LES HOUCHES
Session XXXI
1978

LA MATIÈRE MAL CONDENSÉE

ILL-CONDENSED MATTER

LES HOUCHES, ÉCOLE D'ÉTÉ DE PHYSIQUE THÉORIQUE

ORGANISME D'INTÉRÊT COMMUN DE L'UNIVERSITÉ SCIENTIFIQUE ET MÉDICALE DE GRENOBLE ET DE L'INSTITUT NATIONAL POLYTECHNIQUE DE GRENOBLE

INSTITUT D'ÉTUDES AVANCÉES DE LA DIVISION DES AFFAIRES SCIENTIFIQUES DE L'OTAN

AIDÉ PAR LE COMMISSARIAT À L'ÉNERGIE ATOMIQUE

CONFÉRENCIERS

David J. THOULESS
Jacques JOFFRIN
Philip W. ANDERSON
Valentin POÉNARU
Scott KIRKPATRICK
Tom C. LUBENSKY
Pierre-Gilles de GENNES

Directeurs scientifiques de la session: Roger Maynard, Centre de Recherches sur les Très Basses Températures, BP no. 166, 38041, Grenoble, et Gérard Toulouse, Laboratoire de Physique de l'École Normale Supérieure, 24 rue Lhomond, 75231 Paris

LES HOUCHES

SESSION XXXI

3 Juillet – 18 Août 1978

LA MATIÈRE MAL CONDENSÉE
ILL-CONDENSED MATTER

Édité par

ROGER BALIAN, ROGER MAYNARD, GERARD TOULOUSE

NORTH-HOLLAND PUBLISHING COMPANY
AMSTERDAM · NEW YORK · OXFORD
&
WORLD SCIENTIFIC PUBLISHING CO PTE LTD
SINGAPORE

LES HOUCHES, ÉCOLE D'ÉTÉ DE PHYSIQUE THÉORIQUE

Session		
I	1951	Mécanique quantique. Théorie quantique des champs
II	1952	Quantum mechanics. Mécanique statistique. Physique nucléaire
III	1953	Quantum mechanics. Etat solide. Mécanique statistique. Elementary particles
IV	1954	Mécanique quantique. Théorie des collisions; two-nucleon interaction. Electrodynamique quantique
V	1955	Quantum mechanics. Non-equilibrium phenomena. Réactions nucléaires. Interaction of a nucleus with atomic and molecular fields
VI	1956	Quantum perturbation theory. Low temperature physics. Quantum theory of solids; dislocations and plastic properties. Magnetism; ferromagnetism
VII	1957	Théorie de la diffusion; recent developments in field theory. Interaction nucléaire; interactions fortes. Electrons de haute énergie. Experiments in high energy nuclear physics
VIII	1958	Le problème à *N* corps (Dunod, Wiley, Methuen)
IX	1959	La théorie des gaz neutres et ionisés (Hermann, Wiley)
X	1960	Relations de dispersion et particules élémentaires (Hermann, Wiley)
XI	1961	La physique des basses températures. Low temperature physics (Gordon and Breach, Presses Universitaires)
XII	1962	Géophysique extérieure. Geophysics: the earth's environment (Gordon and Breach)
XIII	1963	Relativité, groupes et topologie. Relativity, groups and topology (Gordon and Breach)
XIV	1964	Optique et électronique quantiques. Quantum optics and electronics (Gordon and Breach)
XV	1965	Physique des hautes énergies. High energy physics (Gordon and Breach)
XVI	1966	Hautes énergies en astrophysique. High energy astrophysics (Gordon and Breach)
XVII	1967	Problème à *N* corps. Many-body physics (Gordon and Breach)
XVIII	1968	Physique nucléaire. Nuclear physics (Gordon and Breach)
XIX	1969	Aspects physiques de quelques problèmes biologiques. Physical problems in biology (Gordon and Breach)
XX	1970	Mécanique statistique et théorie quantique des champs. Statistical mechanics and quantum field theory (Gordon and Breach)
XXI	1971	Physique des particules. Particle physics (Gordon and Breach)
XXII	1972	Physique des plasmas. Plasma physics (Gordon and Breach)
XXIII	1972	Les astres occlus. Black holes (Gordon and Breach)
XXIV	1973	Dynamique des fluides. Fluid dynamics (Gordon and Breach)
XXV	1973	Fluides moléculaires. Molecular fluids (Gordon and Breach)
XXVI	1974	Physique atomique et moléculaire et matière interstellaire. Atomic and molecular physics and the interstellar matter (North-Holland)
June Inst.	1975	Structural analysis of collision amplitudes (North-Holland)
XXVII	1975	Aux frontières de la spectroscopie laser. Frontiers in laser spectroscopy (North-Holland)
XXVIII	1975	Méthodes en théorie des champs. Methods in field theory (North-Holland)
XXIX	1976	Interactions électromagnétiques et faibles à haute énergie. Weak and electromagnetic interactions at high energy (North-Holland)
XXX	1977	Ions lourds et mésons en physique nucléaire. Nuclear physics with mesons and heavy ions (North-Holland)
XXXI	1978	La matière mal condensée. Ill-condensed matter (North-Holland)
En préparation		
XXXII	1979	Cosmologie physique. Physical cosmology
XXXIII	1979	Biophysique des membranes et communication intercellulaire. Biophysics of membranes and intercellular communication
XXXIV	1980	Interaction laser–plasma
XXXV	1980	Physique des défauts

PARTICIPANTS

Axel, Françoise, DPhT, Orme des Merisiers, CEN Saclay, BP2, 91190 Gif-sur-Yvette, France.

Ball, Robin Chris, Cavendish Laboratory, Madingley Road, Cambridge, England.

Bernard, Douglas Alan, Physics Department, Jadwin Hall, Princeton University, Princeton, N.J. 08540, USA.

Calemczuk, Roberto, SBT-CENG, Avenue des Martyrs, Grenoble, France.

Clerc, Jean-Pierre, Université de Provence, Marseille, France.

Dallacasa, Valerio, Physics Department, Via M. d'Azaglio 85, 43100 Parma, Italia.

Dasgupta, Chandan, University of California at San Diego, La Jolla, California, USA.

De Almeida, Jairo, Department of Mathematical Physics, Birmingham University, B15 2TT, Birmingham, England.

Demongeot, Jacques, IRMA 31, B.P. 53, 38041, Grenoble Cedex, France.

Derrida, Bernard, Groupe de Physique des Solides, Ecole Normale Supérieure, 24 rue Lhomond, Paris 75005, France.

Devoret, Michel, S.P.S.R.M., Orme des Merisiers, CEN Saclay, 91190 Gif-sur-Yvette, France.

Ermilov, Alexandr, Chair of Quantum Statistics, Physics Department, Moscow State University, Moscow, 117234, USSR.

Fasol, Gerhard, Cavendish Laboratory, Madingley Road, Cambridge, England.

Fesser, Klaus, Institut für Festkörperforschung der KFA, Jülich D-517 Jülich, Deutschland.

Fishman, Shmuel, Department of Physics and Astronomy, Tel-Aviv University, Ramat-Aviv, Israël.

Flytzanis, Nicolaos, Department of Physics, University of Virginia, McCormick Road, Charlottesville, Virginia 22901, USA.

Franck, *Carl*, Department of Physics, University of Virginia, McCormick Road, Charlottesville, Virginia 22901, USA.

Garel, *Alain*, Laboratoire de Physique des Solides, Bât. 510, Université Paris Sud, 91405 Orsay, France.

Gerl, *Maurice*, Laboratoire de Physique des Solides, Université Nancy I, C.O. 140, 54037 Nancy Cedex, France.

Ginsparg, *Paul*, Department of Physics, Clark Hall, Cornell University, Ithaca, NY 14853, USA.

Hikami, *Shinobu*, Research Institute for Fundamental Physics, Kyoto University, Kyoto, Japan.

Jackson, *Shirley*, Bell Laboratories, ID 337, 600 Mountain Avenue, Murray Hill, NJ 07974, USA.

Joanny, *Jean-François*, Collège de France, Physique de la Matière Condensée, 11 place Marcelin Berthelot, 75005 Paris, France.

Khanna, *Shiv*, GTP, CNRS, avenue des Martyrs, 38042 Grenoble Cedex, France.

Khokhlov, *Alexei*, Physics Department, Moscow State University, Moscow 117234, USSR.

Kühl, *Helmut*, Institut für Theoretische Physik der Universität zu Köln, Zülpicher st. 77, 5000 Köln 41, Deutschland.

Lederer, *Pascal*, Laboratoire de Physique des Solides, Bât. 510, Université Paris Sud, 91405 Orsay, France.

Leibler, *Ludwik*, Collège de France, Physique de la Matière Condensée, 11 pl. Marcellin Berthelot, Paris 75005, France.

Lemaire, *Henri*, 16 chemin St Jean, 38700 La Tronche, France.

Maillard, *Jean-Marie*, Laboratoire de Physique des Solides, Ecole Normale Supérieure, rue Lhomond, 75005 Paris, France.

Moudden, *Abdelhamid*, Laboratoire de Physique des Solides, Université Paris Sud, 91400 Orsay, France.

Parlebas, *Jean-Claude*, Université L. Pasteur, Laboratoire de Structure Electronique des Solides, 4 rue Blaise Pascal, 67000 Strasbourg, France.

Pelcovits, *Robert*, Department of Physics, University of Illinois, Urbana, Illinois 61801, USA.

Peliti, *Luca*, Istituto di Fisica "G. Marconi", Piazzale delle Scienze, 5 I-00185 Roma, Italia.

Rammal, *Rammal*, Centre de recherches sur les très basses températures, CNRS, avenue des Martyrs, 38400 Grenoble, France.

Riedinger, *Roland*, Université de Haute Alsace, Institut des Sciences Exactes et Appliquées, 4 rue des Frères Lumière, 68093 Mulhouse Cedex, France.

Rivier, Nicolas, Department of Physics, Imperial College, Prince Consort Road, London SW7 2BZ, England.

Sarma, Gobalakichena, CEA Saclay, DPh SRM, Orme des Merisiers, B.P. no. 2, 91190 Gif-sur-Yvette, France.

Seabra Lage, Eduardo, Laboratorio de Fisica, Faculdado de Ciencias, Universitade di Porto, Porto, Portugal.

de Sèze, Laurent, CEA Saclay, DPh-G-SRM, Orme des Merisiers, B.P. no. 2, 91190 Gif-sur-Yvette, France.

Shastry, Sri-Ram, School of Physics, University of Hyderabad, Hyderabad, 500001, India.

Sneddon, Leigh, Department of Theoretical Physics, 1 Keble Road, Oxford OX1 3NP, England.

Soukoulis, Costas, The James Franck Institute, University of Chicago, 5640 S. Ellis Avenue, Chicago, Illinois 60637, USA.

Starobinets, Samuel, Weizmann Institute of Science, Chemical Physics Department, Rehovot, Israël.

Szeftel, Jacob, Physique des Solides, Bât. 510, Orsay 91405, France.

Thickstun, Thomas, Departement de Mathématiques, Université Paris Sud, Bât. 425–426, 91405 Orsay, France.

Tremblay, André-Marie, Physics Department, Laboratory for Atomic and Solid State Physics, Cornell University, Ithaca, NY 14853, USA.

Troper, Amos, Centro Brasiliero de Pesquisas Fisicas, Avenida Wenceslau, Braz 71, Botafogo, Rio de Janeiro, Brasil.

Uzelac, Katerina, Laboratoire de Physique des Solides, Université Paris Sud, Bât. 510, 91405 Orsay, France.

Vannimenus, Jean, Laboratoire de Physique de l'Ecole Normale Supérieure, 24 rue Lhomond, 75005 Paris, France.

Velarde, Manuel, Departimento de Fisica-C-3, Universidad Autonoma, Cantoblanco, Madrid, Espagne.

Wilson, Clive, Theoretical Physics, The University, Manchester M13 9PL, England.

Yaouanc, Alain, Departement de Recherche Fondamentale, CENG, Chemin des Martyrs, 38 Grenoble, France.

Yeomans, Julia, Department of Theoretical Physics, 1 Keble Road, Oxford, England.

"...les verres ont des propriétés universelles..." (ce volume, page 107).

PRÉFACE

Cette session de physique de la matière condensée a été consacrée à "la matière mal condensée". Cela veut dire qu'on s'y est intéressé aux effets spécifiques de désordre, par opposition aux effets caractéristiques des matériaux homogènes, bien ordonnés.

Il y a dans le petit adverbe "mal" une nuance péjorative qui ne trompera que les ignorants. Il est de fait qu'abondent dans cette branche de la physique les vocables à signification péjorative dans l'usage vulgaire (désordre, défauts, amorphes, instabilités, bruit,...). Mais, depuis Galilée et son observations de taches (maculae) dans le soleil, à la grande irritation des contemporains, les scientifiques ont pris plaisir à braver les préjugés, jusque dans le choix des mots. Bravade justifiée souvent par l'intérêt des phénomènes et des concepts ainsi nommés.

La recherche des effets spécifiques du désordre a conduit dans les dernières années à une importante avancée, tant expérimentale que théorique. Ce livre en porte témoignage. En vérité, le choix 1977 du jury Nobel de physique, survenu au cours de la préparation de cette session, en avait confirmé l'opportunité.

Au début de la session, J. Joffrin a réussi la tâche difficile de présenter une revue à la fois descriptive et critique de l'étude expérimentale des systèmes désordonnés, principalement des verres et verres de spin. Parallèlement, D. J. Thouless a donné un cours sur les problèmes de percolation et de localisation, problèmes modèles par excellence des effets spécifiques du désordre et pour lesquels de nombreux résultats, parfois anciens d'ailleurs, ont été obtenus. Puis P. W. Anderson a su, malgré sa gloire récente, rester égal à lui-même dans sa présentation des propriétés des verres, verres de Fermi et verres de spin; il a fourni, comme on verra, un texte fourmillant d'idées. S. Kirkpatrick a donné un cours profond et original, soulignant en particulier l'apport important des méthodes de simulation numérique dans l'étude de divers effets de désordre. Dans un passé récent, la richesse du modèle de Potts fournissant des fonctions génératrices pour la statistique de nombreux modèles de désordre est apparue plus pleinement: T. Lubensky en a présenté une revue lucide et détaillée. V. Poénaru a donné un cours d'introduction à la topologie algébrique, avec application aux défauts et aux textures dans les milieux ordonnés. Ce cours, par son originalité et sa qualité pédagogique, a constitué un des points forts de la session. De plus, V. Poénaru, palliant une fâcheuse défaillance, a improvisé une série de cours sur la théorie de l'ergodicité, révélant ainsi à son auditoire un autre pan des mathématiques utiles aux physiciens. Enfin, P.-G. de Gennes a donné un cours d'introduction à un sujet immense: la physique des polymères, traitant successivement de leurs propriétés statiques et dynamiques.

Un certain nombre de conférenciers ont été invités, afin de couvrir avec plus de détails des domaines expérimentaux ou théoriques abordés dans le cours. D'autres, au contraire, introduisaient à des problèmes différents en physique de la matière condensée. Par ordre chronologique, on peut citer: A. Guinier: structure atomique des matériaux amorphes – J. Souletie: phénoménologie des verres de spin – R. Romestain: localisation et diffusion Raman avec "spin flip" dans CdS – M. Devoret: verre quadrupolaire dans l'hydrogène solide – I. Solomon: un semiconducteur amorphe bien condensé: le silicium hydrogéné – D. Sherrington: le modèle de Heisenberg–Mattis: oscillations non triviales sur un modèle plat – P. Nozières: un nouveau matériau quantique: l'hélium 3 polarisé – C. J. Adkins: seuil de conduction dans les couches d'inversion – D. C. Licciardello: défauts et excitations de basse énergie dans les verres – J. Villain: le rôle de la dimensionalité des spins dans le problème des verres de spin – G. Sarma: le passage du régime dilué au régime semi-dilué dans les polymères – J. Joffrin: neutrons et cohérence. Un texte complet de certains de ces séminaires figure dans l'ouvrage, tandis que pour d'autres, une référence principale à un article de revue récent permettra au lecteur de retrouver l'essentiel de l'exposé.

Des groupes d'étude se sont organisés presque spontanément sous la responsabilité d'animateurs dévoués: verres de spin (T. Garel) – verres (N. Rivier) – localisation (G. Sarma) – polymères (L. Leibler) – hydrodynamique et phénomènes de bifurcation (M. Velarde) – bruit en $1/f$ (A.-M. Tremblay) – ondes de densité de charge, etc. Ces groupes d'étude ont offert un cadre sympathique pour les très nombreux séminaires donnés par la plupart des participants et qu'il n'est pas possible de citer entièrement ici.

Parmi les enseignants et les conférenciers, il en est un grand nombre qui ont fourni une participation très féconde, au-delà de leur strict enseignement, en prolongeant leur séjour pour assister aux autres cours et en contribuant par là à susciter des discussions très vivantes. Que ceux-là soient particulièrement remerciés, avec une mention spéciale à V. Poénaru qui assista à presque toute la session, et à S. Kirkpatrick qui en fit quasiment autant.

Remerciements

La réalisation de la session XXXI de l'Ecole des Houches et de ce volume de notes de cours n'a été possible que grâce à de nombreuses contributions:

– le soutien financier de l'Université Scientifique et Médicale de Grenoble, et les subventions de la Division des Affaires Scientifiques de l'OTAN, qui a inclus la session dans son programme d'Instituts d'Etudes Avancées, et du Commissariat à l'Energie Atomique;

– l'orientation et le soutien effectif du Conseil de l'Ecole;

– le soin apporté par Lauris Balian et Thérèse Lécuyer dans la préparation et la frappe des manuscrits;

– la coopération de tous les participants, qui non seulement ont pris une part active dans l'élaboration de nombre de notes de cours, mais ont aussi enrichi les cours eux-mêmes par leurs questions vivantes et leurs commentaires;

– l'aide d'Henri et Nicole Coiffier, de Christiane Rocher, d'Inca San Nicolas et de toute l'équipe pour résoudre les problèmes d'intendance et d'accueil;

– le talent photographique de Françoise Axel et André-Marie Tremblay, qui ont contribué à égayer ce volume.

Nous tenons enfin à remercier les conférenciers, pour les nombreuses heures passées à préparer leurs cours, et leur inépuisable disponibilité envers les participants tout au long de la session.

Roger Balian
Roger Maynard
Gérard Toulouse

"...glasses have universal properties..." (this volume, page 107).

PREFACE

This condensed matter session was devoted to “ill-condensed matter”, where we are interested primarily in disorder effects, as opposed to the study of homogeneous, perfect crystals.

Only the uninitiated will be misled by the use of the word “ill” to arrive at possibly pejorative conclusions, especially as the use of such disparaging words (disorder, defects, amorphous, instabilities, noise, etc.) is rampant in the subject. However, since Galileo’s discovery of sunspots (maculae), much to the irritation of his contemporaries, scientists have revelled in their efforts to overcome prejudice, and this even in their choice of words.

In this spirit, important advances, both experimental and theoretical, have been made in recent years in the study of amorphous systems. The choice of the 1977 Nobel laureates in Physics, and, in more modest fashion, this book, are both tributes to these successes.

At the beginning of the session, a largely empirical and descriptive account of the experimental aspects of the subject was given by J. Joffrin. In parallel, D. J. Thouless gave a series of lectures on percolation and localization problems, typical of amorphous systems and which have led in the more or less recent past to many clear-cut results. Following this, and despite his recent glory, P. W. Anderson remained true to himself in an ingenious discussion of glasses, Fermi glasses and spin glasses; his text, as one will see, is swarming with ideas. S. Kirkpatrick gave an original course, emphasizing the specific approach and interest of the numerical simulation methods in treating the effects of disorder. T. Lubensky gave a lucid and detailed description of the Potts model which furnishes generating functions for statistical treatments of disordered systems. V. Poénaru, in one of the most complete contributions to the school, gave a very pedagogical course on algebraic topology with application to defects and textures in ordered media. Moreover, he filled a breach in giving a brilliant presentation of the ergodic problem. Finally, P.-G. de Gennes gave an introductory course to a huge subject, the physics of polymers, treating both static and dynamic properties.

Some more specialized invited seminars were also given in order to cover with more details some experimental or theoretical aspects evoked during the lectures. In some cases the full text of the seminars has been published and when this is not so, complete references to recent work in the subject are given.

Workshops were organized almost spontaneously by the participants, instigated by several devotees: spin glasses (T. Garel), glasses (N. Rivier), localization (G. Sarma), polymers (L. Leibler), hydrodynamics and bifurcation problems (M. Verlarde), 1/f noise (A.-M. Tremblay), charge density waves, etc. It is impossible here to cite all the seminars that were organized in this framework; suffice it to say that they led to a very agreeable working climate.

Many of the speakers brought a very intangible but greatly appreciated contribution by their participation in innumerable discussions throughout the course. We warmly thank them, especially V. Poénaru who remained almost throughout the whole session, and also S. Kirkpatrick who was a close second.

Acknowledgements

The XXXI Session of the Les Houches Summer School and this volume of lecture notes would not have been possible without:

– the financial support from the Université Scientifique et Médicale de Grenoble, the NATO Scientific Affairs Division (who included this session in its Advanced Study Institutes Programme) and the Commissariat à l'Energie Atomique;

– the guidance of the school board;

– the careful preparation and typing of the manuscripts by Lauris Balian and Thérèse Lécuyer;

– the cooperation of all the participants, who not only took an active part in the preparation of many of the lecture notes but also contributed greatly to the lectures by their lively questions and comments;

– the help of Henri and Nicole Coiffier, of Christiane Rocher, Inca San Nicolas and the whole team, for looking after our material needs;

– the photographic talent of Françoise Axel and André-Marie Tremblay, which helped to enliven this volume.

Finally we wish to thank the lecturers for the many hours of preparation which went into their lectures and the way they tirelessly made themselves available for discussion throughout the school.

Roger Balian
Roger Maynard
Gérard Toulouse

CONTENTS

LECTURES

SEMINARS

CRASH COURSE

COURSE 1

PERCOLATION AND LOCALIZATION*

David J. THOULESS

Physics Department, Queen's University,
Kingston, Canada

* Part of these lectures have also been delivered at the Scottish Universities Summer School of Physics on the metal–non-metal transition in disordered systems (St. Andrews, 1978). The corresponding texts will be included for completeness in the proceedings of both Les Houches and St. Andrews schools.

Contents

R. Balian et al., eds.
Les Houches, Session XXXI, 1978 – La matière mal condensée/Ill-condensed matter

PART I: PERCOLATION

Throughout these lectures we will primarily be concerned with the effects of static disorder in infinite systems. The basic ideas with which we are concerned can be applied to a wide range of phenomena, such as the spread of an epidemic, the reliability of an electrical network, the strength of a composite material such as concrete, or the behaviour of a noisy parametric amplifier, but our attention will be restricted to those problems that arise in condensed matter physics. We will be concerned with the properties of electrons in disordered materials, and to a lesser extent with phonons in disordered systems and with the behaviour of some sorts of random magnetic systems.

For the sake of simplicity a lot of important features of real physical systems will be ignored for most of the time. The disorder will be assumed to be uncorrelated, so that we forget about the possible effects of long wavelength components of the disorder. The disorder will be taken to be static, so that we are dealing with quenched systems with fixed impurities rather than annealed systems in which impurities can move in such a way as to minimize the free energy. This will also prevent us from considering the effects of transient disorder such as is produced by the thermal motion of the lattice unless we make considerable modifications of the theory. We also for the most part ignore many-body effects, on the grounds that the one-body problems have to be properly understood before a many-body theory is developed.

1. What is a percolation problem?

1.1. Definition of the problem

Various examples of percolation problems have been studied over the years, but the formal definition, together with a lot of practical examples, was given in a paper by Broadbent and Hammersley [1]. The subject was reviewed by Frisch and Hammersley [2], and a number of examples are discussed in this review, as well as the relation of percolation problems to other problems such as the self-avoiding random walk. A general review of the subject was given by Shante and Kirkpatrick [3]. The review by Essam [4] is particularly concerned with the combinatorial aspects of the problem and with the behaviour near the critical point. There is also a review by Kirkpatrick [5] which is particularly concerned with the relation between percolation theory and transport problems such as the conductivity of a random network.

Consider a solid which has a distribution of small holes in random positions. If the concentration of holes is low, as in gruyère, the holes will for the most part be isolated, although there will be a few clusters of overlapping holes. As the concentration of holes is increased, the average size of the clusters increases, until at some concentration it becomes infinite. Beyond this concentration there may still be finite clusters, but there is also an infinite cluster, connected to the exterior, as in a sponge or a piece of "vycor" glass. Fluid can now percolate from the exterior to the interior. This is in fact how "vycor" is made; soluble impurities are mixed into the glass when it is formed, and then the impurities are leached out, leaving the porous glass behind. If the concentration of holes is increased still further, a point will be reached when the solid ceases to be connected to itself and will fall to pieces.

From this example the importance of dimensionality can be clearly seen. If we were dealing with a two-dimensional system, a plate with holes punched at random in it, the critical concentration for percolation, the concentration at which the holes first form an infinite cluster, would coincide with the concentration of holes at which the solid ceases to be connected as an infinite cluster. This duality property in two dimensions is exploited in sect. 2.3. In one dimension an arbitrarily small concentration of holes causes the system to disintegrate; the strength of a chain is that of its weakest link. To achieve percolation in one dimension, however, the concentration of holes has to have its maximum value, as any solid can block flow in one dimension.

For formal purposes it is easier to consider percolation processes on a regular lattice. We consider an array of atomic sites connected to their neighbours (usually only the nearest neighbours) by bonds. There are then two classes of percolation problem known as site problems and bond problems. In a site problem the atomic sites are occupied (open, available) with probability p. The percolation probability $P^{\mathrm{S}}(p)$ is the probability that a site is a member of the infinite cluster of occupied sites. The critical probability for site percolation p_c^S is defined as the highest value of p for which $P^S(p)$ is zero. In the bond percolation problem it is the bonds which are open (available) with probability p and closed (unavailable) with probability $q = 1 - p$. Then $P^B(p)$ is the probability that a site is connected to infinitely many other sites by open bonds, and the critical probability p_c^B for bond percolation is the highest value of p for which $P^B(p)$ is zero. Any bond problem can be mapped onto an equivalent site problem.

1.2. Magnetic models

Percolation theory has been used as a model for the behaviour of alloys of magnetic with non-magnetic atoms for a long time [6]. Antiferromagnetic alloys such as $KMn_pMg_{1-p}F_3$ show a percolation behaviour [7,8], while $K_2Mn_pMg_{1-p}F_4$, $K_2Co_pMg_{1-p}F_4$ [7,8] and $Rb_2Mn_pMg_{1-p}F_4$ [9] behave like two-dimensional percolating systems. The model is obviously a site problem, with magnetic atoms corresponding to occupied sites and non-magnetic atoms to unoccupied sites. A magnetic phase transition can only occur if there is an infinite cluster of neighbouring magnetic atoms. The finite clusters can respond to a weak magnetic field and contribute a Curie-like term to the magnetic susceptibility.

A useful contact can be made between the percolation problem and other models of critical phenomena by considering the limit in which the coupling between adjacent magnetic atoms becomes infinite (see ref. [10] for a clear account of this analogy). If the spins are Ising-like, either up or down, then the limit of the magnetic moment per site as the external field tends to zero is $P(p)$, since only the infinite cluster contributes. If the probability that a site belongs to a cluster of n atoms is $f(n, p)$ then the magnetic moment per site is

$$M = P(p) + \sum_n f(n, p) \tanh nH/T \tag{1.1}$$

and the zero field susceptibility is given by

$$\chi = T^{-1} \sum_n nf(n, p) . \tag{1.2}$$

Since an occupied site must either be in a finite cluster or the infinite cluster

$$\sum_{n=1}^{\infty} f(n, p) + P(p) = p , \tag{1.3}$$

so that the mean size of the finite cluster in which a particular occupied site lies is given by

$$S(p) = \sum_n nf(n, p) \Big/ \sum_n f(n, p) = \chi T(p - P)^{-1} . \tag{1.4}$$

In this way further relations may be established between the properties of the clusters in percolation theory and the magnetic properties of a random Ising model in the strong coupling or low temperature limit.

For a planar on Heisenberg spin system there is an additional simple quantity of interest, namely the spin-wave stiffness. If we look at a percolation system close to p_c we can see that although the infinite cluster covers quite a large portion of the system, some parts of it are only tenuously connected with other parts. Thus while the whole of the infinite cluster contributes to the inertia of the spins, the potential energy can be reduced by concentrating the changes in angle in the narrow channels. Thus the spin wave stiffness drops to a very low value close to p_c. This is directly related to the problem of the conductivity of a random network [5] and is discussed further in sect. 5.

1.3. Electrical network problems

For any type of percolation problem an electrical analogy can be constructed. For example the bonds of the percolation problem may be identified with identical resistors, connecting the nodes, or sites, of a network. In a bond percolation problem a fraction $1 - p$ of the resistors is removed at random, whereas in a site percolation problem all those resistors connected with a proportion $1 - p$ of the nodes are removed. Percolation occurs if there is a non-zero probability of a node being electrically connected to other nodes infinitely far away, and the percolation probability is the fraction of nodes which are thus connected. One can also define the resistivity of the network in terms of the resistance across a large cube of the network multiplied by the side of the cube. If the cube is much larger than some characteristic length scale of the percolation problem this quantity should approach a limit. One can also ask what is the probability distribution of the resistance to ground from a node in a network of three dimensions or more. In the case of a two-dimensional network the resistance between one node and a large circle centred on that node increases logarithmically with the radius of the circle, and the coefficient of the logarithm characterizes the macroscopic resistivity of the network.

This is a closely related formulation of the problem in which instead of a fraction $1 - p$ of the resistances being removed, a fraction p is replaced by short-circuits. The resistance is non-zero until the critical probability p_c is reached, when the short-circuit becomes infinite in extent.

The continuum versions of these models are of great practical importance as models of the electrical properties of inhomogeneous materials. Both to calculate the electrical conductivity of a metal with insulating material embedded in it, and to calculate the dielectric properties of an insulator with conducting material embedded in it, such a problem has to be solved; these two cases correspond to the two versions of the network problem. These are two extreme forms of the general problem of the electrical properties of composite materials, which has been studied for more than a hundred years. A recent survey of this problem has been given by Landauer [11].

2. Exact results

2.1. Inequalities for the critical probability

Comparisons between different types of percolation problem and between the percolation problem and similar lattice problems allows certain inequalities to be established both for the percolation probability P and for the critical probability for percolation p_c.

In the first place it can be shown that the percolation probability for the bond problem is greater than the percolation probability for the corresponding site problem [12]. In fact we have

$$pP^{B}(p) \geqslant P^{S}(p)\,, \qquad p_c^{B} \leqslant p_c^{S}\,. \tag{2.1}$$

Consider a particular cluster of n sites. The probability that this cluster is all open in a site problem is p^n. There is at least one subset of the bonds joining the n sites that forms a Cayley tree (a connected graph with no closed loops). This tree has $n - 1$ bonds in it, so the probability that such a cluster is connected in the bond problem is at least p^{n-1}. In this way the inequality (2.1) follows.

Inequalities for the percolation probabilities for different lattices can be established if one lattice can be mapped onto another with a reduction of the number of bonds. For example the triangular lattice can be mapped

Fig. 2.1. Diagram to show how triangular lattice, square lattice and honeycomb lattice can be mapped onto one another. The honeycomb has only the solid bonds, the square lattice has also the dashed bonds, and the triangular lattice has also the dotted bonds.

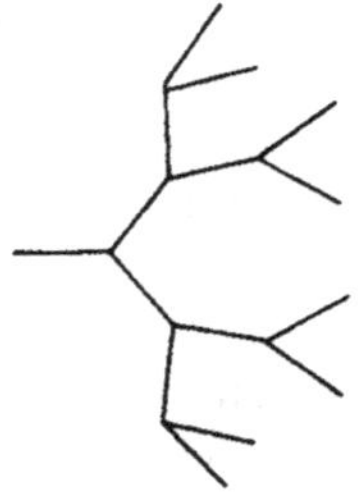

Fig. 2.2. Part of a Bethe lattice with $K = 2$.

onto the square lattice if a third of the bonds are removed, and the square lattice can be mapped onto the honeycomb lattice if a quarter of its bonds are removed (see fig. 2.1), so we have

$$P^{\mathrm{T}}(p) \geqslant P^{\mathrm{SQ}}(p) \geqslant P^{\mathrm{HC}}(p) . \tag{2.2}$$

In three dimensions the face-centred cubic lattice can be mapped onto the body-centred cubic lattice with the removal of bonds, and the body-centred cubic can be mapped onto the simple cubic which can be mapped onto the diamond lattice, so we have

$$P^{\mathrm{FCC}} \geqslant P^{\mathrm{BCC}} \geqslant P^{\mathrm{SC}} \geqslant P^{\mathrm{D}} . \tag{2.3}$$

These results were obtained by Fisher [13].

Further inequalities can be obtained by comparison with the non-intersecting walk problem. If the number of non-intersecting walks of L steps from a point on the lattice is N_L, then the connective constant K is defined by

$$\ln K = \lim_{L\to\infty} L^{-1} \ln N_L . \tag{2.4}$$

K is certainly less than $Z - 1$, where Z is the coordination number of the lattice, since after the first step of the walk there can be at most $Z - 1$ possibilities for the subsequent ones. If $P(p)$ is non-zero there must be a non-zero lower bound for the probability of a non-intersecting path of length L from a site, where the bound is independent of L. The probability that one particular path is open in a site percolation problem is p^{L+1}, and so the probability that one of the N_L paths is open cannot exceed $N_L p^{L+1}$. If Kp is less than unity this tends to zero, so we have

$$p_{\mathrm{c}}^{\mathrm{S}} \geqslant K^{-1} \geqslant (Z - 1)^{-1} . \tag{2.5}$$

This result was obtained by Hammersley [14]. From (2.1) we see that the inequality holds more strongly for the bond problem. Some values for K can be found in an article by Domb [15].

2.2. *The Bethe lattice*

The Bethe lattice is defined as an infinite regular Cayley tree. For the Bethe lattice

$$N_L = Z(Z - 1)^{L-1}, \tag{2.6}$$

so that the connective constant K is $Z - 1$; in fact we shall see that all the inequalities (2.1) and (2.5) become equalities for the Bethe lattice (see fig. 2.2).

The percolation probability was calculated by Fisher and Essam [12]. We consider the bond percolation problem, but very little modification is needed to obtain it for the site problem. We calculate the probability $1 - P$ that a particular site is a member of a finite cluster. This occurs if for each of the $K + 1$ bonds either the bond is closed, or it is open and the site at its end is connected to a finite number of sites beyond it. Writing Q for the probability that a site is connected to a finite number of sites by the remaining K bonds this gives

$$1 - P = (1 - p + pQ)^{K+1}. \tag{2.7}$$

In the same way one of these neighbouring sites is connected to a finite number of other sites by the remaining K bonds if for each bond either the bond is closed, or it is open and the site at its end is connected to a finite number of sites beyond it, so

$$Q = (1 - p + pQ)^K. \tag{2.8}$$

This equation always has the solution $Q = 1$, $P = 0$, but for some values of p it has a second solution which gives a non-zero value of Q. This second solution exists for

$$p > p_c = K^{-1} \tag{2.9}$$

and for values of p slightly larger than this critical value we can substitute $Q = 1 - \varepsilon$ and get

$$1 - \varepsilon \approx 1 - Kp\varepsilon - \tfrac{1}{2}K(K - 1)p^2\varepsilon^2, \tag{2.10}$$

so that we have

$$P(p) \approx 2(1 - p_c/p)(K + 1)/(K - 1) , \tag{2.11}$$

and the percolation probability approaches zero linearly.

The Bethe lattice includes the linear chain as a special case with $K = 1$, and eq. (2.9) shows that the critical probability for the linear chain is unity, so that P is always zero.

Another model which can be solved exactly was considered by Erdös and Rényi [16]. If all sites are connected to all $N - 1$ other sites by bonds which have a probability p of being open, then there is an infinite cluster if pN exceeds unity.

2.3. Duality in two-dimensional systems

Various relations can be obtained for two-dimensional lattices by exploiting duality properties of the lattice. For example the critical probability for a site problem on a triangular lattice can be found. Each finite cluster of occupied sites is surrounded by a perimeter of unoccupied sites which are themselves adjacent, so to each finite cluster of the original problem there corresponds a larger (in the sense of the maximum distance of its members) cluster in the dual problem in which p and $q = 1 - p$ are interchanged. Since this problem is self-matching the critical probability for percolation is $\frac{1}{2}$ (see fig. 2.3a).

For the site problem on a square lattice the perimeter sites to a cluster may be either nearest neighbours or next nearest neighbours, so the matching problem is one in which there are bonds to next nearest sites as well as nearest neighbours. No analytic expression is known for the critical probability in this case, but the critical probabilities for the two matching problems must sum to unity.

For the bond problem on a square lattice, the dual lattice, the lattice in which each loop is replaced by a site and each bond by a bond at right angles, is also a square lattice. If for each closed bond in the original lattice we have a corresponding open bond in the dual lattice again we can see that the perimeter of each cluster in the original lattice is a cluster of open bonds in the dual lattice. Again the critical value for p is $\frac{1}{2}$ (fig. 2.3b).

The dual lattice for the triangular bond problem is a honeycomb lattice, and from this we can deduce that the critical probabilities for the two lattices add up to unity (see fig. 2.4). Consider a honeycomb lattice with a proportion q of open bonds, and compare this with the triangular lattice obtained from the honeycomb lattice by deleting half the sites – the bonds

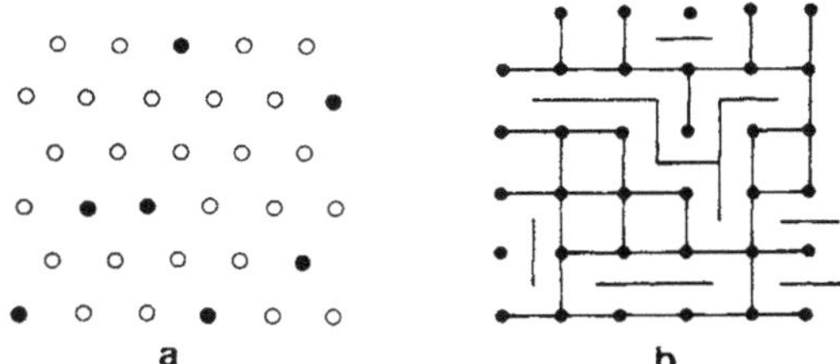

Fig. 2.3. (a) Open and closed sites for a site percolation problem on a triangular lattice. Finite clusters of one type are completely surrounded by a perimeter of nearest neighbours of the other type. (b) Open and closed bonds for a bond percolation problem on a square lattice. The closed bonds are represented as bonds on the dual lattice.

of this new lattice have a probability p of being open. Now consider the three sites surrounding one of the deleted sites of the honeycomb lattice. The probability that these three sites are all connected through the deleted site of the honeycomb lattice is q^3, while the probability that they are all connected by the three triangular bonds is $p^3 + 3p^2q$. The probability that two are connected and one disconnected is q^2p for either problem. The probability that all three are disconnected is $p^3 + 3p^2q$ for the honeycomb lattice and q^3 for the triangular lattice. Since the duality property tells us that either the triangular lattice or the honeycomb lattice, but not both, has an infinite cluster, we see that the condition for the triangular lattice to be better connected than the honeycomb lattice is

$$p^3 + 3p^2q > q^3 . \tag{2.12}$$

The critical probability for the triangular lattice is therefore given by

$$3p_c - p_c^3 = 1 \quad \text{or} \quad p_c = 2\sin(\pi/18) = 0.3473 . \tag{2.13}$$

These results were obtained by Sykes and Essam [17].

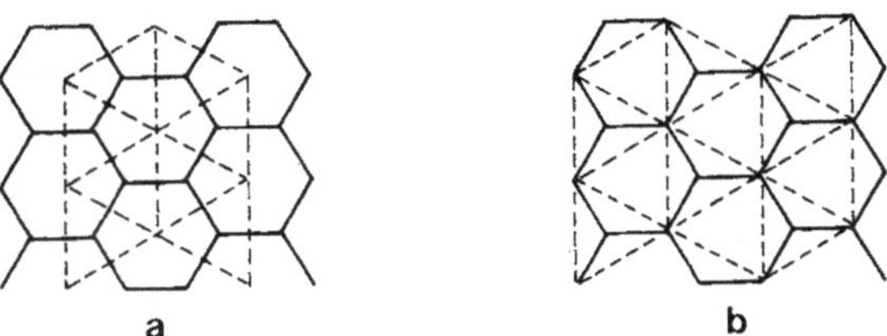

Fig. 2.4. (a) The solid lines are the bonds of a honeycomb lattice, and the dashed lines are the bonds of the dual, triangular, lattice. (b) The dashed lines show the bonds of the triangular lattice obtained by eliminating half the sites of the honeycomb lattice.

Recent work by Straley [18] has exploited the duality property of resistance networks in two dimensions. For any planar resistance network we can construct an equivalent dual network by replacing each resistance by an intersecting element whose conductance has the value of the original resistance. Each voltage becomes a current loop of the dual network and each current loop becomes a voltage. In the context of percolation theory this transformation turns missing bonds into short-circuits, since the infinite resistance is replaced by an infinite conductance. One result that can be deduced from this argument is that if a square lattice contains unit resistances with probability p, and infinite and zero resistances each with the same probability $\frac{1}{2}(1 - p)$, then the resistance of the network is unity. The network is self-dual, and the resistance of the original network is equal to the conductance of the dual. This result has been obtained by Marchant and Gabillard [19].

3. Power series and simulation

3.1. Evaluation of power series

The use of high temperature power series for general magnetic models and of low temperature series for the Ising model has been very successful in calculating the thermodynamic properties of magnetic models in the appropriate temperature range and for finding the critical properties in the neighbourhood of the critical point, which is just where the series diverge. Since the problems of enumerating graphs embedded in a lattice are essentially the same for these magnetic problems and for the percolation problem, the problems have been studied in parallel. It happens that the resulting series are slightly more difficult to analyse for the percolation problem than for the Ising model.

To construct a power series for the calculation of $P(p)$ we calculate the probability that a site is a member of a finite cluster, starting with the simplest clusters and working up to as large clusters as we can manage. For example, on a square lattice site problem we have (cf. fig. 3.1)

$$\begin{aligned} f(1,p) &= pq^4\,, \\ f(2,p) &= 4p^2q^6\,, \\ f(3,p) &= 6p^3q^8 + 12p^3q^7 \\ f(4,p) &= 8p^4q^{10} + 32p^4q^9 + (16 + 16 + 4)p^4q^8\,, \end{aligned} \qquad (3.1)$$

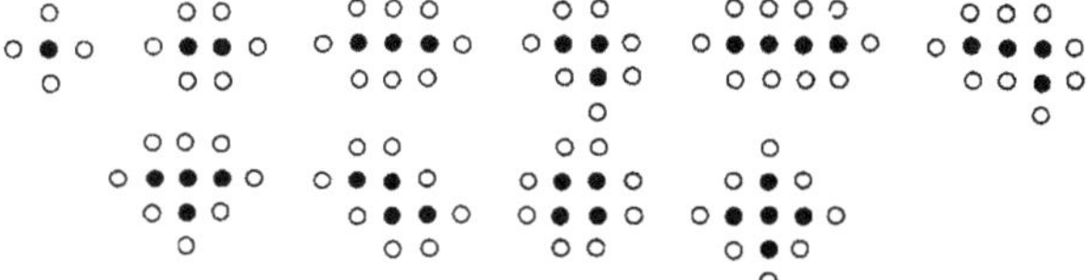

Fig. 3.1. Clusters that contribute to the series (3.1).

where the terms correspond to the different graphs of fig. 3.1. There is a factor p associated with each solid circle and a factor q with each hollow circle. By setting $p = 1 - q$ and rearranging the sum of these terms we get the power series

$$P(p) = 1 - q^4 + q^5 - 4q^6 - 4q^7 - 15q^8 - 69q^9 - \cdots; \tag{3.2}$$

to get the last two terms it is necessary to include also the cluster with five sites arranged on a +. If we take the same series and set $q = 1 - p$ we get identically zero; this we should expect because P is zero for small p. We can however construct series for the mean number of sites per cluster using eq. (1.4), and we get

$$S(p) = 1 + 4p + 12p^2 + 24p^3 + 52p^4 + \cdots. \tag{3.3}$$

In a similar manner a series in q can be constructed for S.

We can write the series (3.1) in the form

$$P(p, q) = p - \sum_{n=1}^{\infty} \sum_{b=1}^{\infty} C(n, b) p^n q^b, \tag{3.4}$$

where n is the number of occupied sites in a cluster and b is the number of vacant sites on the perimeter of the cluster. From this the mean number of sites per cluster can be calculated by taking the partial derivative with respect to p, and other quantities of interest can be obtained in a similar way. It is important in all applications of this method that the quantities $C(n, b)$ be calculated correctly, so to go beyond the lowest orders in the series it is necessary to have a systematic computer program to generate the clusters.

For many purposes it is desirable to have the quantity analogous to the spin–spin correlation function of a magnetic system. In a percolation problem this is the pair-connectedness, which is the probability that two given spins are in the same cluster. Series methods can be developed to calculate this quantity. For large separations the pair-connectedness tends to the limit P^2, and to approach its asymptotic value like $\exp(-r/\xi)$.

It has recently been observed [20–22] that the large clusters which occur close to p_c are ramified, that is they have a surface area proportional to their volume. Since the expression (3.4) is zero for $q = 1 - p$, $p < p_c$, its partial derivative with respect to p is equal to its partial derivative with respect to $q = 1 - p$. Thus we have

$$\frac{\partial P}{\partial p} = -\frac{1 - \langle n \rangle}{P} = \frac{\partial P}{\partial q} = -\frac{\langle b \rangle}{1 - p}. \tag{3.5}$$

The surface to volume ratio of the large clusters tends to $(1 - p_c)/p_c$ as the critical point is approached from below.

Some recent work by Kunz and Souillard [23] on the probability P_n of finding a cluster of size n has some bearing on this problem. These authors have shown that, for p well below p_c, $\ln P_n$ is proportional to n, while, for p well above p_c, $\ln P_n$ is proportional to $n^{1-1/d}$. This is to be expected below p_c if the clusters are ramified, with a surface to volume ratio of order unity since clusters contributing to the pair connectedness at a distance of order n intersite spacings will contain of order n sites, whereas above p_c, if the large finite clusters are more compact, clusters of n sites should extend less far in space.

3.2. Power series calculation of critical properties

Once sufficiently many terms of a power series in p and q have been calculated the ratio method or Padé approximants can be used to calculate critical properties. If the series (3.3) represents a function whose dominant singularity is of the form $(p_c - p)^{-\gamma}$, then the ratio of successive coefficients of the series for large n behaves like

$$a_{n-1}/a_n \sim p_c[1 + (\gamma + 1)/n]. \tag{3.6}$$

A plot of the ratio against n^{-1} should give a line approaching the axis whose intercept gives p_c and whose slope gives the exponent γ. It appears that the series (3.2) is smoother than (3.1), and indeed most of the early work used the series in p rather than the series in q. The process of extrapolation is easier if, as for some of the two-dimensional lattices, the value of p_c is known.

Extensive results for p_c obtained in this way are quoted by Essam [4]. For example p_c for the site problem on a square lattice is given as 0.59, for the site problem on a diamond lattice it is 0.425, for the site and bond problems on a simple cubic lattice it is 0.307 and 0.247, and on a face-centred cubic lattice it is 0.195 and 0.119. Values are also given for problems in which bonds connect also second and third neighbours [24].

The results are consistent with the universality hypothesis that the exponents depend only on the dimensionality of the system and are independent of the nature of the lattice and whether we are dealing with bond or site percolation. In fact the universality hypothesis is used to establish the most probable values of the critical exponents.

If the series is rather irregular or alternating in sign it may be that the dominant singularity is not the one of interest at p_c, and in such a case the ratio method is useless. The series can however be represented by a Padé approximant, a rational fraction for which the leading terms in the power series agree with the power series in question. The singularities of the rational fraction are poles at the zeros of the denominator, so a Padé approximant can reproduce the singularity structure of the original function. Poles of the original function should produce poles in the approximants and cuts should become a sequence of poles. Since a pole is the form of singularity directly given by the approximant, it is advantageous, when a singularity of the form $(p_c - p)^{-\gamma}$ is expected for $S(p)$, to calculate Padé approximants for $d \ln S/dp$, since these should have a simple pole at p_c with residue $-\gamma$.

Padé approximant methods for analysing the series for the pair-connectedness function are used in the paper by Dunn et al. [25]. They have used this analysis to determine values for various exponents including γ, which is found to be 2.38 in two dimensions and 1.70 in three dimensions, and also the exponent ν, which gives the characteristic length scale ξ of connectedness in the form

$$\xi \sim (p_c - p)^{-\nu} . \tag{3.7}$$

They found ν to be 1.34 in two dimensions and 0.82 in three dimensions.

3.3. *Monte Carlo methods*

Many of the early results on percolation theory came from computer simulations of large random systems but these seemed to give somewhat less reliable results than the series methods. Recent calculations [5,26–28] have used very large samples and made careful studies of how the calculated quantities depend on sample size. For comparison with the power series results, Sur et al. [28] find $p_c = 0.3115$ for the simple cubic lattice, and get $\gamma = 1.6 \pm 0.1$, $\nu = 0.8 \pm 0.1$. They also find the exponent β to have the value 0.41 ± 0.01, where β is defined by

$$P(p) \sim (p - p_c)^{\beta} . \tag{3.8}$$

A preliminary report of extensive calculations of critical exponents has been given by Hoshen et al. [50a].

Power series methods have not been used to calculate the conductivity properties of a random network, and mechanical simulations [29–32] and computer simulations [5,18,27,33] have been used to study this problem. These studies show that as the critical probability is approached from above the paths through the system get more and more contorted and restricted. This is of course much easier to see in two dimensions than in three, since a single picture can be used to display the whole system.

4. Scaling and renormalization

4.1. Length scaling

One of the most fruitful ideas in the theory of critical phenomena has been that close to the critical temperature there is a characteristic length scale for correlations, and that as the critical temperature is approached only the length changes, and other quantities scale with this length. The same idea has been adapted to the percolation problem by Dunn et al. [10] and by Levinshtein et al. [26]. It is assumed that for p close to p_c the various sums over clusters are dominated by clusters of a typical size $S(p)$ and linear dimension $L(p)$, and that close to p_c they diverge like

$$S(p) \sim |p - p_c|^{-\Delta} , \qquad L(p) \sim |p - p_c|^{-\nu} , \tag{4.1}$$

while the concentration $N(p)$ per site of such clusters behaves like

$$N(p) \sim |p - p_c|^{2-\alpha} . \tag{4.2}$$

Then the probability that a site is in a finite cluster will have a singular part that behaves as

$$S(p)N(p) \sim |p - p_c|^{2-\alpha-\Delta} , \tag{4.3}$$

so that the exponent for the percolation probability is

$$\beta = 2 - \alpha - \Delta , \tag{4.4}$$

while the mean size of finite clusters behaves like $S^2(p)N(p)$, so the exponent γ is given by

$$\gamma = 2 - \alpha - 2\Delta . \tag{4.5}$$

Similarly higher moments have exponents which differ by multiples of Δ, and this has been confirmed by the work of Dunn et al. [25].

It is assumed that the pair-connectedness at p_c falls off with distance like $r^{-(d-2+\eta)}$, where d is the dimensionality. Close to p_c it should have this form up to $r = L(p)$, and beyond that distance should fall off exponentially. If we integrate this function out to $L(p)$ we should again get the mean size of finite clusters, and this gives

$$\gamma = (2 - \eta)\nu . \tag{4.6}$$

If it is further supposed that the concentration of large clusters is inversely proportional to the volume they occupy then we have the additional hyperscaling relation

$$2 - \alpha = d\nu , \tag{4.7}$$

which implies

$$2\beta + \gamma = d\nu . \tag{4.8}$$

This relation is approximately obeyed by the quoted values of the critical exponents.

4.2. Use of the renormalization group

Harris et al. [34] showed how the relation between the Ashkin–Teller–Potts model and the percolation problem, demonstrated by Kasteleyn and Fortuin [35], can be used to bring the percolation problem into a form in which standard renormalization group ideas can be applied. In the g-level Ashkin–Teller–Potts model there is a variable u_i associated with each site i that can take on any of the integer values from 1 to g. The hamiltonian has the form

$$H = -J \sum_{(ij)} (g\delta_{u_i u_j} - 1) - h \sum_{i} (g\delta_{u_i 1} - 1) . \tag{4.9}$$

The partition function for this hamiltonian can be written as

$$Z = \sum_{u} \prod_{(ij)} [\exp\{-J/T\} + \delta_{u_i u_j}(\exp\{(g-1)J/T\} - \exp\{-J/T\})]$$
$$\times \exp\left\{+h \sum (g\delta_{u_i 1} - 1)T\right\} . \tag{4.10}$$

When the product is expanded out each term contains clusters connected by a Kronecker delta. For each site outside the cluster there is a factor

$$e^{+h(g-1)/T} + (g-1)\,e^{-h/T}, \tag{4.11}$$

while for each connected cluster of n connected sites there is a factor

$$e^{+nh(g-1)/T} + (g-1)\,e^{-nh/T}. \tag{4.12}$$

With each of the bonds connecting the members of a cluster there is a factor

$$e^{(g-1)J/T} - e^{-J/T} \tag{4.13}$$

while for the other bonds there is a factor $\exp(-J/T)$. For $g = 2$ this gives a slightly disguised version of the Ising model, but as g tends to 1 it gives the bond percolation problem, with $q = \exp(-J/T)$.

In this (4.13) is p, and Z tends to unity, since it is just the binominal expansion of $p - q$ to the power of the total number of bonds. If we differentiate with respect to h, divide by $g - 1$ and let h tend to zero, then each finite cluster gives zero and the infinite cluster gives n_∞/T, so that we get

$$NP(p) = T \lim_{h\to 0} \lim_{g\to 1} (g-1)^{-1}\, \partial Z/\partial h. \tag{4.14}$$

Similarly higher derivatives with respect to h give higher moments of the cluster size. Stephen [36] describes how the same method can be used for the bond distribution.

In the mean field theory for this model the free energy per site is given near the critical point by

$$(g-1)^{-1}F = \tfrac{1}{2}(T - T_c)v^2 - \tfrac{1}{6}T(g-2)v^3 + \tfrac{1}{12}T(g^2 - 3g + 3)v^4, \tag{4.15}$$

where v is the (necessarily positive) spontaneous magnetization. The positive cubic term implies that the classical critical behaviour should be attained in six dimensions, rather than in four as for magnetic models with no cubic term. This led to expansions in $\varepsilon = 6 - d$ for the critical exponents, with results such as

$$\gamma = 1 + \varepsilon/7, \qquad \beta = 1 - \varepsilon/7, \qquad \nu = \tfrac{1}{2} + 5\varepsilon/84. \tag{4.16}$$

Kirkpatrick [27] has checked these predictions against Monte Carlo calculations in six, five, four, and three dimensions (see fig. 4.1). It is found that the exponents are at least close to their classical values in six dimen-

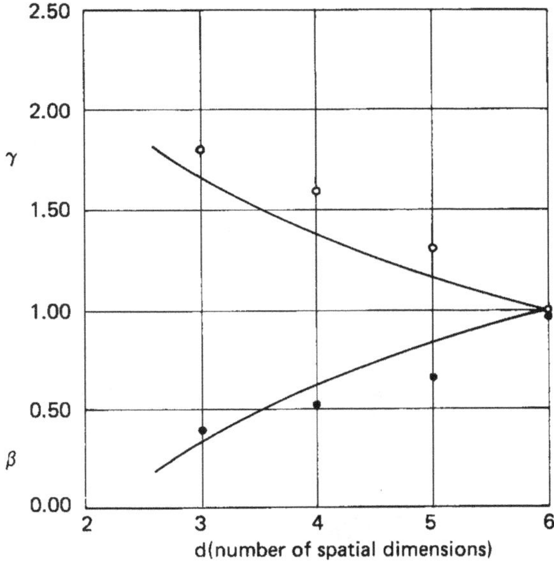

Fig. 4.1. Exponents γ (circles) and β (squares), obtained from analysis of Monte Carlo data, plotted as functions of dimension. The solid lines show predictions of the renormalization-group expansion. This figure is taken from Kirkpatrick [27].

sions, but there is possibly some disagreement on the detailed behaviour. Somewhat different results have been obtained from the renormalization group by Ginzburg [37].

Various methods have been developed for exploiting real space renormalization methods [34,38,39]. To show how these go we consider a very simple-minded approach to site percolation on a triangular lattice. We group together the sites in blocks of three arranged on a triangle, and say that the block is occupied if two or more of its sites are occupied, and vacant otherwise (see fig. 4.2). This leads to a new problem, also on a triangular lattice but with $\sqrt{3}$ times the length scale, and a new probability of occupation

$$p' = p^3 + 3p^2q = 3p^2 - 2p^3\,; \tag{4.17}$$

this has, by chance, a fixed point at the correct value of $p = \frac{1}{2}$. Since we have

$$(p' - \tfrac{1}{2})/(p - \tfrac{1}{2}) \approx \tfrac{3}{2}\,, \tag{4.18}$$

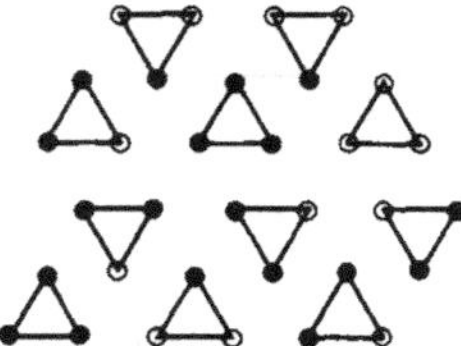

Fig. 4.2. Blocks used for real space renormalization of the site percolation problem on a triangular lattice.

while the length scale changes by $\sqrt{3}$, we get the value of ν as

$$\nu \approx \tfrac{1}{2} \ln 3/\ln \tfrac{3}{2} = 1.35\,, \tag{4.19}$$

which is very close to the right answer. To calculate the percolation probability we argue that a site is connected to the infinite cluster if it is occupied and at least one of its neighbours is occupied, and if the resulting occupied block is a member of the infinite cluster in the rescaled problem. This leads to

$$P(p) = (2p^2q + p^3)P(p')/p' = P(p')(2p^2 - p^3)/(3p^2 - 2p^3)\,, \tag{4.20}$$

which gives $\frac{3}{4}$ for P'/P, and so

$$\beta \approx \ln \tfrac{4}{3}/\ln \tfrac{3}{2} = 0.71\,, \tag{4.21}$$

which is not very good. More careful work than this gives more credible answers.

5. Transport in random networks

5.1. Effective medium theory

For a continuous medium the problem of steady current flow in an inhomogeneous medium is equivalent to the problem of the dielectric response of an inhomogeneous insulating medium – the dielectric constant simply has to be replaced by the conductivity. Such problems have been known for a long time, and the Clausius–Mossotti formula gives a useful approximation if the components do not differ too much in their properties. An effective medium theory which could deal satisfactorily with a percolation problem was developed by Bruggeman [40]. This is similar in spirit to the coherent potential approximation. The inhomogeneous medium is replaced by an effective medium of conductivity σ_m in such a way that the average effect of one of the real components embedded in the effective medium is zero. To show how this works we consider a square lattice bond percolation problem.

The lattice is made up of resistances R with probability p, and infinite resistances with probability $1 - p$. The effective medium consists of a network of resistances which all have the value R_m. If one of these is replaced by a resistance R the current through that link is reduced from I by an amount

$$(R - R_m)I/(R + R_m)\,, \tag{5.1}$$

as can be calculated by standard electrical network theory. This produces a dipolar disturbance in the current flow. Now R_m should be chosen in such a way that the average disturbance produced by replacing the link by R or infinity is zero, so we have

$$p(R - R_m)/(R + R_m) + (1 - p) = 0\,, \tag{5.2}$$

and this gives immediately

$$R_m = R/(2p - 1)\,. \tag{5.3}$$

This is certainly correct for p close to unity, and even gives the correct percolation threshold. In other examples the percolation threshold is not given so accurately by this theory. For example the resistance of the medium becomes infinite for $p = \frac{1}{3}$ for both the triangular lattice and for the simple cubic, but this is slightly too low in the first case and somewhat too high in the second. Effective medium theories have been extensively reviewed by Landauer [11].

5.2. *Critical effects*

Close to the percolation threshold we should not expect the effective medium theory to work, as the assumption that the inhomogeneity can be represented by an average effect is implausible in this region – the conduction paths are undoubtedly very complicated near threshold. It can be seen in the series of pictures shown by Kirkpatrick [41] how complicated these paths are in two dimensions. Close to threshold the majority of the sites in the infinite cluster are on dead ends that make no contribution to the conductivity. The exponent t is related to the conductivity σ by

$$\sigma(p) \sim (p - p_c)^t\,. \tag{5.4}$$

Estimates have been made of t using simulation methods [5,18,42], real space renormalization [39,43], and scaling arguments [44,45]. In two dimensions t comes in the range from 1 to 1.4, with about 1.1 as the preferred value, and in three dimensions it is about 1.7. The scaling relation

$$t = (d - 1)\nu , \tag{5.5}$$

which holds if conduction is essentially by way of straight strands of unit thickness a distance ξ apart is possibly correct in three dimensions, but seems to give too high a value in two. By contrast De Gennes proposed

$$t = 1 + (d - 2)\nu . \tag{5.6}$$

It should be expected that as d tends to 6 the exponent should reach the classical value which it has for the Bethe lattice. Stinchcombe [46] studied the conductivity from one node to ground for the Bethe lattice and found that the exponent was 2. De Gennes [45] pointed out that this is the wrong calculation, and argued that the quantity corresponding to bulk conductivity in real systems should have the exponent 3, in agreement with eq. (5.6) for $d = 6$. This has been shown to be correct by Straley [47].

5.3. *Spin-wave stiffness and related problems*

The problem of the conductivity of a random material has implications for a number of other systems. In particular it gives information about the low frequency spin waves of the corresponding random magnetic system. The velocity of low frequency waves in a random medium is determined by two parameters, an inertial parameter and a stiffness parameter. The inerted parameter for the random mixture of magnetic and non-magnetic atoms is proportional to the number of spins in the infinite cluster, since if the frequency of a disturbance is low all the spins in the infinite cluster will follow the motion. The stiffness parameter is given by the potential energy per unit volume associated with a slow spatial variation of the direction of the spins. This is directly related to the solution of the potential problem for the site percolation system, and so the energy of spin waves of wave-number k is proportional to

$$\sigma(p)k^2/P(p) . \tag{5.7}$$

The exponent associated with the spin wave energy is therefore $t - \beta$. This relation is explained in more detail by Kirkpatrick [5].

5.4. *The Hall coefficient*

In papers by Skal and Shklovskii [48] the scaling argument has been applied to the Hall effect. If we imagine a current I passing through each effective link in the network at a distance ξ apart from one another, the Hall potential across each wire is proportional to I, if the thickness of the link is independent of $p - p_c$, so the average Hall field is proportional to $I\xi^{-1}$. The current density is $I\xi^{1-d}$, and so the Hall coefficient scales as ξ^{d-2}. This argument cannot be correct in detail if eq. (5.5) is not correct.

Levinshtein et al. [44] also made measurements of the Hall constant by punching holes in conducting paper. To make a three-dimensional system they stacked the sheets of paper on top of one another and applied pressure. They found that in two dimensions the Hall coefficient was independent of the number of holes, and that as the percolation threshold was approached the Hall coefficient rose sharply in three dimensions; it was not possible to determine in this experiment if the exponent in three dimensions is really ν.

As Landauer [11] points out it has been known for some time that the Hall constant is strictly independent of the concentration of holes in two dimensions. The argument given by Landauer and Swanson [49] is that, to lowest order in the magnetic field $\boldsymbol{B}$, the current $\boldsymbol{i}_1$ and electric field $\boldsymbol{E}_1$ are related to the current density $\boldsymbol{i}_0$ in the field free medium and the Hall constant R_0 of the homogeneous conducting material by

$$\boldsymbol{i}_1 = \sigma_0 \boldsymbol{E}_1 + R_0 \sigma_0 (\boldsymbol{i}_0 \times \boldsymbol{B}) \,. \tag{5.8}$$

Since $\boldsymbol{i}_0$ is tangential at the surface of the insulating region, $\boldsymbol{i}_0 \times \boldsymbol{B}$ is normal, and so a solution of the equations and boundary conditions can be found in which $\boldsymbol{i}_1$ is zero, $\boldsymbol{E}_1$ is $R_0 \boldsymbol{B} \times \boldsymbol{i}_0$ in the conducting medium and zero in the insulator, and so the average Hall field is $-BR_0$ times the average current density. This argument is independent of the shape and size of the insulating regions.

PART II: LOCALIZATION

6. The Anderson model for disordered solids

6.1. Localized and extended states

Even at the most elementary level some account must be taken of the effects of disorder in solid state physics. The eigenstates for electrons in a perfect crystal are Bloch waves, and for a perfect crystal the electronic specific heat and the shapes of the energy surfaces may be calculated. The theory of transport processes in metals depends crucially on the existence of disorder which causes transitions between Bloch states. The disorder is static in the case of impurity scattering and dynamic in the case of scattering by phonons. If we restrict attention to the impurity scattering, which dominates at low temperatures, we can see that the eigenstates for the electrons in the disordered potential are superpositions of Bloch waves concentrated around an energy surface in a region whose width is of order $\hbar/\tau$, where τ is the collision time for the electron. It is essential for the theory of metallic conductors, with a conductivity that tends to a constant as the temperature tends to zero, that these eigenstates are extended right across the system. If they are not an electron initially in a particular region of space cannot diffuse outwards, and since electron diffusion and electrical conduction are intimately related, there will be no electrical conductivity.

In contrast to this situation Anderson [51] argued in his classic paper on "Absence of Diffusion in Certain Random Lattices" that it was natural to assume that for sufficiently strong disorder the electron eigenstates would be localized in regions where the potential was particularly suitable, and perturbation theory can then be used to calculate the exponential rate at which the wave function falls off from its centre of localization and the degree of disorder necessary to stabilize the localized state. Although the idea of a continuum of localized states is an unfamiliar one to metal physicists it was a natural way of describing the impurity band in a lightly doped semiconductor. In fact the evidence for the existence of localized states in a strongly disordered system is by now much stronger than the evidence for the existence of extended states in a weakly disordered system.

This idea of localization by disorder was developed in a long series of articles by Mott, and its range of application is reviewed in the books by Mott and Davis [52], and Mott [53]. Many of the topics discussed in these lectures are reviewed by Thouless [54].

6.2. *Anderson's model*

The Anderson model is essentially a tight binding model of a regular solid, with the disorder introduced by letting the energy vary from site to site. There is a single electron state per site and the hamiltonian matrix for the electrons has constant elements $-V$ connecting sites to their nearest neighbours, and diagonal elements ε_i where the ε_i are independently distributed. In the notation of second quantization we have

$$H = \sum_i \varepsilon_i a_i^+ a_i - \sum_i \sum_j V_{ij} a_i^+ a_j , \tag{6.1}$$

where V_{ij} is $-V$ for nearest neighbours and zero otherwise. For most of this discussion it will be assumed that the ε_i are uniformly distributed over the range from $-W/2$ to $W/2$. The ratio W/V measures the degree of disorder, or for many purposes $W/2ZV$, where Z is the coordination number of the lattice, is a better measure. There are many possible generalizations of this model which can give a more realistic description of disordered systems, but this is already complicated enough to do for our purposes.

A crude estimate can be made of the rate at which the amplitude falls off from its maximum value in a localized eigenstate. If we consider an electron which in the limit $V \to 0$ would be localized at a site whose energy ε_i is near the centre of the band, then the amplitude on a neighbouring site j is, to lowest order in perturbation theory,

$$-V/(\varepsilon_i - \varepsilon_j)\,. \tag{6.2}$$

Since the denominators can range between $\pm W/2$ a typical value is of order $W/4$. There are however Z neighbours, and the rate of fall-off of the amplitude is determined by the smallest denominator, which will be of order $W/4Z$. The condition for convergence of this perturbation theory is then

$$4ZV/W < 1\,. \tag{6.3}$$

Anderson's analysis was based on a more detailed consideration of the perturbation series for the energy shift $S_i(E)$. According to Feenberg's formulation [55] of perturbation theory this can be expressed as a sum over nonintersecting paths which start from and return to i. Each bond in the path contributes a factor V_{jk} and each site a factor $(E - \varepsilon_j - S_j')^{-1}$, where S_j' is similarly defined in terms of non-intersecting walks that avoid the sites already visited in the earlier walk. It is then necessary to make a statistical analysis of the contributions from these paths.

Two assumptions are made in this analysis. One is concerned with the treatment of the self-energy S_j. Anderson was concerned with the centre of the band, and there he argued that the main effect would be to cut off contribution from sites which lie too close in energy to E. The argument is that if a path passes through such a site then the self-energy of the previous site on the path will have a large contribution from this site close to E, and so will give a large denominator which cancels the small denominator. Anderson argued that therefore sites should only contribute if their energies differed from E by more than $2V^2/W$. It is clear that away from the centre of the band there will be a systematic departure of S_j' from zero since more contributions to its perturbation series will have the same sign as E than the opposite sign, and so some average value of S_j' should be used. Economou and Cohen [56] suggested using the coherent potential approximation, and this seems reasonable.

The second assumption made by Anderson is that all the non-intersecting paths of L steps contribute independently. In order to consider the convergence of the perturbation series or the behaviour of the amplitude at large distances we want to deal with large values of L, and there are of order K^L such paths, where K is the connective constant of the lattice. It then remains to consider the probability distribution of a product of L independent factors of the form $(E - \varepsilon_j - S_j')^{-1}$. Anderson observed correctly that the distribution had a long tail and that the quantities were equally likely to be positive or negative (this is true for large L even if most of the factors have the same sign), and so the sum over independent terms of this sort would be dominated by the largest term. He was able to calculate the asymptotic form, and found that when S_j' was neglected the condition for convergence was

$$W > 2eKV \ln(W/2V) , \tag{6.4}$$

while if the cut-off on small energy denominators was allowed for the result was

$$W > 4KV \ln(W/2V) . \tag{6.5}$$

The logarithmic factor changes the critical value of W considerably. Its presence seems to depend in a rather detailed way on the assumption that the K^L paths make independent contributions. It is correct for the Bethe lattice, as we discuss in sect. 6.4, but it seems to be wrong for two- and three-dimensional lattices. Various alternative assumptions have been made. I do not think the assumptions made by Economou and Cohen [56] are mutually consistent, but the final answer is plausible, which is

$$KV \exp(-\langle \ln|E - \varepsilon_i - \bar{S}|\rangle) < 1 , \tag{6.6}$$

where $\bar{S}$ is a self-energy given by the coherent potential approximation, and the geometric mean of the energy denominators is taken. The results obtained from this by Licciardello and Economou [57] seem to agree well with numerical results.

Anderson's [51] analysis can be extended to obtain more details of the localized wave function. Anderson [58] used the series for the matrix element of the Green function between two distant sites both to show that the wave function falls off exponentially with distance, and to calculate the dependence of the rate of fall off on $W - W_c$. The range of localization is approximately $(W - W_c)^{-3/(d+2)}$ in d dimensions, for $1 < d \leqslant 4$. Last and Thouless [59] calculated the power law controlling the fall-off of the wave function at threshold, but a different result has recently been obtained by Sadovskii [60].

6.3. Localization in one dimension

It was argued by Mott and Twose [61] that however weak the disorder in a one-dimensional system, all eigenstates should be localized. Later work has established this result under fairly general conditions, both for the Schrödinger equation with a random one-dimensional potential and for the tight-binding model with independent random site energies. There is a survey of both theoretical and numerical work on this by Ishii [62]. Most of the discussions of this problem consider the solution of the problem with one-point boundary conditions, that is they start with a given wave function and derivative (on the amplitude at two neighbouring points in the tight binding case), and show that the solution grows exponentially with probability unity. It is then argued that an eigenstate for a long but finite system occurs at just those energies where an exponentially growing wave function from the left matches an exponentially growing wave function from the right. Wegner [63] has demonstrated wave function localization without using this approach.

The most detailed studies of localization in the one-dimensional Anderson model have been made by Papatriantafillou and Economou [64], who have explored a wide range of values of W/V and the whole energy band, using an integral equation that can be derived for the probability density of S_j'.

Herbert and Jones [65] developed a very useful relation for studying one-dimensional systems. They observed that the $1N$ element of the Green function $G = (E - H)^{-1}$ for the one-dimensional Anderson model with N sites is given by $(-V)^{N-1}/\det(E - H)$, from the familiar expression for the inverse of a matrix in terms of its cofactors. Since $G_{jN}(E)a_N$ satisfies the eigenvalue equation for the Anderson model everywhere except at $j = N$, G_{NN}/G_{1N} gives the value of $a_N(E)$ which would be obtained by starting from $a_1 = 1$ and solving iteratively the equations

$$a_{j+1}(E) = (\varepsilon_j - E)a_j/V - a_{j-1}\,. \tag{6.7}$$

We therefore get

$$a_N(E)/a_1(E) = G_{NN}(E)\det(E - H)(-V)^{-N+1}\,. \tag{6.8}$$

In terms of the eigenvalues E_α and normalized eigenvectors, this gives

$$\ln\left|\frac{a_N(E)}{a_1(E)}\right| = \sum_{\alpha=1}^{N} \ln|E - E_\alpha| - (N-1)\ln V + \ln\left[\sum_{\alpha=1}^{N} \frac{|a_N^\alpha|^2}{E - E_\alpha}\right]. \tag{6.9}$$

Unless E is very close to one of the E_α the right side of this equation is approximately $(N-1)\kappa(E)$, where the exponential growth rate of the amplitudes is given by

$$\begin{aligned}\kappa(E) &= \int n(x)\ln|E - x|\,\mathrm{d}x - \ln|V|,\\ &= \mathrm{Re}\int^{E} G_{ii}(E')\,\mathrm{d}E',\end{aligned} \tag{6.10}$$

and $n(x)$ is the density of states per site. It is only when $\ln|E - E_\alpha|$ is of order N, so that E is exponentially close to an eigenvalue, that this expression changes significantly, and in the limit $E = E_\alpha$ we get

$$\ln|a_1^\alpha a_N^\alpha| = -\sum_{\beta\neq\alpha} \ln|E - E_\beta| + (N-1)\ln|V| = -(N-1)\kappa, \tag{6.11}$$

so that κ also gives the rate at which the wave function has fallen off from its maximum of order unity to its value at the boundary.

The expression (6.10) for the localization rate is closely related to eq. (6.6) of the theory of Economou and Cohen [56] for the case $\kappa = 1$. It can be studied in the limit of large W/V, in which case $n(E)$ is W^{-1} for $|E| < \frac{1}{2}W$; and the expression gives

$$\kappa(E) = \tfrac{1}{2}\ln\frac{W^2 - 4E^2}{4e^2V^2} + \frac{E}{W}\ln\frac{W + 2E}{W - 2E}. \tag{6.12}$$

It can also be studied for small W/V, using the perturbation expansion for $G(E)$. We have

$$\begin{aligned}N^{-1}\sum_i G_{ii} = N^{-1}\Bigg[\sum_i G_{ii}^0 &+ \sum_i\sum_j G_{ij}^0\varepsilon_j G_{ji}^0\\ &+ \sum_i\sum_j\sum_k G_{ij}^0\varepsilon_j G_{jk}^0\varepsilon_k G_{ki}^0 + \cdots\Bigg].\end{aligned} \tag{6.13}$$

The second term in the square brackets averages to zero, and the third term averages to

$$\sum_i\sum_j (W^2/12)G_{ij}^0 G_{jj}^0 G_{ji}^0 = -(W^2/12)\sum_j G_{jj}^0\,\mathrm{d}G_{jj}^0/\mathrm{d}E. \tag{6.14}$$

Substitution of this into eq. (6.10), using the value

$$G^0_{ii}(E) = (E^2 - 4V^2)^{-1/2} \tag{6.15}$$

gives the localization rate as

$$\kappa = (W^2/24)(4V^2 - E^2)^{-1}. \tag{6.16}$$

This gives a distance for the exponential fall off which is just twice the mean free path which would be calculated for backward scattering by using the Born approximation.

Abrikosov and Ryzhkin [66,67] have calculated both dc and ac conductivity for a one-dimensional system of finite length. They find the dc conductivity decreases exponentially with length. There appears however, to be a discrepancy of a factor of four between their exponent and the one given in eq. (6.16). This discrepancy also exists in the work of Gogolin et al. [68].

6.4. *Self-consistent theory*

Anderson's theory was established on an apparently new basis by Abou-Chacra et al. [69]. In this theory the Feenberg perturbation series for the self-energy S_i was truncated after the leading term, giving the equation

$$S_i = \sum_j |V_{ij}|^2(E - \varepsilon_j - S_j')^{-1}. \tag{6.17}$$

Instead of the rather casual treatment of the self-energy in the denominators which is used in the original theory, the equation is treated self-consistently, in the sense that the given probability distribution for ε_j and an assumed distribution for S_j' generate a probability distribution for S_i. Since S_j' and S_i are defined by similar equations we ask that the probability density used for S_j' should generate the same probability density for S_i.

This theory is exact for the Bethe lattice, since the only nonintersecting walk which returns to the initial point is the walk with two steps. Except for the first equation, which has $K + 1$ term in the sum instead of K, all the equations defining S_j' in terms of S_k'' and so on have the same structure. Since it is correct for the Bethe lattice we can regard this theory as the high dimensionality limit of localization theory. Unfortunately we do not know what is the critical dimensionality. Many arguments have been advanced to suggest that it is four, but I do not think any of them are relevant – my guess is that six is the right answer. Even if we knew the critical dimensionality the self-consistent theory is sufficiently complicated that an ε expansion would be very hard to do.

We can use S_i to determine whether or not states are localized by giving the energy E a small imaginary part η. Then we write

$$S_i(E + i\eta) = E_i - i\Delta_i \,. \tag{6.18}$$

For extended states Δ_i tends to a constant as η tends to zero – this limiting value gives a measure of the rate at which a particle of energy E at the site i will escape. For localized states Δ_i is proportional to η and the ratio is 1 less than the inverse of the square of the amplitude at i, so that

$$\lim \Delta_i/\eta = \sum_{j \neq i} |a_j|^2/|a_i|^2 \,. \tag{6.19}$$

These relations are in the review by Thouless [54]. If we consider the limit of small η for localized states we can take real and imaginary parts of eq. (6.17) to get

$$\begin{aligned} E_i &= \sum_j |V_{ij}|^2 (E - \varepsilon_j - E_j)^{-1} \,, \\ \Delta_i/\eta &= \sum_j |V_{ij}|^2 (1 + \Delta_j/\eta)(E - \varepsilon_j - E_j)^{-2} \,. \end{aligned} \tag{6.20}$$

The problem is to find out whether these equations have a self-consistent solution. It should be noted that although the distribution of E_i can be found independently of Δ_i, the value of Δ_i/η is correlated with that of E_i – this fact is related to Anderson's argument for introducing a cut-off.

A rather simple result is obtained if we ignore this correlation. If we write $p(x)$ for the probability density of $x = \varepsilon_j + E_j$ we get the equation for the Laplace transform $f(s)$ of Δ_i/η as

$$f(s) = \left[\int p(E - x) f(sV^2/x^2) \exp(-sV^2/x^2) \, dx \right]^K . \tag{6.21}$$

If states are extended, we can expect an attempt to solve this equation iteratively to collapse to the trivial solution which is unity for $s = 0$ and zero everywhere else. For localized states there should be a non-trivial solution. The existence of a solution can be established from the fact that replacing f on the right by a larger (smaller) positive function gives a larger (smaller) positive function on the left. Thus if the first stage of iteration gives a larger (smaller) function f than was used on the right, the iterative sequence is always increasing (decreasing). We can prove the existence of a solution by finding one form of f that generates an increasing sequence, and another, larger, f that generates a decreasing sequence. The solution is then squeezed between the two. The decreasing sequence is generated by taking $f(s) = 1$, and the increasing sequence can be found by taking

$$f(s) = 1 - As^{\beta}, \tag{6.22}$$

where it is positive and zero otherwise, provided A is sufficiently small and β is less than unity and satisfies

$$1 = KV^{2\beta} \int p(E - x)|x|^{-2\beta}\, dx. \tag{6.23}$$

This equation is the consistency condition for a solution of (6.21) which behaves like (6.22) for small s. With Anderson's form of the cut-off a solution of this equation can be found with $\beta > \frac{1}{2}$ provided we have

$$1 > KV \int p(E - x)|x|\, dx, \tag{6.24}$$

and this gives the condition (6.5) with $E = 0$ and a cut-off at $x = 2V^2/W$, so that Anderson's [51] results are recovered, despite the different formulation.

Part of this analysis can be carried out when the correlation between Δ_i and E_i is taken into account, but the theory is much more complicated. It was pointed out by Kumar et al. [70], among various other observations that I think are wrong, that if $p(0)$ is zero without the help of the cut-off then the relevant value of β in eq. (6.22) can be unity, and eq. (6.23) should be replaced by the inhomogeneous equation

$$A = \frac{KV^2 \int x^{-2} p(E - x)\, dx}{[1 - KV^2 \int x^{-2} p(E - x)\, dx]}, \tag{6.25}$$

and (6.24) by the condition

$$1 > KV^2 \int x^{-2} p(E - x)\, dx. \tag{6.26}$$

6.5. *The Ioffe–Regel criterion and minimum metallic conductivity*

For various reasons no satisfactory theory of localization has been developed by means of a study of the stability of extended states. Mott [52,53] has made extensive use of the Ioffe–Regel criterion for the consistency of the extended states. It is argued [71] that since the mean free path is the distance over which the electron wave function loses phase coherence and the wavelength is the distance over which the phase changes by 2π

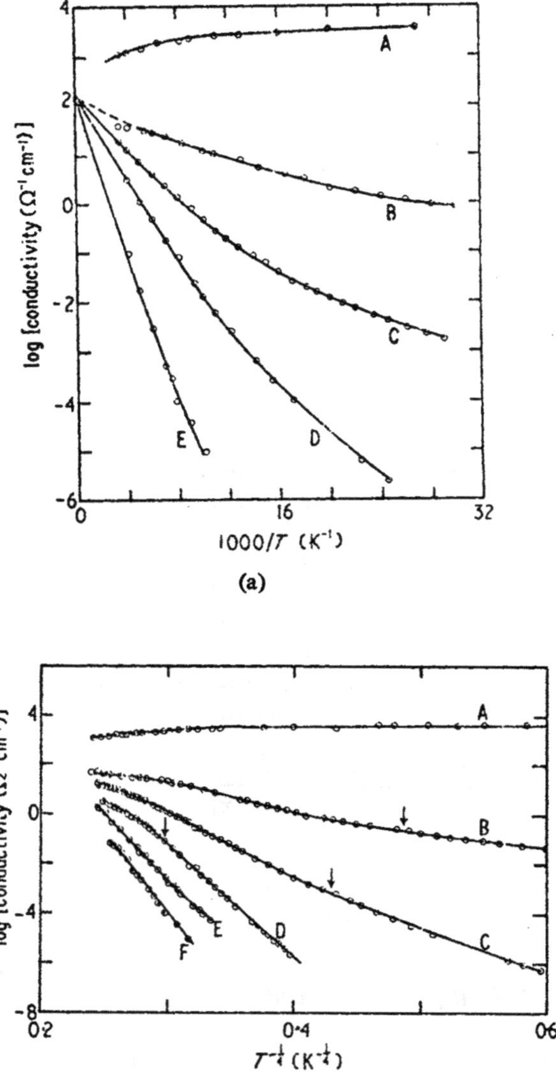

Fig. 6.1. Logarithm of conductivity against (a) $1000/T$ and (b) $T^{-1/4}$ for various compositions of $La_{1-x}Sr_xVO_3$. A: $x = 0.3$; B: $x = 0.2$; C: $x = 0.1$; D: $x = 0.05$; E: $x = 0$ in (a) and $x = 0.02$ in (b); F: $x = 0$. These figures are taken from Sayer et al. [75].

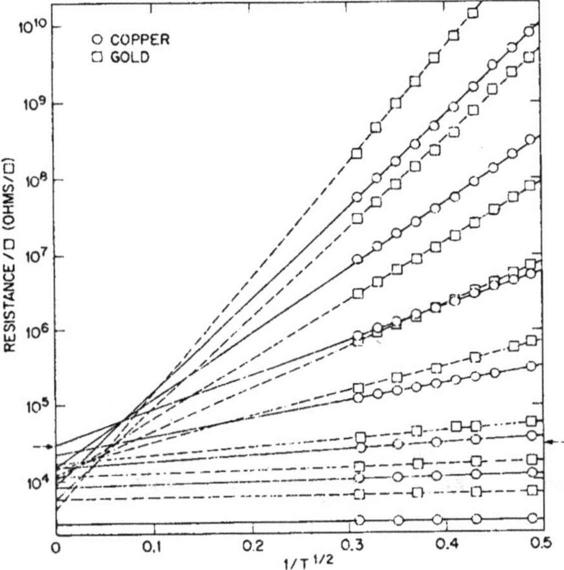

Fig. 6.2. Resistance against $T^{-1/2}$ for very thin films of copper and gold. High-temperature extrapolations for the high-resistance curves all tend to converge to about 30,000 ohm. This figure is taken from Dynes et al. [76].

it does not make sense to have a wavelength which is substantially more than the mean free path. The Ioffe–Regel criterion is therefore

$$\lambda k > 1\,, \tag{6.27}$$

where λ is the mean free path and k the wavenumber. The right side of the inequality is not necessarily exactly unity, but there is evidence that it is close to unity. When this criterion is violated the states are supposed to be localized. It is obviously not a sufficient condition for the existence of extended states, as it is satisfied in one dimension for weak disorder, yet the states are localized.

If this formula is combined with the relation between conductivity and mean free path that comes from kinetic theory an important relation is obtained. We have, for free electrons in three dimensions

$$\sigma_{3\mathrm{D}} = ne^2\tau/m = e^2k_{\mathrm{F}}^2\lambda/3\pi^2\hbar > e^2k_{\mathrm{F}}/3\pi^2\hbar\,, \tag{6.28}$$

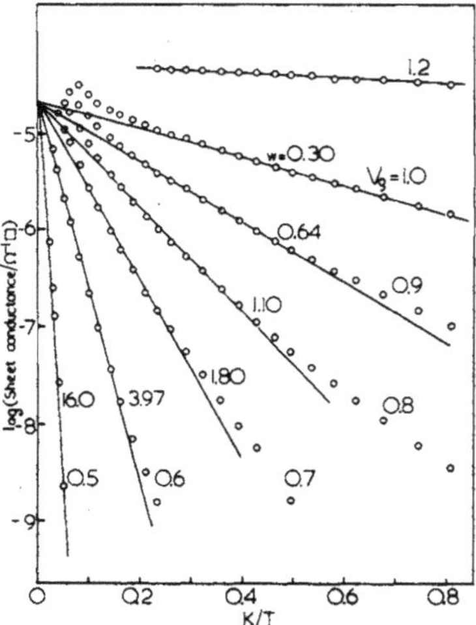

Fig. 6.3. Logarithm of the conductance of an inversion layer plotted against $1/T$. The applied gate voltage is V_g, and the activation energies w ($= E_C - E_F$) in meV are indicated. This figure is taken from Pollitt [77].

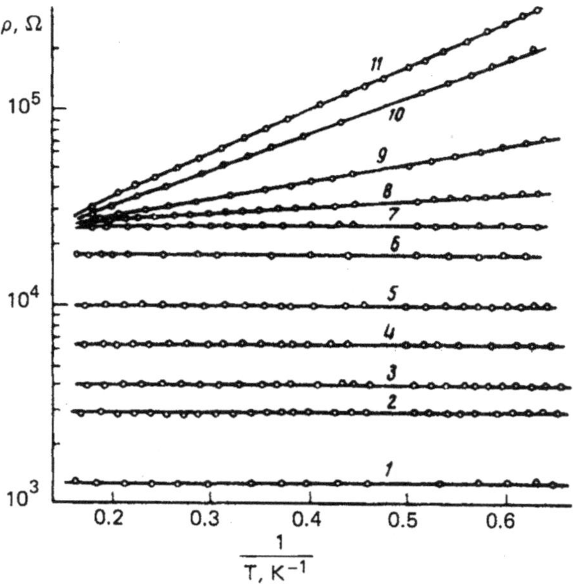

Fig. 6.4. Temperature dependence of the resistivity of the boundary layer between two germanium crystals. The increasing numbers denote a sequence of bicrystals with increasing angle of disorientation. This figure is taken from Vul et al. [74].

so a minimum metallic conductivity is predicted which is inversely proportional to the wavelength. In two dimensions we get

$$\sigma_{2D} = e^2 k_F \lambda / 2\pi\hbar > e^2/2\pi\hbar \,, \tag{6.29}$$

so that the minimum metallic conductivity has a value with no length scale in it, which is close to $4 \times 10^{-5}\ \text{ohm}^{-1}$. The experimental evidence for such a minimum metallic conductivity has been surveyed by Mott et al. [72]. More recent work on two-dimensional minimum metallic conductivity has been described by Adkins et al. [73], who studied the inversion layer, Vul et al. [74] who studied the layer between two germanium crystals grown at different angles, and Dynes et al. [76] who studied thin metallic films (see figs. 6.1–6.4).

7. Simulation and scaling

7.1. Computer studies of the Anderson model

Because good theoretical work has proved so difficult and the interpretation of experimental information is so ambiguous, computer simulations of disordered systems are particularly important. The method used is to consider a lattice of finite size, usually with periodic boundary conditions to avoid surface effects, to choose particular values of the random site energies, and to solve the resulting matrix equations. Once the solutions have been found there are problems of interpretation. For a sufficiently large system there is no difficulty in telling the difference between extended and localized states, but systems of a size that can be handled on a computer are not sufficiently large, and so care must be exercised in distinguishing between the two and in extrapolating results to an infinite system. It is found that quantities of interest have very large fluctuations, and this increases the difficulties enormously; this is probably a reflection of the fact that we are dealing with probability densities with long tails, as Anderson [51] pointed out, and as can be seen from eq. (6.22).

Calculations can most easily be done for two-dimensional systems, since large linear dimensions can be achieved without getting enormous numbers of sites. From the earliest calculations it was apparent that localized states were much more stable than Anderson's criterion (6.5) suggested, and much weaker conditions like (6.3) or (6.6) were appropriate. The most convincing calculations have been performed by Yoshino and Okazaki [78,79], who have found eigenstates on a square lattice of size 100×100 (see fig. 7.1). From (6.5) the critical value of W/V is about 30, but for

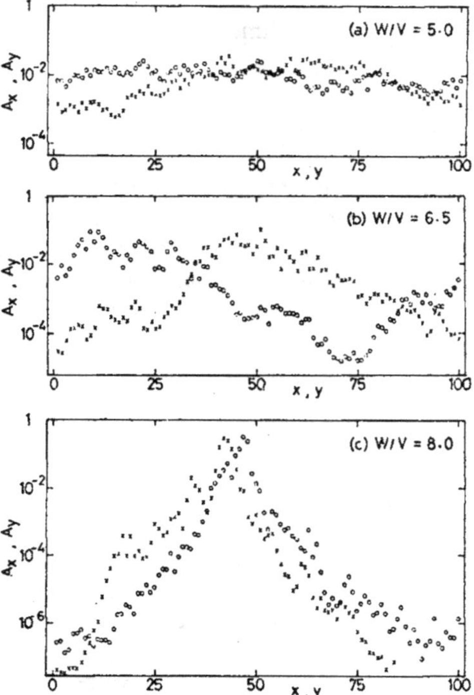

Fig. 7.1. This figure shows for an eigenstate near the centre of the band on a 100 × 100 square lattice with three values of W/V the sum of the squares of the amplitude along a strip of 100 sites parallel to the x or to the y axis as a function of the position of the strip. For the circles A_x is the sum taken over a strip parallel to the y axis, and satisfies periodic boundary conditions, while for the crosses A_y is made to vanish at the edges. This figure is taken from Yoshino and Okazaki [79].

$W/V = 8$ the wave functions in the centre of the band are strongly localized and appear to decay roughly exponentially with a localization length of about 5 atomic spacings. There is, however, a very great variation in decay rate between different states in the same energy range. From these results alone it can be said that the existence of exponentially localized states in two dimensions is clearly established.

The existence of extended states is less clearly established. Yoshino and Okazaki believe that the critical value of W/V in the band centre is about 6.5, and this is in good agreement with earlier results by Edwards and

Thouless [80] and Licciardello and Thouless [81]. However the wave functions shown for $W/V = 5$ seem to have much more spatial variation than one would expect for an extended state, and could be compatible with a localization length comparable with the size of the system. Later work by Licciardello and Thouless [82] suggests that those states which were earlier interpreted as extended are not, but are localized states with a very long localization length. Hower, even if these states are not actually extended, there is a relatively sharp transition between states with a short localization length and those with a very long length, and it may be hard to distinguish experimentally between extended states and large localized states.

For three-dimensional systems the situation is more confused, and the critical value of W/V obtained seems to depend on the criterion used to identify localized and extended states. Weaire and Srivastava [83] have studied the inverse participation ratio

$$\sum_i |a_i|^4 \Big/ \Big(\sum_i |a_i|^2\Big) \tag{7.1}$$

and find a non-zero value for W/V greater than 8 for the diamond lattice and 15 for the simple cubic lattice. The result for the diamond lattice is in agreement with the result obtained by the same method by Licciardello and Thouless [82], but Edwards and Thouless [80] found a considerably higher value using the method of boundary condition sensitivity which is discussed in the next section. Licciardello and Thouless [82] concluded that the wave functions, although apparently localized, did not fall off exponentially with distance for W/V between 9 and 12. This is in accord with an uncomfortable suggestion made earlier by Last and Thouless [59].

Weaire and Srivastava [84] have also calculated the condition for localization in hypercubic lattices in four and five dimensions (fig. 7.2). They find the critical value of W/ZV increases with dimensionality, but it is still a factor six or seven less than the value given by (6.5). It should be borne in mind that the iterative scheme given in eq. (6.21) converges rather slowly, and we have found that very large systems are needed to show that for W close to the critical value states are actually non-localized on the Bethe lattice.

There have also been studies of the effect of "off-diagonal disorder", that is constant ε_i and random V_{ij}. Kikuchi [85,86] and Debney [87] have studied the sort of disorder one gets in a doped semiconductor. For a low concentration of impurity levels there is an enormous variation of V_{ij} which results from the random distances between centres and the exponential dependence of V_{ij} on separation, and this can lead to localization of the

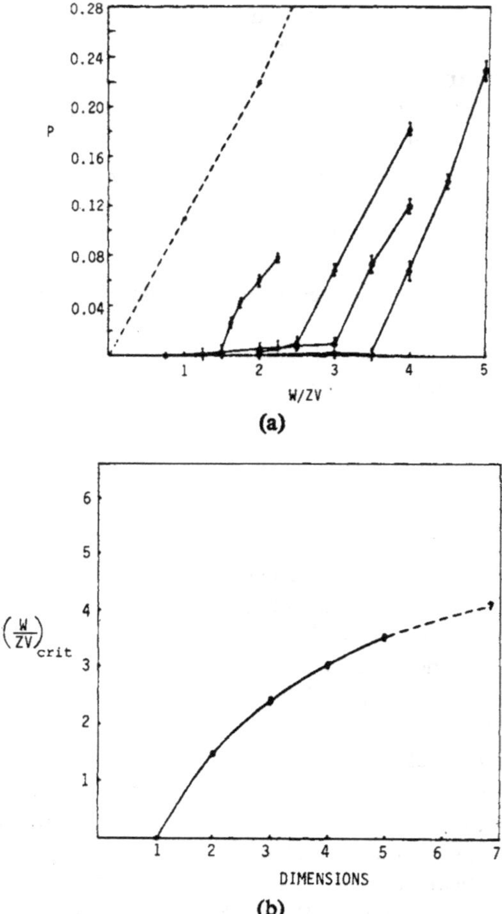

Fig. 7.2. (a) shows the calculated variation of the inverse participation ratio P as a function of W/ZV for a hypercubic lattice in 1, 2, 3, 4, and 5 dimensions; (b) shows the variation with dimensionality of the critical value of W/ZV for localization in the centre of the band. These figures are taken from Weaire and Srivastava [24].

entire band. Weaire and Srivastava [88] have studied a less pathological distribution of V_{ij} and conclude that there are very few localized states. This conclusion is in accord with the arguments of Economou and Antoniou [89] and Hoshino and Watabe [90]. This kind of distribution is likely to be more relevant to the spin glass problem.

7.2. Boundary conditions and resistance

In the work of Edwards and Thouless [80] and Licciardello and Thouless [81] the localized states are recognized by their insensitivity to boundary conditions. If the range of localization is much less than the linear dimensions of the system the amplitude should be exponentially small at the boundaries, and so the shift in energy level due to a change in boundary conditions should be exponentially small. To avoid surface effects periodic and antiperiodic boundary conditions are used.

The sensitivity to boundary conditions of the energy levels can be closely related to the conductance of the system at that energy. One way of showing this is to consider the perturbation used in going from periodic to antiperiodic boundary conditions, which is proportional to the current operator. Second order perturbation theory for the energy shift involves the same matrix elements as the Kubo formula for conductivity, and a direct connection can be made in that way. Alternatively an argument can be made on the basis of the uncertainty relation between time and energy. Suppose that an electron is initially in a wave packet at the centre of the system, which is supposed to be many mean free paths across. Until it has had time to reach the boundary, at a distance $L_1/2$, its motion is insensitive to boundary conditions. Once it has had time to reach the boundary its motion depends strongly on the boundary conditions. If we write T for the time taken to go a distance L_1, and D for the diffusion constant we have

$$\Delta E = \hbar/T = \hbar D/L_1^2 . \tag{7.2}$$

If we combine this with the Einstein relation between conductivity, diffusion constant and density of states

$$\sigma = \tfrac{1}{2}e^2 D \, \mathrm{d}n/\mathrm{d}E , \tag{7.3}$$

we get the sensitivity of energy levels to boundary conditions in a system with sides L_1, L_2, L_3 as

$$\Delta E = \frac{2\hbar}{e^2}\frac{\sigma L_2 L_3}{L_1}\frac{1}{L_1 L_2 L_3}\frac{\mathrm{d}E}{\mathrm{d}n} = \frac{2\hbar}{e^2}\frac{1}{r}\frac{\mathrm{d}E}{\mathrm{d}N}, \tag{7.4}$$

where r is the resistance of the system and $\mathrm{d}E/\mathrm{d}N$ is the spacing between energy levels. Therefore the ratio of the spacing between energy levels to the sensitivity to boundary conditions is proportional to the resistance of the system.

This result leads to a method of interpreting the results and extrapolating to larger systems. We consider initially the square lattice. The real problem does not involve a square of 10 × 10 or 100 × 100 sites with periodic or antiperiodic boundary conditions, but a square surrounded by statistically similar squares. The sensitivity to boundary conditions ΔE gives the strength of coupling V' of a level on one square to its neighbours (see fig. 7.3). The spacing $\mathrm{d}E/\mathrm{d}N$ measures the amount W' by which an energy level on one square fails to match the nearest energy level on its neighbour. We therefore get a new, rescaled, form of the original Anderson problem, with the degree of disorder given by

$$W'/V' = re^2/2\hbar \,. \tag{7.5}$$

It is therefore the resistance per square that determines whether or not states are localized. It was this argument that led to the prediction of a universal value of the maximum metallic resistance per square for disordered systems [91].

A further example of the use of this method is provided by the paper of Thouless and Elzain [92]. This work was prompted by the suggestion of Pepper [93] that the difference between those inversion layers that displayed the expected value of the minimum metallic conductivity and those that do not [94–96] is that the disorder in the former has a short range, while the disorder in the latter has a long range. To study the effect of short range disorder, the white noise limit, the Anderson model was studied with a small value of W/V (actually 2 was sufficiently small) so that the localization edge occurred near the band edge. There the wavelength is much longer than the spacing between sites, so the white noise limit is well approximated. In this limit the density of states can be calculated in the main part of the band by the coherent potential approximation [97], and in the tail of the band by the method of Halperin and Lax [98]; there the density of states falls off exponentially with energy in two

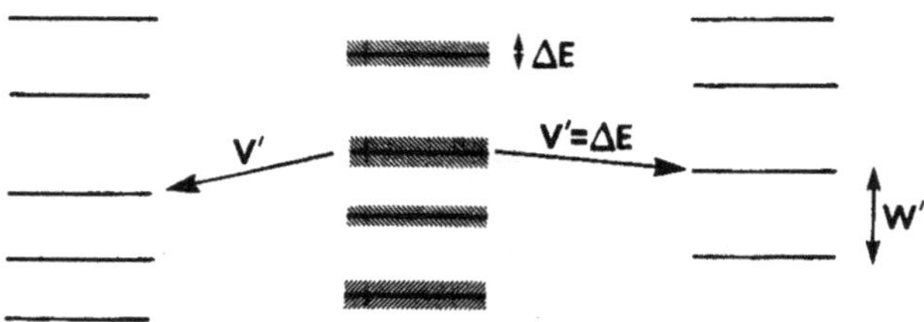

Fig. 7.3. This represents a large cell, with spacing W' between its energy levels, coupled to its neighbouring cells by a coupling V'. The magnitude of V' is of order ΔE, the amount by which a level can be changed in energy by a change of boundary conditions.

dimensions. The calculated density of states matches with these two theories in the appropriate regions (see fig. 7.4). The conductivity was found to be roughly independent of the size of the system well into the band, but to decrease rapidly as the system gets larger further out towards the tail. The mobility edge where this transition occurs is well into the band, where the density of states is more than 0.8 of its maximum value, and where the coherent potential approximation works well. This suggests that localization is not just restricted to those states in the tail sustained by anomalous fluctuations of the potential, but is due to a more subtle interference effect such as occurs in one dimension. In the region of extended states the conductivity can also be calculated by using the coherent potential approximation, but, as has been observed earlier by Hoshino and Watabe [99], this approximation gives results that are higher than the numerical results by a factor of about 2.5.

These results are for the most part in good agreement with the measurements of Pollitt [77] for the inversion layer. However, the Hall effect does not seem to behave in the manner that would be expected for non-interacting electrons, and this has led Adkins [100] to propose a completely different interpretation of these results.

For long thin wires these ideas give an even more surprising result. To build up a large system we put together cells consisting of a given length of the thin wire. So long as the resistance of the cell is sufficiently low the coupling between cells is strong, but if the resistance exceeds $2\hbar/e^2$ or so the coupling is weak, and electron states will remain localized instead of extending from one cell to another. Thus the theorem that states are localized in a one-dimensional system seems to extend also to long thin wires, and the localization length is the length of wire that gives a resistance of 10,000 ohm or so at low temperatures.

Some rather different results for two-dimensional systems have recently been obtained in numerical calculations by Prelovšek [101].

7.3. Scaling theory

The calculation outlined in the previous section suggests that the localization problem could be solved by a renormalization programme very similar to the procedure used by Wilson [102] to solve the Kondo problem. First a density of states and a distribution of coupling strengths (energy shifts) is calculated for a certain cell, large compared with the mean free path. Then attention is focussed on a narrow range of these energy levels, and the problem is solved again for a collection of cells. There may be a fixed point

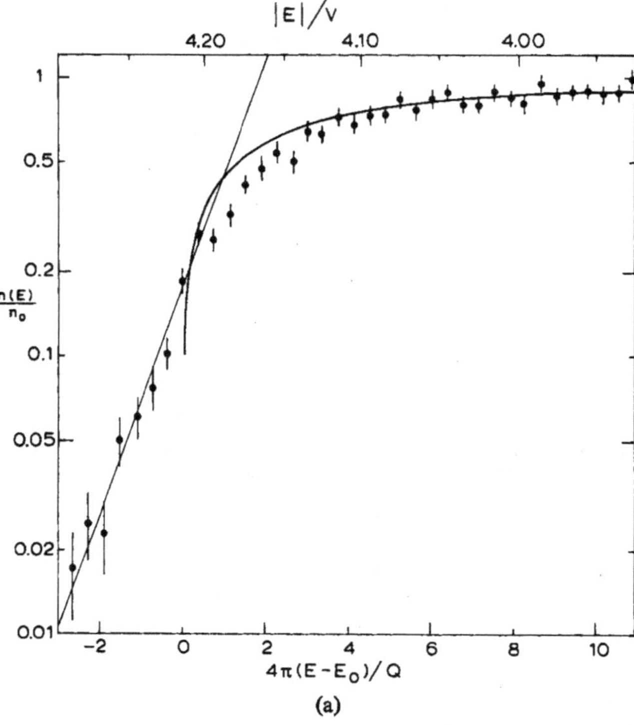

(a)

Fig. 7.4. Density of states and conductivity for a two-dimensional system with short-range disorder (white noise). (a) Density of states $n(E)$ on a logarithmic scale as a function of energy E. The curve gives the CPA density of states and the straight line the exponential tail. Calculated results are shown with error bars to show the standard errors. (b) Conductivity σ and inverse localization length κ as a function of energy E. The curve shows the CPA result for conductivity scaled down by a factor 0.4; the conductivities calculated for the four sample sizes are shown for 28 × 32 by ▽, for 40 × 44 by △, for 58 × 62 by ◇, and for 82 × 86 by □. The circles show estimated results for κ [92].

for which the probability distribution of coupling strengths, trivially rescaled by the energy spacing which is certainly inversely proportional to volume, has a limiting form as the size of the system tends to infinity. For localized states the coupling strengths get weaker and weaker rapidly as the system gets larger. In two dimensions, for extended states, all sufficiently large values of the coupling strength should give a fixed point. Attempts to implement such a programme have not been successful, and so we can only give a qualitative discussion. A closely related renormalization procedure

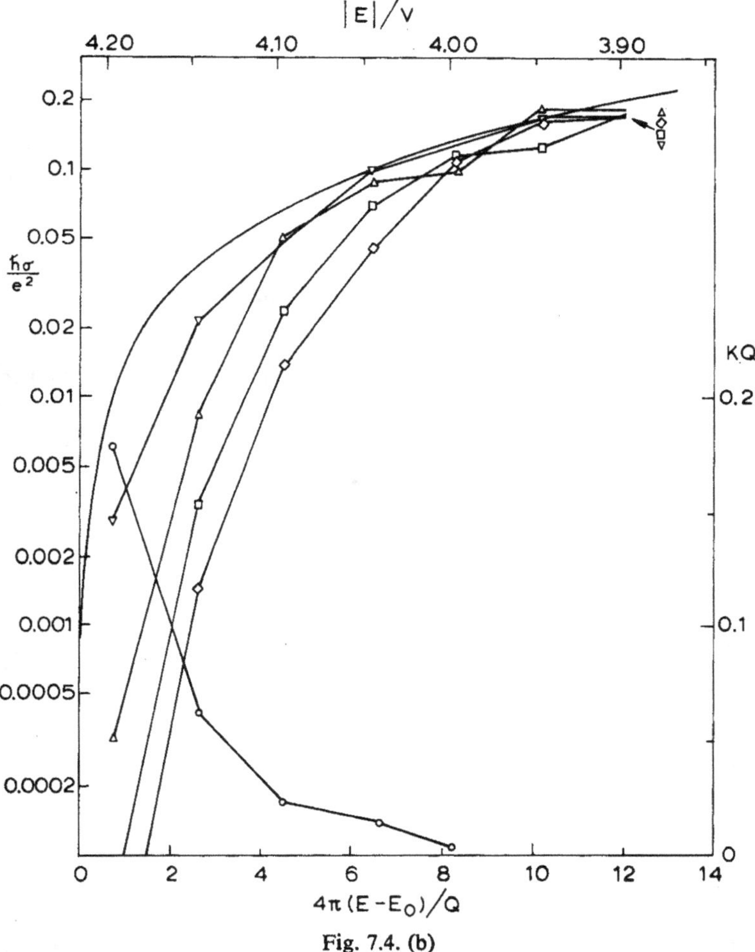

Fig. 7.4. (b)

has been proposed by Wegner [103], but the results are also somewhat inconclusive.

To develop a scaling theory we consider a block of material of dimensions $L_1 \times L_2 \times L_3$, where all the linear dimensions are longer than the mean free path λ and suppose that its properties are entirely characterized by the spacing between energy levels and the resistances in the three directions. The scaling hypothesis implies that the resistance r in the x

direction is a function of the resistance r_0 of a cube of dimensions $\lambda \times \lambda \times \lambda$ and of the ratio of the length L_1 to λ and the cross-sectional area $A = L_2L_3$ to λ^2, so that we have

$$r = g(L_1/\lambda, A/\lambda^2, r_0) . \tag{7.6}$$

Furthermore, from the arguments that have been presented in the previous section we should expect the effect of doubling L_1 to depend only on the value of r, and if this is correct it implies

$$r = f(L_1/\lambda L_0(A/\lambda^2, r_0)) . \tag{7.7}$$

According to the ordinary theory of metallic resistance, we have

$$f(x) \sim x , \qquad L_0(a, r_0) \sim a/r_0 , \tag{7.8}$$

and we should expect this to be the correct behaviour for small x and small r_0. We also expect that in the weak coupling regime, where r is large compared with $2\hbar/e^2$, doubling the length should square the ratio of the resistance to $2\hbar/e^2$, so that we have for sufficiently large x

$$f(x) \sim (2\hbar/e^2) \exp(x/x_0) , \tag{7.9}$$

with some uncertainty about the constant in front. The function $f(x)$ must change smoothly from its small x behaviour to its large x behaviour.

Some assumption must also be made about the behaviour of $L_0(a, r_0)$. If it has the form a/r_0 for all r_0 then there is no maximum resistance per square and no minimum metallic conductivity in three dimensions. This part of the theory still has to be worked out. Further work along these lines has been done by Abrahams et al. [140], and a numerical calculation of the scaling function in two dimensions has been made by Lee [141].

8. Electrical conductivity

8.1. Conduction in a static lattice

For non-interacting electrons in a static lattice the conductivity can in principle be calculated from the Kubo formula, but since it is usually very difficult to evaluate the matrix elements involved in the formula we make use only of the general structure. We can write this formula in the two equivalent forms

$$\begin{aligned}\sigma(\omega) &= \frac{e^2\pi}{m^2\omega\Omega} \sum_{\alpha} \sum_{\beta\neq\alpha} |\hat{p}_{\alpha\beta}|^2 (f_\beta - f_\alpha)\delta(\varepsilon_\alpha - \varepsilon_\beta - \hbar\omega) \\ &= \frac{e^2\pi}{\Omega} \sum_{\alpha} \sum_{\beta\neq\alpha} |\hat{x}_{\alpha\beta}|^2 (f_\beta - f_\alpha)\omega\delta(\varepsilon_\alpha - \varepsilon_\beta - \hbar\omega) ,\end{aligned} \tag{8.1}$$

where Ω is the volume of the system, ω is the frequency, $\hat{p}$ and $\hat{x}$ are components of the momentum and position operators between states of energy ε_α, ε_β, whose probabilities of occupation in thermal equilibrium are given by f_α, f_β. At low temperatures the first form of this result can be written as

$$\sigma(\omega) = \frac{\pi e^2 \hbar \Omega}{2m^2} \int_{E_F - \hbar\omega}^{E_F} \frac{|\hat{p}|^2_{\text{av}} n(E) n(E + \hbar\omega)\, \mathrm{d}E}{\hbar\omega} . \tag{8.2}$$

The standard results of kinetic theory can be recovered if we assume that the eigenstates can be written in the form

$$|\alpha\rangle = \sum_k a_k^\alpha |k\rangle \tag{8.3}$$

where the a_k^α are independent random Gaussian variables whose variance is approximately

$$\langle |a_k^\alpha|^2 \rangle = (\pi/\lambda k^2 \Omega)[(k - k_\alpha)^2 + \tfrac{1}{4}\lambda^2] , \tag{8.4}$$

and $\varepsilon_\alpha = \hbar^2 k_\alpha^2/2m$. Substitution of this form gives

$$\begin{aligned} |\hat{p}|^2_{\text{av}} &= \frac{1}{6\lambda^2\Omega} \int \frac{\mathrm{d}k}{[(k - k_\alpha)^2 + \frac{1}{4}\lambda^2][(k - k_\beta)^2 + \frac{1}{4}\lambda^2]} \\ &= (2\pi/3\lambda\Omega)[(k_\beta - k_\alpha)^2 + \tfrac{1}{4}\lambda^2]^{-1} \end{aligned} \tag{8.5}$$

and this yields the Drude formula

$$\sigma(\omega) = (k_F^3 e^2 \tau / 3\pi^2 m)(1 + \omega^2 \tau^2)^{-1} . \tag{8.6}$$

For localized states the operator $\hat{x}$ has finite matrix elements, and therefore $\hat{p}$ has elements that tend to zero as the energy difference between the states goes to zero. The second form of the expression (8.1) is then proportional to ω^2, as there is one explicit factor of ω and one implied by the factor $f_\beta - f_\alpha$. In fact for small ω the states which dominate the sum will be those in which there are two localized levels separated by the distance such that the tunnelling frequency between them is of order ω. For exponentially localized states this distance is proportional to $\ln \omega$, and as a result the frequency dependence for low frequencies is

$$\sigma(\omega) \simeq \omega^2 (\ln \omega)^{d+1} . \tag{8.7}$$

In the one-dimensional case Berezinskii [104] and Abrikosov and Ryzhkin [66] have shown that the exact solution has this form at low frequencies.

If the localization occurs in a region where the distortion from the free electron wave function is not very great, as it seems to in the two-dimensional white noise calculations of Thouless and Elzain [92], then it is reasonable to expect the approximation (8.3) for the wave function to be usable for fairly large values of ω, so that the correlations that give rise to localization are unimportant. One would in that case get a return to the Drude form (8.6) for large frequencies. In any case, from the dipole sum rule, the integral of $\sigma(\omega)$ must be the same for localized and extended states.

At non-zero temperatures the conductivity is given by a thermal average. For the dc case we have, from eq. (8.1),

$$\sigma(0) = \int \sigma_E(0)\, \mathrm{d}f/\mathrm{d}E\, \mathrm{d}E\,. \tag{8.8}$$

The localized states contribute nothing to this integral. If the Fermi energy is in a metallic region we get a conductivity which is independent of temperature. If the Fermi energy is in the region of localized states, then this integral should be dominated by the metallic states lowest in energy, at the mobility edge E_c, and the integral gives

$$\sigma(0) = \sigma_{\min} \exp[(E_F - E_c)/kT]\,. \tag{8.9}$$

This formula is widely used in determining the minimum metallic conductivity $\sigma_{\min}$. A plot of $\ln \sigma$ against $1/T$ should give a straight line whose intercept is $\ln \sigma_{\min}$.

Further details of these arguments about the conductivity can be found in Mott and Davis [52].

8.2. *Hopping conductivity*

In the metallic region of a highly disordered material the presence of phonons is not expected to have a major influence on the conductivity, since at low temperatures impurity scattering should be much more effective than phonon scattering. In the region of localized states current can only flow if electrons can hop from one level to an overlapping level, and for this to occur energy must be supplied, since neighbouring levels must have a different energy. This energy can come from the phonons, or it can also come from other electrons through the electron–electron interaction. This problem of hopping conductivity for localized states is also discussed in detail by Mott and Davis [52] and by Mott [53].

At low temperatures one might expect dc hopping conductivity to be controlled by a term of the form $\exp(-W'/kT)$, where W' is a typical energy separation between neighbouring localized states. Mott [105] however argued that at low temperatures there would be a tendency for longer range hopping, since although the rate for hopping a distance R decreases exponentially as $\exp(-2\alpha R)$, yet a longer hop will enable the electron to go to a state closer in energy. Since the number of states in a range R is proportional to R^d, the activation energy needed should be proportional to R^{-d}. The product of these two exponentials is greatest for

$$\frac{1}{kT}\frac{9}{4\pi R^4}\frac{\mathrm{d}n}{\mathrm{d}E} - 2\alpha = 0\,, \tag{8.10}$$

so that the hopping distance is proportional to $T^{-1/4}$, and the conductivity is given by

$$\sigma \simeq A\exp[-(T_0/T)^{1/4}]\,, \tag{8.11}$$

where T_0 is given roughly by

$$T_0 \simeq (512\alpha^3/9\pi k)\,\mathrm{d}n/\mathrm{d}E\,. \tag{8.12}$$

A similar argument in two dimensions replaces the exponent $\frac{1}{4}$ by $\frac{1}{3}$. This form of the conductivity has been observed in many systems (see, e.g., fig. 6.1), and gives a clear indication of the existence of a continuum of localized states near the Fermi level, although other mechanisms may give similar behaviour, at least over a limited temperature range (see, e.g., ref. [106]).

An important aspect of the problem is left out in this treatment, as was pointed out by Ambegaokar et al. [107] and by Pollak [108]. We have calculated the easiest hopping process from a given site, but dc conduction can only occur if there is a continuous chain of possible hops across the system. For this reason we have to consider a percolation process. Since there is an exponential variability of the individual links the conductivity will be dominated by the threshold, the path for which the maximum resistance is as low as possible, and the power law variation of percolation conductivity with $p - p_c$ should be unimportant.

If we consider two localized states with energies $\varepsilon_\alpha > \varepsilon_\beta$ a distance R apart, the rate of transitions from α to β is $W^0_{\beta\alpha}$, and the rate of transitions from β to α is

$$W^0_{\alpha\beta} = W^0_{\beta\alpha}\exp[(\varepsilon_\beta - \varepsilon_\alpha)/kT]\,, \tag{8.13}$$

so that there is detailed balance in thermal equilibrium. If now electrostatic potentials V_α, V_β are applied to the levels, there will be a net transition rate

$$(1 - f_\beta)W_{\beta\alpha}f_\alpha - (1 - f_\alpha)W_{\alpha\beta}f_\beta = e(V_\alpha - V_\beta)W^0_{\beta\alpha}f_\alpha(1 - f_\beta)\,. \tag{8.14}$$

This allows a resistance given by

$$r_{\alpha\beta}^{-1} = e^2 W_{\beta\alpha}^0 f_\alpha^0 (1 - f_\beta^0) \tag{8.15}$$

to be associated with each pair of localized states.

Various attempts have been made to solve this percolation problem, and the results are tabulated by Seager and Pike [109]. The results for $T_0^{1/4}$ do not differ much from Mott's value given in eq. (8.12). One important consequence of the percolation argument, pointed out by Kurkijärvi [110], is that since there is no percolation in one dimension the formula cannot be extended to give a $T^{-1/2}$ exponent, but the correct exponent is T^{-1}.

An aspect of eq. (8.15) that must be regarded with suspicion is that for a given energy difference the exponential factors are largest for two levels on opposite sides of the Fermi surface, so that hops from hole states to particle states and back are favoured. This comes about because in the derivation of eq. (8.15) it is assumed that the probability of occupation of a site is given by the Fermi function. Now in fact if an electron hops from a state below the Fermi surface to one above, the hole it leaves behind remains empty unless a further hopping process occurs. The most likely outcome is that the electron will return to the hole it vacated, but by associating a Fermi factor with the hole we make this unlikely. It would probably be better to treat electron hopping and hole hopping independently, and not to consider these transitions across the Fermi surface.

For ac hopping conductivity the percolation problem does not arise. The theory is based on Debye's theory of dielectric relaxation. For a dipole with an energy difference ΔW between its states and a relaxation time τ this gives

$$\sigma(\omega) = \frac{nD^2}{3kT} \frac{1}{1 + \exp(\Delta W/kT)} \frac{\omega^2 \tau}{1 + \omega^2 \tau^2}, \tag{8.16}$$

where n is the concentration of dipoles and D is their moment. In a disordered material there will be a distribution of such dipoles corresponding to all pairs of neighbouring localized states, and so a distribution of values of τ. If it is assumed that we are dealing with exponentially localized states, then τ will depend exponentially on the separation, and this formula yields a conductivity form

$$\sigma(\omega) \simeq A(\ln \omega)^4 kT\omega \,. \tag{8.17}$$

In practice $\sigma(\omega)$ usually depends on ω with a power that varies in the range from 0.6 to 1.

8.3. One-dimensional conductivity

In sect. 7.2 it was argued that all eigenstates in a long thin wire, like those in a true one-dimensional system, should be localized. I believe that Mott's [111] argument for localization in one dimension leads to the same conclusion. It is obviously important to understand why this does not lead to paradoxical results such as that, in accordance with eq. (7.8), the resistance of a wire whose resistance is in excess of 10^4 ohm should increase exponentially with its length. Recent discussions of this problem have been given by Thouless [112] and Adkins [113].

A partial answer to this paradox is that the effects of thermally activated transitions from one level to another ensure that only at very low temperatures can the localization be seen. It is possible that the effects of electron–electron interactions at zero temperature are also important, but this has not been explored. At first sight the conventional description of the hopping process in terms of a transition rate from one localized state to another seems to give a completely different result from what would be expected from extended states. If, as seems reasonable, the lifetime of an electron in a localized state is more or less independent of the localization length, then the electron can hop a distance equal to the localization length in this lifetime, and so the resistance is inversely proportional to the square of the localization length. This description in the conduction process is however, inadmissible if the lifetime is less than $\hbar$ divided by the spacing between overlapping levels, and the description has to be reformulated in terms of the diffusion of electron wavepackets (see fig. 8.1). If the electron is initially localized in a region much smaller than the localization distance the wavefunction is made up of a wavepacket of these localized states. The motion of this wavepacket will initially be just the same as if it were made up of extended states (incoherent superposition of Bloch waves as in eq. (8.3)). If there is a collision with a phonon the wavepacket is relocalized and the diffusion process starts again. It is only if the electron can diffuse a distance comparable with the localization length without a transfer of energy that the localization can be effective. Then the electron has to wait

Fig. 8.1. Overlapping localized energy levels.

for a process that transfers energy before it can go further. It is estimated that at sufficiently low temperatures the electron–electron collisions will be the most effective mechanism, and these should lead to a resistance which increases as T^{-2}.

To observe these effects in a wire of diameter 50 nm with a localization length of 10 μm it has been estimated that a temperature of order 1 K will be necessary. Since the spacing between overlapping localized levels is then of order 10^{-8} eV it can be argued that electric fields of less than 10^{-3} V m^{-1} should be used to avoid mixing of the localized levels. Such weak fields, which imply current of the order of 1 nA, should not be necessary, as there should be localized states, different ones, in the presence of a uniform field, and the only condition on the field is that it should not disturb the thermal equilibrium of the electrons, so that the voltage drop in a localization length should be much less than kT.

If preliminary experimental results that suggest this effect is not to be found are confirmed, they will indicate that something important is missing in the theory of electrons in disordered systems.

9. Other theories of disorder

9.1. Long-range disorder

It was pointed out by Ziman [114] that in a potential that varied slowly in relation to the electron wavelength electrons would only be able to penetrate into regions in which the potential energy was less than the electron energy. At low energies there would be separated regions of space where the electron could exist, and as the energy was raised these regions would grow and merge until eventually at some threshold energy the electron could move through the whole system. Only if the Fermi energy was higher than this threshold energy for percolation would the system behave as a metal.

These ideas have been developed in much more detail by Cohen et al. [115]. In this review there is a summary of the work and references to the original papers. In this work the ideas developed for inhomogeneities sufficiently broad that a classical description of the electrons is appropriate are extended to a regime in which the inhomogeneities are on a much smaller scale.

Pepper [93] has argued that those measurements on inversion layers which give a higher than expected value of the minimum metallic conductivity are due to the existence of long-range fluctuations of the potential. If these fluctuations are Gaussian, or if they are due to the symmetrical effects of positive and negative charges, the percolation threshold comes at an energy such that the potential is equally likely to be higher or lower, than the median energy. This implies that the semiclassical density of states is half of its maximum value.

In such a model some care has to be exercised in determining the behaviour of the conductivity at the threshold. Two considerations have to be borne in mind. Firstly there can be tunnelling between classically inaccessible regions. Secondly, even above the classical percolation threshold there can be narrow channels which act essentially as tunnelling barriers in quantum theory. Extended states therefore cannot exist until the tunnelling rate is comparable with the spacing between energy levels within well-connected regions. This argument suggests that the minimum metallic conductivity should be unchanged by the long-range potential. However, the transition from the classical insulator to the classically connected conductor could be sufficiently abrupt that the true minimum metallic conductivity was not seen, but a higher conductivity was apparent. Alternatively the temperature might be sufficiently high that the localization of the states was not seen – the observed conductivity would come from the conducting regions, and the tunnelling between the barriers would be too rapid to affect it.

9.2. Analogy with phase transitions

There has been a considerable amount of work which has exploited the analogy between the problem of an electron in a random potential and the statistical mechanics of a polymer [116–118]. Since the polymer problem can be regarded as the limit of a spin system with n degrees of freedom when n tends to zero this analogy has enabled many of the techniques of the renormalization group to be applied to the problem. Some recent papers on this subject are by Nitzan et al. [119], by Aharony and Imry [120], by Sadovskii [121] and by Cardy [122].

A common feature of these papers is that they study the white noise potential which is indeed analogous to the Landau–Ginzburg–Wilson free energy in phase transition theory. There is no doubt that this does some peculiar things in four dimensions. If we consider a Gaussian potential fluctuation, its strength is measured by the correlation strength.

$$\Phi = \int \langle (v(\boldsymbol{r}) - v)(v(\boldsymbol{r}') - v) \rangle \, d^3\boldsymbol{r}' \, . \tag{9.1}$$

In d dimensions this defines a characteristic length by

$$L^{4-d} = (\hbar^2/2m)^2/\Phi\,, \tag{9.2}$$

and we have a white noise problem if the correlation length for the potential fluctuation is much less than this.

In the four-dimensional case there is no scaling length, but the right-hand side reduces to a dimensionless measure of the reciprocal of the degree of disorder. There is no white noise limit, but the correlation length always affects the form of the spectrum however short it is. The tail of the energy band is indeed given by semiclassical considerations, in that the wave-function can be confined within a single correlation length without the kinetic energy overwhelming the potential energy. If the disorder is sufficiently strong one should expect all states whose wavelength is greater than the correlation length to be localized. These arguments are given in more detail by Thouless [123], and the existence of localization in four and five dimensions has been shown by Weaire and Srivastava [83,84,88]. It is not clear to me that any of the works cited that exploit the analogy with phase transitions are really describing localization rather than the anomalous behaviour of the tail of the density of states. If the critical dimensionality is actually greater than four some way must be found to get round this awkward behaviour of the tail.

10. Correlation and disorder

10.1. The Mott–Hubbard transition

In this chapter a brief discussion is given of some of the ways in which the mutual interactions of the electrons and the dynamic interaction with the lattice can modify the effects of disorder. Our understanding of this subject is far from complete, but there are several aspects of the problem that have been recognized as important.

Both the Mott transition and Wigner crystallization, which is discussed in the next section are described in detail in chapter 4 of Mott's book [53]. It is argued that if the atoms of a metal were constrained to be separated by a much larger distance than usual the Coulomb interaction between electrons and positive ions would trap the electrons and an insulator would be formed. Only when there are enough free electrons present to screen this interaction can the system be a metal. Correspondingly it is expected that when hydrogen is subjected to very high pressure it will become metallic.

The Hubbard [124] model also predicts a transition between a metallic and an insulating state, but the mechanism does not involve the long-range effects of the Coulomb force. The Hubbard hamiltonian describes electrons that are free except that there is an interaction U between two electrons on the same atom. Since two s-electrons with the same spin direction cannot be on the same atom the interaction is between electrons of opposite spin, and so electrons of one spin move in a potential determined by the instantaneous position of the electron of the opposite spin. If U is larger than the bandwidth the spin-up electrons can split the band for the spin-down electrons into two separate sub-bands, and similarly the spin-down electron will split the spin-up band. If the number of electrons is equal to the number of sites both lower Hubbard sub-bands will then be full, and the system will be an insulator. However if there are somewhat less or somewhat more electrons than sites, or if a few electrons are excited to the upper sub-band, then the electrons in the upper sub-band and the holes in the lower sub-band will be in extended states and fully mobile in contrast to the situation that exists with Anderson localization. The ground state of this system is antiferromagnetic, but the arguments for reduced conductivity apply even in the paramagnetic state.

At least in the antiferromagnetic state the presence of disorder should modify this picture considerably. An electron added to the upper sub-band will, like any other electron in a band tail, be localized by the disorder, so that there can be a region between the sub-bands in which the density of states is non-zero but the mobility is zero. However, a short ranged random potential has more effect on the position of the band edge than on the position of the mobility edge relative to the band edge. In three dimensions for the Anderson model with small W/V the shift in the band edge is proportional to W^2/V, as can be seen from elementary perturbation theory, whereas the shift in the mobility edge relative to this is proportional to W^4/V^3, as can be seen either by using the Ioffe–Regel criterion or from the theory of Economou and Cohen [56], so that weak disorder does more to broaden the sub-bands than to narrow the mobility sub-bands. Therefore it would seem that a larger ratio of U to unperturbed bandwidth is needed to produce an insulating state in the presence of disorder than in its absence.

For the Mott transition similar considerations should apply. The presence of disorder broadens the band, but the existence of localized states reduces screening, and therefore helps the transition to occur.

The problem can also be approached from the other end. If we imagine interacting electrons being placed in the lowest energy states in a random potential, those which go into the strongly localized states will themselves produce a localized potential which enhances the disorder in the potential. Therefore from either point of view Anderson localization and the Mott–Hubbard insulating mechanism reinforce one another.

10.2. Wigner crystallization

Wigner [125] pointed out that the lowest energy state for a low density of electrons in a positive background would be a regular lattice, a body-centred cubic lattice, so that if the Coulomb energy is sufficiently much more important than the Fermi kinetic energy and thermal energy, crystallization should occur. The mechanism is obviously very similar to the Mott transition except that there is no crystal lattice to determine the positions of the electrons and therefore the transition occurs much less readily than the Mott transition. On the other hand it can occur for any number of carriers. Once the lattice is formed it is presumably locked into position either by the boundaries or by pinning of the lattice by defects in the background, and so conduction can occur by vacancies or by interstitials in the Wigner lattice – these do not have to be in equal numbers since the lattice constant can adjust itself to keep the electron density constant. The interplay between the effects of disorder and Wigner crystallization should be very similar to the interplay between disorder and the Mott transition, so that we should expect the disorder to help the crystallization. Aoki [126] has made a study of the effect of Wigner crystallization in the presence of disorder for a two-dimensional electron gas in a strong magnetic field. This system is a particularly good candidate for the Wigner transition because the magnetic field freezes the transverse kinetic energy of the electrons [127]. The concept of a "Wigner liquid" has also been used by Adkins [100] in his interpretation of inversion layer experiment in the absence of a strong magnetic field.

10.3. The interaction of localized electrons

The localization of electrons should profoundly modify their interactions, and this in turn should alter the localization properties, as Anderson [128] has observed. Firstly the Coulomb interaction between localized electrons is repulsive, and so a second electron on a particular localization site has more energy than the first. This implies that if the Fermi energy lies in the region of localized states, or even just above it, there will be unpaired electrons in these states which can give a Curie-like magnetic susceptibility and a paramagnetic resonance signal. This seems to be the situation in some doped semiconductors and in amorphous Si and Ge.

Secondly the localization of the electrons implies that the Coulomb interaction will not be screened at low temperatures, although the dielectric constant will be high if the localized states near the Fermi surface are big. This is obviously important for the Mott and Wigner transitions. There is a detailed discussion of screening in the one-dimensional system by Bush [129].

Thirdly it is possible for the lattice to mediate a short-range attractive force between electrons which can overcome the Coulomb repulsion between them and produce a negative Hubbard U. This idea was proposed by Anderson [130,131] and developed by Street and Mott [132]. A single electron may find an energetically favourable location, and the atoms will then adjust themselves so as to lower the energy. A second electron will then be more tightly bound and cause an even greater adjustment of the atoms. In this way a pair of electrons can form something like a covalent bond in the band gap of a disordered solid. Anderson suggested that in materials such as chalcogenide glasses only paired electrons would be localized, and unpaired electrons would be mobile. The Fermi energy should be pinned by the pairs to the middle of the gap, and no paramagnetic effects should be observed.

10.4. Hopping conductivity and the Coulomb interaction

It has been argued by Efros and Shklovskii [133] that the Coulomb interaction between localized states causes a reduction in the density of states which are close in energy and space in the neighbourhood of the Fermi surface. If there is a state at the position $\boldsymbol{r}_i$ just below the Fermi surface there cannot be another state at the position $\boldsymbol{r}_j$ whose energy is greater than that of the first by an amount less than $e^2/\varepsilon|\boldsymbol{r}_i - \boldsymbol{r}_j|$, because if there was the electron in the first state would be able to move to the second and lose energy because of its interaction with the hole it left behind. It is claimed that such a local reduction in the density of states leads to an activation energy for a hop of length R proportional to $e^2/\varepsilon R$, and minimization of

$$e^2/\varepsilon RkT - 2\alpha R\,, \tag{10.1}$$

leads to an $\exp(-T^{-1/2})$ conductivity. Mott [134] disagrees with this, and claims that when adjustment of the other electrons in the neighbourhood is allowed for no such reduction of the density of states occurs.

I agree with Mott's conclusions. As electrons are added to the system they will go into the lowest available levels, and the addition of each extra electron must cause some adjustment of the others, in such a way as to preserve a more or less uniform density of electrons. As in any other normal many-fermion system the chemical potential – the energy of the last electron – should be a continuous function of the number of electrons. As we have discussed in sect. 8.2 the main contribution to dc hopping conductivity does not come from transitions between hole and particle states, but from transitions between an extended network of possible particle or hole states. The spatial distribution of these unoccupied states (or of the hole states) is not affected by the argument of Efros and Shklovskii, and so the $T^{-1/4}$ law should not be altered by the Coulomb interaction.

In very thin metallic films, which seem to consist of metallic islands separated by tunnelling barriers, the conductivity does seem to follow an $\exp(-T^{-1/2})$ law at low temperatures [76,135]. The derivation of such a law depends on assumptions made about the statistics of size and separation of the metallic islands [136,137].

Recently Fleishman et al. [138] and Fleishman and Anderson [139] have studied the conditions under which electron–electron interactions can produce hopping. If the interaction is short range no transitions can be produced, since energy conservation is not generally possible for transitions between localized electron states in a finite volume. The phonons can induce hopping because they have a continuous spectrum. The matrix elements involved when two distant electrons change their state simultaneously are dipolar, and these fall off sufficiently slowly that in three dimensions energy can be conserved, and the electron–electron interaction causes hopping. In two dimensions the dipolar force falls off too fast to produce hopping.

References for sections 1–5 (Percolation)

[1] S. R. Broadbent and J. M. Hammersley, Proc. Camb. Phil. Soc. 53 (1957) 629.
[2] H. L. Frisch and J. M. Hammersley, J. Soc. Indust. Appl. Math. 11 (1963) 894.
[3] V. K. S. Shante and S. Kirkpatrick, Adv. Phys. 20 (1971) 325.
[4] J. W. Essam, in: Phase Transitions and Critical Phenomena, Vol. 2, eds. C. Domb and M. S. Green (Academic Press, New York, 1972) p. 197.
[5] S. Kirkpatrick, Rev. Mod. Phys. 45 (1973) 574.
[6] R. J. Elliott, B. R. Heap, D. J. Morgan and G. S. Rushbrooke, Phys. Rev. Letters 3 (1960), 366.
[7] D. J. Breed, K. Gilijamse, J. W. E. Sterkenburg and A. R. Miedema, J. Appl. Phys. 41 (1970) 1267.

[8] D. J. Breed, Physica 68 (1973) 303.
[9] R. A. Cowley, G. Shirane, R. J. Birgenau and H. J. Guggenheim, Phys. Rev. B15 (1977) 4292.
[10] A. G. Dunn, J. W. Essam and J. M. Loveluck, J. Phys. C8 (1975) 743.
[11] R. Landauer, in: Electrical Transport and Optical Properties of Inhomogeneous Media, eds. J. C. Garland and D. B. Tanner (American Institute of Physics, 1978) p. 2.
[12] M. E. Fisher and J. W. Essam, J. Math. Phys. 2 (1961) 609.
[13] M. E. Fisher, J. Math. Phys. 2 (1961) 620.
[14] J. M. Hammersley, Ann. Math. Statist. 28 (1957) 790.
[15] C. Domb, J. Phys. C3 (1970) 256.
[16] P. Erdös and A. Rényi, Magyar Tud. Akad. Mat. Kut. Int. Közl. 5 (1960) 17. Reprinted in: P. Erdös, The Art of Counting (MIT Press, Cambridge, Mass., 1973) p. 574.
[17] M. F. Sykes and J. W. Essam, J. Math. Phys. 5 (1964) 1117.
[18] J. P. Straley, Phys. Rev. B15 (1977) 5733.
[19] J. Marchant and R. Gabillard, C.R. Acad. Sci. B281 (1975) 261.
[20] C. Domb, J. Phys. C7 (1974) 2677.
[21] D. Stauffer, J. Phys. C8 (1975) L172.
[22] P. L. Leath, Phys. Rev. B14 (1976) 5046.
[23] H. Kunz and B. Souillard, Phys. Rev. Letters 40 (1978) 133.
[24] C. Domb and N. W. Dalton, Proc. Phys. Soc. 89 (1966) 859.
[25] A. G. Dunn, J. W. Essam and D. S. Ritchie, J. Phys. C8 (1975) 4219.
[26] M. E. Levinshtein, B. I. Shklovskii, M. S. Shur and A. L. Efros, Soviet Phys. JETP 42 (1975) 197.
[27] S. Kirkpatrick, Phys. Rev. Letters 36 (1976) 69.
[28] A. Sur, J. L. Lebowitz, J. Marro, M. H. Kalos and S. Kirkpatrick, J. Stat. Phys. 15 (1976) 345.
[29] B. J. Last and D. J. Thouless, Phys. Rev. Letters 27 (1971) 1719.
[30] D. Adler, L. P. Flora and S. D. Senturia, Solid State Commun. 12 (1973) 9.
[31] J. P. Fitzpatrick, R. B. Malt and F. Spaepen, Phys. Letters 47A (1974) 207.
[32] B. P. Watson and P. L. Leath, Phys. Rev. B9 (1974) 4893.
[33] S. Kirkpatrick, Phys. Rev. Letters 27 (1971) 1722.
[34] A. B. Harris, T. C. Lubensky, W. K. Holcomb and C. Dasgupta, Phys. Rev. Letters 35 (1975) 327.
[35] P. W. Kasteleyn and C. M. Fortuin, J. Phys. Soc. Japan (Suppl.) 26 (1969) 11.
[36] M. J. Stephen, Phys. Rev. B15 (1977) 5674.
[37] S. L. Ginzburg, Soviet Phys. JETP 44 (1976) 599.
[38] A. P. Young and R. B. Stinchcombe, J. Phys. C8 (1975) L535.
[39] S. Kirkpatrick, Phys. Rev. B15 (1977) 1533.
[40] D. A. G. Bruggeman, Ann. Physik 24 (1935) 636.
[41] S. Kirkpatrick, in: Electrical Transport and Optical Properties of Inhomogeneous Media, eds. J. C. Garland and D. B. Tanner (American Institute of Physics, 1978) p. 99.
[42] C. D. Mitescu, H. Ottavi and J. Roussenq, in: Electrical Transport and Optical Properties of Inhomogeneous Media, eds. J. C. Garland and D. B. Tanner (American Institute of Physics, 1978) p. 377.
[43] P. M. Kogut and J. Straley, in: Electrical Transport and Optical Properties of

Inhomogeneous Media, eds. J. C. Garland and D. B. Tanner (American Institute of Physics, 1978) p. 382.
[44] M. E. Levinshtein, M. S. Shur and A. L. Efros, Soviet Phys. JETP 42 (1975) 1120.
[45] P. G. DeGennes, J. Physique Lettres 37 (1976) L1.
[46] R. B. Stinchcombe, J. Phys. C7 (1974) 179.
[47] J. P. Straley, J. Phys. C10 (1977) 3009.
[48] A. S. Skal and B. I. Shklovskii, Sov. Phys. Semicond. 8 (1974) 1029.
[49] R. Landauer and J. A. Swanson, I.B.M. Technical Report, unpublished.
[50] B. I. Shklovskii, Sov. Phys. JETP 45 (1977) 152.
[50a] J. Hoshen, R. Kopelman and E. M. Monberg, J. Stat. Phys. 19 (1978) 219.

References for sections 6–10 (Localization)

[51] P. W. Anderson, Phys. Rev. 109 (1958) 1492.
[52] N. F. Mott and E. A. Davis, Electronic Processes in Non-Crystalline Materials (Clarendon Press, Oxford, 1971).
[53] N. F. Mott, Metal-Insulator Transitions (Taylor and Francis, London, 1974).
[54] D. J. Thouless, Phys. Reports C13 (1974) 94.
[55] E. Feenberg, Phys. Rev. 74 (1948) 206.
[56] E. N. Economou and M. H. Cohen, Phys. Rev. B5 (1972) 2931.
[57] D. C. Licciardello and E. N. Economou, Phys. Rev. B11 (1975) 3697.
[58] P. W. Anderson, Proc. Nat. Acad. Sci. US 69 (1972) 1097.
[59] B. J. Last and D. J. Thouless, J. Phys. C7 (1974) 715.
[60] M. V. Sadovskii, Soviet Phys. JETP 43 (1976) 1008.
[61] N. F. Mott and W. D. Twose, Adv. Phys. 10 (1960) 107.
[62] K. Ishii, Supp. Progr. Theor. Phys. 53 (1973) 77.
[63] F. J. Wegner, Z. Physik B22 (1975) 273.
[64] C. Papatriantafillou and E. N. Economou, Phys. Rev. B13 (1976) 920.
[65] D. C. Herbert and R. Jones, J. Phys. C4 (1971) 1145.
[66] A. A. Abrikosov and I. A. Ryzhkin, Soviet Phys. JETP 44 (1976) 630.
[67] A. A. Abrikosov and I. A. Ryzhkin, Adv. Phys. 27 (1978) 147
[68] A. A. Gogolin, V. I. Melnikov and E. I. Rashba, Sov. Phys. JETP 42 (1976) 168.
[69] R. Abou-Chacra, P. W. Anderson and D. J. Thouless, J. Phys. C6 (1973) 1734.
[70] N. Kumar, J. Heinrichs and A. A. Kumar, Solid State Commun. 17 (1975) 541.
[71] A. F. Ioffe and A. R. Regel, Progr. Semiconductors 4 (1960) 237.
[72] N. F. Mott, M. Pepper, S. Pollitt, R. H. Wallis and C. J. Adkins, Proc. Roy. Soc. A345 (1975) 169.
[73] C. J. Adkins, S. Pollitt and M. Pepper, J. Physique Colloq. 37, C4 (1976) 343.
[74] B. M. Vul, E. I. Zavaritskaya, Yu. A. Bashkirov and V. M. Vinogradova, JETP Letters 25 (1977) 187.
[75] M. Sayer, R. Chen, R. Fletcher and A. Mansingh, J. Phys. C8 (1975) 2059.
[76] R. C. Dynes, J. P. Garno and J. M. Rowell, Phys. Rev. Letters 40 (1978) 479.
[77] S. Pollitt, Commun. Phys. 1 (1976) 207.
[78] S. Yoshino and M. Okazaki, Solid State Commun. 20 (1976) 81.
[79] S. Yoshino and M. Okazaki, J. Phys. Soc. Japan 43 (1977) 415.
[80] J. T. Edwards and D. J. Thouless, J. Phys. C5 (1972) 807.

[81] D. C. Licciardello and D. J. Thouless, J. Phys. C8 (1975) 4157.
[82] D. C. Licciardello and D. J. Thouless, J. Phys. C11 (1978) 925.
[83] D. Weaire and V. Srivastava, J. Phys. C10 (1977) 4309.
[84] D. Weaire and V. Srivastava, in: Amorphous and Liquid Semiconductors, ed. W. E. Spear (1977) p. 286.
[85] M. Kikuchi, J. Phys. Soc. Japan 37 (1974) 904.
[86] M. Kikuchi, J. Phys. Soc. Japan 41 (1976) 1459.
[87] B. T. Debney, J. Phys. C10 (1977) 4719.
[88] D. Weaire and V. Srivastava, Solid State Commun. 23 (1977) 863.
[89] E. N. Economou and F. D. Antoniou, Solid State Commun. 21 (1977) 285.
[90] K. Hoshino and M. Watabe, J. Phys. Soc. Japan 43 (1977) 583.
[91] D. C. Licciardello and D. J. Thouless, Phys. Rev. Letters 35 (1975) 1475.
[92] D. J. Thouless and M. E. Elzain, J. Phys. C11 (1978) 3425.
[93] M. Pepper, Proc. Roy. Soc. A353 (1977) 225.
[94] D. C. Tsui and S. J. Allen, Phys. Rev. Letters 32 (1974) 1200.
[95] D. C. Tsui and S. J. Allen, Phys. Rev. Letters 34 (1975) 1293.
[96] A. Hartstein and A. B. Fowler, J. Phys. C8 (1975) L249; Phys. Rev. Letters 34 (1975) 1435.
[97] P. Soven, Phys. Rev. 156 (1967) 809.
[98] B. I. Halperin and M. Lax, Phys. Rev. 148 (1966) 722.
[99] K. Hoshino and M. Watabe, Solid State Commun. 18 (1976) 1111.
[100] C. J. Adkins, J. Phys. C11 (1978) 851.
[101] P. Prelovšek, Phys. Rev. Letters 40 (1978) 1596.
[102] K. G. Wilson, in: Nobel Symposia – Medicine and Natural Sciences, eds. B. and S. Lundqvist (Academic Press, New York, 1973) p. 68.
[103] F. J. Wegner, Z. Physik B25 (1976) 327.
[104] V. L. Berezinskii, Soviet Phys. JETP 38 (1974) 620.
[105] N. F. Mott, J. Non-Cryst. Solids 1 (1968) 1.
[106] D. Adler, L. P. Flora and S. D. Senturia, Solid State Commun. 12 (1973) 9.
[107] V. Ambegaokar, B. I. Halperin and J. S. Langer, Phys. Rev. B4 (1971) 2612.
[108] M. Pollak, J. Non-Cryst. Solids 11 (1972) 1.
[109] C. H. Seager and G. E. Pike, Phys. Rev. B10 (1974) 1435.
[110] J. Kurkijärvi, Phys. Rev. B8 (1973) 922.
[111] N. F. Mott, Adv. Phys. 16 (1967) 49.
[112] D. J. Thouless, Phys. Rev. Letters 39 (1977) 1167.
[113] C. J. Adkins, Phil. Mag. 36 (1977) 1285.
[114] J. M. Ziman, J. Phys. C1 (1968) 1532.
[115] M. H. Cohen, J. Jortner and I. Webman, in: Electrical Transport and Optical Properties of Inhomogeneous Media, eds. J. C. Garland and D. B. Tanner (American Institute of Physics, 1978) p. 63.
[116] S. F. Edwards, J. Non-Cryst. Solids 4 (1970) 417.
[117] R. A. Abram and S. F. Edwards, J. Phys. C5 (1972) 1183.
[118] K. F. Freed, Phys. Rev. B5 (1972) 4802.
[119] A. Nitzan, K. F. Freed and M. H. Cohen, Phys. Rev. B15 (1977) 4476.
[120] A. Aharony and Y. Imry, J. Phys. C10 (1977) L487.
[121] M. V. Sadovskii, Soviet Phys. Solid State 19 (1977) 1366.
[122] J. L. Cardy, J. Phys. C11 (1978) L321.
[123] D. J. Thouless, J. Phys. C9 (1976) L603.

[124] J. Hubbard, Proc. Roy. Soc. A281 (1964) 401.
[125] E. Wigner, Trans. Faraday Soc. 34 (1938) 678.
[126] H. Aoki, Surface Sci. 73 (1978) 281.
[127] Yu. E. Lozovik and V. I. Yudson, JETP Letters 22 (1975) 11.
[128] P. W. Anderson, Nature Phys. Sci. 235 (1972) 163.
[129] R. L. Bush, Phys. Rev. B13 (1976) 805.
[130] P. W. Anderson, Phys. Rev. Letters 34 (1975) 953.
[131] P. W. Anderson, J. Physique Colloq. 37, C4 (1976) 339.
[132] R. A. Street and N. F. Mott, Phys. Rev. Letters 35 (1975) 1293.
[133] A. L. Efros and B. I. Shklovskii, J. Phys. C8 (1975) L49.
[134] N. F. Mott, Phil. Mag. 34 (1976) 643.
[135] P. Sheng and B. Abeles, Thin Solid Films 41 (1977) L39.
[136] P. Sheng, B. Abeles and Arie, Phys. Rev. Letters 31 (1973) 44.
[137] J. Heinrichs, A. A. Kumar and N. Kumar, J. Phys. C9 (1976) 3249.
[138] L. Fleishman, D. C. Licciardello and P. W. Anderson, Phys. Rev. Letters 40 (1978) 1340.
[139] L. Fleishman and P. W. Anderson, to be published (1978).
[140] E. Abrahams, P. W. Anderson, D. C. Licciardello and T. V. Ramakrishnan, Phys. Rev. Letters 42 (1979) 673.
[141] P. A. Lee, to be published.

COURS 2

LES SYSTÈMES DÉSORDONNÉS

Aspect expérimental

Jacques JOFFRIN

Institut Laue–Langevin, 156 X, 38042 Grenoble Cedex, France

Contents

Table

R. Balian et al., eds.
Les Houches, Session XXXI, 1978 – La matière mal condensée/Ill-condensed matter

Introduction

Les systèmes désordonnés.... Le physicien fait semblant de les découvrir; mais celui qui feuillette un livre d'histoire, une encyclopédie et même sa mémoire, s'aperçoit que c'est tout au plus une redécouverte.

Si le physicien n'est pas insensible, et s'il admet que l'esthétique fait partie de la qualité de la vie, alors le plus beau des systèmes désordonnés du côté de la réalité est sans conteste le vitrail; pur produit de la péninsule européenne, il est le résultat d'une préparation qui a duré des millénaires; désordonné dans ses formes, divers dans ses teintes et ses couleurs, c'est par une alchimie qui traite de cette multiplicité d'un coup qu'un autre état a été reconstruit, produisant l'effet "rosace" qui s'étend de Chartres à Rouault.

Du coté du concept, le verre de spin avec toutes ses variantes a une autre perfection: celle du modèle; il n'est limité que par l'imagination du théoricien qui a trouvé un concurrent récent dans celui qui "simule". Tout cela pour dire que les physiciens n'ont pas inventé le sujet des systèmes désordonnés même s'ils en parlent d'une autre manière.

Les systèmes désordonnés du physicien ont une grande variété et il n'est pas simple de discerner quelles propriétés ils ont en commun. Tout au moins, il n'est pas évident que s'ils ont une chaleur spécifique linéaire en température, ce soit pour des raisons identiques, de sorte qu'il vaudra mieux procéder de manière analytique, décrivant chacun d'eux successivement, et réservant pour une dernière partie les concordances, les similitudes et peut-être les convergences.

Il y a deux grandes classes de systèmes désordonnés:
- les verres dont la grande variété a attiré l'attention des expérimentateurs depuis longtemps;
- les verres de spin qui ont focalisé l'intérêt des théoriciens en raison de la richesse et de la nouveauté des modèles qui leur sont associés.

Ce sont ces deux catégories qui formeront l'essentiel des chapitres suivants. Pourtant, d'autres systèmes désordonnés sont justiciables d'une étude similaire. Il y a d'abord le cas des métaux amorphes qu'on sait de mieux en mieux préparer sous forme de matériau massif; leurs propriétés, à basse température au moins, rappellent fortement celles des verres isolants; cela aidera à simplifier l'exposé.

Il y a ensuite le cas des matériaux magnétiques amorphes isolants ou métalliques de concentration égale à un. C'est un domaine qui se développe assez vite et dont on se contentera de faire un rapide survol en cherchant simplement à dresser une liste des cas possibles résultant des grandeurs relatives des forces d'échange, des forces d'anisotropie cristalline et des forces dipolaires.

Un inventaire des systèmes désordonnés, de ceux à propos desquels il existe un minimum d'informations expérimentales, doit aussi faire une place aux cristaux qui contiennent des impuretés moléculaires substitutionnelles distribuées au hasard; on décrira les particularités d'un tel système en cherchant à mettre en évidence ce qui le distingue des verres ou des verres de spin.

Enfin, dernier exemple et non des moindres: celui des systèmes polymériques; sous forme solide, et pour leurs propriétés au voisinage de la transition vitreuse, on verra comment ils se rapprochent des autres verres; en solution plus ou moins concentrée, un cours spécial de cette école leur est réservé.

1. Verres de spin

1.1. Définition

On appelle verre de spin un système obtenu en diluant dans une matrice simple monocristalline d'un métal noble (Cu, Ag, Au, . . .) des impuretés de transition (Mn, Fe, Cr, . . .). En concentration assez faible, c'est-à-dire pour environ 1%, on peut penser que les impuretés se substituent de manière aléatoire; on a un système à désordre de position; les fluctuations statistiques de concentration donnent naissance à des régions microscopiques où la concentration des atomes dilués est plus forte que dans d'autres; un bon verre de spin correspond à une distribution aléatoire des atomes substitués. Pour préparer un verre de spin, les impuretés introduites dans la matrice sont choisies pour conserver un moment magnétique; celui-ci a pour effet de polariser les électrons de conduction de la matrice. Les autres ions magnétiques perçoivent cette polarisation par l'intermédiaire d'une interaction indirecte entre impuretés dont la forme analytique est donnée dans tous les cours de physique des solides des universités; c'est celle de Rudermann–Kittel–Kasuya–Yosida. Pour deux impuretés i et j à la distance r_{ij} l'interaction effective s'écrit:

$$H_{ij} = -\frac{A}{r_{ij}^3}\cos(2k_F r_{ij} + \phi)\boldsymbol{S}_i\cdot\boldsymbol{S}_j\,. \tag{1.1}$$

On remarquera sur (1.1) que, pour les concentrations annoncées, $2k_F r_{ij}$ atteint très facilement la valeur de π, de sorte que le signe de H_{ij} est positif ou négatif suivant la distance des atomes de la paire; cette propriété est essentielle pour avoir un verre de spin. De plus, en raison du caractère aléatoire des distances entre spins, chacun d'eux est soumis par ses voisins à des sollicitations compétitives. Comme bien des interactions indirectes, celle-ci a un caractère dipolaire et décroit en $1/r^3$ à longue distance.

Si l'on réécrit H_{ij} sous la forme

$$H_{ij} = -J_{ij}\boldsymbol{S}_i\cdot\boldsymbol{S}_j\,. \tag{1.2}$$

J_{ij} a donc une distribution de valeurs en module et en signe; c'est l'origine de toute une série de "modèles". Enfin, on se souviendra que dans les cas simples cette interaction est du type Heisenberg (par opposition au cas Ising où l'on aurait un terme $S_i^z S_j^z$). L'ordre de grandeur de J dans les matrices comme le Cu est tel que:

$$J/K_B \simeq 10^3 c \text{ en degrés}\,, \tag{1.3}$$

où c est la concentration des impuretés.

L'expression (1.1) est intéressante à un autre titre: elle permet de s'affranchir du système de référence habituel pour adopter celui du système d'impuretés c'est à dire pour bâtir des lois de correspondance dans les propriétés thermodynamiques des alliages de concentration différente. Tout alliage de concentration c peut en effect servir de modèle pour un alliage de concentration c' pourvu que dans (1.1) on fasse la substitution

$$\begin{cases} c \to c' \\ r_{ij} \to r'_{ij} \end{cases} \text{avec} \quad c r_{ij}^3 = c' r_{ij}'^3\,. \tag{1.4}$$

La partie sinusoïdale de (1.1) oscille vite et lorsque l'on effectue des moyennes sur toutes les impuretés, sa valeur moyenne donnera toujours le même résultat. Les fonctions thermodynamiques rapportées à une impureté s'exprimeront donc comme des fonctions uniques des variables H/c, h/c, T/c où H est le champ moléculaire local, h le champ extérieur appliqué et T la température. Ces lois de correspondance ont été vérifiées pour la chaleur spécifique ou l'aimantation par impureté [1.1]. Elles autorisent en principe à ne considérer les propriétés que d'un seul alliage.

Dans le cadre d'une théorie de champ moyen, valable dans la limite des basses températures, on peut se demander à quelle loi statistique doit obéir la distribution des champs magnétiques locaux H_i appliqués à chaque spin i du fait de (1.1). Si μ_j est le moment magnétique du spin j on aura suivant (1.1)

$$H_i = \sum_j \mu_j \frac{\cos(\ )}{r_{ij}^3},$$

où μ_j, à basse température au moins, varie comme un moment de Brillouin:

$$\mu_j(T) = \mu_j^0 B\left(\frac{H_j + h}{KT}\right).$$

Les deux formules précédentes montrent que les quantités réduites qui gouvernent la thermodynamique du verre sont H/c, h/c, T/c puisque

$$\frac{H_i}{c} = \sum_j \mu_j^0 B\left(\frac{H_j/c + h/c}{KT/c}\right)\frac{\cos(\ )}{r_{ij}^3 c}.$$

Ainsi, l'énergie libre par unité de volume du verre de spin sera

$$F(T, h, c) = -KT \log Z^{Nc},$$

avec

$$Z = \exp\left(-\frac{\mu^0 H}{KT}\right) = \exp\left(-\frac{\mu^0 H/c}{KT/c}\right),$$

soit encore

$$F(T, h, c) = \cdot c^2 f(T/c, h/c),$$

d'où il résulte immédiatement

$$C_v(T, h, c) = -Tf''(T/c, h/c). \tag{1.5}$$

(1.5) implique qu'une chaleur spécifique linéaire en T est compatible avec ces lois de correspondance; ce sera le cas à basse température. De même, une loi du type c^2 est compatible avec (1.5); c'est l'expression de C_v dans la limite des hautes températures.

En conclusion, on parlera d'un verre de spin dans deux limites de concentration assez bien définies: il faut qu'elle soit assez élevée pour que les manifestations de l'effet Kondo, effet à un seul spin, n'interfèrent pas avec celles qui sont caractéristiques des verres de spin; pratiquement, il faut que c soit supérieure à quelques centaines de ppm. D'un autre côté, c doit être assez petite pour que des effets de précipités des impuretés ne

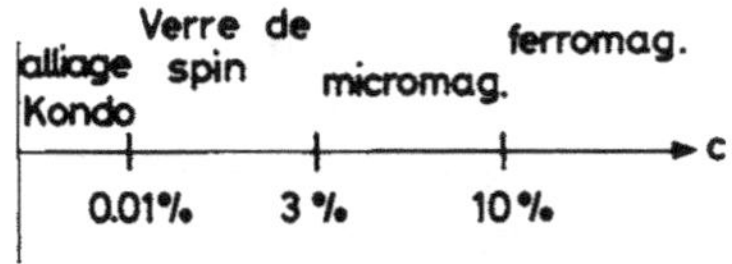

Fig. 1.1.

détruisent pas leur répartition aléatoire; $c = 10^{-2}$ est une limite qu'il ne faut pas dépasser.

Au-delà, on tombe dans un domaine que l'on appelle quelquefois mictomagnétisme: le nombre des premiers voisins est plus important et il faut tenir compte des forces d'échange; des phénomènes de ségrégation peuvent se produire. Assez souvent, les lois de correspondance ne sont plus vérifiées en raison d'une importance accrue des effets à courte distance puisque les interactions d'échange varient bien plus rapidement qu'en $1/r^3$. On peut donc résumer la situation par le graphique de la fig. 1.1.

En quoi le verre de spin doit-il être distingué très précisément des autres systèmes magnétiques? En plus du désordre de position, on le définira comme n'ayant pas d'ordre magnétique à longue distance quoiqu'il existe un blocage du spin de chaque impureté en dessous d'une température de transition T_{SG}; en d'autres termes, il n'apparait pas de moment magnétique macroscopique même si, localement, on décèle une polarisation permanente; on ajoutera qu'un verre de spin doit obéir aux lois de correspondance déduites de (1.5) dans la région de basse température.

Cette définition, on va le voir dans la suite soulève bien des questions: la température de transition peut-elle être caractérisée comme on le fait pour les transitions de phase? Quel est le paramètre d'ordre? Quelles quantités ont un comportement critique? De même, a-t-on le droit d'affirmer qu'il s'agit d'un phénomène coopératif où tous les spins seraient bloqués simultanément à la même température? Il n'est pas assuré, du point de vue expérimental qu'il existe un système qui réponde aux définitions données plus haut; c'est ce que l'on examinera dans les paragraphes suivants.

1.2. *Propriétés statiques*

1.2.1. *Chaleur spécifique*

La mesure de la chaleur spécifique est probablement celle qui prête le moins à la controverse car elle ne s'attaque pas directement à l'aimantation du système. La perturbation ne modifie pas la symétrie du système.

On note qu'à basse température ΔC est approximativement linéaire en température et présente un maximum à une température T_m. ΔC est l'excès de chaleur spécifique par rapport à la matrice pure en faisant l'hypothèse que la partie magnétique est simplement additive. On a montré que cet excès de chaleur spécifique obéit bien aux lois de correspondance de même que la température du maximum T_m [1.2, 1.3]

$$\Delta C/c \simeq T/c\,, \tag{1.6}$$

$$T_m/c \simeq \text{constante}\,. \tag{1.7}$$

La loi (1.6) est trop importante pour qu'on n'y revienne pas. Si la chaleur spécifique traduit la densité d'excitations $n(E)$ du système, la linéarité en température de ΔC implique que $n(E)$ soit à peu près constant en énergie dans la gamme explorée: il y a une densité uniforme d'excitations. Par ailleurs, aucun effet dépendant du temps n'a été observé sur ces mesures contrairement à d'autres systèmes désordonnés tels que les verres de polymères.

A haute température au contraire, la chaleur spécifique est du type

$$\Delta C/c \simeq (T/c)^{-1}\,. \tag{1.8}$$

Cet excès peut être observé relativement haut en température (100 K). Le fait que ΔC augmente quand on s'approche de T_m est souvent interprété comme la signature d'une augmentation de la corrélation des spins (fig. 1.2).

C'est bien le maximum de ce que l'on peut dire tant le "pic" de chaleur spécifique, quand il existe, est arrondi. En particulier, ΔC ne présente aucun saut, et aucune variation critique qui puisse faire penser à un phénomène de changement de phase, de sorte que si c'est à cette conclusion que l'on était pourtant conduit il resterait à expliquer pourquoi C n'a aucune singularité visible.

1.2.2. Susceptibilité magnétique

Les systèmes avec impuretés magnétiques étaient étudiés depuis bien longtemps lorsque les premières mesures de susceptibilité montrant un maximum aigu au voisinage d'une température qu'on notera T_{SG} jetèrent le trouble, faisant irrésistiblement penser à un phénomène de transition de phase. Autant la chaleur spécifique laisse quelquefois des doutes sur l'identification d'une vraie température de transition, autant la susceptibilité statique, ou celle mesurée à basse fréquence, présente un maximum aigu pourvu que les expériences soient faites en champ faible (quelques

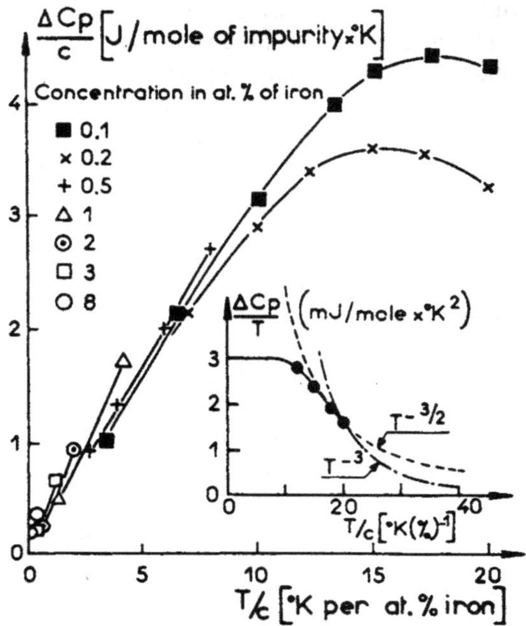

Fig. 1.2. Dépendance en température de l'excès de chaleur spécifique pour des alliages Au–Fe; diagramme en quantités réduites: $\Delta C_p/c(T/c)$.

gauss). L'ordre de grandeur de cette susceptibilité reste petit: 10^{-3} environ de celle que présente une impureté isolée qui polarise son environnement. La forme de la courbe de susceptibilité en fonction de la température ressemble à celle d'un anti-ferromagnétique.

A titre d'exemple, mentionnons le cas de Au–Fe dont la susceptibilité a été examinée dans une gamme de concentrations allant de 1 à 13%, concentration au-delà de laquelle on voit apparaitre le ferromagnétisme. La grandeur du pic de susceptibilité croît très vite avec la concentration. Typiquement, pour $c = 1\%$, on a $T_{SG} \simeq 8$ K et pour $c \simeq 12\%$, $T_{SG} \simeq 36$ K [1.4].

La température de transition T_{SG} est assez exactement proportionnelle à c. La susceptibilité dont il est question ici est la partie réversible; on peut la mesurer par l'un des deux processus suivants:

$$\chi_{rev}(T) = \left.\frac{M_{\text{macroscopique}}}{h}\right|_{h\to 0} = \left.\frac{\partial M}{\partial h}\right|_{h\to 0}.$$

Dans la limite des basses températures, elle obéit aux lois de correspondance:

$$\chi_{rev}(T, h) = f(T/c, h/c) .$$

On retiendra que T_{SG} est très inférieur à T_m température déduite des mesures de chaleur spécifique.

L'application d'un champ magnétique de 100 G suffit à effacer le pic ou tout au moins à arrondir les formes au point qu'il devient difficile de parler de température de transition (fig. 1.3). C'est d'ailleurs une des raisons pour lesquelles les premiers expérimentateurs avaient manqué cette observation. Retenons en tous cas que la susceptibilité est divergente ou tout au moins qu'elle présente un maximum aigu qui permet de définir sans ambiguité T_{SG}.

1.2.3. Mesure de l'aimantation locale

S'il y a naissance d'un ordre ou tout au moins s'il y a un phénomène coopératif qui vise à bloquer les spins, en dessous d'une certaine tempé-

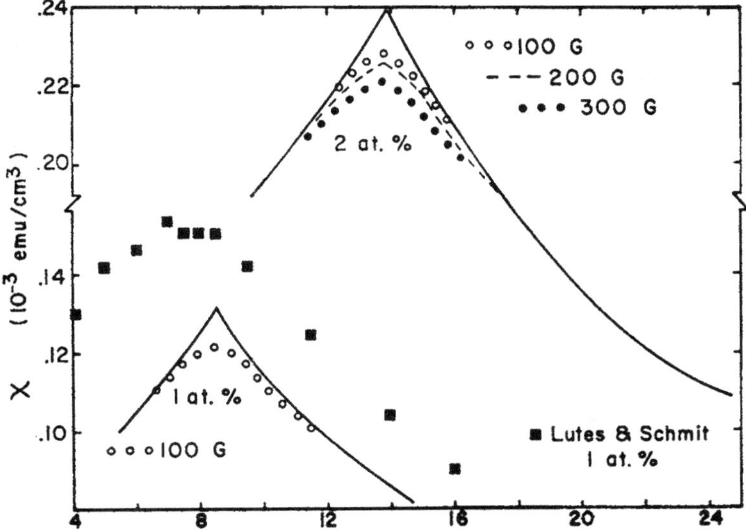

Fig. 1.3a. Susceptibilité en fonction de la température T(K) pour des échantillons de concentrations 1 et 2%. Dans ce graphique, on a donné des courbes pour un champ nul et pour diverses valeurs du champ appliqué; les résultats de Lutes et Schmit correspondent à une concentration de 1%.

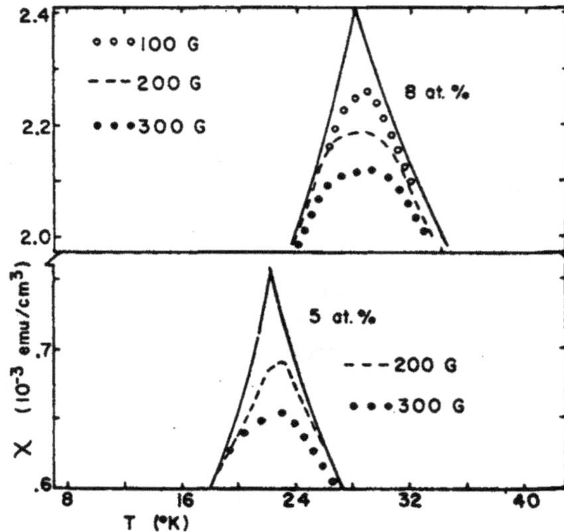

Fig. 1.3b. Susceptibilité pour des échantillons de concentration 5 et 8% pour diverses valeurs du champ appliqué.

rature dite de transition, on doit pourvoir mesurer une aimantation locale. Les spectres d'effets Mössbauer permettent de s'en rendre compte.

Si pour un spin, on peut définir une direction moyenne $\langle \boldsymbol{S} \rangle$ à une échelle de temps supérieure à celle qui correspond à une transition Mössbauer ($\tau = 10^{-8}$ s) il en résulte un champ hyperfin qui lève la dégénérescence des niveaux nucléaires d'une quantité proportionnelle à $\langle \boldsymbol{S} \rangle$. A la limite, même si le spin, ou le groupe de spins relaxe d'une direction à une autre, d'une configuration à une autre dans un temps supérieur à τ, tout se passe comme si le système était statique; on pourrait même imaginer qu'il y ait des oscillations autour d'une position moyenne. Ce qui est important ici, c'est l'existence d'une orientation moyenne presque stable.

Expérimentalement, ce n'est pas aussi clair; si les spectres Mössbauer permettent d'identifier une température de transition, celle-ci, en fait, est souvent plus élevée que celle obtenue par analyse de la susceptibilité. Ou plutôt, une analyse assez précise des résultats montre que cette transition n'est pas très bien définie comme si certains spins ou groupes de spins s'"ordonnaient" avant que la transition ne se soit produite [1.5] dans tout l'échantillon. Il faut insister sur cet aspect; l'échelle de mesure du temps, soit τ, correspond à la largeur naturelle Γ_0 de la transition nucléaire γ du noyau. Tout élargissement de la raie, peut avoir une origine statique (inhomogénéité du champ cristallin et donc du champ quadrupolaire ou distribution du champ magnétique hyperfin) ou dynamique (mouvement de l'aimantation électronique à une fréquence plus grande que Γ_0^{-1}). Une expérience de spectroscopie Mössbauer fournit donc à la fois des informations sur la dynamique des spins par mesure de Γ et sur la distribution de l'aimantation locale par observation de la forme et de la distance en champs des six raies de résonance.

Dans le cas des verres de spins, deux quantités retiennent l'attention: l'aimantation dont on a montré qu'elle obéit approximativement à une loi de Brillouin en fonction de la température, quelle que soit C; un temps caractéristique du mouvement des spins dont on décèle le ralentissement à une température supérieure à T_{SG}; on l'explique en disant qu'il y a formation de "nuages" ou de groupes de spins.

Cette image, on parle de "nuages" ou de cluster, est d'ailleurs en accord avec une interprétation plus dynamique, qui veut que ces groupes de spins aient un comportement lent, propre à produire cette levée de dégénérescence, mais sans que cela implique que les mêmes spins restent liés entre eux à plus long terme. Retenons malgré tout que la transition qui est observée n'est pas très bien définie en température.

Une autre méthode pour se faire une idée de l'aimantation locale consiste à mesurer la précession de spin des muons polarisés implantés dans les alliages [1.6]. En effet, il précessent autour du champ magnétique local et émettent des positrons dans une direction préférentielle parallèle au spin; l'asymétrie de l'émission dans le temps est une mesure du champ magnétique local. Un muon peut être considéré comme un proton radioactif occupant de manière aléatoire les sites de l'alliage. La fréquence de précession du muon est déterminée par la valeur du champ magnétique local autour duquel il précesse sur un cône d'angle au sommet fixé; par décomposition radioactive, le muon donne un positron qui est détecté avec une intensité proportionnelle à l'angle entre le spin et la direction du compteur.

$$\mu \rightarrow e^+ + \delta + \bar{\delta}\,.$$

Si tous les muons étaient placés dans le même champ, le signal mesuré oscillerait avec une période égale à l'inverse de la fréquence de Larmor des muons. Dans un verre de spin, il y a a au contraite une distribution de champs magnétiques locaux qui se superpose en général au champ magnétique extérieur.

Ou encore, pour employer le langage de la résonance nucléaire, le temps de dépolarisation T_2 de l'émission des muons est une mesure de la largeur effective de la distribution des champs locaux par la relation

$$\Delta = (\gamma_\mu T_2)^{-1} ,$$

où

$$\gamma_\mu = 8.5 \times 10^4 \text{ G}^{-1} \text{ s}^{-1} .$$

Au-dessus de la température de transition de verre de spin, cette largeur est faible; au-dessous de la transition, au contraire, cette largeur est grande.

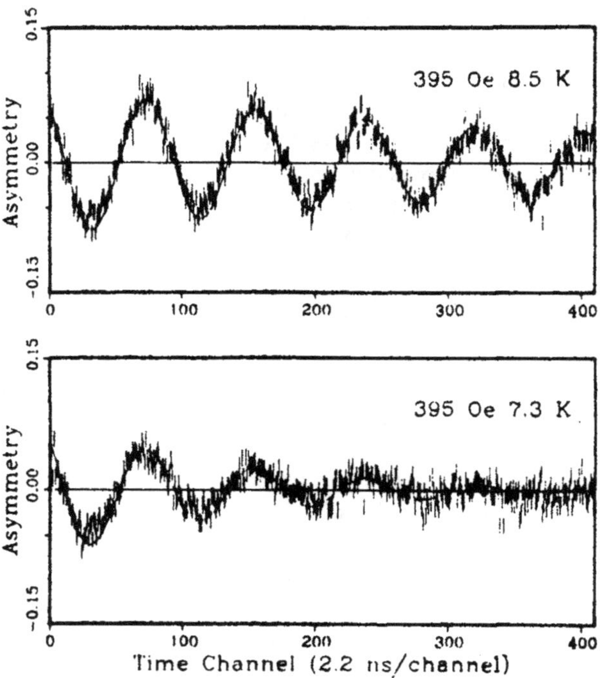

Fig. 1.4a. Asymétrie du signal de positrons en fonction du temps de stationnement des muons dans un alliage Cu–Mn à 0.7‰; T_0 = 7.7 K. Les barres d'erreurs figurent les erreurs statistiques. L'échantillon était refroidi sous champ depuis 20 K jusqu'à la température de mesure. La ligne continue est obtenue par une méthode de moindre carré pour figurer une décroissance exponentielle du signal.

Fig. 1.4b. Taux de dépolarisation exprimé en largeur de raie en gauss en fonction de la température de l'échantillon pour diverses valeurs du champ magnétique appliqué. T_0 est égal à 7.7 K pour Cu–Mn et 11.6 K pour Au–Fe. fc correspond à un refroidissement sous champ, zfc à un refroidissement en champ nul. Les barres d'erreurs résultent des erreurs statistiques standard et des erreurs résultent de mesures indépendantes répétées.

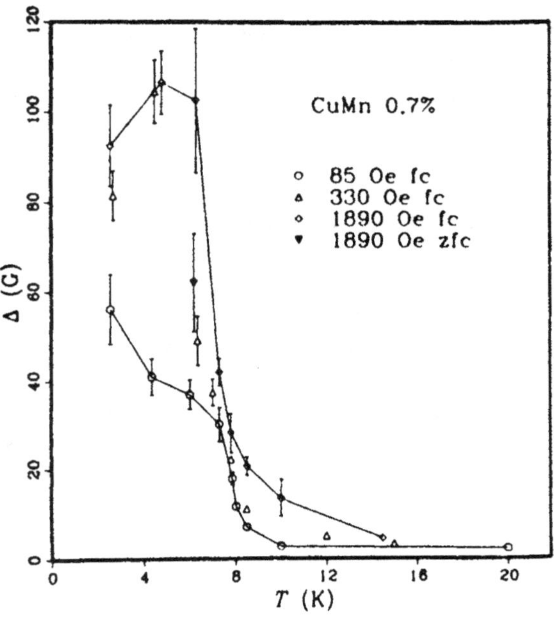
CuMn 0.7%
85 Oe fc
330 Oe fc
1890 Oe fc
1890 Oe zfc
Δ (G)
T (K)
0
20
40
60
80
100
120
0
4
8
12
16
20

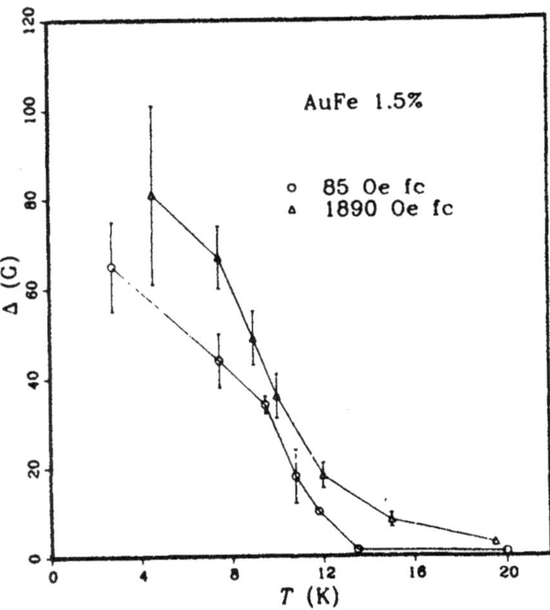
AuFe 1.5%
85 Oe fc
1890 Oe fc
Δ (G)
T (K)
0
20
40
60
80
100
120
0
4
8
12
16
20

Fig. 1.4b.

Dans Cu-Mu à 0.7% par exemple, elle passe de quelques gauss à une centaine de gauss; et la transition apparait assez aigüe. En champ appliqué plus élevé (200 à 300 G), on note l'apparition de champs locaux plus grands; un refroidissement sous champ favorise cette tendance sur laquelle on s'expliquera plus loin (fig. 1.4).

En définitive, la quantité qui est mesurée présente une variation nette à T_{SG}; mais un désavantage de cette expérience tient à l'obligation d'appliquer en même temps un champ magnétique.

1.2.4. Mise en évidence des nuages: diffraction des neutrons aux petits angles

On vient de le voir, les mesures de chaleur spécifique montrent qu'au-dessus de T_{SG} il subsiste une bonne partie de l'entropie magnétique, suggérant que des nuages de spins continuent d'avoir une existence.

Une expérience de diffusion de neutrons aux petits angles permet de mettre ce fait en évidence et de voir leur évolution en température dans le cas un peu particulier des alliages Au–Fe proches de la limite de percolation [1.7]. Partant des hautes températures où les spins sont isolés, le système évolue par formation d'unités correspondant à des spins correlés. Comme dans tous les cas semblables, lorsque le temps de relaxation de ces unités devient plus long que le temps de mesure ces unités apparaissent statiques; c'est tout au moins la première interprétation des résultats.

Si elle est vraie, on s'attend à mettre en évidence toute une série de températures de transition T_{SG} correspondant à des nuages de dimensions différentes; plus le nuage est grand, plus il a un temps de relaxation long, plus haut en température il peut être considéré comme statique. Il y a là une nette similitude avec les résultats des mesures d'effet Mössbauer. L'observation se fait par diffusion de neutrons aux petits angles. Soit en effet un ensemble de N spins répartis en nuages identiques notés i, de rayons R_i, et comprenant n_i spins de moment μ alignés ferromagnétiquement (on sait que cet alliage a cette propriété). Le facteur de forme magnétique est alors:

$$\begin{aligned} F_i(q) &= A \int_0^{R_i} \frac{n_i \mu}{R_i^3} \frac{\sin qr}{qr} r^2 \, \mathrm{d}r \\ &= A \frac{n_i \mu}{R_{iq}^{33}} (\sin qR_i - qR_i \cos qR_i) \,. \end{aligned} \tag{1.9}$$

On voit alors comment dans ce cas la formation de nuages favorise la diffusion aux petits angles; c'est le premier point. On a en effet

$$\frac{\mathrm{d}\sigma}{\mathrm{d}\Omega} \simeq \sum_i F_i^2(q) \,. \tag{1.10}$$

Pour $q \to 0$ le gain de diffusion est donné par

$$\frac{\text{nuages}}{\text{pas de nuages}} \simeq \frac{N}{n}(n\mu)^2/N\mu^2 \simeq n\,.$$

On voit l'intérêt ici du groupement ferromagnétique; dans un autre alliage, CuMn par exemple, on aurait un gain de 1 seulement car le moment magnétique de chaque nuage vaudrait $(\sqrt{n}\mu)$ et le gain serait

$$\frac{N}{n}(\sqrt{n}\mu)^2/N^2 \simeq 1\,. \tag{1.11}$$

L'équation (1.9) montre ainsi que la diffusion par un nuage est fortement anisotrope et centrée vers les q petits; la plus grande part de la diffusion est quasiélastique et ses variations en température reflètent les variations de $\chi(q)$ (fig. 1.5).

Pour une température inférieure à la température de blocage $T_{SG}(q)$ des nuages de dimension π/q le facteur de structure des corrélations dynamiques est absorbé par le facteur de structure statique qui donne la diffusion de Bragg. Expérimentalement, on vérifie bien que la diffusion aux petits angles passe par un maximum aigu pour une température $T_{SG}(q)$ et que celle-ci décroît quand q augmente (fig. 1.6). Ainsi, dans le cas particulier d'un alliage à corrélations ferromagnétiques, on décèle la formation

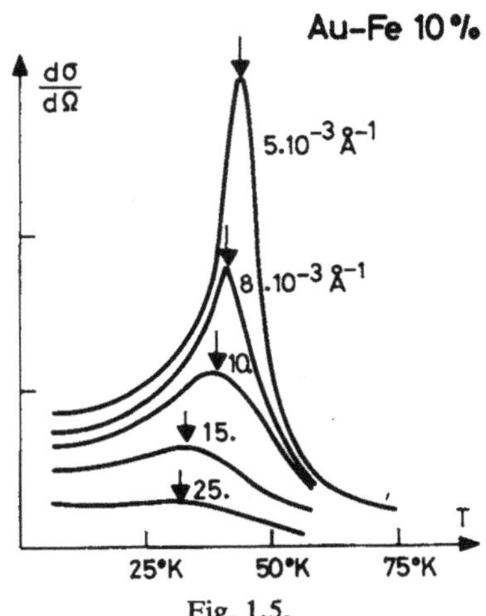

Fig. 1.5.

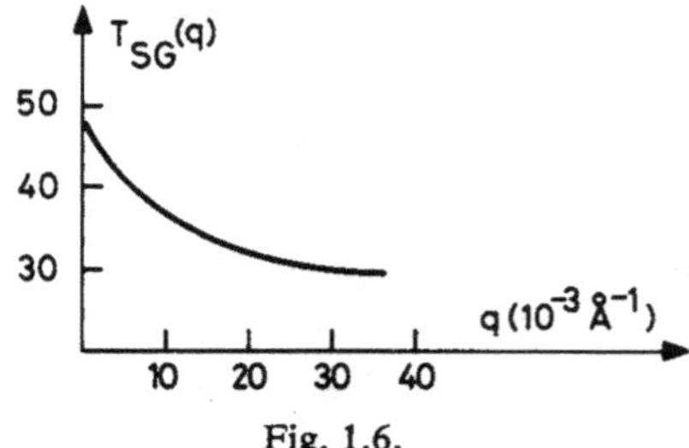

Fig. 1.6.

de nuages; la susceptibilité magnétique classique, mesurée sur le même alliage correspond bien aux mesures extrapolées des neutrons. C'est une confirmation qui fortifie le point de vue développé ci-dessus.

Mais ce n'est à mon avis qu'une interprétation: car on peut quand même se demander si dans les courbes expérimentales il n'y aurait pas lieu de distinguer deux contributions; la première qui est bien la susceptibilité

$$S(q) = KT\chi(q)$$
$$= \sum_{ij} \{\langle S_i S_j \rangle - \langle S_i \rangle \langle S_j \rangle\} \exp\{i\boldsymbol{q}\cdot(\boldsymbol{R}_i - \boldsymbol{R}_j)\} F^2(q) \frac{1}{N\pi}, \tag{1.12}$$

et la deuxième qui est une contribution élastique du type Bragg mais qui, dans un verre de spin n'est jamais nulle quel que soit $\boldsymbol{q}$

$$S_2(q) = \text{“Bragg”} = \sum_{ij} \{\langle S_i \rangle \langle S_j \rangle\} \exp\{i\boldsymbol{q}\cdot(\boldsymbol{R}_i - \boldsymbol{R}_j)\} F^2(q) \frac{2}{N}. \tag{1.13}$$

Pour des i et des j relativement voisins, et en particulier si le paramètre d'ordre dans la phase basse température a une extension spatiale, la contribution de $S_2(q)$ devrait croître lorsque la température décroît en dessous de T_{SG} puisqu'elle est proportionnelle à la valeur de ce paramètre d'ordre; elle a donc tendance à déplacer le maximum de $S(q)$ vers des températures inférieures à T_{SG} $(q = 0)$; en faisant cette correction, il n'est donc pas exclu que l'on restaure pour $S_1(q) = KT\chi(q)$ un maximum en température, indépendant du vecteur d'onde q et dont le maximum coïncide avec celui mesuré de manière plus traditionnelle. Ce point de vue vient d'ailleurs d'être mis en bonne forme par un groupe de l'Université de Chicago et sera soumis à publication dans l'année (fig. 1.7).

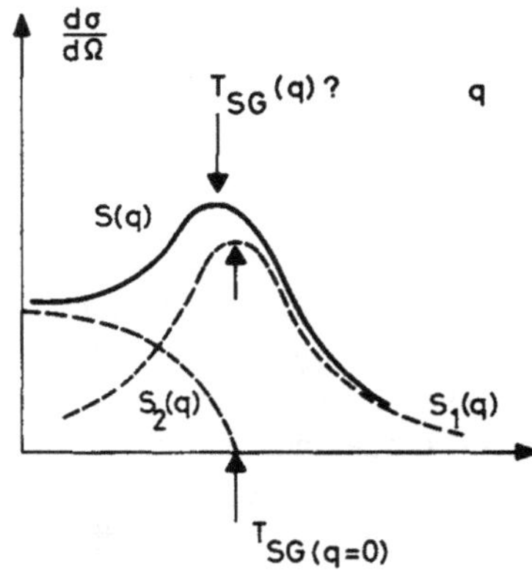

Fig. 1.7.

1.2.5. Premières conclusions

Les mesures précédentes permettent de se faire une première représentation d'un verre de spin. Au fur et à mesure que la température du système est abaissée, beaucoup de ces spins tendent à correler leurs directions pour former des "nuages" de tailles de plus en plus grandes; ces nuages sont en interaction les uns avec les autres par l'intermédiaire des spins qui subsistent dans leur intervalle; le temps caractéristique du mouvement des paquets de spins s'allonge quand on se rapproche de T_{SG} ou apparemment tout le verre de spin se fige.

Le détail du processus de "prise en masse reste inconnu; mais à ce stade, on doit éviter de penser à ces nuages comme étant de composition fixe; en particulier, rien n'indique que l'on puisse leur appliquer le terme de grain fin de Néel car leur géométrie n'est pas stable; aucune mesure à $T > T_{SG}$ ne confirme une identité fixe. Par contre, une question est légitime: quel est le rôle de l'anisotropie qui est de toute façon présente en raison de l'énergie dipolaire des spins de chaque nuage. L'importance de l'échelle de temps qui repère les phénomènes doit être soulignée.

Il n'en reste pas moins troublant de voir que certaines expériences conduisent à penser que la transition est facile à identifier (susceptibilité magnétique, dépolarisation des muons, effet Mössbauer) alors que d'autres suggèrent l'apparition d'une polarisation progressive (chaleur spécifique) sans qu'aucune anomalie ou sans qu'un saut n'ait lieu.

A mon avis, la température de transition est bien identifiée; il ne me parait pas excessif de parler de "transition" au sens des changements de phase; il reste toutefois à expliquer que la chaleur spécifique ne présente pas de maximum net à T_{SG}. On trouvera dans la suite d'autres expériences qui appuient ce point de vue.

1.3. Phénomènes de transport

On connait bien l'influence des impuretés magnétiques très diluées sur les propriétés de transport des alliages; cependant, comme dans un système non dilué les interactions indirectes sont assurées par le gaz d'électrons, il est raisonnable d'attendre quelques manifestations de l'état verre de spin sur toutes les propriétés où les électrons sont concernés au premier chef. En particulier, les mesures de résistivité, d'effet Hall anormal, d'atténuation acoustique et de pouvoir thermo-électrique seront sensibles à l'état magnétique du système et en constitueront un révélateur possible.

1.3.1. Effet Hall anormal

Quand le potentiel d'interaction entre un atome paramagnétique et un électron dépend du spin (couplage spin–orbite par exemple) les diffusions droite–gauche pour un électron polarisé ne sont plus symétriques. Dans un système d'impuretés polarisées, cette asymétrie engendre un effet Hall anormal dont la grandeur dépasse la valeur de l'effet Hall normal dès que la concentration dépasse quelques 10^{-3} dans des alliages comme Au–Fe.

Dans l'effet Hall normal [1.8], c'est la force de Lorentz classique qui engendre une tension ou un courant. Dans l'effet Hall anormal, le courant électronique peut être divisé en deux parties: celle des spins + et celle des spins −; chacune est diffusée différemment par une impureté magnétique polarisée. Dans un verre de spin, en l'absence de polarisation macroscopique induite, on n'observe aucun effet Hall anormal: les deux composantes de courant sont diffusées identiquement, car statistiquement, il y a une distribution d'impuretés; en présence d'un faible champ au contraire, les "nuages" ont, en moyenne, une certaine aimantation; elle est d'autant plus grande que l'on est près de T_{SG}: pour $T > T_{SG}$, l'aimantation a une composante moyenne non nulle suivant $\bar{h}$ même s'il y a des fluctuations; pour $T \leqslant T_{SG}$ le champ appliqué est assez grand pour découpler quelques spins, ou quelques nuages les uns des autres et pour imposer une certaine aimantation statique.

En principe, l'effet Hall anormal est un effet à un spin; il est additif et proportionnel à l'aimantation. Dans le cas du verre de spin au contraire, il sert à manifester l'état collectif du système; mais il faut pour le mettre en évidence briser la symétrie du système de spins. C'est le gros ennui de ce

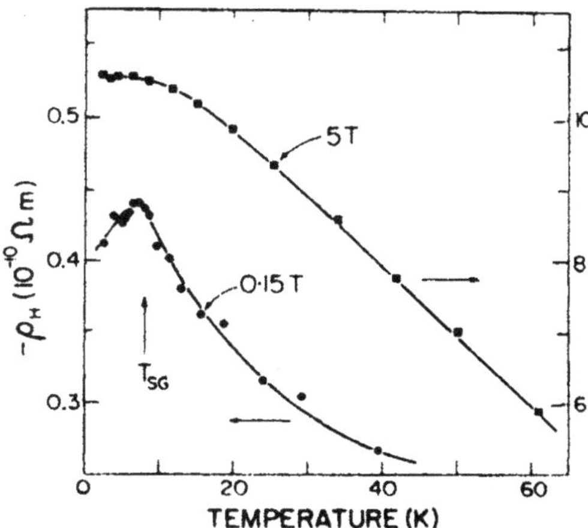

Fig. 1.8. Dépendance en température de la résistivité de Hall pour un alliage Au–Fe à 0.98% dans les champs indiqués. Le maximum local observé près de la température de transition ($T_{SG} = 8$ K) est modifié quand l'intensité du champ appliqué est augmentée.

type d'expérience: il faut appliquer un champ extérieur pour briser l'isotropie du verre de spin: isotropie dynamique à $T > T_{SG}$; isotropie statique à $T < T_{SG}$.

La dépendance en température de cet effet est assez analogue à celle que présente la susceptibilité; en bas champ, le maximum est très prononcé et situé à la même température T_{SG}; en champ élevé, il s'arrondit (fig. 1.8). Mais tous les essais pour vérifier des lois de correspondance se sont révélés infructueux.

1.3.2. Résistivité

Plusieurs alliages ont été étudiés dont le comportement est assez voisin; à titre d'exemple, on citera Au–Fe, Au–Cr, Au–Mn, Ag–Mn, Cu–Mn [1.9]. En général, dans le régime basse température, la loi en température de la résistivité est assez bien représentée par une variation du type $\rho \simeq T^{3/2}$ ou même en T^2 [1.10] (et quoi de plus normal?!). En s'approchant de T_{SG} elle tend à devenir du type $\rho \simeq T$. Les coefficients de ces deux lois décroissent comme la concentration sans que l'on puisse établir une loi de correspondance. La résistivité présente un maximum mais il serait malaisé de pouvoir identifier une température de transition avec cette seule mesure; d'ailleurs le maximum de ρ se produit à des températures T_ρ bien supérieures à T_{SG}. Pour $T \simeq T_{SG}$ la résistivité décroît probablement comme $1/T$: tout au

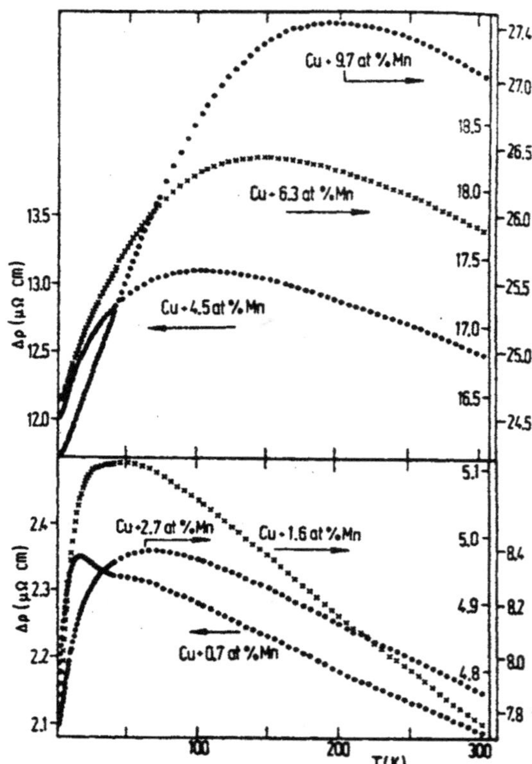

Fig. 1.9. Dépendance en température du changement de résistivité $\Delta\rho(\mu\Omega$ cm) pour des alliages Cu–Mn dont les concentrations sont comprises entre 0.7 et 9.7%. Notez le changement d'échelle lorsque la concentration augmente.

moins la contribution qui vient de l'effet verre de spin (fig. 1.9). T_ρ obéit grossièrement à la relation $T_\rho \simeq c$. Comme pour la chaleur spécifique, T_ρ diffère de T_{SG} de plusieurs degrés.

Ces résultats laissent perplexes. Aussi, certains auteurs ont cherché à porter $d\rho/dT$; ce coefficient est souvent plus significatif (rappelons-nous par exemple le cas des diodes supraconductrices). Cette fois-ci, le maximum de $d\rho/dT$ se situe nettement en dessous de T_{SG}, mais il présente un maximum très net (fig. 1.10).

Comment interpréter ces résultats ?

A basse température, le comportement en $T^{3/2}$ a été relié à l'existence de modes collectifs diffusifs du type "ondes de spins" [1.11]; un régime

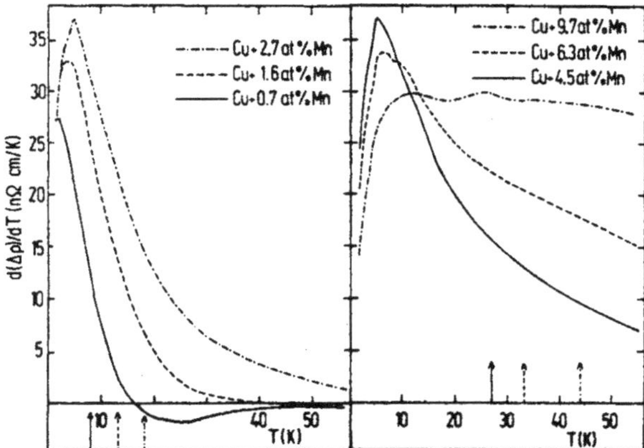

Fig. 1.10. Dépendance en température de la dérivée de la résistivité par rapport à la température $d(\Delta\rho)/dT$ ($n\Omega$ cm/K) pour des alliages de Cu–Mn. Les flèches représentent la température de transition déterminée par des mesures de susceptibilité.

différent serait même atteint avant que T soit de l'ordre de T_{SG} réconciliant à la fois les mesures de $\rho(T)$ et de $d\rho/dT$. Visiblement, cette interprétation pose la question des modes collectifs du type ondes de spin; il faudra y revenir ultérieurement. Le comportement en T^2 semble, lui, mieux compris [1.12].

Il reste malgré tout à comprendre pourquoi on ne perçoit aucune signature de T_{SG} sur la résistivité; faut-il invoquer le fait que même pour $T > T_{SG}$ la dimension d des nuages de spins serait plus grande que le libre parcours moyen l des électrons de sorte que le maximum de ρ se produirait justement pour la condition $l \simeq d(T)$ qui n'a bien sûr rien à voir avec T_{SG}? Visiblement, une meilleure compréhension de ce phénomène est nécessaire.

Mais l'opération sera difficile si l'on se souvient que les mesures de résistivité semblent bien plus affectées par le traitement thermique de l'alliage que par la concentration d'impuretés magnétiques [1.13].

1.3.3. Atténuation acoustique

Les ondes acoustiques ont souvent prouvé leur aptitude à explorer les propriétés statiques ou dynamiques des changements de phases. La mesure de leur vitesse dans la limite des basses fréquences fournit en effet quelques coefficients d'une équation d'état du type énergie libre; inversement, la mesure de la vitesse et de l'absorption à fréquence finie fournissent des indications sur les valeurs et la dépendance en température des temps caractéristiques des fluctuations critiques.

Les essais n'ont pas tous été fructueux; un example de verre de spin où un effet a été décelé est celui de l'alliage Au–Cr [1.14]. On mesure alors un écart de vitesse par rapport à la matrice qui se situe à une température très exactement égale à celle mesurée par susceptibilité; sa variation est très aigüe (fig. 1.11), même si l'ordre de grandeur (10^{-4}) reste faible.

Pour le cas présent, les auteurs ont suggéré que la variation de constante élastique était reliée à la modification de la chaleur spécifique au voisinage de T_{SG}. En l'absence de toute autre mesure, il est difficile d'en décider mais j'ai tendance à ne pas croire à cette explication; il serait en effet plus convaincant d'opérer aussi avec des ondes de polarisation transversale pour pouvoir montrer que dans ce cas il n'y a aucun effet, de disposer d'un monocristal pour éviter les effets de moyenne sur les orientations des grains, enfin d'examiner la dépendance en fréquence. Le résultat expérimental annoncé est donc seul indicatif. D'autres explications peuvent être proposées: le mécanisme de couplage entre l'onde acoustique et le système de spin me parait en effet plutôt devoir être de type magneto-élastique; probablement pas une modulation du champ cristallin qui est un effet à un

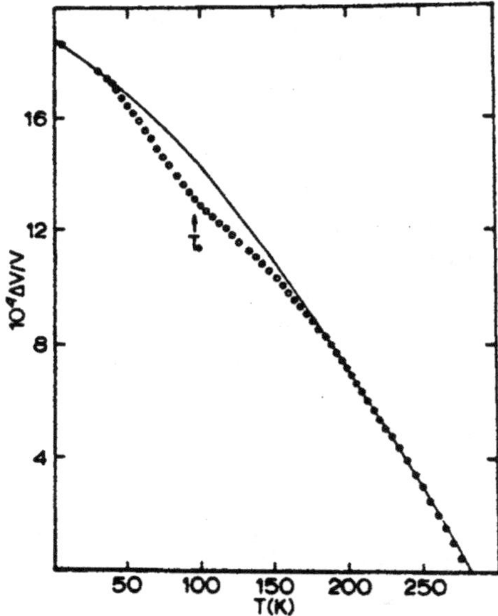

Fig. 1.11a. Dépendance en température de la vitesse du son d'ondes longitudinales de fréquence 13 MHz pour un alliage Au–Cr à 10.8%. La ligne en trait continu est une estimation des variations lentes. La température du pic de susceptibilité T_0 est indiquée sur le graphique.

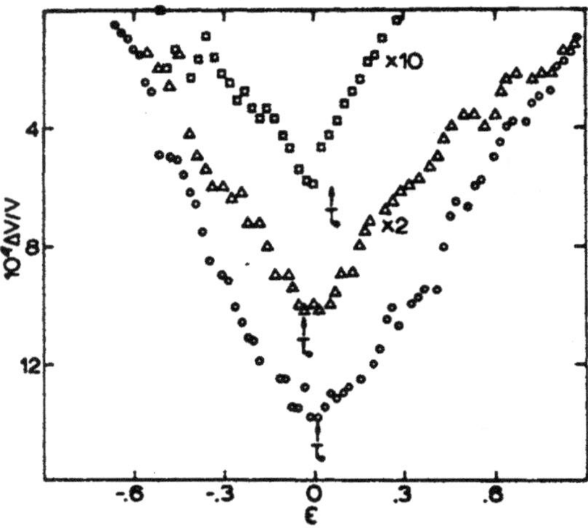

Fig. 1.11b. Variations de la vitesse du son en fonction de la température réduite

$\square C = 3.7$ at.%, $T_m = 39$ K, $T_0 = 41.5$ K
$\triangle C = 6.1$ at.%, $T_m = 58$ K, $T_0 = 56.2$ K
$\bigcirc C = 10.8$ at.%, $T_m = 95$ K, $T_0 = 96$ K.

spin et qui implique un fort moment orbital; plus certainement une modulation des interactions entre spins, en particulier des interactions RKKY; la forme de $\Delta v(T)$ indiquerait que c'est un effet dynamique qui est observé et non un effet statique de l'aimantation locale car ce dernier ne serait visible que pour $T < T_{SG}$.

Dernière remarque: Ces expériences me paraissent présenter l'intérêt que la déformation élastique qui entraîne une variation du champ magnétique local (hypothèse de champ moyen) le fait comme il se doit pour être couplée directement avec le paramètre d'ordre (qui est non uniforme).

A mon avis, ces expériences mériteraient d'être reprises d'autant que les dernières qui ont été publiées [1.15] sur $\underline{\text{Cu}}$-Mn ajoutent au mystère!

1.4. Rémanence

Le phénomène de rémanence, ou plus généralement le phénomène de mémoire n'est pas toujours très bien accepté par le physicien qui le considère comme un peu "sale". En magnétisme, comme en ferroélectricité, le problème des parois et de leur déplacement plus ou moins irréversible n'est pas encore parfaitement élucidé.

Pour les verres de spins, la rémanence a des caractéristiques qui la mettent pourtant en étroite relation avec les propriétés déjà mentionnées. C'est pour cette raison qu'il faut s'y attarder.

En principe, les quantités qui peuvent être soumises aux effets de rémanence sont toutes celles qui mettent en jeu l'aimantation locale dans la phase basse température.

A ce titre, on imagine aisément que la rémanence doit être affectée par le processus de passage à travers T_{SG} (vitesse de refroidissement, champ magnétique appliqué); de même, si la rémanence est associée à des irréversibilités, la grandeur des perturbations appliquées au verre de spin conditionnera sa réponse; on a déjà mentionné cet effet en analysant les measures de susceptibilités magnétiques.

1.4.1. Grandeurs affectées par la rémanence

L'aimantation est la première grandeur qui fasse preuve de rémanence; comme sa partie rémanente obéit elle aussi aux lois de correspondance, elle semble liée aux interactions du type RKKY ou tout au moins à des interactions en $1/r^3$. On observe la loi suivante:

$$m_r/c = f(R/c) = \exp(-\alpha T/c)\,, \tag{1.14}$$

où α est une constante [1.16]. Cette loi rappelle celle que l'on a pour tous les systèmes classiques qui ont une barrière de potentiel à franchir et qui sont activés thermiquement. On y reviendra plus tard (figs. 1.12a et 1.12b).

Il y a en fait deux moyens de mesurer l'aimantation rémanente: une mesure isotherme à $T < T_{SG}$ en appliquant un champ magnétique jusqu'à saturation; en refroidissant sous champ de $T > T_{SG}$ à $T < T_{SG}$. Dans les deux cas, la valeur à saturation est identique et vaut quelques pour cent de l'aimantation calculée pour l'alliage considéré sans interaction entre spins (fig. 1.13). Le système ne s'est donc pas ordonné ferromagnétiquement sous l'influence du champ, mais il y a eu orientation préférentielle de quelques "nuages" ou groupes de spins. Sans être trop précis, disons que certaines barrières de rotation ont été franchies; on trouverait le même comportement dans le cas des grains fins ferro ou antiferromagnétiques [1.17].

Tout comme l'aimantation, la susceptibilité, définie comme

$$\chi = \left.\frac{M(H)}{H}\right|_{H\to 0} \tag{1.15}$$

présente une rémanence. On distingue la susceptibilité réversible dont on a donné les propriétés dans la première partie et la susceptibilité rémanente

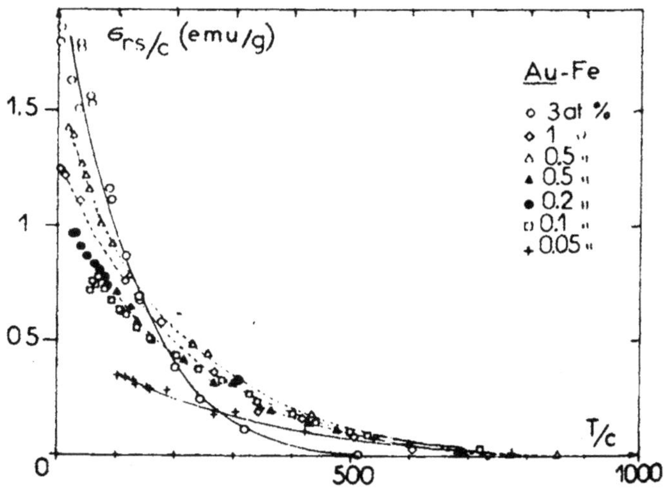

Fig. 1.12. Diagramme en coordonnées réduites de la variation avec la température de l'aimantation rémanente à saturation pour un alliage Au–Fe.

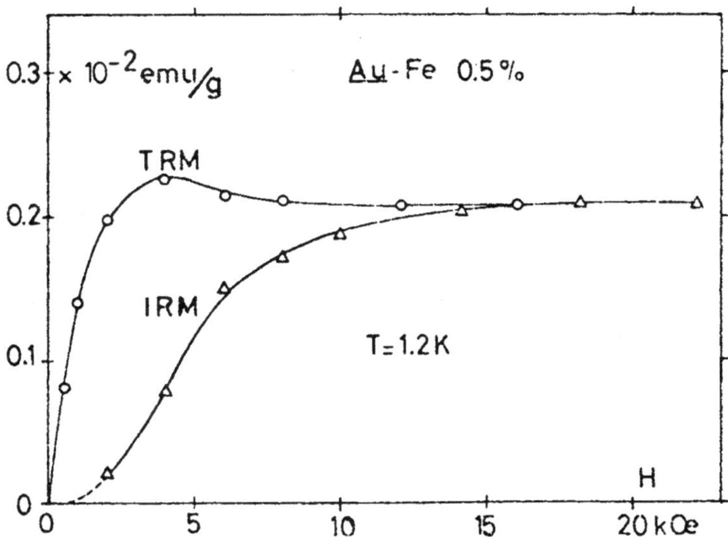

Fig. 1.13. Dépendance en fonction du champ de l'aimantation thermorémanente (TRM) obtenue après refroidissement depuis une température supérieure à T_m jusqu'à T = 1.2 K ; aimantation isotherme rémanente (IRM) obtenue quand le champ appliqué à 1.2 K est supprimé.

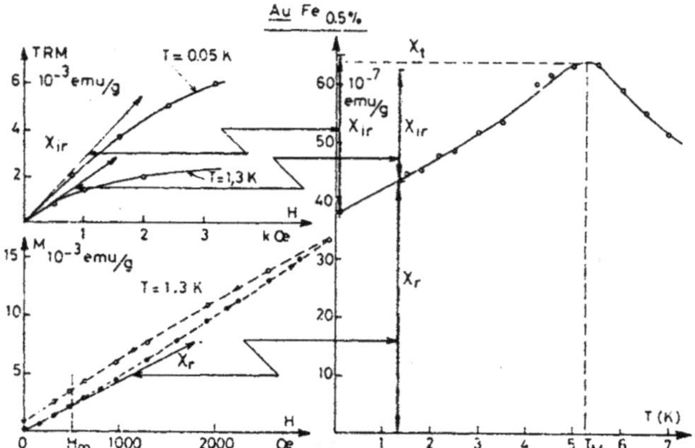

Fig. 1.14. Méthodes expérimentales pour obtenir respectivement la susceptibilité réversible (χ_r) et irréversible (χ_{ir}); variation thermique de χ_r et $\chi_t = \chi_r + \chi_{ir}$.

mesurée par exemple dans le processus de refroidissement sous champ [1.18].

Pour des températures inférieures à T_{SG}, on observe que la susceptibilité totale, $\chi_{total} = \chi_{réversible} + \chi_{rémanente}$ reste constante. C'est dire que la partie irréversible augmente lorsque la température baisse (fig. 1.14). Puisqu'il est question de susceptibilité irréversible, notons que des mesures d'effet Hall anormal avec un verre de spin préparé dans des conditions similaires, pourraient être instructives.

1.4.2. Evolution temporelle

Pour une aimantation de départ donnée et une température donnée l'aimantation évolue dans le temps dès que l'on supprime le champ appliqué. On observe que

$$m_r \simeq -\log \Delta t\,, \tag{1.16}$$

où Δt est l'intervalle de temps que le système a passé en champ nul. Le coefficient de la loi (15) est à peu près indépendant de la température. Les échelles de temps qui interviennent sont extrêmement longues et s'expriment en heures; c'est le signe d'un phénomène qui se développe à l'échelle macroscopique; et nous voici ramenés au problèmes des "nuages" (fig. 1.15a).

Il faut probablement rattacher à ces mesures celles qui correspondent à l'évolution de la susceptibilité en fonction de la fréquence, dans le régime

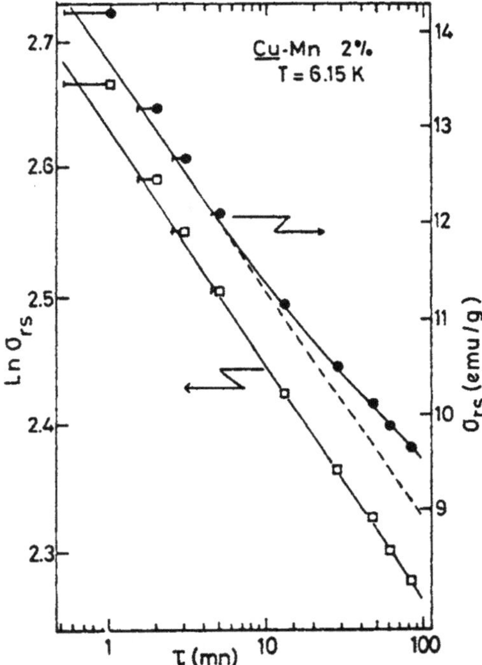

Fig. 1.15a. Dépendance en fonction du temps à 6.15 K de l'aimantation rémanente à saturation pour un alliage Cu–Mn à 2%.

des très basses fréquences (0.01 Hz à 10 Hz) car c'est une échelle de temps très en rapport avec celle des expériences de rémanence; dans des alliages comme $(La_{1-x}, Gd_x)Al_2$ l'influence de la fréquence de la mesure, dans un domaine qui va de 1 c/s à 10 c/s est forte, elle semble supporter là encore le modèle des nuages [1.19] le temps caractéristique de chacun d'eux suit une loi d'Arrhénius (voir figs. 1.15b et 1.15c).

Notons que, curieusement, la chaleur spécifique ne semble pas dépendre très fortement de phénomènes de mémoire [1.20]; "curieusement" puisque c'est le cas pour les verres de polymères qui ont par ailleurs tant d'analogies avec les verres de spins.

1.4.3. Seconde conclusion provisoire

La transition que l'on observe aux alentours de T_{SG}, le phénomène de rémanence rappellent irrésistiblement les propriétés de la transition

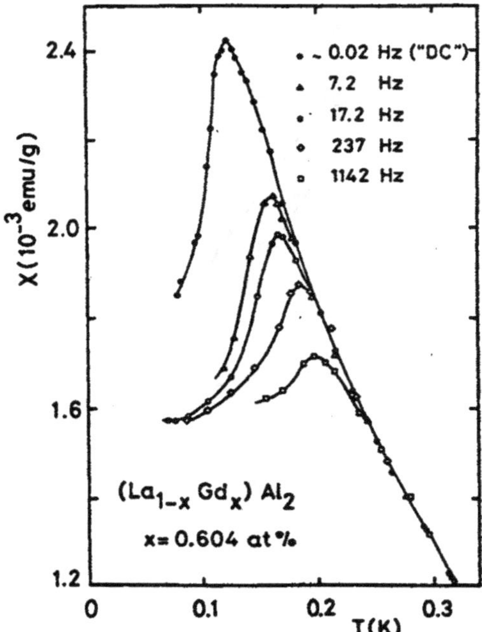

Fig. 1.15b. Susceptibilité en fonction de la température pour un alliage de (La, Gd)Al_2 à différentes fréquences.

vitreuse pour laquelle la cinétique de la transition joue un rôle essentiel; la température de transition dépend de la rapidité du passage, la viscosité croît exponentiellement.

On connait depuis longtemps dans la littérature l'exact analogue de ces effets dans le cas des grains fins ferromagnétiques; ce n'est donc pas un hasard si le modèle, appliqué aux verres de spins, a été développé à Grenoble; il faut en dire quelques mots au moins.

Première idée: à chaque "nuage" ou monodomaine est associé une énergie d'anisotropie E_a et le temps caractéristique de relaxation en direction de l'aimantation locale de ce monodomaine est donné par un processus d'activation thermique sous la forme

$$\tau = \tau_0 \exp(E_a/2KT)\,, \tag{1.17}$$

τ_0 est un temps microscopique.

Deuxième idée: pour chaque type de mesure m, on définit une échelle de

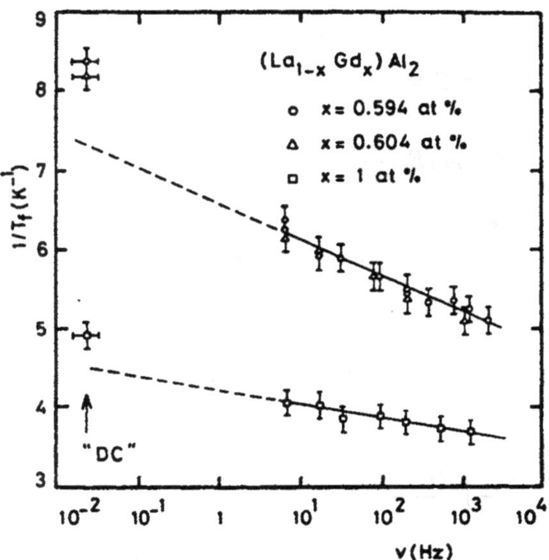

Fig. 1.15c. Variation en fonction de la fréquence de l'inverse de la température du maximum de susceptibilité.

temps τ_m (voir par exemple effet Mössbauer, susceptibilité dynamique, . . .); par la formule (1.16) on lui associe une température de blocage T_{bm} des monodomaines m d'énergie d'anisotropie E_a:

$$2KT_{bm} = \frac{E_a}{Q + \log \tau_m}, \qquad Q = -\log \tau_0 . \tag{1.18}$$

Troisième idée: il y a une distribution des énergies d'anisotropie $p(E_a)$: elle entraîne une distribution des temps de relaxation.

Il en résulte que si une grandeur M macroscopique fait preuve de rémanence et si pour chaque "nuage" m est suffisamment défini par la seule donnée de E_a on a:

$$dM = p(E_a)m(E_a)\, dE_a ,$$

où $p(E_a)$ est la probabilité de distribution des E_a. En utilisant (1.17) on écrit

$$dE_a = \frac{E_a}{\log \tau}\, d \log \tau + \frac{E_a}{T}\, dT ,$$

d'où il résulte:

$$\left.\frac{\partial M}{\partial \log \tau}\right|_T = p(E_a)m(E_a)\frac{\partial E_a}{\partial \log \tau},$$
$$\left.\frac{\partial M}{\partial T}\right|_\tau = p(E_a)m(E_a)\frac{\partial E_a}{\partial T}. \tag{1.19}$$

Des mesures de relaxation de l'aimantation rémanente (fig. 1.15a) on tire deux résultats: le fait que pour l'aimantation rémanente m_r

$$\left.\frac{\partial m_r}{\partial \log \tau}\right|_T = \text{constante}$$

implique que la quantité $p(E_a)\, m_r(E_a)$ varie lentement avec E_a; de plus, le rapport des deux expressions (18) donne une mesure de τ_0. On a en effet [1.21]:

$$\left.\frac{\partial m_r}{\partial T}\Big/\frac{\partial m_r}{\partial \log \tau}\right|_T \frac{(Q + \log \tau m_r)}{T_{bm}}. \tag{1.20}$$

On trouve pour τ_0 des valeurs de l'ordre de 10^{-7} s à 2 K dans les alliages de type Au–Fe; c'est une valeur en accord avec le caractère macroscopique des monodomaines; τ_0, on le verra plus loin, est ainsi beaucoup plus grand que le temps de relaxation d'un spin isolé.

Le même modèle peut être exploité pour déterminer le nombre moyen n de spins appartenant à un nuage; il faut s'y attarder.

Supposons en effet que les moments de chaque spin d'un nuage soient répartis au hasard; l'aimantation d'un paquet de n spins sera de l'ordre de

$$m = \sqrt{n}\mu\,,$$

où μ est le moment magnétique d'une impureté isolée. Si tous les spins étaient découplés et alignés, on aurait l'aimantation à saturation m_s

$$m_s = Nc\mu = pn\mu\,,$$

où N est le nombre de sites possibles et p le nombre de nuages créés par unité de volume. L'aimantation rémanente à saturation au contraire est donnée par

$$m_{rs} = p\sqrt{n}\mu\,.$$

Le rapport m_{rs}/m_s fournit donc une valeur de n; on trouve que n est de quelques centaines. Mais on peut faire mieux et chercher la distribution des n, moyennant un modèle conçu pour retrouver le résultat des figures 12a et 12b.

Supposons que l'énergie d'anisotropie E_a de chaque nuage soit proportionnelle au nombre de spins et à un champ d'anisotropie h_a qui croit comme c. On aura

$$E_a = h_a n\mu \,.$$

Ces deux hypothèses, notons le, supposent que l'énergie d'anisotropie a toutes les chances d'être de nature dipolaire magnétique ($1/r^3 \to c$) et qu'elle agit dans le volume du nuage; les interactions de frontière aves les autres nuages ne sont donc pas impliquées. Grâce à (1.17), on peut encore écrire

$$h_a n\mu = KT \log(\tau/\tau_0) \,.$$

Si pr(n) est la probabilité de distribution des n, on aura, à toute température

$$m_{rs}(T) = \int_{n(T)}^{\infty} m(n)\mathrm{pr}(n)\,\mathrm{d}n \,.$$

En effet, la sommation ne peut s'étendre qu'aux "nuages" qui ont un nombre assez grand de spins pour qu'à la température T, et à une échelle de temps τ, ils apparaissent bloqués.

A la probabilité de n, on associe une probabilité de m

$$\mathrm{pr}(m) = \mathrm{pr}(n)(\mathrm{d}n/\mathrm{d}m) \,.$$

En choisissant

$$\mathrm{pr}(m) = \frac{2}{\pi}\frac{1}{m_0}\exp\left(-\frac{m^2}{2m_0^2}\right),$$

on trouve que

$$\mathrm{pr}(n) = \frac{1}{2\pi}\frac{1}{nn_0}\exp\left(-\frac{n}{2n_0}\right),$$

avec $n_0\mu = m_0$.

La seule preuve que ce choix pour pr(m) est bon, c'est qu'il permet de calculer correctement l'aimantation rémanente à saturation et de retrouver les résultats de la figure 1.15a; en effet:

$$\begin{aligned} m_{rs}(T) &= \int_{n(T)}^{\infty} (n\mu)\frac{1}{(2\pi nn_0)^{1/2}}\exp\left(-\frac{n}{2n_0}\right)\mathrm{d}n \\ &= \left(\frac{2}{\pi}\right)^{1/2} m_0 \exp\left[-\frac{n(T)}{2n_0}\right] = \left(\frac{2}{\pi}\right)^{1/2} m_0 \exp\left[-\frac{KT\log(\tau/\tau_0)}{2n_0 h_a \mu}\right]. \end{aligned}$$

Ou encore, en rassemblant ces résultats

$$m_{rs}(T)/m_{rs}(T = 0) = \exp(-\alpha T/c) .$$

C'est cette dernière formule, en accord avec l'expérience, qui justifie seule la loi de distribution de n ou de m. Le modèle est donc cohérent; on peut encore ajouter que les valeurs de E_a trouvées expérimentalement sont en accord avec les calculs de h_a d'origine dipolaire.

Pour conclure, disons que ce n'est pas mal trouvé, mais qu'à mon avis c'est un peu tarabiscoté, comme on dit en français! Ce qui me parait le plus hasardeux, c'est de supposer que l'énergie d'anisotropie est proportionnelle à n; en effet, les énergies de type RKKY dominent les énergies dipolaires; pour certains spins, celles-ci ne sont donc pas dans leur configuration minimum et toute modification d'orientation qu'on impose au nuage doit se traduire pour certains spins par une augmentation d'énergie dipolaire, par une diminution pour d'autres. J'aurais donc plus volontiers considéré une formule du type

$$E_a = h_a \mu \sqrt{n} .$$

Par ailleurs, certains résultats expérimentaux sur l'évolution temporelle de l'aimantation s'expliquent mieux en donnant une distribution de probabilité pour E_a qui ne soit pas uniforme mais lorentzienne [1.22]. Bref, il y a matière à raffiner le modèle.

Finalement, la question est celle-ci: cette image des nuages est séduisante; mais elle rend compte des seules mesures portant sur l'aimantation; il serait donc intéressant de disposer de mesures qui ne portent pas que sur cette quantité; on aimerait voir des expériences portant sur la chaleur spécifique ou la dilatation. Par ailleurs, une donnée du problème est insuffisamment prise en compte; c'est celle qui concerne la taille des domaines encore que dans le cadre de la théorie de Néel, on puisse estimer la taille moyenne d'un nuage. Un début d'interprétation se développe qui tendrait à assimiler la longueur de cohérence ζ relative au changement de phase qui a lieu à T_{SG} avec la taille des "nuages"; il y a là matière à réflexion. Il serait également instructif de connaître la loi de distribution des temps de relaxation des nuages.

En tout cas, ce modèle, qui rappelle celui des grains fins développés par Néel, est un bon cadre d'explication pour les phénomènes où la rémanence entre un peu dans la phase basse température; il serait souhaitable de faire joint avec les propriétés autour de T_{SG}.

1.5. Propriétés dynamiques

On prétend dans tous les cours de physique que la connaissance des propriétés dynamiques permet de lever bien des ambiguités des modèles qui décrivent un système; je crains bien qu'ici elles n'ajoutent au mystère.

Dans ce paragraphe, on examinera successivement des expériences de diffusion de neutrons et de susceptibilité dynamique.

1.5.1. Diffusion inélastique cohérente de neutrons: un mode collectif d'ondes de spin?

Ce n'est que récemment qu'un article très complet a développé une théorie hydrodynamique des ondes de spin pour un système où les positions d'équilibre des spins en chaque site ne sont pas nécessairement colinéaires [1.23], comme c'est le cas pour les verres de spins.

Comme dans toute théorie hydrodynamique, il faut supposer que les fréquences du mode collectif sont faibles devant les temps de relaxation de tous les degrés de liberté ou tout au moins que leur couplage avec eux est assez faible. Or, dans un verre de spin, la difficulté tient à l'existence d'excitations de basse fréquence, celles qui sont révélées par les mesures de chaleur spécifique et dont on retrouvera la présence également dans les verres. Enfin, pour parler de mode collectif, il importe qu'il existe une constante de rappel aux mouvements impliqués: rien n'est moins sûr.

Toujours est-il que des raisonnements de type hydrodynamique, dans le cadre ci-dessus, assortis de considérations de symétrie, supposant en particulier l'isotropie du système invitent à conclure à l'existence d'ondes de spin, dont la dispersion serait linéaire en vecteur d'onde.

Qu'en est-il expérimentalement? Comme souvent dans de tels cas, la diffusion inélastique cohérente de neutrons peut fournir une réponse. En règle générale, la section efficace de diffusion est donnée par la formule:

$$\frac{d^2\sigma}{d\Omega\, d\omega} = N\left(\frac{\gamma e^2}{mc^2}\right)^2 \frac{\kappa'}{\kappa_0}\, S(q\omega)\,, \tag{1.21}$$

où N est le nombre de spins, κ' le vecteur d'onde des neutrons diffusés, κ_0 celui des neutrons incidents; $\gamma e^2/mc^2$ est la constante d'interaction spin–neutron et $S(q\omega)$ la fonction de diffusion reliée à la susceptibilité dynamique $\chi(q\omega)$ par l'expression

$$S(q\omega) = 2\,\frac{\hbar\omega}{1 - \exp(-\hbar\omega/KT)}\, F^2(q)\, \frac{1}{g^2\mu^2}\, \chi(q\omega)\,. \tag{1.22}$$

$F(q)$ enfin est le facteur de forme magnétique [1.24]. Tout mode collectif se traduit par la présence dans $\chi(q\omega)$ de deux pics centrés en $\omega = \pm\omega(q)$ et munis chacun d'une largeur $\Gamma(q)$

$$\chi(q\omega) = \chi(q)\left\{\frac{\Gamma(q)}{[\omega - \omega(q)]^2 + \Gamma(q)^2} + \frac{\Gamma(q)}{[\omega + \omega(q)]^2 + \Gamma^2(q)}\right\}\frac{1}{2\pi}. \tag{1.23}$$

C'est la mesure de $\Gamma(q)$ et de $\omega(q)$ qui a été tentée à l'aide d'un spectromètre à temps de vol sur un alliage Mn–Cu à 5% et à une température voisine de 5 K [1.25]. Apparemment, grand soin a été pris pour s'affranchir des diffusions par les phonons de la matrice, pour comparer la diffusion avec gain ou perte d'énergie, pour mesurer les intensités du pic élastique ou inélastique incohérent.

Le résultat après traitement des données montre au dire des auteurs qu'il existe un pic inélastique, à environ 4 meV pour $q = 3$ Å^{-1} dont la disper-

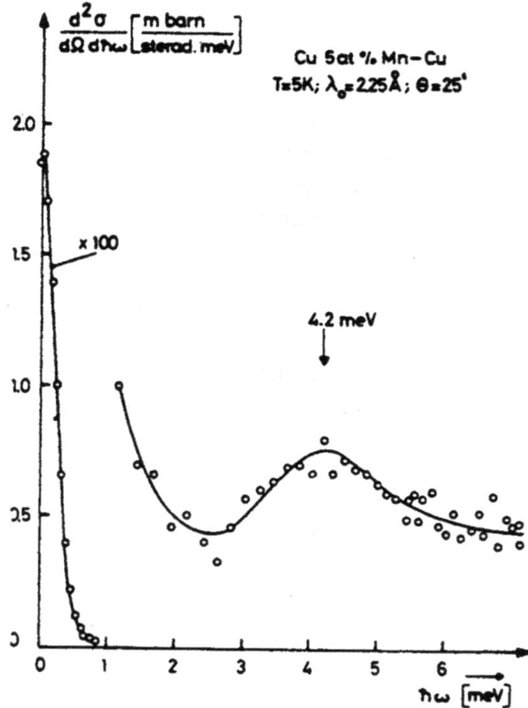

Fig. 1.16. Spectre en temps de vol obtenu par différence entre un alliage Cu–Mn à 5% et du cuivre pur. Les unités sont notées en mbarn/sr, meV et $Cu_{0.95}$–$Mn_{0.05}$.

sion est nulle dans la zone explorée, contrairement aux prédictions théoriques (fig. 1.16).

Faut-il croire à ce résultat? Une étude plus systématique serait nécessaire car une impureté produirait le même résultat ou même une mauvaise correction de bruit de fond; une évolution en température de la raie serait donc un bon critère de crédibilité. Bref, on gardera ce résultat en mémoire sans trop insister; pour y croire définitivement il faudrait d'abord que la statistique de comptage soit améliorée, et qu'une étude plus soigneuse en température soit poursuivie, notamment près de T_{SG}, au besoin en variant la concentration.

1.5.2. Diffusion magnétique diffuse de neutrons: temps de relaxation des spins

Au-dessus de la température T_{SG} on s'attend à ce que les spins soient partiellement découplés les uns des autres: certains relaxent en principe par un processus de Korringa (électrons libres du métal); leur temps de relaxation doit être inversement proportionnel à la température.

D'autres qui font partie des "nuages" non dissociés peuvent avoir un temps de relaxation τ plus long.

Là aussi, la mesure de la section efficace de diffusion fournit une mesure de τ, leur temps de relaxation, puisque cela se traduit par un élargissement de la raie de diffusion des neutrons monochromatiques incidents; on doit observer une raie quasi-élastique dont la dépendance en $\boldsymbol{q}$ dépend uniquement du facteur de forme magnétique $F(\boldsymbol{q})$ de l'atome paramagnétique. Des expériences ont été faites dans cet esprit sur des alliages du type Au–Fe dans une gamme de concentration de 5 à 15% et une zone de température allant de 30 K à 300 K [1.26] (fig. 1.17).

On observe en fait deux contributions: la première, indépendante de la concentration, relativement large et centrée en zéro d'énergie, est attribuable aux relaxations des spins individuels; à 300 K la demi-largeur est de 10 meV; elle correspond à des τ de l'ordre de 10^{-13} s ce qui est correct pour un métal. Son facteur de forme $F(\boldsymbol{q})$ est bien celui d'un moment 3d.

La deuxième raie, malheureusement, n'est pas interprétable aussi aisément; des expériences supplémentaires sont nécessaires pour en fournir une interprétation satisfaisante.

Plus récemment, une étude systématique [1.27] des mêmes relaxations dans les alliages Cu–Mn a fourni une vision plus précise de ce qui se passe.

A faible concentration, et à température assez haute, les spins sont découplés; ils relaxent uniquement par un processus de Korringa; la largeur de raie est du type

$$h\tau^{-1} \simeq \Gamma_0 \simeq \pi[J_{s\text{-}d}\rho(E_F)]^2 KT,$$

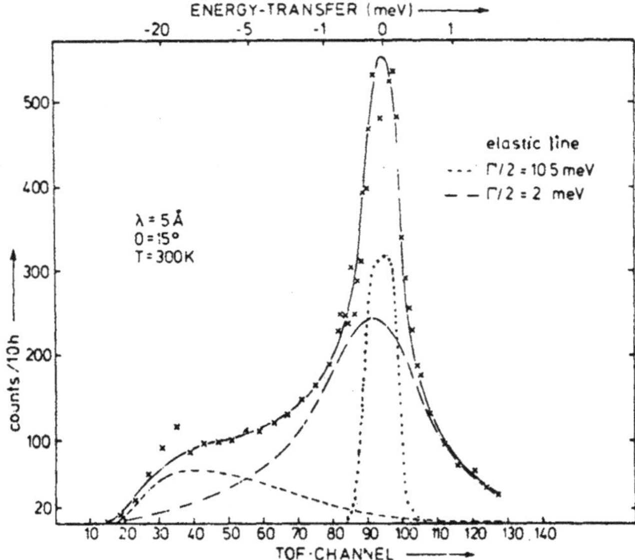

Fig. 1.17. Spectre en temps de vol obtenu par différence entre un alliage Au–Fe à 15% et de l'Au pur pour un angle de diffusion constant de 15° à une température de 300 K.

où J_{s-d} est l'intégrale d'échange entre les électrons libres et les impuretés magnétiques; $\rho(E_F)$ est la densité des électrons au niveau de fermi. L'ordre de grandeur ($\Gamma_0 \simeq 1$ meV) est correct et bien sûr Γ_0 est indépendant de q.

A plus forte concentration, il faut faire entrer en ligne de compte l'influence des impuretés entre elles; à Γ_0 s'ajoute une contribution du type

$$\Gamma^2(q) \simeq \tfrac{8}{3}s(s + i)\sum_j (J_{ij}^2 r_{ij}^2)q^2 ,$$

qui pour q grand donne [1.28]

$$\Gamma^2 \simeq \tfrac{8}{3}s(s + i)\sum_j J_{ij}^2 .$$

C'est approximativement ce qui est vérifié compte tenu du fait que la raie passe progressivement d'une forme lorentzienne à une gaussienne et que dans la limite $q \to 0$ il semble exister un rétrécissement par le mouvement. Bref, je crois que la situation est assez claire cette fois.

Quoiqu'il en soit, il faut prendre pour certain qu'il est possible d'appréhender le mouvement individuel des spins dans un verre; il reste à explorer leur comportement près de T_{SG}.

1.5.3. Susceptibilité dynamique et susceptibilité statique dépendant de q; diffusion de neutrons: expériences en temps de vol

La formule (1.22) assure la liaison entre la section efficace de diffusion et la susceptibilité dynamique $\chi(q\omega)$. Dans un paragraphe précédent, on a analysé le cas d'un mode d'onde de spin; on voudrait ici décrire des expériences qui portent sur une exploration systématique de $\chi(q\omega)$ dans un domaine d'énergie beaucoup plus petit où il apparait que $\chi(q\omega)$ a la forme suivante:

$$\chi(q\omega) = \chi(q)\frac{1}{\pi}\frac{\Gamma(q)}{\omega^2 + \Gamma^2(q)} . \tag{1.24}$$

Tous les résultats expérimentaux [1.29] qui ont porté sur les alliages Cu–Mn ou Au–Fe à quelques $\%_0$, montrent que la diffusion peut s'analyser en deux contributions: une partie élastique, qui a une largeur en énergie égale à la fonction d'appareil, et une partie quasi-élastique de la forme (1.24).

Pour cette dernière contribution, on note que $\Gamma(q)$ décroit comme la température mais ne s'annule pas lorsque l'on se rapproche T_{SG}; cela n'est pas déraisonnable si l'on combine deux idées: dans le cadre d'une théorie de champ moyen on sait calculer [1.30], quel que soit T, la fonction de correlation $\langle S_i(0)S_i(t)\rangle$ qui joue le rôle de paramètre d'ordre pour $T < T_{SG}$. On devrait observer que

$$\Gamma(q) \simeq T - T_{SG} .$$

En fait, suivant la première interprétation des auteurs, cette singularité est "lavée" s'il y a une distribution de T_{SG}. Toujours la même idée est sous-jacente: il y a des "nuages" qui ont leur propre température de transition: le fameux $T_{SG}(q)$ que l'on a vu précédemment et qu'il faut, je crois, répéter. Par ailleurs, et cela est compatible avec le même calcul, on observe une augmentation de l'intensité contenue dans la partie élastique; cette augmentation prend naissance à des températures très supérieures à T_{SG} car une partie des spins sont bloqués déjà à haute température.

D'ailleurs, la température à laquelle on détecte une augmentation significative de la diffusion élastique est très dépendante de la résolution de la fonction d'appareil $\Delta\omega$, faisant irrésistiblement penser à une distribution des temps de relaxation; dans des systèmes comme les verres de spin, on ne peut donc espérer voir des singularités aussi franches que celles qu'on observe pour les changements de phase ordinaires.

Le dernier renseignement que l'on peut extraire de ces mesures est la valeur de $\chi(q)$ par intégration de la partie quasi-élastique $\chi(q\omega)$. Le maximum de $\chi(q)$ se produit à des températures supérieures à celle observée pour $\chi(q = 0)$, en sens inverse donc du comportement observé par diffusion aux petits angles.

Ces différences s'expliquent aisément si on tient compte une fois encore du temps caractéristique de la mesure. Ici, la séparation entre partie élastique et quasi-élastique est fixée par $\Delta\omega$ la résolution de l'appareil; ne sont compris dans $\chi(q\omega)$ que des contributions correspondant à des mouvements plus rapides que $1/\Delta\omega \simeq 10^{-11}$. Le calcul de $\chi(q)$ dans ces conditions ne prend en compte qu'une partie des fluctuations. Inversement, dans une expérience de diffusion aux petits angles, c'est $\chi(q)$ total qui est mesuré directement et dont la valeur n'est pas grevée par la séparation élastique–quasi-élastique.

Une dernière expérience [1.31] mérite d'être décrite, qui a été conçue dans le but de conforter cette lecture; elle me parait souligner l'importance de l'échelle de temps de la mesure dont on a repéré les effets à plusieurs moments; elle me semble aussi propre à éliminer l'idée d'une transition progressive de différents groupes de spins à des températures supérieures à T_{SG}.

C'est une expérience de diffusion inélastique de neutrons, où l'on mesure la quantité $S_2(q)$ définie par la formule (1.13) avec des résolutions de plus en plus grandes; c'est, en d'autres termes, faire une expérience avec des échelles de temps variables. $S_2(q)$ comme on l'a vu est proportionnel au paramètre d'ordre η défini dans le modèle de Edwards et Anderson

$$S_2(q) = 2F^2(q)\eta .$$

Mais pour un instrument donné, on intègre toujours dans le pic de diffusion élastique une somme de contributions dynamiques qui tombent à l'intérieur de la résolution δE. Si la fonction de correlation temporelle des spins est

$$\langle S_i(0)\, S_i(t)\rangle = \sum_{\tau} \mathrm{pr}(T, \tau) \exp(-t/\tau) , \qquad (1.25)$$

où $\mathrm{pr}(T, \tau)$ est la probabilité, à la température T, d'avoir un temps de relaxation τ, le pic élastique incluera non seulement $S_2(q)$ mais aussi toutes les fluctuations pour lesquelles

$$\tau > h/E .$$

A température supérieure à T_{SG} il y a un ralentissement progressif des spins lorsque $T \to T_{SG}$; le pic "élastique" doit croître et l'on vérifie bien que plus la résolution est mauvaise (δE grand) plus il prend naissance à température élevée; en utilisant le même instrument, le même échantillon, Cu–Mn à 8%, on a montré que pour $\delta E \simeq 250$ eV, on voit croître le pic élastique à 75 K, pour $\delta E = 25\ \mu$eV à 55 K, et pour $\delta E = 1.5\ \mu$eV à 47 K, alors que $T_{SG} \simeq 40$ K.

Cette expérience confirme donc bien ce qu'on a souligné précédemment: l'échelle de temps entre en ligne de compte pour expliquer des résultats comme ceux obtenus par effet Mössbauer; par ailleurs, tout semble compatible avec l'existence d'un transition de phase à T_{SG} qui se manifeste par un véritable ralentissement.

2. Les verres

2.1. Introduction

"Le verre, vieux de 4 ou 5,000 ans, est probablement un des premiers matériaux synthétisés par l'homme. Contrairement à d'autres produits de l'artisanat préhistorique, il n'en reste aucun vestige. Cependant, les écrits sont là; les scribes Sémites (Mésopotamie) sous la dictée des Sumériens gravent dans l'argile le premier traité sur les méthodes de fabrication du verre (XVIIème siècle av. J.C.). C'est ensuite dans les pays du Levant que se développe l'art du verrier et la Conquête Romaine le diffuse aux quatre coins de l'Empire. Après la chute de l'administration romaine, l'usage du verre semble se restreindre. Ce n'est que beaucoup plus tard, au Moyen Age, sous la poussée du Mysticisme Chrétien que la verrerie est de nouveau appréciée et que l'on voit apparaître les verres teintés des grands vitraux. Avec l'expansion colonialiste du XVième siècle on exporte le verre en Chine et on en fait une monnaie d'échange avec Montezuma et les Indiens d'Amérique. Puis la Rennaissance Italienne avec les maîtres de Venise lui redonne un peu de sa gloire: plus qu'un objet d'usage domestique, il redevient précieux comme chez les Romains. Finalement, la Bohème, en inventant le "verre qui se taille comme un cristal", devient le centre verrier de l'Europe.

A la même époque, le chimiste Johann Kunckel publie à Potsdam (1679) son "Ars Vetraria Experimentalis". Cette première approach scientifique au problème fit que l'art du verrier devint l'industrie du verre. En fait les techniques demeurent assez empiriques. Le verre est un matériau secret pour le fabricant mais aussi pour le chercheur qui pendant longtemps le considère comme peu attrayant et lui préfère la "simplicité" du cristal. Malgré tout, on entreprend des recherches destinées à établir les relations entre les propriétés et la composition. Les résultats ont conduit à la spécialisation, à l'amélioration et à l'invention de produits nouveaux; les domaines d'utilisation s'élargissent; le verre devient un matériau esthétique et peu coûteux. Demain, on le rencontrera dans le domaine des communications; les fibres optiques permettant de véhiculer une densité maximum d'information avec un minimum d'affaiblissement remplaceront avantageusement les cables classiques; les semi-conducteurs amorphes donneront peut-être naissance aux prochaines générations d'ordinateurs" [2.1].

Pour revenir à la physique des systèmes désordonnés et en particulier aux verres, disons que deux aspects méritent de retenir l'attention: la préparation du verre, c'est-à-dire l'ensemble des propriétés qui gouvernent la transition vitreuse et l'état qui en résulte; le verre apparait alors comme un système non ergodique, au sens où la moyenne temporelle d'une grandeur est différente de la moyenne d'ensemble. Placé à une température inférieure à la température de transition vitreuse T_G le verre est figé dans un état métastable: sa structure locale, sa viscosité, son volume libre, sont le résultat et d'une histoire thermique et des conditions qu'on lui impose.

Le deuxième aspect des verres qui retiendra l'attention est celui qui couvre leurs propriétés thermodynamiques et dynamiques à basse température; leur similarité avec celles des métaux amorphes ou des verres de spins posera d'ailleurs plus de problèmes qu'elle n'en résoudra. Une explication a été proposée qui rend compte tout à la fois de la chaleur spécifique, de la conductibilité thermique, des propriétés acoustiques... tout tiendrait à l'existence de systèmes à deux niveaux: S2N (TLS: "two level systems" en anglais) résultant de la présence de doubles puits de potentiel entre lesquels un paramètre de configuration peut osciller par effet tunnel ou par activation thermique. Ce concept de S2N a été fortement exploité; mais ce qui reste étrange c'est que fort peu de choses soient connues sur la nature de ces S2N ou sur ce qu'implique ce paramètre de configuration: correspond-il à un mouvement atomique ou à un mouvement électronique? aucune expérience jusqu'à maintenant n'a fourni un commencement de réponse. Nous pourrons donc spéculer à l'envi!

2.2. Les verres: les différents types de verres

On le verra, les verres ont des propriétés "universelles" c'est-à-dire que, quelle que soit leur nature, les variations avec la température des coefficients des équations d'états, les ordres de grandeur des coefficients de transport sont à peu près identiques.

Raison de plus pour faire ici l'inventaire de leur nature diverse et pour l'oublier ensuite une fois débarrasés de cette nomenclature.

Le plus simple et le plus commun des verres est la silice fondue dont les formes cristallines sont le quartz, la cristobalite et la tridynite. L'unité de composition est le tétraèdre SiO_4; on connait plusieurs variétés de silice, très souvent d'origines commerciales diverses qui ne diffèrent que par le procédé de préparation et qui possèdent donc des impuretés métalliques, ioniques, ou même moléculaires en proportion variable.

Dans la même ligne, il existe une autre famille de verres dont les proportions d'oxydes SiO_2, B_2O_3, K_2O, Na_2O varient mais dont les propriétés là aussi sont voisines; moins usuels sont ceux qui comportent des oxydes de P_2O_5, ou des oxydes de terres rares: La_2O_3, ThO_2, Nb_2O_5. Beaucoup de ces oxydes ont comme élément structurel de base un tétraèdre type MO_4. La silice pure, ou la silice contenant des impuretés moléculaires type OH^- étant un matériau relativement bien identifié et reproductible, la majorité des expériences ont été faites sur elle; c'est d'ailleurs un matériau très industriel. On n'évoquera pas plus dans la suite les silices obtenues par densification sous haute pression au voisinage de T_G; ce sont à l'opposé des matériaux fragiles dont on a guère exploré plus que les propriétés optiques et les caractéristiques de structures [2.2].

Une deuxième catégorie de verres correspond aux verres organiques dont l'élément de base est une molécule organique, alcool ou cétone; un exemple fameux: le glycerol $C_3O_3H_8$ dont la température de transition vitreuse est de 170 K. D'autres exemples correspondant au PMMA (polymethylmethorylate) ou au PVA (polyvinyl acetate); dans ce dernier cas la répétition du même motif de nombreuses fois conduit à la constitution de chaines qui confèrent à ce système des propriétés de viscosité et d'élasticité dont on trouvera des éléments dans le cours qui porte sur les polymères.

Une troisième catégorie de verres est celle des semiconducteurs type Ge ou Se, le S, ou même les chalcogénures dont certains ont une structure en couche désordonnée, tel As_2Te_3.

Pour être complet il faut ajouter à cette liste quelques sels ioniques ou partiellement ioniques comme $ZrCl_2$ ou même quelques solutions electrolytes près de la composition ou elles forment un entectique, type $HCl\text{–}H_2O$.

Tous ces exemples ont en commun de posséder des éléments constituants ayant une certaine versatilité dans leur aptitude à former des liaisons chimiques. Mais la préparation du verre, on le sait depuis les premiers cours de chimie-physique enseignés à l'Université, est une question de cinétique; si un germe est donné, ou tout au moins si une position de germination est acquise, un cristal se développe si les processus de diffusion dans le liquide sont assez rapides. Inversement, un verre est favorisé si la structure du cristal qui devait se former est compliquée et implique une trop grande spécificité des liaisons à établir.

2.3. La transition vitreuse

Le passage de l'état liquide surfondu à l'état vitreux s'appelle la transition vitreuse. En fait, la transition est mal définie ou tout au moins, elle réclame pour l'être une procédure très stricte; on parle plutôt d'une région de transition.

Lorsque les conditions de cristallisation ne sont pas réunies, soit qu'il n'y ait pas de germes, soit parce que les temps caractéristiques nécessaires à l'établissement de l'équilibre thermodynamique sont plus longs que l'évolution imposée au liquide, on obtient finalement un verre c'est-à-dire un état métastable dont la conformation elle-même évolue à des vitesses qui dépendent de l'histoire du verre et des conditions qui lui sont imposées.

La formation d'un verre s'observe aisément, parce que diverses grandeurs macroscopiques présentent au moins un changement de pente en fonction de la température; il définit T_G. Mais T_G dépend beaucoup de la vitesse de refroidissement: plus la vitesse de refroidissement est grande, plus T_G est voisin de T_c, la température de cristallisation (fig. 2.1).

Sur cette figure, on a porté le volume spécifique du verre (ici PVA) en suivant le schéma suivant. On part d'une situation où le liquide est surfondu et où il a des propriétés reproductibles. On le refroidit rapidement à une température T où l'on mesure son volume à un temps t postérieur à cette opération; suivant que t est petit ou grand (cent heures) on vérifie que les points expérimentaux s'alignent sur deux droites dont l'intersection avec celle relevée dans l'état surfondu définit une température de transition vitreuse T_G dépendant de t. Il est bien sûr essentiel de toujours partir d'une situation de départ reproductible.

Les mêmes mesures pourraient être faites en observant le volume spécifique au même instant t mais après avoir refroidi le liquide surfondu avec des vitesses différentes; on obtiendrait une autre série de T_G. Les verres obtenus à des vitesses différentes ont des densités ou des entropies différentes. C'est donc par préciser quelques aspects de la transition vitreuse

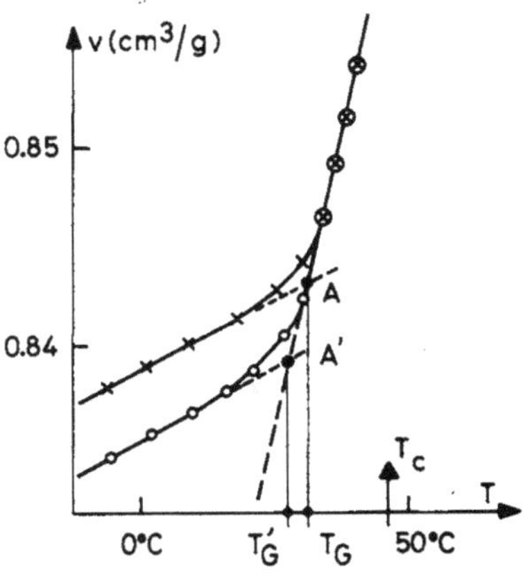

Fig. 2.1.

qu'il faut commencer en retenant que T_G n'est pas définie de manière unique; la procédure expérimentale doit être toujours soigneusement précisée.

2.3.1. Description de la transition vitreuse

Les particularités de la transition vitreuse sont illustrées par trois types de mesures: des mesures avec évolution en température: chaleur spécifique ou analyse thermique différentielle. Des mesures de relaxation isotherme et

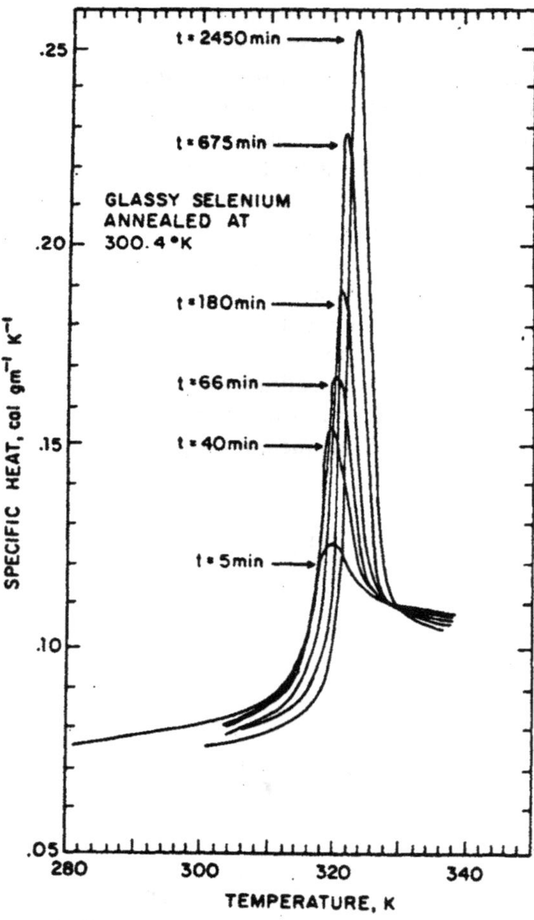

Fig. 2.2. Chaleur spécifique en fonction de la température pour le Se dans la région de la transition vitreuse après recuit à 300.4 K (27.2°C) pendant différents temps.

isobarique qui portent sur des grandeurs comme le volume spécifique, l'enthalpie, ou la constante dielectrique. Enfin, des mesures de coefficient de transport, la viscosité essentiellement, et l'atténuation ultrasonore accessoirement.

(a) La chaleur spécifique est singulière: elle présente des pics au voisinage de T_G qui dépendent de manière spectaculaire de la rapidité des évolutions en température que l'on impose et de l'histoire thermique précédente du matériau. Il est assez difficile d'en extraire des informations précises sur la cinétique de la transition ou sur les temps caractéristiques des relaxations de conformation, mais c'est un bon procédé pour déceler l'influence sur T_G de la concentration des solutions ou de la masse moléculaire dans le cas des polymères. A titre d'exemple, les courbes de la figure 2.2 donnent l'allure du phénomène pour le glycérol.

La procédure expérimentale d'enregistrement des courbes est la suivante: à partir d'une même température où le glycérol est liquide, l'échantillon est refroidi dans des conditions identiques jusqu'à une température T_R inférieure à T_G; là, il est stabilisé pendant des durées variables puis il est réchauffé à vitesse constante. Pour un échantillon recuit peu de temps, courbe (1), le pic de chaleur spécifique est petit. Plus le recuit est long, plus le pic est grand et plus il se déplace vers les hautes températures, courbe (3) [2.3, 2.4].

Le schéma de cette expérience est le suivant dans un diagramme où l'on porte les températures en abcisse et les temps de stabilisation $\bar{t}$ en ordonnée:

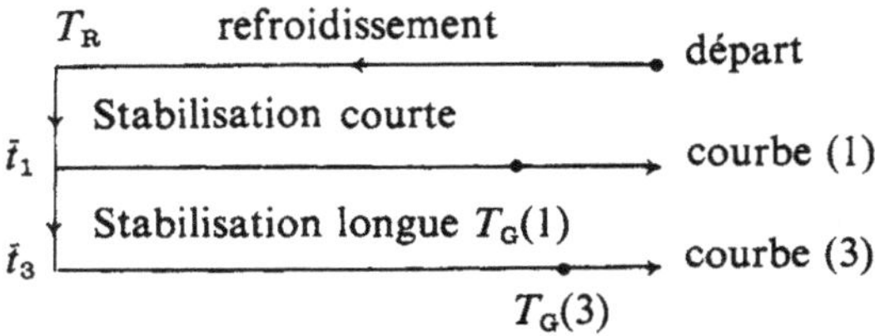

Dans tous les cas, on vérifie que

$$dT_G(E)/d\log \bar{t} \simeq RT_G^2/E_a\,,$$

où E_a est une énergie d'activation qui ne dépend presque pas de la grandeur qui est examinée; en particulier, elle reste vraie pour T_G défini à partir des mesures de volume spécifique du paragraphe précédent.

Peut-on expliquer ce comportement? Qualitativement, on peut raisonner ainsi: plus longtemps le glycérol est stabilisé à $T_R < T_G$ plus il a le temps de relaxer vers la configuration d'équilibre à T_R; ce faisant, il s'écarte d'autant plus de la configuration d'équilibre qu'il devrait avoir en passant par T_G; les barrières de potentiel qui gouvernent les réarrangements moléculaires sont plus élevées et le pic de chaleur spécifique se déplace vers les hautes températures.

(b) Un modèle existe-t-il qui rende compte de ces comportements? Oui et non; on va le voir en traitant des relaxations isothermes.

Ces dernières sont en effet fort instructives car elles donnent accès à la cinétique du retour à l'équilibre; les expériences ont porté sur l'enthalpie, l'indice de réfraction et le volume spécifique.

Prenons comme premier exemple le volume spécifique et sa dépendance temporelle après une trempe de l'état liquide en équilibre jusqu'à la température T en dessous de T_G. Un réseau de courbes typiques est donné dans la figure 2.3 pour différentes valeurs de T. Les points à retenir sont les suivants:

– L'évolution en température constante dépend fortement de la température de trempe, mais peu ou pas de la température de départ.

– L'évolution du rapport $[v(t) - v(\infty)]/v(\infty) = \delta(t)$ est proportionnelle à $\log t$ sur quelques décades de temps:

$$\delta(t) \simeq \log(t - t_{\text{initial}}) .$$

– Aucune des courbes données ici ne peut être ramenée à une loi unique par une méthode faisant apparaître des correspondances ou des quantités réduites; on est loin de pouvoir envisager une description à l'intérieur des lois linéaires; en particulier, ces mesures ne peuvent être normalisées par δ initial.

– Ces courbes suggèrent immédiatement l'idée d'une distribution de temps de relaxation très dépendante de la température à laquelle l'expérience est faite; on peut toujours définir un temps de relaxation effectif à chaque instant par la formule

$$\frac{1}{\delta(t)} \frac{d\delta}{dt} = -\tau_{\text{eff}}(t) .$$

Si la relaxation de δ est exponentielle, τ_{eff} est une constante; dans le cas d'un verre, τ_{eff} varie avec t, avec la configuration du verre (temps de stabilisation) et avec la température (voir fig. 2.3).

Toutes ces caractéristiques se retrouvent pour des verres aussi différents dans leur nature que le selenium amorphe, As_2O_3, PVA [2.5]. On observe donc là une propriété relativement universelle qui ne peut tenir qu'à la cinétique de réarrangements de configuration et non à leur nature chimique.

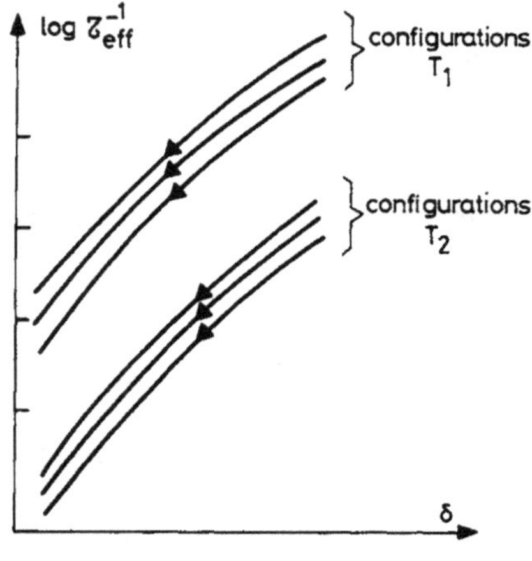

Fig. 2.3.

Par ailleurs, si l'on fait l'opération en sens inverse, c'est-à-dire si, après stabilisation, on réchauffe le verre, sa dilatation se fait plus rapidement que par une lois exponentielle.

En d'autres termes, le retour à l'équilibre se fait de manière progressivement accélérée lors de la dilatation, mais de manière ralentie lors de la contraction [2.6].

Cette dernière observation suffit à montrer que la distribution des temps de relaxation n'est pas suffisante à expliquer le comportement du verre mais qu'il faut dire en plus que cette distribution elle-même dépend de la structure du verre à la température où l'expérience est faite; cet effet est bien plus important que l'histoire du verre; toujours est-il que cela interdit de vouloir représenter l'évolution du verre avec une équation différentielle impliquant les seules quantités $(v(t), \dot{v}(t), v_\infty)$; les conditions initiales elles-mêmes jouent un rôle.

Les expériences qui portent sur la relaxation isotherme de l'enthalpie obéissent aux mêmes remarques.

(c) A ce stade, il est utile de décrire une expérience où l'on mesure des courants de dépolarisation thermique. Elle met en évidence l'existence d'une température singulière loin en dessous de T_G qui a quelquefois été appelée température de changement de phase, On procède ainsi: l'échantillon est polarisé par un fort champ électrique (10 kV/cm) à une température supérieure à T_G; on attend que l'équilibre soit atteint. Puis, on trempe l'échantillon et on coupe le champ électrique. On réchauffe enfin en enregistrant à la fois la polarisation $\boldsymbol{P}(T)$ et le courant de dépolarisation $\boldsymbol{j}(T)$ à chaque instant en imposant une histoire thermique $T(t)$.

A tout instant, on peut écrire

$$\mathrm{d}P(t)/\mathrm{d}t = j(t)\,.$$

On définit alors un temps de relaxation par l'équation

$$\mathrm{d}P(t)/\mathrm{d}t = -P(t)/\tau(t)\,.$$

A supposer que le verre relaxe avec ce seul temps caractéristique $\tau(T)$ on montre aisément que $\tau(T) = P(T)/j(T)$: c'est le rapport de deux quantités mesurées indépendamment.

Le résultat est tout à fait étonnant: on trouve que $\log[\tau(T)]$ a un comportement critique; "critique" au sens des changements de phase comme si loin en dessous de T_G il existait une température de transition vers une autre phase. Dans la figure 2.4 on a porté les variations de $\tau(T)$ en fonction du paramètre $10^3/(T - T_\infty)$ pour un polymère: le polyethylène isophtalamide [2.7]. Il est remarquable de voir comment la loi

$$\log[\tau(T)] = A/(T - T_\infty) \tag{2.1}$$

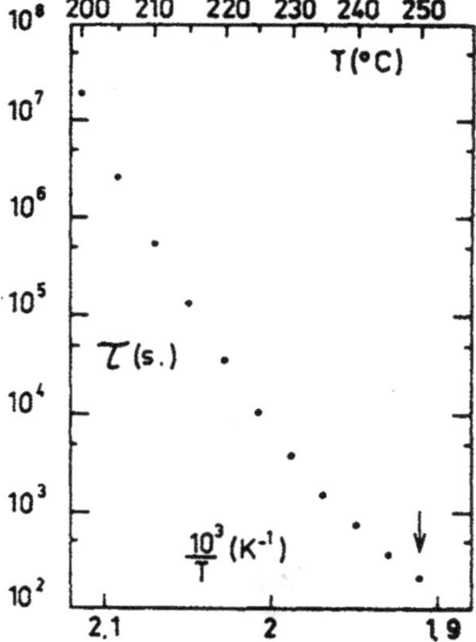

Fig. 2.4a. Diagramme de ln τ en fonction de $10^3/T$ à partir de pic de DTC obtenu dans le poly(ethyleneisophtalamide) au-dessus de la température de transition vitreuse. La flèche indique la position pour le maximum de thermocourant.

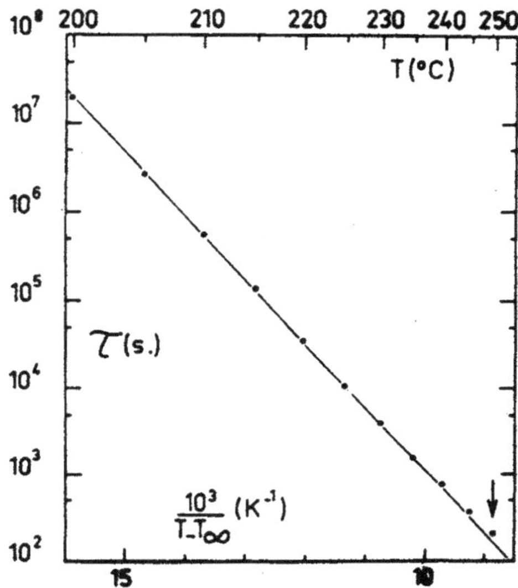

Fig. 2.4b. Graphique donnant ln τ en fonction de $10^3/(T - T_\infty)$ où $T_\infty = 410$ K partir des mêmes données que dans la figure ci-dessus.

est suivie sur un grand nombre de décades en τ. Par comparaison avec les mesures de quantités critiques autour d'un point de changement de phase, la gamme de température n'est certes pas très grande; mais la précision est suffisante pour distinguer (2.1) d'une loi d'activation ordinaire du type

$$\log[\tau(T)] \simeq 1/T\,. \tag{2.2}$$

Plus surprenant encore: sur le glycérol, des expériences poursuivies soit par relaxation de l'enthalpie [2.8], soit par relaxation diélectrique [2.9], soit par viscosité [2.10] dans une grande gamme de température ont montré que le temps de relaxation $\tau(T)$ obéit à une seule et même loi du type (2.1); cela indique pour le moins que la même évolution de la configuration du verre gouverne toutes ces propriétés. Mais quels paramètres de configuration entrent en jeu? La loi empirique (2.1) a été proposée par Vogel–Fulcher; elle se révèle assez générale. Son développement autour de T_G fournit des expressions analytiques qui reproduisent tous les modèles particuliers qui ont été proposés pour représenter l'évolution d'un verre: modèle dit du volume libre [2.11], description empirique de WLF [2.12] ou théorie thermodynamique statistique [2.13]. Quand on y regarde d'un peu plus près, on a l'impression que la loi de Vogel–Fulcher les contient toutes: elle a un caractère plus fondamental.

Par ailleurs, les coefficients de (2.1), c'est-à-dire A et T_∞ ont un caractère universel: leurs valeurs numériques sont très voisines quel que soit le verre considéré.

De sorte que le problème est à l'heure actuelle de tenter d'expliquer pourquoi on doit s'attendre à un tel comportement, quelle structure microscopique en est à l'origine; c'est une question aux théoriciens. Aux expérimentateurs, on peut demander de ne pas s'obnubiler sur la forme analytique de (2.1); après tout dans une gamme de température assez large, une autre expression que (2.1) pourrait aussi bien coller aux résultats expérimentaux. D'ailleurs (2.1) a pu être remplacée dans certains cas par une expression du genre:

$$\log[\tau(T)] \simeq [T \log(T/T_0)]^{-1} . \tag{2.3}$$

On a plus besoin en ce domaine de mesures méticuleuses, réalisées avec une procédure bien définie, que de résultats ajustés sur une loi proposée au terme d'une théorie chevelue et douteuse.

2.3.2. Quelques idées pour la description de la transition vitreuse

Une grande quantité de modèles tentent de rendre compte des principaux aspects de la transition vitreuse; toute une littérature s'efforce de décrire les avatars de l'une ou l'autre des quantités étudiées: volume spécifique, enthalpie, constante diélectrique, diffusivité thermique, viscosité.

Disons le tout de suite: aucun n'explique vraiment ce qui se passe pour que l'on puisse dire que ce problème est classé; pourtant, plusieurs idées existent qui, combinées ensemble, traduisent à peu près la réalité. Ce qui manque, c'est un fondement statistique plus fermé; un effort comme celui qui a été développé à propos des verres de spins.

Rassemblons donc les éléments qui doivent être pris en compte:

(1) Le verre a une structure compliquée; sa configuration est associée à toute une série de paramètres de configuration ξ_i. Pour T et p donnés, chacun d'eux relaxe vers une situation d'équilibre avec un temps caractéristique τ_i. On admettra que chaque paramètre ξ_i contribue pour δ_i à une quantité macroscopique δ, seule atteinte expérimentalement. On écrit à tout instant:

$$\delta = \sum_i \delta_i . \tag{2.4}$$

Les δ_i sont construits pour être nuls à l'équilibre. L'évolution de $\delta(t)$ obéit donc à l'équation:

$$\frac{\mathrm{d}\delta(t)}{\mathrm{d}t} = \sum_i \frac{\mathrm{d}\delta_i(t)}{\mathrm{d}t} = -\sum_i \frac{\delta_i(t)}{\tau_i} . \tag{2.5}$$

Bref, on admettra qu'il y a une distribution de temps de relaxation. Le choix des $\delta_i(0)$ dans (2.25) permet de rendre compte des effets de mémoire: la préparation du système avant son évolution ultérieure peut être différente.

(2) Le retour à l'équilibre se fait parfois par une évolution non monotone [2.14] (voir fig. 2.5). On pourrait en rendre compte en choisissant les $\delta_i(0)$ positifs ou négatifs. Mais c'est assez peu vraisemblable. Une idée plus fructueuse consiste à admettre que les divers τ_i dépendent à chaque instant de la valeur de tous les ξ_j.

C'est prendre en compte l'aspect coopératif de la transition vitreuse, et se donner la possibilité de décrire l'évolution du système en fonction de son état structurel; en particulier, le retour à l'équilibre n'a aucune raison de se faire de la même manière à partir d'une situation avec $\delta > 0$ ou avec $\delta < 0$. Les divers ξ_j sont donc couplés.

La difficulté, ou la diversité des modèles tient au grand choix qui existe pour donner une forme analytique à ces couplages. Donnons en quelques exemples.

Une forme simple consiste à n'envisager qu'un seul temps de relaxation et à décrire l'évolution de δ avec un seul paramètre ξ:

$$\tau(T, \delta) = \tau_0 \exp\left(\frac{A}{T - T_\infty}\right) \exp(-\alpha\delta) . \tag{2.6}$$

En portant (2.6) en (2.5), on a une équation non linéaire avec mémoire en

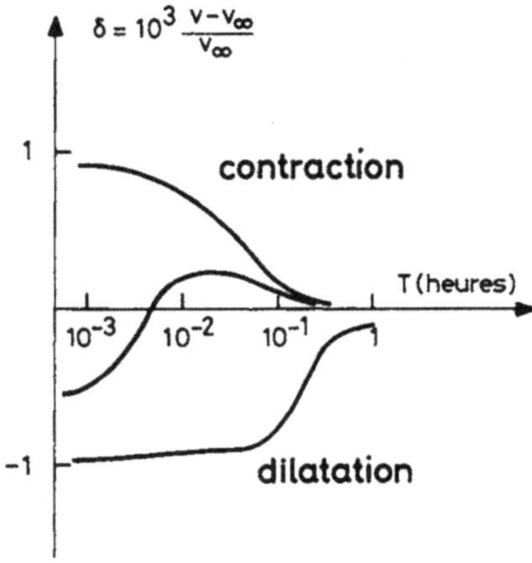

Fig. 2.5.

raison du terme coopératif en $\alpha\delta$. On sait en extraire une solution analytique dont le comportement asymptotique dépend du signe de $\delta(t = 0)$.

La même formule (2.6) peut être utilisable dans un modèle à deux paramètres (A_1, α_1 ; A_2, α_2). Pour des constantes 1 et 2 très différentes on simule des évolutions non monotones [2.15] analogues à celles relevées à propos du volume spécifique [2.14]. Les mêmes formules fournissent également une description de l'histoire thermique du verre, correspondant à un chemin imposé $T(t)$.

Ces résultats prouvent que les ingrédients d'une théorie valable sont bien ceux qui ont été mentionnés. Mais ces formules restent à mon avis du bricolage tant qu'elles ne sont pas mieux fondées. On recherche des idées nouvelles!

2.4. *Quelques particularités de la phase vitreuse*

A ce stade, il est utile de signaler quelques particularités de la phase vitreuse et de les distinguer de celles d'un cristal. On ne peut passer sous silence par exemple la structure du verre et ses conséquences sur la dynamique des mouvements collectifs.

2.4.1. *Structure statique d'un verre*

Loin en dessous de T_G, le verre n'évolue plus; les positions moyennes des atomes sont fixes, à l'agitation thermique près. Certes, les dispositions réciproques des atomes sont dépendantes du processus de préparation du verre: mais il y a cependant quelques remarques générales qui peuvent être articulées.

Pour déterminer la structure d'un verre, on cherche à mesurer le facteur de structure $S(\boldsymbol{k})$ soit par diffraction des RX soit par diffraction des neutrons. Pour un système monoatomique (Ge ou Se, . . .), on définit une fonction de paire $\rho(\boldsymbol{r}) = \rho(\boldsymbol{r})/\rho_0$, où ρ_0 est la densité moyenne du verre et $\rho(\boldsymbol{r})$ la densité à la distance $\boldsymbol{r}$ d'un atome pris comme référence. En définissant

$$S(\boldsymbol{k}) = \frac{1}{N}\left\langle \left|\sum_{i=1}^{N} \exp(\mathrm{i}\,\boldsymbol{k}\cdot\boldsymbol{r}_i)\right|^2 \right\rangle, \tag{2.7}$$

on montre aisément que

$$\rho(\boldsymbol{r}) = 1 + \frac{1}{2\pi^2\rho_0 r}\int_0^\infty k[S(\boldsymbol{k}) - 1]\sin kr\,\mathrm{d}k\,. \tag{2.8}$$

Avoir une bonne idée de $g(\boldsymbol{r})$ consiste donc à explorer $S(\boldsymbol{k})$ dans une zone allant de k tout petit jusqu'à k maximum. Pour avoir une bonne précision pour des valeurs de r petites, il faut mesurer $S(\boldsymbol{k})$ pour des grands vecteurs d'ondes.

Dans la figure 2.6 sont données les allures des fonctions $S(\boldsymbol{k})$ et $g(r)$ pour le sélénium amorphe. Les ordres de grandeur y sont. Que constate-t-on? En règle générale, la distance des premiers voisins, liés de manière covalente, n'est pas affectée par le passage de l'état amorphe à l'état liquide; par contre, le deuxième pic de la fonction $4\pi r^2\rho_0[\rho(\boldsymbol{r}) - 1]$ qui donne la distribution radiale du deuxième voisin est beaucoup plus large que le premier semblant indiquer que l'angle des liaisons est largement distribué autour d'une valeur moyenne. C'est une observation assez générale dans les verres. Par ailleurs, le nombre des premiers voisins reste limité à deux alors que suivant les interprétations, le nombre des seconds voisins varie de 6 à 8 [2.16]; c'est un effet des modèles utilisés pour décrire le sélénium amorphe: chaîne plane formée d'éléments qui sont des triangles isocèles analogues à ceux du sélénium cristallisé dans la forme hexagonale; anneaux

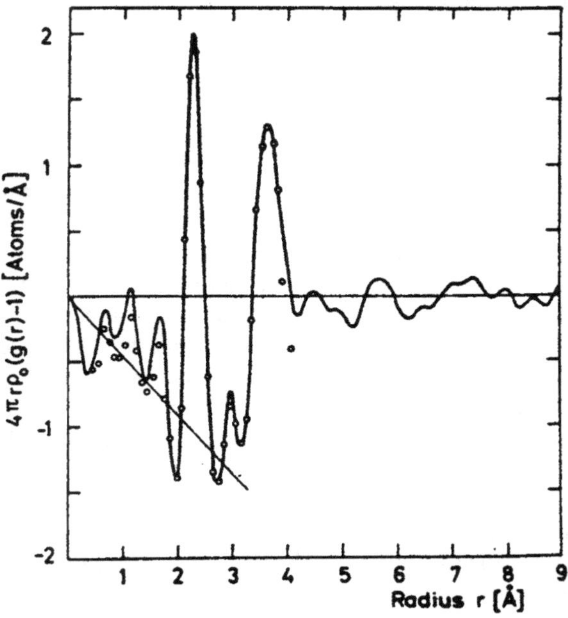

Fig. 2.6. $4\pi r\rho_0(g(r) - 1)$ obtenu à partir d'une transformation de Fourier directe des valeurs expérimentales de $S(k)$ (ligne continue du graphique) et pour le modèle proposé (ronds ouverts).

composés de huit atomes, mémoire de la forme monoclinique du sélénium; parmi les 6 ou 8 premiers voisins, 2 le sont de manière covalente.

Une remarque générale: aucune étude de structure n'existe qui fasse une liaison systématique entre le processus de passage de la transition vitreuse et structure de la phase amorphe; les précisions expérimentales seraient-elles encore insuffisantes pour explorer plus en détail ce qu'implique au niveau microscopique la théorie du volume libre, quantité macroscopique?

On retrouve dans le cas de la silice les mêmes caractéristiques que pour le sélénium: il y a un enchainement tridimensionnel de tétraèdres SiO_4 à peu près réguliers; chacun d'eux peut tourner librement autour de la liaison Si–O qui le lie aux tétraèdres voisins; mais l'angle des liaisons Si–O–Si peut varier de 120° à 180° avec une certaine distribution [2.17]. Lorsque l'on passe du verre normal au verre dense, on observe un déplacement de cette distribution vers les petits angles, correspondant à un rapprochement des atomes Si [2.2] mais sans changement topologique dans l'arrangement structural et sans altération du tétraèdre élémentaire. On notera que l'angle a priori quelconque de la liaison Si–O–Si a des conséquences sur la structure électronique du matériau: la covalence normale du silicium est de 4 et elle conduit à une hybridation de type sp^3. Pour l'oxygène, lié à deux silicium, son hybridation passe progressivement de l'état sp^2 à l'état sp lorsque l'angle de la liaison varie de 120° à 180°. Un modèle de la silice amorphe qui permet de retrouver à la fois les variations de volume, de compressibilité et de dilatation fait jouer un rôle essentiel à l'énergie requise pour modifier l'angle Si–O–Si et passer de la phase α à la phase β par-dessus une barrière de potentiel [2.18]. La structure de la silice et toutes ses propriétés thermodynamiques ou même dynamiques pourraient s'expliquer avec ces seuls ingrédients: des doubles puits de potentiel dont les caractéristiques (profondeur, hauteur de barrière, . . .) seraient largement distribuées. Ces excitations sont-elles celles que l'on observe à basse température? Notons incidemment qu'il y a une autre possibilité de préparer de la silice amorphe: par implantation ionique; on remarquera que ses propriétés s'éloignent beaucoup de celles de la silice qui est préparée par densification.

2.4.2. Modes collectifs

Tous les physiciens du solide sont habitués à la notion de mode collectif et plus particulièrement, ils savent exprimer comment les propriétés de symétrie de translation du solide cristallin conditionnent l'identification des modes propres: chacun d'eux est caractérisé par une fréquence ω et un vecteur d'onde $\boldsymbol{k}$; la fonction $\omega(\boldsymbol{k})$ est la loi de dispersion; elle dépend des forces interatomiques. Muni d'un peu de patience, on sait calculer exactement tous les modes propres; il "suffit" de diagonaliser une matrice $(3n - 3) \times (3n - 3)$, où n est le nombre d'atomes par maille.

On sait encore calculer exactement les modes propres d'un solide lorsque l'on a introduit une seule impureté dans la matrice; mais les problèmes à deux ou plusieurs impuretés deviennent insolubles à moins de recourir à des méthodes d'approximation (CPA, ...) dont la complexité oblige vite à en faire une vocation si l'on veut en tirer quelque chose! Pour un verre, système désordonné parfait, il n'y a aucun espoir de s'en tirer par des approximations. Les modes propres ne sont plus des ondes planes.

Plutôt que de raffiner encore des méthodes d'approximation, il est préférable de faire face directement au système désordonné; traiter par exemples des propriétés en termes de probabilité de distribution des caractéristiques du verres; on en verra plusieurs exemples dans la suite. Pour revenir au problème des modes propres, on peut pourtant explorer leur densité d'état. Des expériences de diffusion inélastique cohérente de neutrons, aptes à mesurer les lois de dispersion $\omega(\boldsymbol{k})$, montrent en effet que pour $\boldsymbol{k}$ donné, les modes explorés couvrent une large bande fréquence mais que cette largeur ne dépend pas de la température; cela s'explique aisément; le vecteur d'onde n'est pas un bon nombre quantique; pour $\boldsymbol{k}$ fixé, il y a toute une série de modes de fréquences diverses qui ont des "composantes" non nulles sur la base des $\omega(\boldsymbol{k})$. On observe un élargissement "inhomogène".

En fait, des expériences de temps de vol des neutrons, qui reviennent à mesurer des densités d'état en fonction de la fréquence, sont plus aptes à donner une information directe. Pour la silice par exemple, on observe que cette densité d'état est bien plus voisine de celle de la cristobalite que de celle du quartz, ce qui est normal si on prend en compte le seul critère de densité, c'est-à-dire de distance entre les atomes [2.19]. Une différence pourtant: dans les régions des basses fréquences, cette densité d'état est toujours supérieure à celle du cristal le plus voisin; on a pu exploiter ce résultat pour rendre compte de la valeur de la chaleur spécifique élevée du verre dans la région de 5 K à 20 K [2.20] sans que le résultat s'accorde avec celui qui est déduit des mesures de vitesse du son. Cette observation est plus générale et on retrouve le même comportement pour à peu près tous les amorphes.

Dans une analyse de modes propres, deux cas limites sont clairs cependant. Une analyse des modes collectifs pour des vecteurs d'onde de plus en plus grands montre que tout effet de structure statique disparait alors et que tout atome diffuse indépendamment des autres [2.21]; on peut mesurer en principe très exactement la densité d'état; l'expérience reste à faire et le mieux serait d'y procéder sur un verre monoatomique tel que Ge amorphe.

L'autre cas limite correspond à une situation où le vecteur d'onde tend vers zéro: c'est l'hypothèse des ondes longues ou des modes hydrodynamiques.

2.4.3. Atténuation ultrasonore – modes basse fréquence

Dans un cristal, l'atténuations des modes hydrodynamiques ou modes sonores est à peu près constante avec la température; son origine est bien connue: elle provient de la modulation de la population des modes thermiques par le champ de déformation.

Dans un verre comme la silice, le comportement de cette atténuation α est "anormal"; α présente un maximum vers 50 K que l'on retrouve d'ailleurs dans les mesures de constante diélectrique (fig. 2.7) et dans bien d'autres verres. On rend compte de α en revenant à l'hypothèse esquissée plus haut: il y a une relaxation de la structure. Une grande quantité de degrés de liberté (les liaisons Si–O–Si) ont un potentiel à deux puits entre lesquels ils relaxent; le temps caractéristique à ces températures est du type Arrhenius

$$\tau = \tau_0 \exp(E_a/KT)\,, \tag{2.9}$$

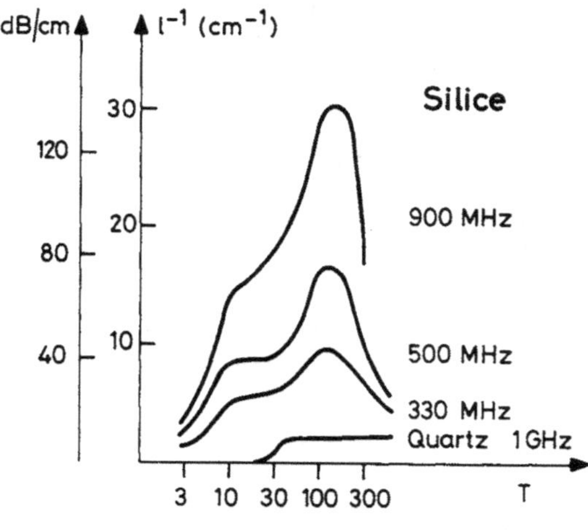

Fig. 2.7.

et l'atténuation du son résultante est

$$\alpha \simeq \frac{1}{T}\int f(E_a)\,\frac{\omega^2\tau}{1+\omega^2\tau^2}\,d(E_a)\,, \tag{2.10}$$

où $f(E_a)$ est la probabilité de distribution de E_a, énergie d'activation. La formule (2.10) convient très bien à expliquer les résultats dans une grande gamme de température et de fréquence [2.22].

Les résultats de diffusion Brillouin qui étendent la fréquence des mesures jusqu'à 30 GHz confirment tout à fait ce point de vue [2.23] que des expériences d'effet Raman viennent encore conforter [2.24].

2.5. Les verres: propriétés de basse température

Si des modes acoustiques de grande longueur d'onde existent dans les verres, leurs propriétés thermodynamiques devraient, comme dans un cristal être déterminées par leur densité d'états, c'est-à-dire par la vitesse du son; la chaleur spécifique devrait être en T^3 comme dans un solde cristallin.

Or, les premières mesures de chaleur spécifique, de conductibilité thermique et d'atténuation du son ont montré un comportement tout à fait différent du cristal qui a beaucoup intrigué avant qu'une première explication satisfaisante [2.25,2.26] ne soit proposée et qu'elle fournisse au moins un cadre de compréhension.

2.5.1. Modèle des systèmes à deux niveaux: S2N

Les auteurs [2.25,2.26] partent de l'idée qu'il existe dans le verre des systèmes à deux niveaux qui ont une position stable et une position métastable dans un double puits de potentiel. Le paramètre de configuration qui a donc deux positions possibles (fig. 2.8) n'est pas nécessairement celui auquel il a été fait allusion dans le paragraphe précédent. Au contraire même. Toujours est-il que la nature de ce système à deux niveaux reste obscure; on a pensé à un groupe d'atomes, un tétraèdre dans la silice par exemple – on a pensé à une liaison covalente ou bipolaron [2.27], mais à l'heure actuelle, rien n'est prouvé. Toujours est-il que ce double puits a des caractéristiques qui varient d'un site à un autre: on est dans un système désordonné; la hauteur de la barrière, la profondeur du puits, les "distances" entre les deux minimum sont donc largement distribuées.

Pour l'explication des propriétés de basse température, la possibilité du passage par effet tunnel d'un puits à l'autre est essentielle; elle implique que

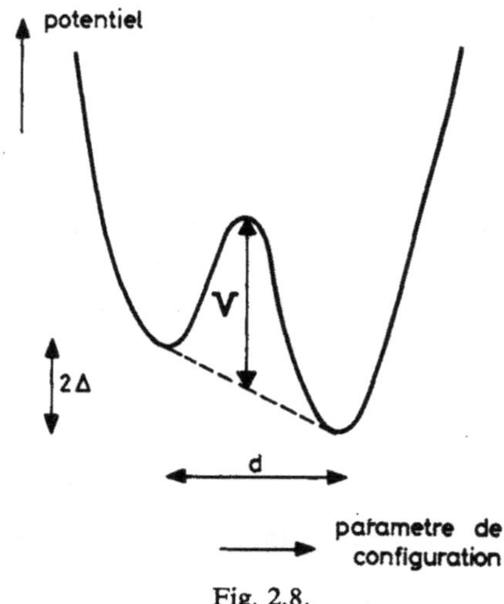

Fig. 2.8.

les barrières ne soient pas trop hautes pour que ces systèmes à deux niveaux puissent donner lieu à des effets résonnants par interaction avec les phonons thermiques; inversement, les systèmes à deux niveaux dont les barrières seront grandes ne seront peut-être pas tous atteints dans des expériences où ils ne peuvent être excités que par des processus par activation; à basse température donc, c'est-à-dire pour $T < 10$ K, la hauteur des barrières qui est à peu près uniformément distribuée entre 0 et 1 eV limite donc le nombre des S2N susceptibles d'intervenir à une faible partie de ceux qui existent dans le verre.

Quels sont les ingrédients importants d'une description d'un système à deux niveaux? La masse m qui tunnelle, la hauteur de barrière V, la distance d entre les deux puits, une énergie $\hbar\omega_0$ qui est une énergie de point zéro dans l'un ou l'autre des puits. Il est classique de construire le paramètre

$$\lambda = \hbar^{-1} d \sqrt{(2mV)}\,, \tag{2.11}$$

et d'en déduire la quantité

$$\Delta_0 = \hbar\omega_0\, e^{-\lambda}\,, \tag{2.12}$$

qui donne l'énergie de couplage tunnel. L'hamiltonien d'un tel système dans la base des états propres dans chaque puits, en l'absence d'effet tunnel, est

$$H_0 = \begin{pmatrix} \Delta & \Delta_0 \\ \Delta_0 & -\Delta \end{pmatrix}, \tag{2.13}$$

où Δ est l'asymétrie du système à deux puits. Les énergies propres de (2.13) sont donc

$$E = \pm\sqrt{(\Delta^2 + \Delta_0^2)}\,. \tag{2.14}$$

Le couplage avec un champ extérieur (élastique ou électrique) dans la même base que (2.13) peut être décrit sous la forme:

$$H_1 = \varepsilon G \begin{pmatrix} 1 & 0 \\ 0 & -1 \end{pmatrix}; \tag{2.15}$$

G est la constante de couplage avec le champ extérieur ε. Si ε est une déformation élastique, G est une constante de couplage spin–phonons; elle est de l'ordre de l'électron volt: c'est un potentiel de déformation; si ε est le champ électrique, G est le moment dipolaire électrique; typiquement une fraction de debye. Dans la base qui diagonalise (2.13), le couplage s'écrit alors

$$H_1 = \varepsilon G \begin{pmatrix} \Delta/E & -\Delta_0/E \\ \Delta_0/E & -\Delta/E \end{pmatrix}. \tag{2.16}$$

Pour plusieurs applications et aussi pour pouvoir employer le langage du paramagnétisme, on fera une stricte analogie entre ces systèmes à deux niveaux et des spins $\frac{1}{2}$; en particulier dans la base qui diagonalise (2.13) on pourra écrire

$$H_0 = ES_z\,,$$

avec

$$S_z = \frac{1}{2}\begin{pmatrix} 1 & 0 \\ 0 & -1 \end{pmatrix}. \tag{2.17}$$

Cette analogie n'implique pas grand'chose sinon que tout opérateur qui agit sur un système deux niveaux est une matrice 2×2. Les quatre opérateurs S_x, S_y, S_z, et I forment une base suffisante. La séduction de cette analogie tient à sa simplicité et à la possibilité qu'elle donne d'utiliser tous les traitements connus en paramagnétisme. Nous sommes dans un système désordonné; pour poursuivre l'analogie: chaque spin i aura une énergie E_i et on admettra que le champ magnétique local est aléatoire en direction et en grandeur; il ne peut être controllé dans une expérience dans la mesure où il est intrinsèque au verre.

De la même manière, l'hamiltonien de couplage (2.16) peut être exprimé à l'aide des opérateurs de spin

$$H_c = \varepsilon_i G_i^x S_i^x + \varepsilon_i G_i^z S_i^z , \tag{2.18}$$

avec

$$S_i^x = \frac{1}{2}\begin{pmatrix} 0 & 1 \\ 1 & 0 \end{pmatrix},$$

et

$$G_i^x = 2G\Delta_0/E , \qquad G_i^z = 2G\Delta/E . \tag{2.19}$$

Pour achever de préciser ce modèle, il reste à se donner une statistique sur les quantités Δ, λ, ω_0, G, ... Il n'y a pas, à vrai dire, d'argument très convaincant pour choisir une distribution plutôt qu'une autre, hormis les conséquences qu'on peut en tirer. On admettra donc que

$$\text{prob.}(\Delta, \lambda) = \text{constante} , \tag{2.20}$$

étant entendu que les variations de λ dominent celles de Δ_0 et que celles de ω_0 peuvent éventuellement être absorbées dans λ.

Une remarque: pour les spins d'énergie E, λ a une valeur minimum donnée par

$$\lambda_{\min}^{(E)} = \log(\hbar\omega_0/E) ; \tag{2.21}$$

pour ces spins là, qui correspondent à $\Delta = 0$ on a $\Delta_0 = E$ et la constante de couplage G^x est maximum; inversement leur couplage G^x est minimum.

C'est donc de ce modèle qu'on se propose de tirer une explication du comportement expérimental des verres à basse température.

2.5.2. *Justifications du modèle des systèmes à deux niveaux*

Il faut à la vérité dire que ce modèle est apparu après que bien des faits aient troublé les expérimentateurs; il a coordonné des éléments éparses qui n'avaient que peu de liens entre eux.

Chaleur spécifique. Expérimentalement, la chaleur spécifique, dans la limite des basses températures, est à peu près linéaire en température contrairement au cas du cristal où elle est cubique (voir fig. 2.9). S'il n'y avait qu'un seul type d'impuretés d'énergie E_i fixée, on observerait une anomalie de type Schottky pour C à T donné par $KT \simeq E_i$. Si l'on suppose

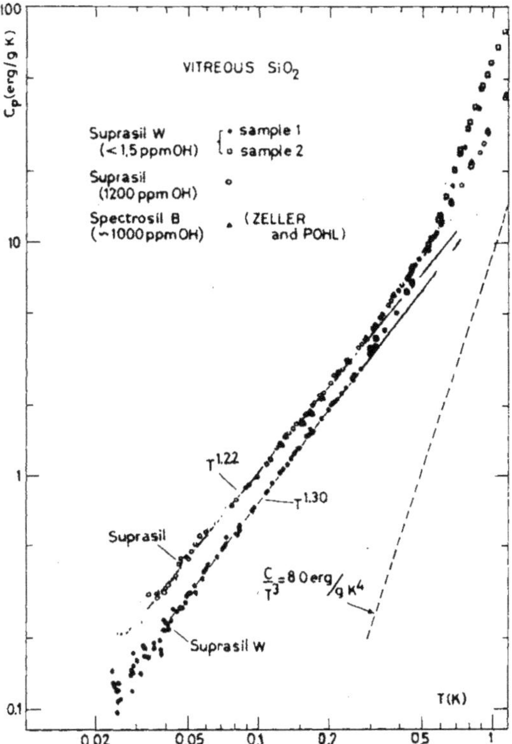

Fig. 2.9. Chaleur spécifique de la silice vitreuse pour différentes concentrations d'OH. La ligne en tirets représente la contribution calculée à partir de la densité (2.20 g/cm^3) et de la vitesse du son (v_ℓ = 5.80 × 10^5 cm/s et v_t = 3.75 × 10^5 cm/s).

que la distribution des E_i, soit $n(2E)$, est constante en énergie et égale à n_0, on trouve bien que C est linéaire en température; en effet, l'énergie libre pour un système à deux niveaux d'énergie $2E$ est

$$F(2E) = -KT\log[2\mathrm{ch}(E/KT)]\,,$$

et la contribution à la chaleur spécifique correspondante est

$$C(2E) = -T\frac{\partial^2 F(2E)}{\partial T^2} = \frac{E^2}{K^2T^2}K\frac{1}{\mathrm{ch}^2(E/KT)}\,.$$

La chaleur spécifique du verre est alors

$$C = \int C(2E)n_0\,\mathrm{d}E = \tfrac{1}{6}\pi^2 n_0 K^2 T\,. \tag{2.22}$$

n_0, pour tous les verres est de l'ordre de 5×10^{32} erg^{-1} cm^{-3} [2.28,2.29]. Expérimentalement, le coefficient de température est légèrement supérieur à 1; on peut l'expliquer en considérant que $n(2E)$ n'est pas tout à fait constante (et pourquoi le serait-elle?); on peut aussi supposer que $n(2E)$ présente vers les basses énergies un cut-off qui produit le même effet [2.30]. Quoiqu'il en soit $n(2E)$ est à peu près constante dans la gamme du degré K; un ennui: la chaleur spécifique ne donne aucune information sur la distribution $n(2E)$ pour $E > 10$ K: le terme en T^3 des solides devient dominant.

Conductibilité thermique. Dans la gamme des basses températures ($T <$ 1 K) la conductibilité thermique [2.28,2.31] obéit à la loi expérimentale

$$K \simeq T^2 . \tag{2.23}$$

On explique ce résultat (voir fig. 2.10) en disant que les systèmes à deux niveaux sont capables d'émettre et d'absorber de manière résonnante des phonons de même fréquence; en partant de la formule classique

$$K = \tfrac{1}{3} C_{\text{ph}} v^2 \tau ,$$

où C_{ph} est la chaleur spécifique des phonons, v leur vitesse et τ leur temps de relaxation, on en déduit que, si $K \simeq T^2$, $C_{\text{ph}} \simeq T^3$, $v =$ constante, on doit avoir τ^{-1}, où

$$l_{\text{ph}}^{-1} = v^{-1} \tau^{-1} \simeq T . \tag{2.24}$$

C'est un point qui mérite d'être expliqué.

Lorsqu'un phonon d'énergie $\hbar\omega$ est en interaction avec des spins dont la séparation des niveaux est égale à $\hbar\omega$ il y a une probabilité d'absorption qui limite son libre parcours moyen même si son énergie est ultérieurement thermalisée dans un autre mode; il y a diffusion résonnante. En fonction des constantes du modèle on calcule que le libre parcours moyen est égal à

$$l^{-1} = \pi \frac{(G^x)^2}{4\rho v^3} n_0 \omega \operatorname{th}\left(\frac{\hbar\omega}{2KT}\right) . \tag{2.25}$$

Pour les phonons thermiques, c'est-à-dire pour ceux où $\hbar\omega \simeq KT$, on déduit immédiatement que

$$l^{-1} \simeq \frac{(G^x)^2}{4\rho h v^3} n_0 KT .$$

L'ordre de grandeur de la conductibilité, sa dépendance en température sont en accord avec ce résultat en prenant pour G^x une valeur d'une fraction d'électron-volt et pour n_0 la valeur déduite de la mesure de chaleur spécifique. Retenons que, expérimentalement, on ne mesure que le produit

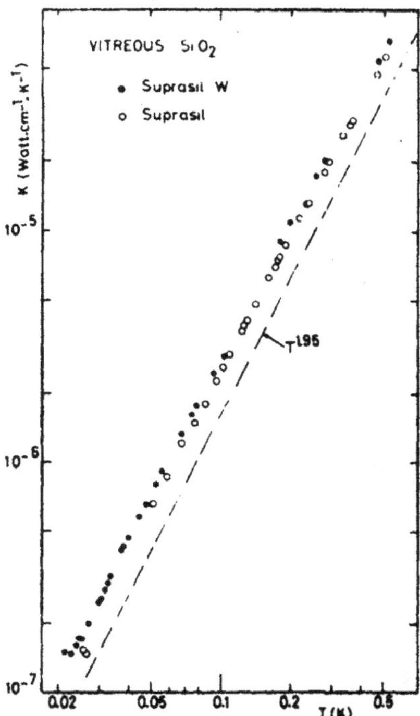

Fig. 2.10. Conductivité thermique mesurée sur le même échantillon de Suprasil que celui utilisé dans la figure précédente pour représenter les valeurs de chaleur spécifique. La ligne en tirets correspond à une variation en température $T^{1.95}$.

$n_0(G^x)^2$ et non chaque terme séparément, et encore, il ne s'agit que d'une moyenne compliquée qu'il faudrait relier au modèle développé plus haut.

Pour le moment, on notera que si K est si petit par rapport au cristal, c'est que les systèmes à deux niveaux diffusent fortement les phonons; leur couplage G^x avec les déformations est grand et les valeurs de K sont à peu près "universelles", c'est-à-dire que tous les verres fournissent des ordres de grandeur identiques. En regardant d'un peu plus près, on trouve que l'exposant de K^0 est égal à 1.9; on peut faire les mêmes remarques qu'à propos de l'exposant de la chaleur spécifique; elles sont concourantes.

2.5.3. Preuve du modèle des systèmes à deux niveaux

On peut discuter longtemps ce qu'est une justification ou une preuve; convenons qu'il sera question ici d'une expérience d'atténuation ultrasonore qu'aucun autre modèle ne pourrait expliquer: elle emporte l'évidence: c'est cela une preuve!

Si l'on part de l'idée que les systèmes à deux niveaux sont couplés fortement avec les phonons, il est aisé de faire un calcul de libre parcours moyen en considérant l'interaction élémentaire un phonon $\rightleftharpoons$ un spin; le résultat quantitatif est la formule (2.25).

La contradiction (provisoire), mais instructive est que des mesures d'atténuation ultrasonore, propices à l'exploration d'un seul paquet de spins, conduisaient à des libres parcours moyens très grands, incompatibles avec (2.25), jusqu'à ce que l'on réalise que, justement avec G^x grand, on pouvait très aisément avoir de la saturation d'absorption comme en résonance paramagnétique électronique ou nucléaire; la difficulté était expérimentale et il fallait opérer à un niveau de puissance suffisamment bas pour observer (2.33) [2.34]; le phénomène de saturation est observable: en fonction de la puissance acoustique, elle varie suivant la loi [2.33]

$$\alpha = \alpha_{\text{non saturé}}(1 + \phi/\phi_c)^{-1/2}\,, \tag{2.26}$$

où ϕ est le flux acoustique et ϕ_c un flux critique qui dépend de la dynamique du système de spin (voir plus loin). C'est ce qui a pu être vérifié avec précision [2.24].

A bas niveau acoustique, l'atténuation croît donc lorsque T décroît; à haut niveau, seule l'atténuation de relaxation résultant de la modulation de la population de spin subsiste, donnant au contraire une atténuation croissante avec la température de la forme

$$\alpha_{\text{relax}} = \frac{n_0(G^z)^2}{gv^3}\int_0^\infty \frac{\mathrm{d}E}{KT}\frac{\exp(E/KT)}{[1+\exp(E/KT)]^2}\frac{\omega^2\tau(E)}{1+\omega^2\tau^2(E)}\,, \tag{2.27}$$

où $\tau(E)$ est le temps de relaxation à la température T des spins d'énergie E; $\tau(E)$ est déterminé, comme en paramagnétisme, par un processus de relaxation directe à un phonon thermique; ce temps sera ultérieurement appelé $T_1(E)$

$$\tau^{-1}(E) = \frac{1}{8\pi\rho h^4}\left(\frac{(G_l^x)^2}{v_l^5} + 2\,\frac{(G_t^x)^2}{v_t^5}\right)E^3\coth\left(\frac{E}{2KT}\right)\cdot \tag{2.28}$$

En reportant (2.36) dans (2.25), on vérifie que pour $\omega\tau(E) \ll 1$ l'atténuation de relaxation croît comme T^3; elle est responsable du maximum d'atténuation observé vers 5 K dans à peu près tous les verres [2.35]. Rappelons qu'à plus haute température, l'atténuation est déterminée par un processus d'activation thermique.

En conclusion, il semble que les mesures de saturation ultrasonore soient importantes; elles prouvent que l'on a à faire à des systèmes qui ne sont pas des oscillateurs harmoniques susceptibles d'être excités indéfiniment sans saturation; ces S2N sont donc l'équivalent de spins ou plutôt, ils ont un nombre fini de niveaux. C'est pour la commodité que l'on a adopté le langage des spins; mais rien ne prouve que ces niveaux soient au nombre de deux seulement; il pourrait y en avoir trois inégalement espacés [2.36].

Dernière remarque: qu'adviendrait-il si ces systèmes étaient des oscillateurs très anharmoniques?

2.5.4. Premières conséquences du modèle

Après que ce cadre explicatif ait vu le jour, diverses expériences ont été développées qui ont encore renforcé la confiance qu'on pouvait avoir en ce modèle par comparaison avec les autres théories.

Vitesse du son. Une mesure précise de la vitesse du son à très basse température constitue en effet un excellent test de ce qui a été proposé. Cette vitesse est la transformée de Hilbert de l'atténuation non saturée. On doit écrire:

$$\frac{\Delta v(\omega T)}{v} = \text{p.p.} \int_0^\infty \mathrm{d}\omega' \,\frac{v}{\pi}\, \frac{l^{-1}(\omega, T) - l^{-1}(\omega', T_0)}{\omega^2 - \omega'^2},$$

où p.p. signifie partie principale de l'intégrale. Elle contient en principe les mêmes informations mais elle est plus facile à mesurer et surtout elle est moins sensible à la puissance acoustique incidente puisque tous les spins dans une largeur de bande égale à KT y contribuent; on observe à la fois la contribution résonnante et la contribution qui provient de la relaxation de la population [2.37] (voir fig. 2.11). La variation de vitesse à très basse température est la suivante

$$\frac{\Delta v}{v} = \frac{(G^x)^2}{4\rho v^2}\, n_0 \log\left(\frac{T}{T_0}\right), \tag{2.29}$$

v et ρ étant connus, la mesure de Δv fournit une valeur du produit $n_0(G^x)^2$. Si la quantité n_0 écrite dans (2.29) est la même que celle estimée par les mesures de chaleurs spécifiques, on obtient alors des évaluations indépendantes des quantités n_0 et $(G^x)^2$; elles peuvent être corroborées par comparaison avec la conductibilité thermique; à un facteur deux près, il y a concordance. Ce n'est pas si mal!

Pour résumer les informations glanées par les méthodes acoustiques disons que l'on accède à un coefficient de couplage, sans dimension de la forme

$$k^2 = n_0(G^x)^2/4\rho v^2\,. \tag{2.30}$$

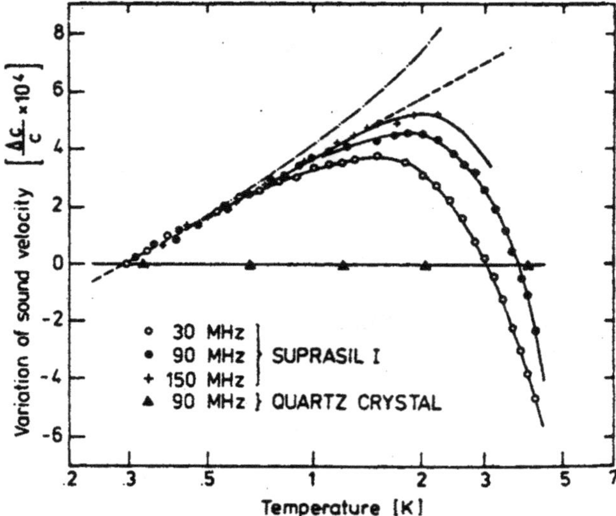

Fig. 2.11. Variation relative à la vitesse du son longitudinal $\Delta c/c$ en fonction de la température. La ligne droite en traits pointillés correspond à une variation proportionnelle à $\ln T$; la ligne en traits pleins correspond à la somme des contributions calculées pour l'interaction résonnante et de relaxation; la ligne points-traits donne la contribution des processus résonnants seuls en supposant une densité d'état $n(E) = n_0[1 + a(E/k_B)^2]$.

Dans tous les verres (et aussi dans les verres métalliques), il a une valeur "universelle": $k^2 \simeq 5 \times 10^{-4}$. Un effet de désordre de structure? Qu'il s'agisse des formules (2.28), (2.29) ou (2.30), là aussi le résultat est en fait une moyenne sur tous les spins qui ont même énergie E; il y a moyenne car dans le modèle G^x dépend de la distribution des deux paramètres indépendants Δ et λ (voir formule (2.20)).

Constante diélectrique du verre. Ces études ont sans doute été inspirées par celles qui ont été réalisées il y a une dizaine d'années à propos des impuretés ioniques ou moléculaires type OH^- introduites de manière substitutionnelle dans les cristaux de la série KCl et dont le moment dipolaire est de l'ordre de 1 debye, ce qui est aisément mesurable (1 debye = 0.2 e × A). Et puis il était tout naturel de se demander si les systèmes à deux niveaux avaient une polarisabilité ou pas.

La réponse est la suivante: un verre de silice pure ne présente aucune variation mesurable de constante diélectrique à très basse température; la polarisabilité électrique des "spins", si elle existe, est donc très faible.

L'application d'un champ électrique continu ne modifie pas non plus les propriétés de transport [2.38]. Par contre, le même verre préparé pour que, après passage de la transition vitreuse, il contienne des ions moléculaires OH^-, a une constante diélectrique qui varie fortement avec la température. Rien d'étonnant dira-t-on! Mais ce qui est plus intéressant c'est qu'elle varie comme $\log T$ [2.39], c'est-à-dire comme la vitesse du son (vitesse de la lumière et constante diélectrique c'est tout comme). Cela prouve que sur la même échelle d'énergie, les impuretés moléculaires ont elles aussi deux niveaux tunnels, dont la séparation E_j a une distribution très plate sur une échelle de l'ordre de KT; elles n'ont pas de caractère intrinsèque mais, elles aussi, peuvent être décrites dans le cadre du formalisme du paramagnétisme. La formule qui donne la constante diélectrique est analogue à (2.29) (voir fig. 2.12):

$$\frac{\Delta c}{c} = \frac{1}{2}\frac{\Delta \varepsilon}{\varepsilon} = \frac{p^2}{\varepsilon_0} n_0' \log\left(\frac{T}{T_0}\right), \tag{2.31}$$

où p est le moment dipolaire transversal des "spins", n_0' leur concentration par unité de volume et d'énergie, c la vitesse de la lumière dans le verre et

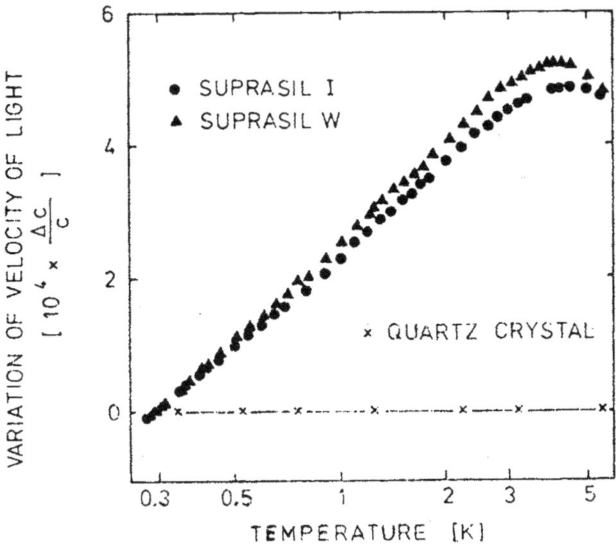

Fig. 2.12a. Variation relative à la vitesse de la lumière $\Delta c/c$ à 1 GHz dans la silice vitreuse en fonction de la température. Le Suprasil W contient 1.5 ppm d'impuretés OH et le Suprasil I 1200 ppm.

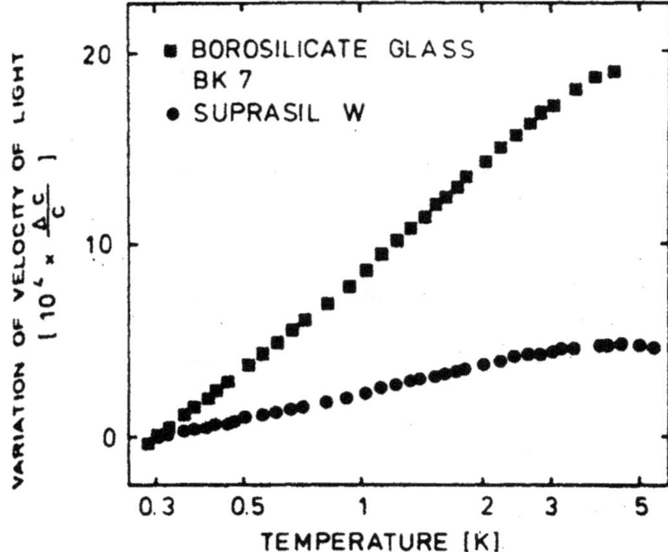

Fig. 2.12b. Variation relative à la vitesse de la lumière $\Delta c/c$ à 1.1 GHz dans le verre BK 7 en fonction de la température. La courbe correspondant au Suprasil W est reportée pour comparaison.

ε_0 la constante diélectrique du vide dans le système MKS. Expérimentalement, on ne mesure que la quantité $n_0'p^2$; comme on n'accède pas indépendamment à n_0' il y a une indétermination qui n'est levée que par des méthodes plus sophistiquées.

Enfin, des expériences qui combinent à la fois un champ électrique et champ de déformation élastique [2.40] montrent que les mêmes impuretés sont couplées à l'un et à l'autre champ; on est donc naturellement conduit à penser que l'assemblée des OH^- est une image fidèle de l'assemblée des systèmes à deux niveaux intrinsèques; les OH^- pourraient jouer dans le cas de la silice le rôle de marqueurs apportant leurs propriétés de polarisation aux défauts tunnels.

Cette propriété sera utilisée ultérieurement pour étudier les propriétés du verre à basse température par des méthodes d'échos dipolaires électriques.

Chaleur spécifique dépendant du temps. Une prédiction immédiate du modèle est que la chaleur spécifique dépend du temps; on retrouve là une idée que l'on a déjà considérée à propos des verres de spin et qui est liée à l'inhomogénéité de la distribution des systèmes à deux niveaux; non seulement des spins d'énergie fixée ont des caractéristiques différentes (distribution des Δ et λ) mais encore dans une expérience de chaleur spécifique, les spins concernés s'étalent sur une bande de largeur au moins égale à KT.

Si le temps de la mesure expérimentale est t, tous les spins qui peuvent relaxer dans un temps inférieur à t participeront au phénomène; les autres au contraire seront considérés comme fixes.

Très naturellement, l'inverse du temps caractéristique T_1 pour faire une transition d'un état à un autre dans une expérience de chaleur spécifique est proportionnel à $(G^x)^2$; $T_1^{-1} \simeq (G^x)^2$, c'est-à-dire à $(G\Delta_0/E)^2 \simeq e^{-2\lambda}$ conformément à la formule (2.19); une distribution des λ entraine une distribution des T_1^{-1}.

On doit se souvenir ici que dans le modèle proposé jusqu'à maintenant λ est équiprobable pour des valeurs supérieures à λ_{mini}. Si donc il faut affiner une formule comme (2.22) pour tenir compte de la répartition des temps caractéristiques, on est amené à écrire, pour une expérience faite dans un temps inférieur à t:

$$C(t) = \frac{\partial}{\partial T}\int \frac{E\,\mathrm{d}E}{1 + \exp(E/KT)}\int_{\lambda_{\text{mini}}\text{ ou }\lambda(t)}^{\lambda_{\text{maxi}}} n(E, \lambda)\,\mathrm{d}\lambda\,;$$

comme $n(E, \lambda)\,\mathrm{d}\lambda = n_0$, $\mathrm{d}\lambda = n_0\,\mathrm{d}T_1/T_1$, on a aussi

$$C(t) = \frac{\partial}{\partial T}\int \frac{E\,\mathrm{d}E\,n_0}{1 + \exp(E/KT)}\int_{T_{1\,\text{mini}}}^{t} \frac{\mathrm{d}T_1}{T_1}\,.$$

Au total, on obtient une dépendance logarithmique de $C(t)$ soit

$$C(t) \simeq \log(t/T_{1\,\text{mini}})\,. \tag{2.32}$$

Malheureusement, aucune expérience n'a vraiment mis cette loi en évidence [2.41]. Cela tient à ce que pour décrire un tel phénomène, il faut tenir compte de toutes les équations de transport d'un système qui comprend non seulement les systèmes à deux niveaux mais le bain des phonons thermiques qui peut lui aussi se trouver hors de l'équilibre thermique. Finalement, des expériences de diffusivité thermique devraient être le mieux à même de trancher cette question. Mon sentiment est que c'est beaucoup demander à de telles mesures de pouvoir tester un modèle; elles sont trop macroscopiques.

Des expériences récentes, analysant les profils de température en réponse à des impulsions de chaleur semblent avoir mis en évidence un écart aux lois "normales", c'est-à-dire à celles qui impliquent une chaleur spécifique constante [2.42]; le temps caractéristique de relaxation des TLS varie en T^{-3}, ce qui est raisonnable; il apparait plus court que ce qui peut être déduit par atténuation acoustique.

Inversement, des expériences d'impulsion de chaleur [2.43] suggèrent que les lois normales suffisent; les explications fournies dans ce dernier cas ne me paraissent pas claires; bref, attendons!

2.5.5. Dynamique des systèmes à deux niveaux; échos dipolaires électriques

Poursuivant l'analogie avec le paramagnétisme, il a été proposé d'analyser la dynamique d'un paquet de "spins" ou de systèmes à deux niveaux par des méthodes beaucoup plus raffinées d'"échos de spins".

Pour décrire l'évolution d'un paquet de spins, on cherche, quand c'est possible, à le faire dans le cadre des équations de Bloch: la composante longitudinale de l'aimantation relaxe avec un temps T_1 et la composante transversale relaxe avec un temps T_2^*:

$$\frac{\partial S_z(t)}{\partial t} = -\frac{S_z(t) - \langle S_z \rangle}{T_1}, \qquad \frac{\partial S_x}{\partial t} = -\frac{S_x}{T_2^*}.$$

Une expérience d'échos de spin se ramène à appliquer une série de deux ou trois impulsions de même fréquence et à observer l'amplitude d'un écho cohérent: avec une séquence de deux impulsions, on peut mesurer le temps de relaxation T_2, qui est la partie non réversible de T_2^*; avec une séquence de trois impulsions on mesure T_1.

Les systèmes à deux niveaux intrinsèques sont couplés seulement avec le champ de déformation; par des méthodes acoustiques, on a pu [2.44] observer des échos et estimer ces temps de relaxation dans une gamme de température de l'ordre de 20 mK à 50 mK. En fait, ces expériences souffrent de difficultés pratiques et, même si elles ont montré la voie, elles n'ont pas apporté grande information physique.

La même tentative a été reprise sur les OH^- dans la silice, en les couplant avec un champ électrique [2.45]. Plus simples à mettre en œuvre ces dernières expériences ont permis de mesurer tout à la fois mais de manière indépendante, le moment électrique dipolaire transversal moyen, qui est de l'ordre de 1 debye, la densité d'état n_0' dans une gamme d'énergie correspondant à quelques centaines de mégacycles; on trouve que $n_0' \simeq cn_0$ où c est la concentration des OH^-; ce résultat renforce l'idée que les OH^- sont une image fidèle de la population des systèmes intrinsèques; la distribution du moment électrique longitudinal est très proche d'une lorentzienne de largeur voisine de 0.2 D [2.46].

En même temps, on a montré que le temps de relaxation longitudinal T_1 n'est pas unique, mais qu'il est largement distribué pour une population d'énergie E fixée; T_1 est limité par les processus directs d'interaction entre un spin et un phonon thermique de même fréquence; un calcul d'ordre de grandeur montre que la constante de couplage OH^- – phonons G^x est encore un peu plus grande que celle des systèmes intrinsèques.

Enfin, T_2 a été correctement mesuré dans une gamme de température s'étendant de 3 mK jusqu'à 30 mK. T_2^{-1} est proportionnel à T [2.47] (voir fig. 2.13). Ce résultat reste inexpliqué en dépit d'une proposition différente [2.48] qui avait un caractère séduisant et qu'il est instructif de reproduire.

Deux idées successives: parce que les OH^- sont fortement couplés avec les déformations locales, il y a une interaction indirecte entre elles par échange de phonons virtuels; parce que le nombre des OH^- "unlike" est beaucoup plus grand que celui des OH^- "like", cette interaction est du type Ising par opposition à Heisenberg. Quantitativement, on obtient, pour deux OH^- notés i et j, à la distance R_{ij} une interaction de la forme

$$H_{\text{int}} = J_{ij}S_i^z S_j^z = \frac{(G^z)^2}{4\pi\rho v^2}\frac{1}{R_{ij}^3}S_i^z S_j^z\,; \tag{2.33}$$

avec des valeurs $G^z = 1$ eV, $v = 5.10^3$ m/s, $R = 10$ Å, on a J de l'ordre de quelques degrés K [2.33].

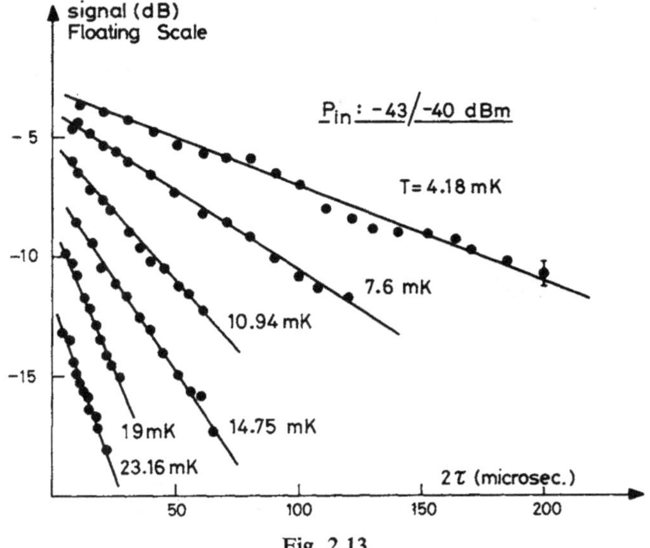

Fig. 2.13.

Nantis de cette interaction relativement forte, en tous cas du même ordre que celle que l'on rencontre dans les verres de spins pour des concentrations analogues, on raisonne ainsi: supposons que les spins unlike fluctuent en raison de l'agitation thermique; par l'intermédiaire de J_{ij}, il en résulte une variation du champ local au niveau du spin qui est observé; bien sûr cette agitation thermique avec son caractère aléatoire n'est pas compensée par renversement du sens du temps et elle pourrait être à l'origine de T_2; on retrouve alors un problème classique en paramagnétisme: celui de la diffusion spectrale d'un paquet de spin sous l'influence de l'agitation brownienne du champ local [2.48].

Le résultat du calcul implique que l'écho observé doit obéir à la relation suivante

$$\text{écho} \simeq \exp[-A(2\tau)^2T^4]\,, \tag{2.34}$$

où 2τ est l'instant où se produit l'écho après la première impulsion. Cette formule requiert que la décroissance soit plus rapide qu'exponentielle et que le temps caractéristique varie comme T^{-2}. Ces deux prédictions sont

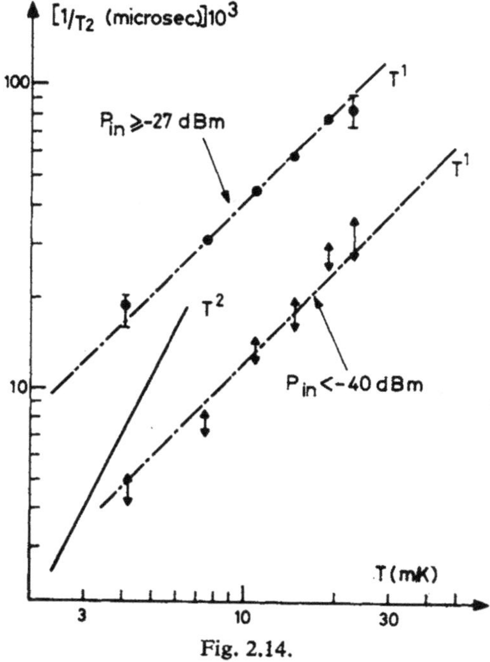

Fig. 2.14.

en contradiction avec les résultats expérimentaux (voir figs. 2.13 et 2.14); de sorte que c'est un problème qui reste ouvert.

Toujours est-il que ces méthodes d'écho ouvrent la possibilité de bien préciser les différents aspects du modèle des systèmes à deux niveaux; il serait souhaitable qu'elles soient poursuivies.

A titre de conclusion provisoire, notons que la question qui reste ouverte est celle de la nature de ces systèmes à deux niveaux dont personne ne sait s'ils impliquent un mouvement atomique ou électronique. Il est paradoxal mais après tout assez fréquent en physique, de faire des expériences sur des entités dont on connait mieux les manifestations que la vraie nature.

Il est également souhaitable de faire une comparaison entre la situation qui prévaut dans un verre de spin et dans un verre en dressant une liste des caractéristiques de symétrie, de champ local, . . . (tableau 2.1).

La leçon du tableau 2.1 est simple: l'analogie entre ces deux ensembles de propriétés est très formelle; elle tient uniquement au mot "verre"! et ne vaut donc que pour le caractère désordonné de la position des spins ou des systèmes à deux niveaux. Comme on l'a vu aussi dans d'autres cours de cette école, pour les verres, le problème qui présente le plus d'analogie avec les verres de spins est celui de la transition vitreuse. J'ai essayé de montrer comment la loi empirique qui paraissait avoir le plus de "virtualités" était la loi de Vogel–Fulcher. Mais il reste à lui trouver un fondement; il reste également à comprendre ce qu'est en général l'état verre que l'on prépare:

Tableau 2.1

Verre de spin	Verre–verre
Spin d'impuretés magnétiques: i	Systèmes à deux niveaux, nature inconnue: i
Dégénérescence de rotation en l'absence d'interactions $H_i(\text{ext}) = 0$	Symétrie du champ cristallin local $H_i(\text{ext}) = E/g\beta$ $p(H_i) = \text{cte}$
Interactions indirectes entre spins du type RKKY de signe aléatoire: modèle de Heisenberg $J_{ij}\mathbf{S}_i \cdot \mathbf{S}_j$	Interactions élastiques indirectes de signe aléatoire: modèle d'Ising $J_{ij}S_i^z S_j^z$
Changement de phase à T_G	Pas de changement de phase dû aux interactions
Chaleur spécifique linéaire à basse température	Chaleur spécifique linéaire: densité d'excitations $n_0 \simeq$ cte
"Nuages" ou paquets de spin ? rémanence	Distribution des T_1 et T_2

quels "défauts", quelle structure se cachent derrière le faisceau de propriétés et de comportement universels que l'on observe. Si cette question était résolue, la nature des excitations des verres à basse température trouverait probablement une explication: simple conséquence d'une meilleure connaissance de la transition vitreuse.

3. Autres systèmes désordonnés

Les verres de spin et les verres sont les deux exemples les plus explorés des systèmes désordonnés. Il en existe pourtant d'autres dont on fera ici un rapide survol et qui se rattachent à l'un ou l'autre des deux précédents pour certaines de leurs propriétés. Il faut mentionner:

- les métaux amorphes;
- les isolants magnétiques amorphes;
- les systèmes d'impuretés moléculaires dans les cristaux de type KCl;
- les assemblées d'impuretés non ionisées dans les semi-conducteurs.

Il faudrait les passer tous en revue; en réalité, la tâche sera simplifiée car bien de leurs propriétés sont identiques à celles que l'on a déjà vu pour les verres de spins ou les verres; on n'insistera guère, sinon pour souligner différences là ou cela sera nécessaire et pour donner souligner les ressemblances troublantes qui s'étendent jusqu'aux ordres de grandeur et soulignent le caractère "universel" des systèmes désordonnés.

3.1. Verres métalliques

3.1.1. Des verres métalliques: lesquels?

Il n'est pas aisé de préparer un verre métallique qui ne présente aucune trace de recristallisation; il y a vingt ou trente ans, on a commencé de dépenser beaucoup d'argent pour préparer un "bon" cristal; on en connait maintenant des presque parfaits: voyez le silicium ou le germanium, et quelques métaux comme le Cu.

On pourrait dire que pour les verres métalliques, on en est encore à la méthode des expédients grossiers; pour parvenir à faire ces verres, on est parfois obligé d'ajouter des ingrédients supplémentaires pour empêcher les traces de cristallisation. Bref, cette opération n'est pas simple et pourtant on a autant besoin maintenant de matériaux parfaitement désordonnés qu'autrefois on avait besoin de matériaux parfaitement cristallisés.

On peut classer les verres métalliques en deux catégories importantes. La première correspond à l'assemblage d'un métal (Ni, Pd, Fe, Co,...)

avec un métalloïde (Si, P, Ge, C, B, . . .); ils se forment facilement lorsque la composition du liquide de départ est voisine de celle d'un eutectique qui pourrait s'obtenir par refroidissement lent. Les exemples les plus fameux sont Pd–Si, Ni–P, Fe–B. La deuxième catégorie est celle des verres où il y a deux métaux qui sont alliés. Ces métaux sont Ni, Co, Fe, Pd, Cu et Zr, Ta, Nb, Y. Les meilleurs exemples sont Zr–Cu, Y–Cu, ou Ti–Cu. Tous ces verres s'obtiennent par trempe à partir d'une solution liquide.

On ne sait pas bien ce qui favorise la formation d'un verre; mais, empiriquement, deux facteurs semblent jouer un rôle. La différence de taille entre les constituants d'abord devrait être assez grande pour favoriser l'apparition de structures compactes plus denses que celles qui seraient obtenues avec des atomes de même diamètre; il en résulterait un assemblage stable face aux fluctuations de composition nécessaires à la formation d'un cristal; une différence de diamètre supérieure à 1.2 est souhaitable. Le deuxième paramètre qui joue un rôle est la différence de valence entre les composants élémentaires; il en résulte une modification importante de l'hybridation des électrons de valence, apparemment favorable à la formation d'un verre.

Un mot enfin sur l'intérêt des verres métalliques avant d'embrayer sur leurs propriétés; cet intérêt est évidemment technologique. Du point de vue mécanique, on notera qu'ils sont très ductiles et que la résistance à l'allongement est grande, en proportion de la valeur des constantes élastiques.

Une seconde caractéristique est le ferromagnétisme faible des verres à base de Fe, Co, Ni: le fait que les anisotropies disparaissent dans un système qui retrouve son invariance par rotation entraîne une grande perméabilité et aussi de faibles pertes.

Un troisième intérêt est lié aux propriétés de corrosion; l'absence de joints de grain ou de défauts qui favorisent la migration et l'infiltration des ions dans un cristal, rend le verre résistant à la corrosion; éventuellement les potentiels électro-chimiques peuvent être modifiés.

Un quatrième intérêt est assez évident: on connait les défauts produits par irradiations dans un cristal; ici, que ferait une irradiation? Le verre métallique pourrait donc être un matériau de choix pour des éléments qui devraient avoir des propriétés mécaniques constantes dans un environnement difficile: pièces mécaniques soumises à de fortes doses d'irradiation.

Enfin, signalons un dernier avantage: la possibilité de stocker une plus grande quantité d'hydrogène dans l'alliage Pd–Si que dans un métal monocristallin comme le Nb. Ainsi, les verres métalliques ont quelque séduction.

3.1.2. Comportement à basse température

On va retrouver ici l'ensemble des propriétés qui ont caractérisé les verres-verres à basse température: les systèmes à deux niveaux conditionnent la vitesse du son, l'atténuation, la chaleur spécifique, la conductibilité thermique.

Mais il est instructif de donner quelques ordres de grandeur. Par ailleurs, on sera amené à s'interroger sur l'influence des systèmes à deux niveaux sur les propriétés électroniques des verres métalliques.

Chaleur spécifique. La chose est entendue; il y a sans ambiguité possible un terme linéaire en chaleur spécifique [3.1].

Un alliage comme Pd–Si a été fréquemment employé car il est aisé à préparer sous forme amorphe ou cristalline et que la comparaison, de ce fait, est aisée; de plus, il contient généralement peu d'impuretés magnétiques. Mais autant il est clair que ce terme linéaire en T existe, autant il n'est pas facile d'extraire de cette mesure la contribution qui correspond aux S2N [3.1] car les électrons masquent leur influence. De sorte qu'une comparaison systématique doit être faite après recristallisations partielles [3.2].

Un ordre de grandeur du terme linéaire en température dû aux S2N ($C \simeq aT$) est donné par a:

$$a = 0.1 \text{ mJ/mole}\cdot\text{K}^2 . \tag{3.1}$$

En comparant ce résultat avec la formule (2.22), on en déduit $n_0 \simeq 10^{33}$ erg^{-1} cm^{-3}; c'est une valeur voisine de la silice.

Pour confirmer ces mesures, il était intéressant de faire une expérience dans un amorphe devenant supraconducteur à une température assez haute de sorte que les électrons ne participent plus en principe à la chaleur spécifique [3.3]. C'est le cas de $Zr_{0.7}Pd_{0.3}$ pour lequel T_c vaut environ 2.5 K; on trouve très exactement (voir fig. 3.1):

$$n_0 = 2.7 \times 10^{33} \text{ erg}^{-1} \text{ cm}^{-3} .$$

Comme dans le cas de Pd–Si, cette valeur reste sujette à controverse dans la mesure où l'on peut s'interroger sur l'existence de zones partiellement recristallisées, ou normales; c'est en particulier le cas de ceux qui parmi les physiciens ne sont pas encore convaincus de l'existence de ces S2N. J'ai tendance à penser que ce résultat est propre.

Conductibilité thermique K; vitesse du son v. On n'observe aucune différence avec le cas des verres. Pour des raisons déjà données, on vérifie que

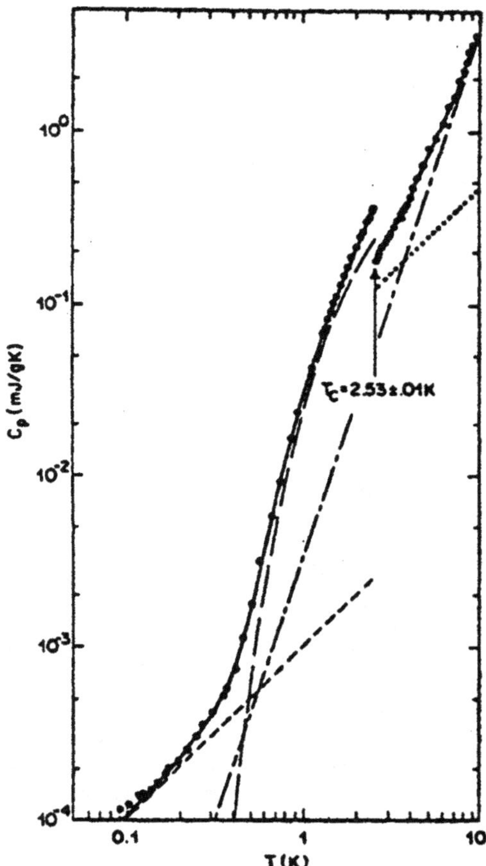

Fig. 3.1. Chaleur spécifique du métal amorphe supraconducteur $Zr_{0.7}$–$Pd_{0.3}$. La ligne solide est ajustée sur les résultats expérimentaux. Les diverses contributions correspondent: ····: électrons au-dessus de T_c; ——: électrons en dessous de T_c; –·–·–: phonons; ----: contribution supplémentaire linéaire en température.

$K \simeq T^2$ [3.3,3.4] et que le coefficient de ce terme en T^2 est très comparable à celui de la silice. Cela a été vérifié pour Zr–Pd, Ni–P, Fe–P, et Pd–Si, que ces métaux soient préparés par trempe ou par électrodéposition.

La vitesse du son suit une loi en log T (fig. 3.2) et obéit bien à la formule (2.29), conduisant à un coefficient de couplage k^2 [3.5] dont la valeur est un ordre de grandeur plus petit que dans la silice soit 5×10^{-5}.

Ces résultats sont à priori surprenants; en effet, si la silice a une structure

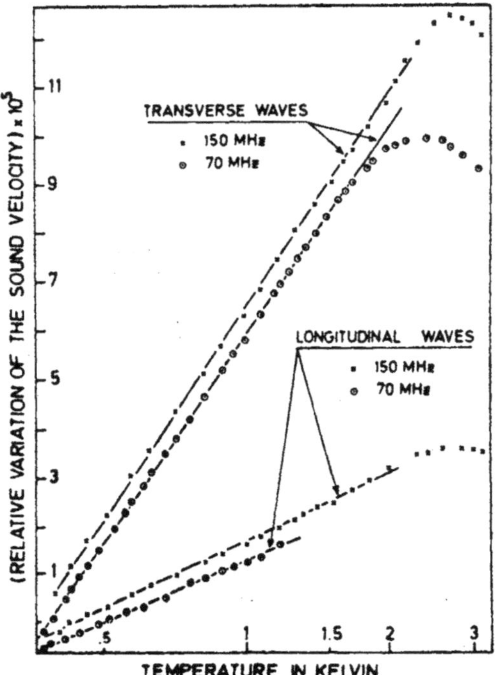

Fig. 3.2. Variation relative de la vitesse du son en fonction de la température pour l'alliage amorphe Ni–P. La température de référence est égale à 0.37 K. L'origine de l'axe vertical est arbitraire. La précision de mesure est meilleure que $\pm 2 \times 10^{-6}$. La vitesse longitudinale est égale à 5.2×10^5 cm/s. La vitesse des ondes transversales est 2.3×10^5 cm/s.

relativement "aérée", les amorphes métalliques ont au contraire un empilement compact. Si donc les S2N correspondent à un mouvement atomique, on comprend mal que k^2 soit approximativement le même. Aussi, plutôt que de s'étonner il vaut mieux renverser la proposition et prendre cette coïncidence comme une indication que les S2N sont plus probablement de nature électronique et qu'ils sont localisés.

Saturation de l'atténuation ultrasonore. Ce n'est que récemment que la "preuve" [3.6] de l'existence de ces S2N a été apportée par une mesure précise de la saturation de l'atténuation ultrasonore; elle obéit très exactement à la loi (2.26) (fig. 3.3).

Quoiqu'aucune mesure "d'écho" n'ait été effectuée, mais là-aussi, il

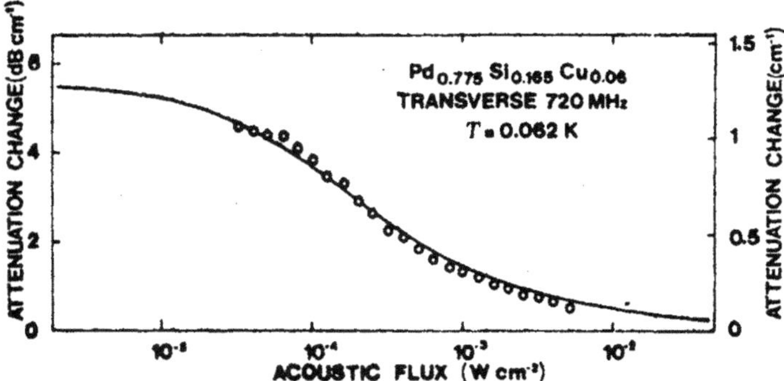

Fig. 3.3. Variation relative de la vitesse du son transversal dans un alliage amorphe Pd–Si–Cu en fonction de l'intensité acoustique de l'échantillon. La ligne continue est calculée à l'aide de la formule:

$$\alpha_R = \alpha_R^{uns}\left(1 + \frac{\Phi}{\Phi_c}\right)^{-1/2}$$

faudrait probablement travailler à une température voisine de quelques mK, les valeurs du flux de saturation acoustique ϕ_c sont grandes; cela indique que les temps de relaxation T_1 et T_2 sont très courts.

Il y a donc une certaine contradiction dans les résultats expérimentaux: on trouve que k^2 n'est pas très grand [3.5] en même temps que la relaxation est rapide. La réponse est probablement la suivante: les électrons participent à cette relaxation de manière plus efficace que les phonons. Inversement, on peut se demander quelle est l'influence des S2N sur le libre parcours moyen des électrons; c'est ainsi que l'on est amené à s'interesser aux mesures de résistivité à basse température.

3.1.3. Résistivité des verres métalliques

Il existe relativement peu d'expériences très propres sur la résistivité et qui ne mélangent pas magnétisme, impuretés, S2N; mais il faut bien faire avec les matériaux dont on dispose!

Grossièrement, il existe un minimum de résistivité aux environs de quelques degrés K; pour une température inférieure ρ a tendance à croître (fig. 3.4). On cherche souvent à interpréter ces résultats [3.7] en reliant les mesures à une formule du type $\log(T^2 + \Delta^2)$. L'accord n'est pas excellent; mais la formule elle-même est sujette à caution [3.8].

Quoiqu'il en soit, il y a là une voie qui mérite de se développer pour réconcilier mesures acoustiques et résistivité dans le cadre d'un modèle

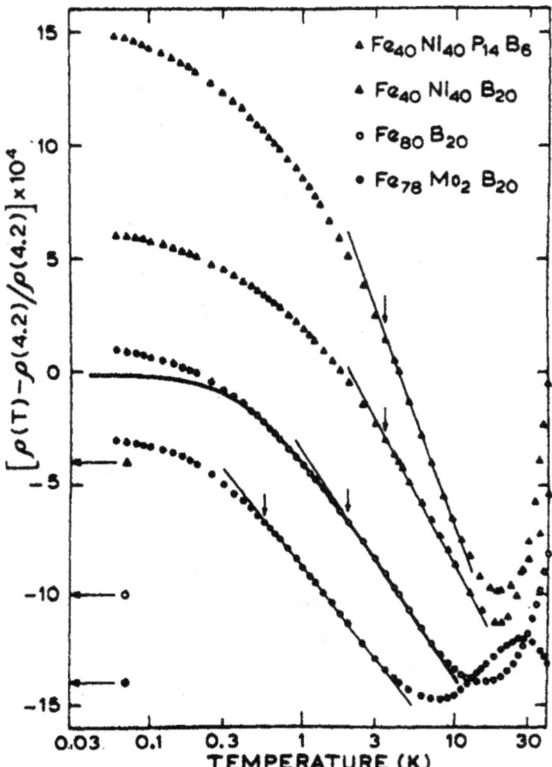

Fig. 3.4. Résistivité normalisée en fonction de log T. Les trois courbes du bas ont été décalées comme l'indiquent les flèches horizontales. La courbe en traits continus est ajustée avec une formule $\ln(T^2 + \Delta^2)$ avec $\Delta = 0.45$ K.

théorique dont les arguments sont similaires à ceux utilisés dans l'explication de l'effet Kondo.

En conclusion, il semble bien que les amorphes métalliques à basse température soient gouvernés, comme les verres isolants, par des excitations dont il reste à préciser la nature.

3.2. *Impuretés moléculaires dans les cristaux ioniques*

Lorsque l'on introduit en faible concentration certaines molécules ou certains ions moléculaires dans des matrices cristallines du genre KCl, ils ont tendance à sa placer en position substitutionnelle; le dipole ou le quadripole électrique qu'ils portent s'oriente alors suivant les minima du potentiel cristallin qui joue le rôle de champ d'anisotropie local. Si la concentration de ces dipoles augmente, on imagine aisément que leur énergie d'interaction peut devenir prépondérante; on retrouve une situation qui s'apparente beaucoup à celle d'un verre de spin et il est naturel de se poser la question: une transition est-elle possible? T_G a-t-il un sens et une réalité? Pour y répondre, il est utile de commencer par décrire les propriétés d'une seule impureté moléculaire.

3.2.1. Propriétés d'une seule impureté

Un ion moléculaire comme OH^- a un moment dipolaire électrique permanent; il se substitue aisément aux ions Cl^- de KCl dans une proportion qui peut aller jusqu'à 1%. On a montré que, dans une première approximation, il s'oriente dans le champ cristallin suivant six directions équivalentes parallèles aux axes 1.0.0 du cristal, l'oxygène occupant le centre d'une cage aménagée par le départ des ions Cl^-. Entre ces six vallées, OH^- exécute des mouvements tunnels de sorte que, en deuxième approximation. la dégénérescence des six niveaux est levée et conduit à une décomposition illustrée sur la figure 3.5. En champ électrique E ou sous déformation élastique ε nuls, on observe en partant des énergies les plus basses un singulet, à 2Δ au-dessus un triplet, et enfin à une distance Δ, un doublet. Pour cet exemple, Δ est égal à 6200 Mc/s soit 0.3 K. L'application d'un champ électrique statique ou d'une déformation élastique permet de mesurer le moment électrique longitudinal permanent de OH^- soit $p_{\mathrm{long}} = 4.0$ D. Pour la constante de couplage de la molécule avec le réseau ou pour le potentiel de déformation, on retrouve des ordres de grandeur de 1 eV.

On comprend assez bien ce qui se passe et des expériences de chaleur spécifique de conductibilité thermique [3.10] laissent clairement voir cette structure sous forme d'anomalies de Schottky; des expériences de résonance paraélectrique ont montré qu'on pouvait induire aisément des transitions entre les niveaux [3.9,3.11].

Incidemment, notons que la raie est très large, non pas qu'il y ait interaction entre ces impuretés mais parce qu'il y a une distribution de champ local: le cristal est inhomogène très probablement à cause des contraintes locales.

A des énergies très supérieures à Δ existent les niveaux de rotation ou de libration [3.10] de la molécule, mais on ne les fera pas intervenir dans la suite.

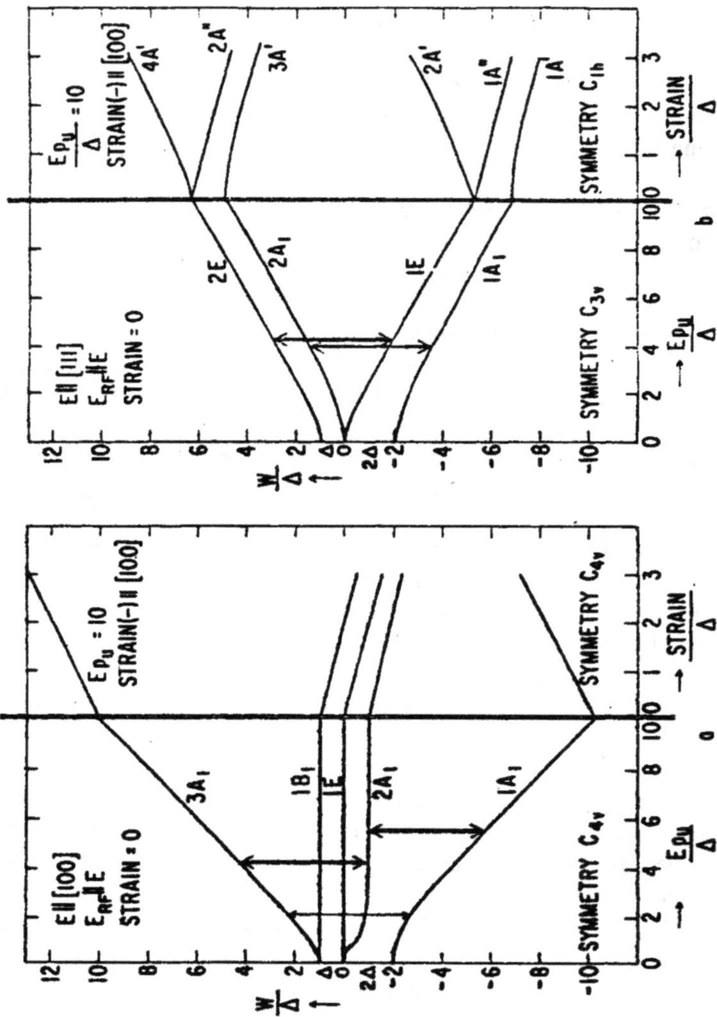

Fig. 3.5. Schéma des niveaux d'énergie d'un dipole OH^- dans un champ cristallin de symétrie octahédrique pour deux orientations du champ électrique appliqué. Les flèches à gros traits indiquent les transitions permises, les flèches à petits traits les transitions interdites. L'effet d'une distorsion uniaxiale positive dans la direction (100) est indiquée dans la partie droite des diagrammes.

3.2.2. Interactions entre impuretés; verre de spin dipolaire

Il y a une quinzaine d'années, alors que l'on s'interrogeait sur les propriétés des impuretés isolées leurs interactions avaient déjà été considérées; le maximum de la constante diélectrique, en fonction de la température [3.12] avait suggéré l'existence d'une transition de phase vers un état ferroélectrique. Ce point de vue fut vite contesté et la suggestion était que la distribution de la polarisation correspondait plus probablement à un antiferroélectrique aléatoire. Par ailleurs l'auteur [3.13] faisait remarquer qu'un traitement du genre Clausius–Mossotti ne tenait pas compte de l'organisation régulière des dipoles dans une sphère de rayon R_c; comme pour les verres de spins [3.14], une correlation devait exister pour $r < R_c$ alors qu'aucun ordre à longue distance ne pouvait subsister.

Chaque volume ainsi correllé devait avoir une température de transition de l'ordre de

$$kT_{max} \simeq \frac{p_0^2}{\varepsilon_0} \frac{1}{R_c}, \tag{3.2}$$

où R_c, à un coefficient numérique près est donné par:

$$R_c \simeq a/c^{1/3}; \tag{3.3}$$

a est la maille cristalline et c la concentration.

Depuis cette première approche, la situation expérimentale pour ces systèmes dipolaires électriques a été considérablement éclaircie.

On retrouve avec quelques variantes qu'il faut souligner une situation assez semblable à celle des verres de spin, mais avec une collection de résultats bien moins grande.

La chaleur spécifique présente à basse température ($T < T_G$) une variation en $T^{3/2}$ [3.15]. Des essais récents non publiés confirment cet exposant étrange qu'on n'attribuera pas à des "ondes de spin" comme on l'a suggéré, mais qui néanmoins est bien 3/2; curieusement d'ailleurs, les ferroélectriques à basse température semblent affublés de la même indignité [3.16]. Le seul reproche qu'on pourrait faire à ces expériences c'est de n'avoir pas été effectuées à suffisamment basse température, mais...

Pour ce qui est des propriétés de polarisation électrique [3.15]: la constante diélectrique passe par un maximum qui dépend bien de la concentration (fig. 3.6). L'absence de "cusp" se comprend si l'on se rappelle que le niveau fondamental est ici non dégéneré. La polarisation rémanente (voir fig. 3.7) a les caractéristiques qu'on s'attend à lui voir et elle a bien une signification macroscopique: relaxation sur une grande échelle de temps (voir fig. 3.8).

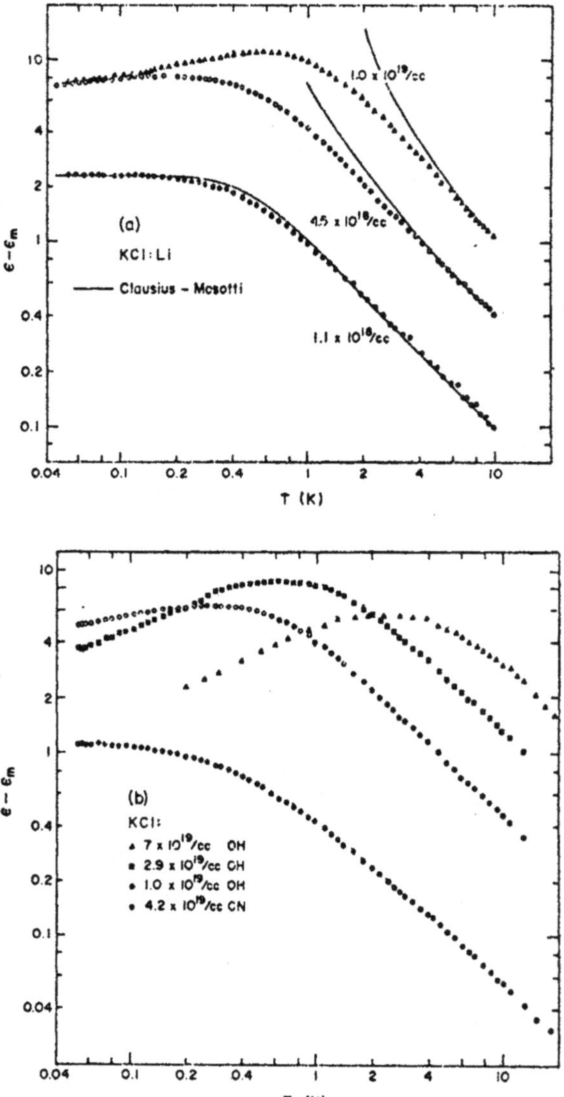
(a)
KCl:Li
Clausius – Mosotti
1.0 x 10^19/cc
4.5 x 10^18/cc
1.1 x 10^18/cc
ε − ε_m
T (K)
(b)
KCl:
7 x 10^19/cc OH
2.9 x 10^19/cc OH
1.0 x 10^19/cc OH
4.2 x 10^19/cc CN
ε − ε_m
T (K)

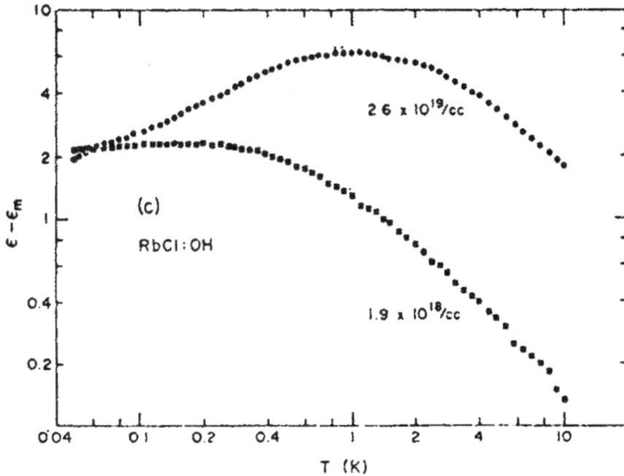

Fig. 3.6. Excès de constante diélectrique $\varepsilon - \varepsilon_m$ en fonction de la température pour différents types de défauts. Les courbes en traits continus de la figure du haut sont calculées à l'aide d'une fonction de Clausius Mosotti.

Tous ces comportements sont propres; ils ont été attribués à un phénomène de changement de phase qui s'apparente à celui des verres de spins. Attendons la suite. Je crois que le caractère séduisant de ce système c'est que l'on connait bien les forces d'interaction élémentaires entre dipoles (si l'on devait, comme dans les verres, introduire les interactions élastiques, on retomberait dans un cas moins trivial). Inversement, la structure non

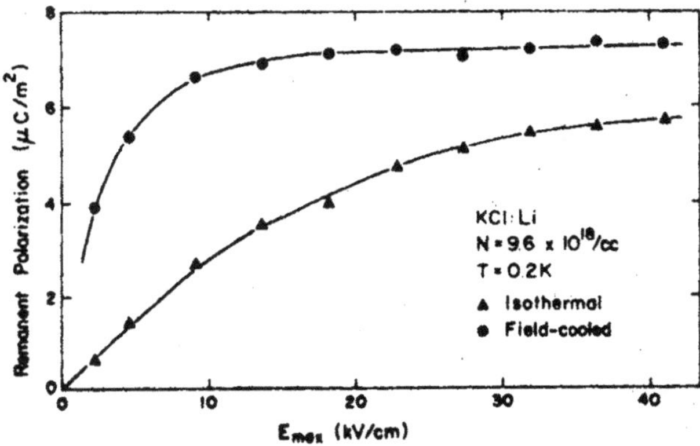

Fig. 3.7. Polarisation rémanente en fonction du champ de polarisation mesurée une minute après la coupure du champ électrique.

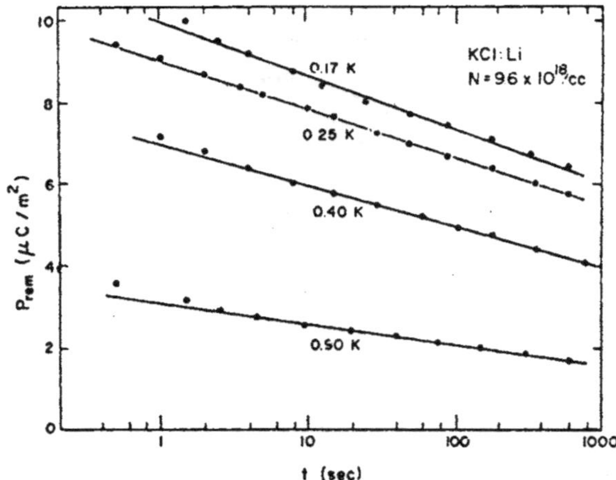

Fig. 3.8. Polarisation rémanente de KCl:Li en fonction du temps (échelle logarithmique pour différentes températures).

dégénérée des six niveaux les plus bas risque de compliquer l'apparence des phénomènes à T_G lorsque

$$\Delta \simeq KT_G\,. \tag{3.4}$$

Dans les expériences mentionnées ci-dessus, c'est un peu le cas.

3.3. Derniers exemples pour mémoire

Pour achever cette revue, je voudrais encore mentionner par titre deux systèmes qui sous certains aspects s'apparentent aux verres.

3.3.1. Le quartz fumé

Lorsqu'un quartz qui contient des impuretés d'aluminium est soumis à des rayonnements ionisants, des centres colorés sont formés d'un trou électronique qui se trouve piégé dans un potentiel à deux puits.

Il existe en fait une distribution de niveaux d'énergie de ces trous qui a été observée par chaleur spécifique (linéaire en T à basse température) [3.17], par atténuation ultrasonore [3.18] (saturation et relaxation, voir fig. 3.9), ou par relaxation diélectrique [3.19]. On met là-aussi en évidence des systèmes à deux niveaux dont on connait mieux cette fois la nature puisqu'on sait "guérir" le quartz par exemple. L'intérêt de ce système vient peut-être de ce que l'on sait doser la densité n_0 de ces excitations en changeant la

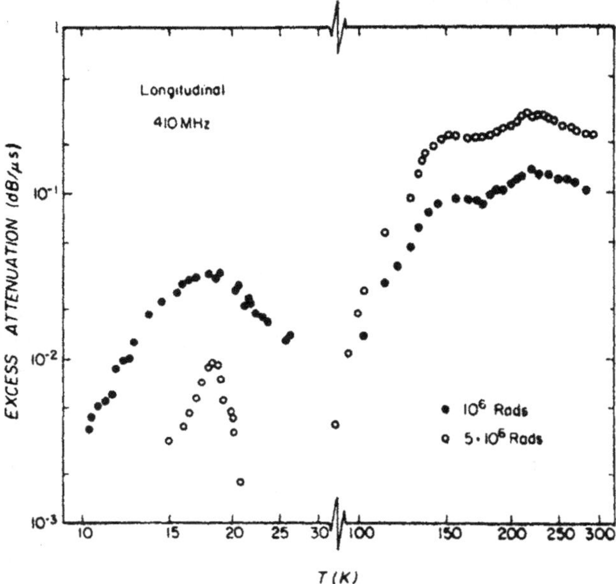

Fig. 3.9. Dépendance en température de l'excès d'atténuation pour des échantillons de quartz irradiés aux rayons γ à basses et hautes températures.

dose d'irradiation; c'est une souplesse dont on ne dispose pas dans les verres.

3.3.2. Accepteurs et donneurs dans un semiconducteur

Les trous liés à un accepteur ou les électrons liés à un donneur ont dans un semiconducteur un grand rayon de l'ordre d'une quinzaine d'ångström. On sait qu'ils sont très couplés aux phonons: des mesures d'absorption acoustique résonnantes ont été effectuées il y a fort longtemps.

On observe par ailleurs que la densité des niveaux est une fonction relativement large en énergie $\simeq$ 0.1 meV [3.20]; toutes les caractéristiques acoustiques d'un verre sont aisément retrouvées. Quelle est l'origine de cette distribution? Comme d'habitude, on pense aux inhomogénéités du cristal; plus probablement, il faut les rattacher à la présence de défauts résiduels.

L'intérêt de ce système ne tient pas à l'objet même qui est étudié; on en sait assez sur les impuretés des semiconducteurs! mais il tient à sa grande souplesse pour servir de modèle à ce qui passe dans un verre; on peut en effet aisément modifier la concentration des trous liés, donc leurs interactions, relier leur distribution avec la présence d'autres impuretés et voir comment l'intensité critique ϕ_c est liée à la concentration des systèmes à deux niveaux [3.21].

Conclusions

Cette présentation de l'aspect expérimental des systèmes désordonnés a voulu attirer l'attention sur quelques propriétés "universelles" des amorphes. Il n'était pas dans mon intention d'en faire une présentation exhaustive. De même, les références que je donne correspondent à une sélection effectuée à mes risques et périls; j'espère en particulier ne pas éveiller trop de susceptibilités!

L'un ou l'autre des exemples mentionnés ici renferme encore bien des obscurités même si ces dernières années ont apporté pour certains d'entre eux, des idées nouvelles. Reste en particulier très ouverte la question de T_G des verres: expérimentalement et théoriquement. Les verres de spins restent l'objet d'apres controverses. Les exemples moins standards dont j'ai parlé ont pour eux une plus grande versatilité. Nous verrons!

A la fin de la rédaction de la deuxième version du cours, je voudrais remercier les élèves de cette session qui par leurs questions ont suggéré une amélioration ou des compléments en plusieurs endroits. Je voudrais enfin dire que j'ai bénéficié des remarques de mes collègues du C.R.T.B.T. (R. Maynard, L. Piché, J. Tholence, R. Rammal, J. Souletie), et de l'I.L.L. (A. Murani, G. Wright) qui m'ont éclairé sur bien des points.

Références

Verres de spin

[1.1] J. Souletie et R. Tournier, J. Low Temp. Phys. 1 (1969) 95.
[1.2] L. Wenger et P. Keesom, Phys. Rev. B11 (1975) 3497.
[1.3] L. Wenger et P. Keesom, Phys. Rev. B13 (1976) 4053.
[1.4] V. Cannela et J. Mydosh, Phys. Rev. B6 (1972) 4220.
[1.5] C. Violet et R. Borg, Phys. Rev. 149 (1966) 540; 162 (1967) 608;
B. Window, C. Longworth et C. Johnson, J. Phys. C3 (1970) 2156.
[1.6] D. Murnick, A. Fiory, et W. Kossler, Phys. Rev. Letters 36 (1976) 100.
[1.7] A. P. Murani, Intern. Symp. on Neutron Scattering, Vienna, 1977.
[1.8] S. P. McAlister et C. M. Hurd, Solid State Commun. 19 (1976) 881.

[1.9] P. J. Ford et J. A. Mydosh, Phys. Rev. B14 (1976) 2057.
[1.10] A. Laborde, Thèse, Grenoble (1977).
[1.11] N. Rivier et K. Adkins, J. Phys. F5 (1975) 1745.
[1.12] K. Matho, communication privée.
[1.13] R. Buckmann, H. P. Falke, H. P. Jablonski, et E. F. Wassermann, Physica 86–88B (1977) 835.
[1.14] E. Hawkins, T. Moran, et R. L. Thomas, dans: Amorphous Magnetism II, eds. R. A. Levy et R. Hasegawa (Plenum, New York, 1977).
[1.15] G. Hawkins et R. L. Thomas, J. Appl. Phys. 49 (1978) 1627.
[1.16] J. Tholence et R. Tournier, J. Physique 34, C4 (1974) 229.
[1.17] L. Néel, Ann. Géophysique 5 (1949) 99.
[1.18] E. Hirschkoff, E. Shanaberger, M. Symko, et O. Wheatley, J. Low Temp. Phys. 5 (1971) 545;
J. Tholence, Thèse, Grenoble (1974).
[1.19] H. Löhneysen, J. Tholence, et R. Tournier, J. Physique 39, C6 (1978) 922.
[1.20] G. J. Nieuwenhuys et J. A. Mydosh, Physica 86–88B (1977) 880.
[1.21] F. Holtzberg, J. Tholence, et R. Tournier, dans: Amorphous Magnetism II, eds. R. A. Levy et R. Hasegawa (Plenum, New York, 1977).
[1.22] H. Rechenberg, L. Bieman, F. Huang, et A. de Graaf, J. Appl. Phys. 49 (1978) 1638.
[1.23] B. Halperin et W. Saslow, Phys. Rev. B16 (1977) 2154.
[1.24] W. Marshall et R. Lowde, Rept. Progr. Phys. 31 (1968) 705.
[1.25] H. Scheuer, M. Loewenhaupt, et J. B. Suck, J. Magnetism Magnetic Mater. 6 (1977) 100.
[1.26] H. Scheuer, M. Loewenhaupt, et W. Schmatz, Physica 86–88B (1977) 842.
[1.27] A. P. Murani, à paraître.
[1.28] P. G. de Gennes, J. Phys. Chem. Solids 4 (1958) 223.
[1.29] A. P. Murani et J. Tholence, Solid State Commun. 22 (1977) 25;
A. P. Murani, IAEA Conf., Vienna, 1977.
[1.30] S. Edwards et P. W. Anderson, J. Phys. F6 (1976) 1361.
[1.31] A. P. Murani et A. Heidemann, à paraître.

Les verres

[2.1] Citation extraite de la thèse de L. Piché, Grenoble (1976), numéro d'ordre: AO-12002.
[2.2] R. Couty, Thèse, Paris (1977).
[2.3] R. Calemczuk, Thèse de 3ème Cycle, Grenoble (1977).
[2.4] R. B. Stephens, J. Non-Crystalline Solids 20 (1976) 75.
[2.5] A. Kovacs, J. Polymer Sci. 30 (1958) 131.
[2.6] A. Kovacs, R. Stratton, et J. Ferry, J. Phys. Chem. 67 (1965) 152.
[2.7] C. Lacabbe et D. Chatain, J. Phys. Chem. 79 (1975) 283.
[2.8] R. Calemczuk, Thèse, Grenoble (1977).
[2.9] D. Davidson et R. Cole, J. Chem. Phys. 19 (1951) 1484;
G. Duffie et T. Litowitz, J. Chem. Phys. 37 (1962) 1699.
[2.10] G. Tammann et W. Hesse, Z. Anorg. Allgem. Chem. 156 (1926) 245.
[2.11] D. Turnbull et M. Cohen, J. Chem. Phys. 34 (1961) 120.
[2.12] M. Williams, R. Landel, et J. Ferry, J. Am. Chem. Soc. 77 (1955) 3701.
[2.13] G. Adam et J. Gibbs, J. Chem. Phys. 43 (1965) 139.

[2.14] A. Kovacs, Fortschr. Hochpolymer Forsch. 3 (1963) 394.
[2.15] R. Rammal, Thèse 3ème Cycle, Grenoble (1977).
[2.16] F. Hansen et T. Knudsen, J. Chem. Phys. 62 (1975) 1556;
R. Bellissent et G. Tourand, dans: Proc. Conf. on Amorphous and Liquid Semiconductors, Edinburgh, 1977.
[2.17] R. Mozzi et B. Warren, J. Appl. Cryst. 2 (1969) 164;
A. J. Leadbetter et A. C. Wright, J. Non-Crystalline Solids 7 (1972) 141.
[2.18] M. R. Vukcevich, J. Non-Crystalline Solids 11 (1972) 25.
[2.19] R. Mozzi et B. Warren, J. Appl. Cryst. 2 (1969) 669.
[2.20] A. Leadbetter, J. Chem. Phys. 51 (1969) 779.
[2.21] J. Axe, dans: Physics of Structurally Disordered Solids, ed. Shashanka (Plenum, New York, 1976).
[2.22] S. Hunklinger, Ultrasonic Symp., Milwaukee, PA, IEEE 74-CHO-896-ISU.
[2.23] J. Pelous et R. Vacher, Solid State Commun. 16 (1975) 279.
[2.24] W. Bagdade et R. Stolen, J. Phys. Chem. Solids 29 (1962) 2001.
[2.25] P. W. Anderson, B. Halperin, et C. Varma, Phil. Mag. 25 (1971) 1.
[2.26] W. Phillips, J. Low Temp. Phys. 7 (1972) 351.
[2.27] P. W. Anderson, Phys. Rev. Letters (1975).
[2.28] R. Zeller et R. Pohl, Phys. Rev. B4 (1971) 2029.
[2.29] J. Lasjaunias, A. Ravex, D. Thoulouze, et M. Vandorpe, dans: Proc. Intern. Conf. on Phonon Scattering in Solids (Plenum, New York, 1975) p. 135.
[2.30] R. Maynard, communication privée.
[2.31] R. Stephens, Phys. Rev. B8 (1973) 2896;
M. Zaitlin et A. Anderson, Phys. Rev. Letters 33 (1974) 1158.
[2.32] W. Arnold, S. Hunklinger, S. Stein, et K. Dransfeld, J. Non-Crystalline Solids 14 (1974) 192.
[2.33] J. Joffrin et A. Levelut, J. Physique 36 (1975) 811.
[2.34] B. Golding, J. Graebner, et R. Schutz, Phys. Rev. B14 (1976) 1660.
[2.35] J. Jäckle, L. Piché, W. Arnold, et S. Hunklinger, J. Non-Crystalline Solids 20 (1976) 365.
[2.36] K. Mon, Y. Chabal, et A. Sievers, Phys. Rev. Letters 35 (1975) 1352.
[2.37] L. Piché, R. Maynard, S. Hunklinger, et J. Jäckle, Phys. Rev. Letters 32 (1974) 1426.
[2.38] R. Stephens, Phys. Rev. B14 (1976) 754.
[2.39] M. von Schickfus, S. Hunklinger, et L. Piché, Phys. Rev. Letters 35 (1975) 876.
[2.40] P. Doussineau, A. Levelut et T.-T. Ta, J. Physique 38L (1977) 37.
[2.41] W. Goubau et R. Tait, Phys. Rev. Letters 34 (1975) 1220.
[2.42] J. Lewis et J. Lasjaunias, J. Physique, Colloq. LT 15, Grenoble (1978);
R. Rammal et R. Maynard, J. Physique, Colloq. LT 15, Grenoble (1978).
[2.43] R. B. Kummer, R. Dynes, et V. Narayanamurti, Phys. Rev. Letters 40 (1978) 1187.
[2.44] B. Golding et J. Graebner, Phys. Rev. Letters 37 (1976) 852.
[2.45] L. Bernard, L. Piché, G. Schumacher, J. Joffrin, et J. Graebner, J. Physique 39L (1978) 126.
[2.46] L. Bernard, L. Piché, G. Schumacher, et J. Joffrin, J. Physique C6 (1978) 957.
[2.47] L. Bernard, L. Piché, G. Schumacher, et J. Joffrin, J. Low Temp. Phys., submitted.
[2.48] J. Black et B. Halperin, Phys. Rev. B16 (1977) 2879.

Autres systèmes désordonnés

[3.1] H. Chen et W. Haemmerle, J. Non-Crystalline Solids 11 (1972) 161; B. Golding, B. Bagley, et F. Hsu, Phys. Rev. Letters 29 (1972) 68.
[3.2] J. Lasjaunias, A. Ravex, et D. Thoulouse, J. Phys. F, to be published.
[3.3] J. Graebner et al., Phys. Rev. Letters 39 (1977) 1480.
[3.4] J. R. Matey et A. C. Anderson, J. Non-Crystalline Solids 23 (1977) 129.
[3.5] G. Belessa, P. Doussineau, et A. Levelut, J. Physique 38L (1977) 65.
[3.6] P. Doussineau, P. Legros, A. Levelut, et A. Robin, J. Physique 39L (1978) 265.
[3.7] R. Cochrane, F. Hedgcock, B. Hästner, et W. Muir, J. Physique 39, C6 (1978) 939.
[3.8] J. Kondo, Physica 84B (1976) 40.
[3.9] G. Feher, I. Shepherd, et H. Shore, Phys. Rev. Letters 16 (1966) 500.
[3.10] C. Chau, M. Klein, et B. Wedding, Phys. Rev. Letters 17 (1966) 521.
[3.11] W. Bron et R. Dreyfus, Phys. Rev. Letters 16 (1966) 165.
[3.12] W. Kanzig, H. Hart, et S. Roberts, Phys. Rev. Letters 13 (1964) 543.
[3.13] R. Brout, Phys. Rev. Letters 14 (1965) 176.
[3.14] R. Brout et W. Klein, Phys. Rev. 132 (1963) 2412.
[3.15] A. Fiory, Phys. Rev. B4 (1971) 614.
[3.16] W. Lawless et A. Morrow, Ferroelectrics 15 (1977) 167.
[3.17] M. Saint-Paul, B. Picot, et R. Nava, à paraître.
[3.18] M. Saint-Paul et R. Nava, J. Physique 39 (1978) 787.
[3.19] W. J. de Vos et J. Volger, Physica 47 (1970) 13.
[3.20] E. Ortlieb, H. Schad, et K. Lassmann, Solid State Commun. 19 (1976) 599.
[3.21] H. Zeile, O. Mathuni, et K. Lassmann, à paraître.

COURSE 3

LECTURES ON AMORPHOUS SYSTEMS*

Philip W. ANDERSON

Bell Laboratories, Murray Hill, New Jersey 07974, USA
and
Physics Department, Princeton University, Princeton, New Jersey 08540, USA

* Work supported in part by NSF DMR 78-03015 and ONR N00014-77-C-0711.

Contents

R. Balian et al., eds.
Les Houches, Session XXXI, 1978 – La matière mal condensée/Ill-condensed matter

1. Introduction

1.1. General principles (if any)

If I had been giving a set of lectures on the problem of disorder a few years ago, and even more surely if any of the other obvious lecturers had been, I should almost certainly have devoted the majority of my time to a set of methods and a subject which, for practical purposes, has been ignored and will be ignored here at Les Houches: one type or another of multiple scattering theory. There has in fact been one of Thomas Kuhn's "revolutions" in this field, which has taken place rather quietly as revolutions go, but is nonetheless radical.

In the multiple scattering approach, we desperately attempt to replace a disorderly or random system by an equivalent clean or regular or smooth one. The very names of the approximations: "effective medium" theory, "coherent potential" approximation, reflect this hope. The latter is a spectacularly successful way of doing the very best job one can of this. One invents an effective medium in which the *average* scattering due to the random objects one is looking at is exactly zero, the remaining scattering then being at least incoherent and often small: it is a clever scheme and an effective one, if what one is concerned to do is to sweep the dirt in our dirty system under the rug and think of it almost entirely as a perturbed clean one. Even the basic question of how an amorphous material could develop a gap and become insulating was discussed by Edwards [1], Beeby [2], and by Mott [3], in terms of a strong multiple scattering phenomenon.

Multiple scattering is a paradigm (in the Kuhn sense) of the old attitude; *localization* and *percolation* are the corresponding paradigms of the new. These are phenomena which are *specific* to disordered and random systems; and they require a finite randomness before they become manifest. There are several other ways in which they typify the new kind of interest in disordered systems in themselves, as a problem genuinely distinct from the regular or "clean" system. One is that they require a treatment in terms of *distributions* rather than averages: one studies the probability distribution of a physical quantity at a given point in the system – connectedness, energy, amplitude, etc. – rather than the average. It has always been true that no point in a real system is the same as any other, and that experiments are done on specific samples, not on ensembles, but we have only realized recently that it is not safe to ignore that fact and treat everything in a conventional statistical way: the most probable and the average are not always the same nor do they even give sufficient information to describe the system.

Closely related is the concept of "*non-ergodicity*". In the end, the use of averages in conventional statistics is justified by the ergodic "theorem" which Boltzmann was cagey enough to express as a "hypothesis". It was assumed that systems run through all possible states as a function of time, and thus that averaging over an ensemble of systems is the same as a time average and hence gives the physical behavior of an individual system. Random systems contravene this argument at several points, but the most typical is their non-ergodic behavior: they do not, for practical purposes, move uniformly through phase space, even given infinite time.

Localization and percolation are specially direct and simple examples of this behavior, which is also characteristic of glass, spin glass, gels, and various other amorphous systems. Localization in particular starts from a Hamiltonian which is identical in form with that for Van Hove transport theory, a specially simple example often used to illustrate conventional non-equilibrium statistical mechanics and how averaging works in that case [4]. Finally, both phenomena exhibit critical points – singularities – which are not a consequence of broken symmetry, as are almost all the singularities of regular systems. It seems to be characteristic of irregular systems that they can go from ergodic to non-ergodic behavior in an apparently discontinuous fashion without any perceptible structural change.

The revolution, then, has left us studying a whole new category of phenomena and asking new questions, as proper scientific revolutions always do. (1) How do disordered systems *differ* from regular systems? (not "how may they be reduced to them?"); (2) What is the *distribution* of the objects of interest: energy states, "tunneling states", clusters, etc.? (3) What is the nature of ergodic–non-ergodic transitions? (such as mobility edges, glass transitions, critical percolation, etc.).

1.2. Problems not solutions

Briefly, what I have to say here is remarkably little about the answers to these various questions, a great deal about the questions themselves. The subject of our consideration here is a vital branch of science in the sense that its deep questions are mostly open.* Thanks to the last few years, we have come to a new realization of what our problems are; but as David Thouless' lectures will have made clear, we are not close, even in the paradigmatic example of the rather straightforward problem of localization, to a final solution as to the nature of the mobility edge and the exponents of the important physical parameters, much less to any precise quantitative solution of individual cases.

In localization we at least know there exists a mobility edge (pure mathematicians would probably not even give us that) but in the more physical question of the glass, the Fermi glass, the spin glass, we are not even sure of the existence of a sharp transition, although in each case experiment confronts us with two clearly distinguishable phases. It is not at all clear why glass itself seems to have no sharp transition, spin glasses and gels are uncertain, while the Fermi glass can hardly be other than a sharp transition at least at $T = 0$; but of course in each case the non-ergodic phase is clearly defined experimentally.

In the following lectures, I will try to clarify the nature, as I see it, of these various questions. The coverage will be extremely uneven; partly because it will reflect a very personal attention span, partly because most of the history of amorphous systems has been one of experimentalists groping in the dark, doing well phenomenologically – as they do – without serious theoretical help, but measuring those phenomena which relate to the specific interests of the technologist or the specific instrumental capability of the experimenter. As theorists it is usually our own fault if we find that the most relevant experimental measurement has not been made.

I hope very much that the lectures which follow will be out of date almost as soon as they are published. There has been a rapid escalation of the quantity and quality of work in this field, and in the new intellectual climate of understanding of the real problems, there is every reason to hope many of the problems I discuss will be soon solved, possibly by some of you.

2. Problems in glass proper

2.1. *The glass transition*

A wide variety of materials can become glasses, and the majority of those which do, exhibit a moderately sharp, characteristic kind of transition into the glassy state from the liquid.

* I remember that after a day's conversation with David Thouless discussing our respective lectures, we realized that we must have touched a dozen questions each of which remained open, and none of which were settled.

Of the elements only sulfur and selenium form glasses, at least easily. The most characteristic glasses are mixtures of covalent elements from 4th, 5th, and 6th columns: Si, Ge, Sn, Pb; P, As, Sb; O, S, Se, Te. These may contain quite large percentages of almost any ion as additions – column 1, 2, and 3 ions in SiO_2 help with glass formation possibly because they allow edge- and corner-sharing tetrahedra (see later). SiO_2 is thought of as the most characteristic glass but it in fact has a number of anomalous properties, possibly because it forms at such a high temperature. Many organic polymers and large molecules also exhibit glassy phases. Much studied glasses are that of glycerin and of sugar solution (hard candy). The canonical glasses are all characterized chemically by flexibility of bonding structure, e.g., as in S or Se, the possibility of forming chains and rings with arbitrary configurations. SiO_4 tetrahedron are known to connect by corners, edges, or faces; corner-sharing tetrahedra like to exhibit an Si–O–Si angle of about 150° rather than 180° which allows rotation of the second tetrahedron about the Si–O bond in the first. Phillips has suggested as a rule of thumb that an average number of bonds between 2 and 3 per atom tends to form a glass: 2 tends to lead to polymer chains, 3 or greater is "overconstrained" and doesn't allow enough flexibility of choice.

Metallic glasses are becoming quite common with the advent of rapid ("splat") cooling techniques. The easiest to form tend to be of the rough formula $(\text{transition metal})_{0.7-0.8}(\text{Si, P, C, B, Ge})_{0.3-0.2}$. (These second atoms are called "metalloids".) The question of the electronic structure of these glasses and the bonding is still up in the air; I believe they exhibit strongly directed bonding of the metalloid atom but possibly not of the metal. It is very noticeable that they are softer to shear than even a typical metal. It is not possible to follow out the glass transition in these because of the ease of recrystallization.

Ease of glass formation* is probably almost entirely a kinetic question. Glass is never, to my knowledge, the equilibrium phase at any temperature. Thus it must form from a supercooled liquid, which can in principle crystallize instead; and the glass itself can crystallize at elevated temperatures. Glasses are easiest to form near eutectics where the liquid has extra entropy of mixing and the solid is a mixture of crystals, so the melting point is depressed and the liquid is less supercooled at T_g (the glass transition). Glasses are also favored if the actual crystal structure is very complicated (as in all SiO_2 phases, GeS_2, S, Se, etc.) and by specificity of bonding in the crystal (if the mixed crystal is stable, the liquid gains no entropy of mixing, and kinetically, segregation – hence the slow process of

* Note 1 added in proof.

diffusion – does not have to precede crystallization, so the mixed crystal forms easily.) Some rather complete discussions exist of these kinetic questions [5].

Many other materials can be obtained in "amorphous" form by such means as evaporation on cold substrates, sputtering on the same, and other techniques of deposition or preparation in situ in such a way that crystallization does not easily occur because of restricted atomic motion. Amorphous Si, Ge, Bi, and C can be prepared in this way, as undoubtedly can many other materials – e.g., almost all transition metals. When enough H is present, a-Si apparently becomes quite stable; this may or may not be a true glass in some sense. Of course, these are all solid; liquids are normally amorphous as well. It is not clear at all what, if anything, precisely distinguishes a glass from either of the other two amorphous states, but in practice we define a glass as a rigid phase formed by cooling from the liquid, and we suspect that its stability is caused by the large size and free energy behavior of crystalline nuclei of critical size. Other amorphous phases may have locally unstable regions, i.e., types of "defects" not present in the glass. A good early review of glass science is Kauzmann [6].

Figure 2.1 shows a schematic presentation of the typical glass transition. We present it on a V–T plot which makes relationships the clearest. The supercooled liquid has a much more rapid thermal expansion than either

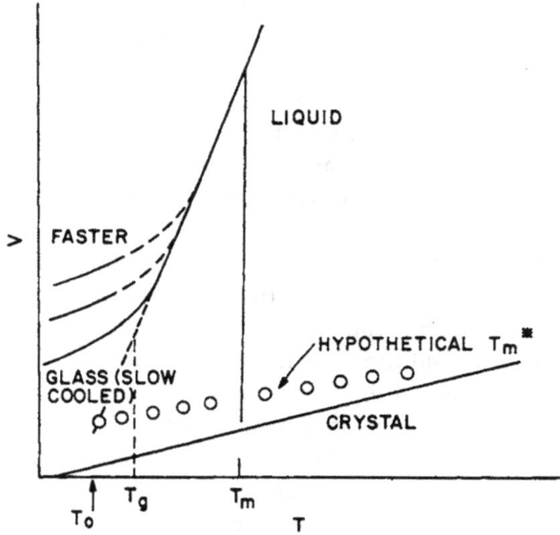

Fig. 2.1. Glass transitions.

the solid or the glass, which are very comparable, so the glass transition shows up as a rapid decrease of the thermal contraction rate. This occurs over a 10–20° range out of several hundred degrees C, i.e., of order 1–5%. The transition is time dependent; it may be delayed by 10° or more by slow cooling, as shown; and a glass annealed near T_g will contract and harden if within 10–20° of the usual T_g. All of this occurs well below the melting point at which liquid and solid are in thermal equilibrium at the same pressure.

Specific heat measured at the same point – on cooling – shows simply a smooth drop from one roughly constant value to a lower one, more like that of the solid (see fig. 2.2) [7,8]. There is no entropy release indicating a configurational change. Incidentally, after cooling and reheating through the glass transition, there often is a peak of heat absorption looking like a specific heat peak, but this is a purely hysteretic, *non*-equilibrium phenomena: we will discuss it when we come to talk about tunneling centers. This represents entropy lost while the glass is at low temperatures.

As the glass cools further, it is now rigid and behaves mechanically and thermally rather like the corresponding solid. The specific heat and thermal expansion of an insulating glass follows a typical Debye curve down to less than 1 K, where the linear specific heat which we discuss in the next

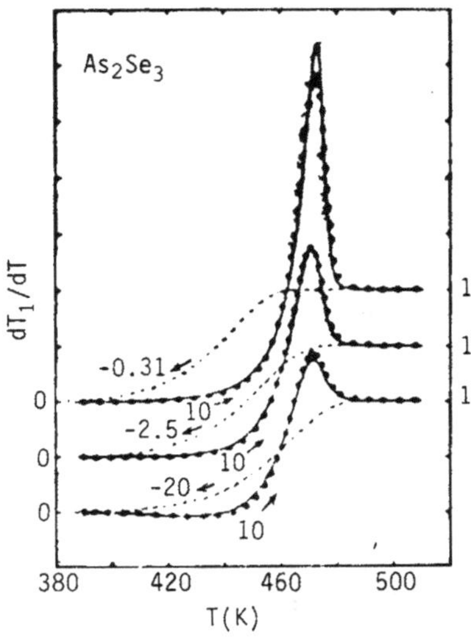

Fig. 2.2.

section shows up. It is notable, however, that the T^3 term characteristically does *not* agree with sound velocity measurements.

Simon and Giauque, both in 1926 [9,10] integrated out the specific heat curve of glass to check Nernst's "law" at absolute zero. In fact, the entropy of glass at $T = 0$ is finite and large, of order unity per molecular unit (say Si atom). Thus the entropy of melting is never made up for by the rather larger specific heat of supercooled liquid and glass – much of the configurational disorder of the liquid relative to the solid is still there. As far as can be told from these ancient measurements, there is no "ideal" glassy phase, with a unique structure; the structure is whatever the liquid finds itself trapped in. Figure 2.3 is an attempt to describe the implications of this remark with regard to the potential energy $V(x)$ as a function in configuration space. We have a large number of more or less equivalent valleys in the energy diagram, separated by mountain passes whose height is variable within a certain range, but almost all of which become essentially impassable much below T_g.

The real essence of the glass transition, and also a strong support for the interpretation of fig. 2.3, comes when we study transport properties in the neighborhood of T_g. It is universally found that the fluidity (inverse viscosity), diffusion rate, nuclear spin relaxation rate, and all other transport rates, decrease *very* rapidly but smoothly as we approach T_g in a supercooled liquid. There seems to be some universal mechanism which is

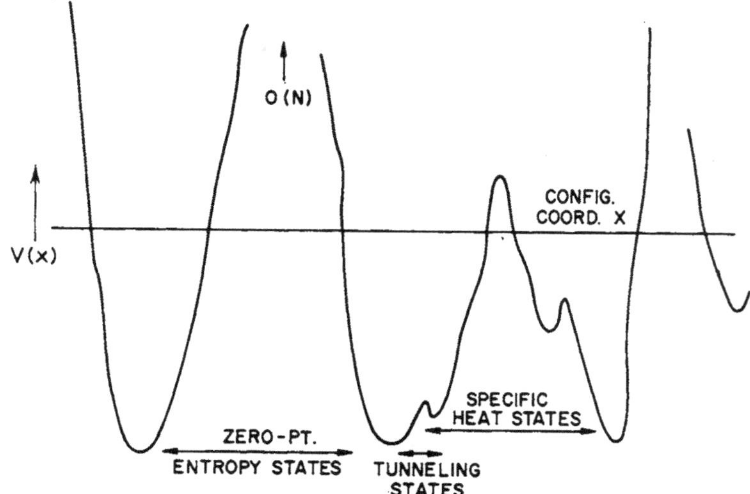

Fig. 2.3. Configuration energy as a "random mountain range".

common to all of the transport processes going on, including even the rate of thermal contraction, and which drops out smoothly at a certain temperature in every supercooled liquid. The functional form of this dependence is simple, unique, unusual, and for practical purposes universal:

$$D, \tau^{-1}, \eta^{-1} = \text{const} \times \exp\left(-\frac{A}{T - T_0}\right). \qquad (2.1.1)$$

Here T_0 is a temperature rather below the glass transition temperature T_g, by between 10 and 100° in general. The "const" in the case of η corresponds to a reasonable viscosity, so that with A of the order of typical total activation energies (0.1–1 eV), $\eta \sim 10^{14}$ poise at T_g which is a good definition of T_g. The form of (2.1.1) is of course so spectacularly steep that all physical processes are frozen out for the age of the universe before we reach T_0. T_0 is a totally inaccessible point in the liquid state. Figure 2.4 shows some examples of (2.1.1), which is almost universal to cold liquids. Data on supercooled water had been fitted to 10-parameter empirical formulas, before it was realized that the 3-parameter (2.1.1) was much better. One must allow small but insignificant temperature variations of "const" and A if one wishes to fit viscosity over ranges of several hundred degrees; but in the T_g range this is never the case.

Vogel [11] seems to have first proposed the law (2.1.1) (Fulcher [12] came a few years later). Later Doolittle [13] taking advantage of the linear thermal expansion, expressed it as

$$C \exp\left(-\frac{A^1}{V - V_0}\right) = C \exp\left(-\frac{V_0}{V_f}\right), \qquad (2.1.2)$$

where V_f is called the "free volume". V_0 is perhaps less pressure-dependent than T_0, as might be expected: with chemical bonding forces very steeply dependent on atomic distances, a change in volume has quite a bit more effect on the energy than a change in T. But I do not believe (as opposed to Cohen and Turnbull) that (2.1.2) is any more fundamental than (2.1.1) and in fact (2.1.2) does not fit the data without some temperature-dependence of V_0. This law has been studied extensively by Williams et al. [14], by Cohen and Turnbull [15], and by Barlow et al. [16].

It is quite important, as especially Cohen emphasizes, that in addition to η such measures of relaxation as D and the spin lattice relaxation process for nuclear spins follow the same law with the same exponential. Thus all essentially configurational – or topological – changes in the atomic relationships, as opposed to distortions without broken bonds, are mediated by some common mechanism which obeys Vogel's law. This situation

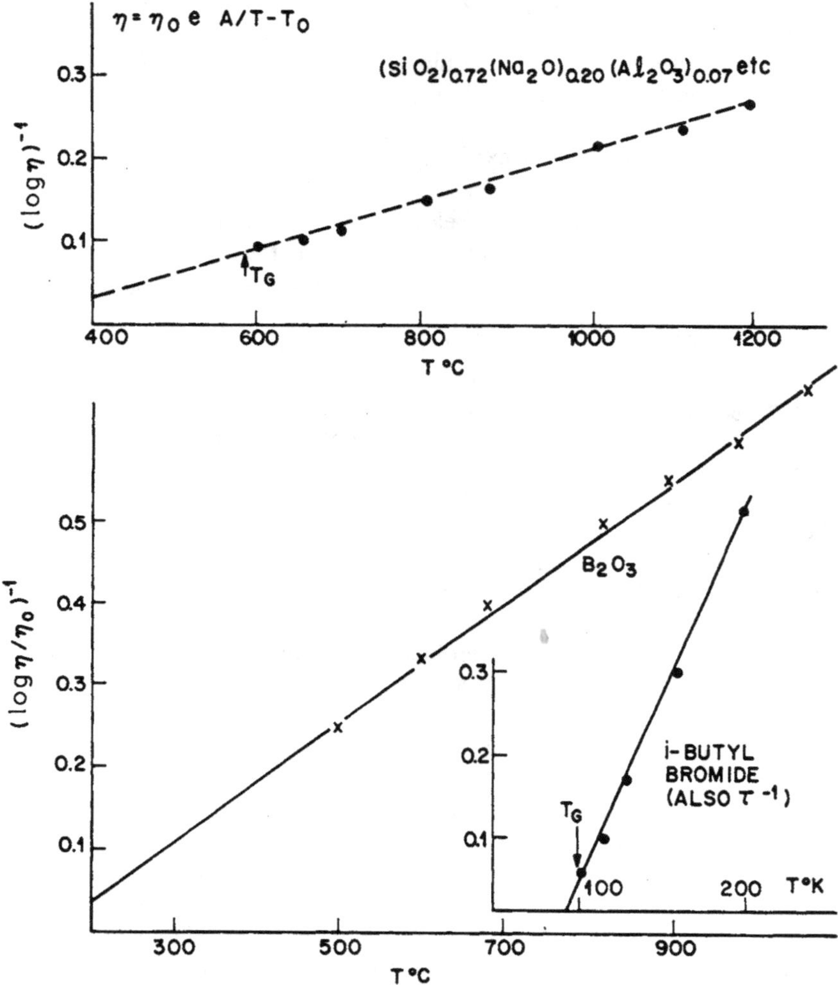

Fig. 2.4. Examples of Fulcher's (Doolittle's) law.

immediately strikes the imagination of anyone familiar with the Kosterlitz–Thouless theory and other examples of defect theories of phase transitions and of transport properties.

The functional form of Vogel's law is also striking and unusual. It is simply not possible to reasonably modify a conventional Arrhenius activation energy law $e^{-E/T}$ by allowing E to be a function of T, or by averaging over a distribution of E's, in such a way as to arrive at a Vogel expression, since the Vogel expression has an essential singularity at T_0 and the Arrhenius expression is too smooth in T. We are thus left with two difficult questions: (1) What is the significance of T_0? and (2) How can we arrive at a Vogel expression for the rates? Clearly T_g is not a singular temperature or transition point in any sense, since all that happens there is that transport processes become so slow as to be imperceptible; but by no means is the same true of T_0; T_0 is at least formally a true transition point or singularity. The problem of the nature of glass can be argued to be identical with the question of what T_0 means.

One answer to these questions is given by Cohen and Turnbull, harking back to much earlier ideas in the theory of liquids. In their theory one emphasizes the expansion (2.1.2) in terms of volume, not the pseudo-Arrhenius form (2.1.1). It is suggested that the basic difference between liquid and solid is the existence of extra volume in the liquid: more or less a picture of the liquid as hard spheres confined to cells by the surrounding configuration of atoms, but the cells are quite a bit bigger than the corresponding cells in the solid. The fractional extra volume per cell is $v_f = (V - V_0)/V_0$.

The idea then is that genuine topological transport cannot take place unless there is an essentially empty cell: a hole the size of a whole atom. The probability of the existence of such a hole is calculated straightforwardly as

$$\sim \exp(-V_0/v_F) \tag{2.1.3}$$

(the probability that n increments of free volume of the average size v_F will accumulate in a given cell is $\sim (1/e)^n$; if $n = V_0/v_F$, this is (2.1.3)).

There are a number of obvious objections to this oversimplified picture. To my mind the most severe is the perfectly linear behavior of the thermal expansion as v_F becomes quite small. One is asked to believe that all motion stops when $v_F \to 0$, but that the thermal contraction continues perfectly smoothly up to that point (even though it then ceases, indicating that all holes of all sizes are frozen in). The Van der Waals equation, for instance, would predict (using $(p + a/v^2)(V - b) = T$)

$$V - V_0 = v_F \propto T ,$$

not $\propto T - T_0$. I do not see how one can have it both ways.

A second serious problem for the free volume theory is the experimental dynamic visco-elastic behavior. (For the references here, I am indebted to J. C. Phillips.) Supercooled liquids respond to shear vibrations with a certain degree of elasticity as well as viscous flow, and with a characteristic frequency dependence not like that typical of a simple relaxation process. It appears that there is a certain delay time during which the liquid behaves like an elastic solid and not a liquid, and that relaxation, when it occurs, does not take place uniformly at all sites. The best fit to the behavior is a model called the "Extended Glarum Model" due to Phillips et al. [17], following Glarum [18]. This model (we will not work out the rather tedious mathematical details) assumes that relaxation takes place at any "site" only when a "defect" has succeeded in diffusing to it. The number of "defects" is rather low, so that the "sites" have a wide distribution of relaxation times depending on whether they are close to or far from the initial position of a "defect".

This behavior tells us that whatever the relaxing mechanism or entity is, it is not a thing like a "hole" which can appear and disappear at any given site and has no permanent identity – in the free volume theory, the "hole" is simply a statistical accumulation of extra volume, and occurs with equal probability anywhere. Instead, some specific type of defect which is quasipermanent in its identity is doing the viscoelastic relaxation, and presumably also all the other transport processes. Whatever the defects are, they are frozen in the glass and they contain some extra volume.

A last argument against the "hole" theory is that it stems from a random hard sphere model of a liquid which is hard to reconcile with the obviously covalent bonded compounds which form glass. It seems most likely that glass-forming liquids already have a fairly extensive random network of directed bonds above T_g.

It seems much more likely that the "defects" of the Glarum model and the Vogel–Doolittle equation are defects of some sort in the covalent network. I have suggested several possible scenarios for this idea, but must confess that as yet I am completely at sea for a really satisfactory theory. There are two choices of what T_0 might be, but all my scenarios depend on the idea that whatever object is responsible for transport is a defect which has properties like the 2d vortices or dislocations of the Kosterlitz–Thouless theory [19].

In the K–T theory, it is observed that the energy of an isolated point defect in a 2d system is infinite with the logarithm of the volume of the sample. The dislocation causes a strain field

$$\nabla u_i \sim 1/r$$

and

$$\text{Total Elastic Energy} \propto \int c(\nabla u)^2 r\,\mathrm{d}r$$

$$\sim c\int_{\text{core}}^{\text{boundary}} \frac{\mathrm{d}r}{r} \sim c \ln \frac{r}{a_0}\,.$$

If the dislocation encounters others of different Burgers vector, the long-range strain fields can be screened out and one gets an energy proportional only to the log of the distance between them: $E \propto \ln r_s = \ln(N)^{-1/2}$, where N is the density. Thus the energy of a number N of dislocations in unit area is of order (a_0 is a core radius cutoff)

$$E = NE_0 + Nc\ln(a_0^2 N^{-1})\,.$$

The free energy has added to this the entropy term due to the random position of the defects:

$$F(N) = NE_0 - Nc\ln N + NT\ln N\,, \tag{2.1.4}$$

and clearly when T is sufficiently high the entropy term wins and there is a finite density of dislocations *in equilibrium.* At $T = T_0 = c$ there is a melting transition above which one finds

$$N = a_0^{-2}\exp\left(-\frac{E_0}{T - T_0}\right), \tag{2.1.5}$$

while below it there are no dislocations in thermal equilibrium at all.

This is, of course, the naive version of K–T theory; in a real system there is a density of bound pairs of dislocations which partially screen the long-range energies. The effective E_0 and T_0 become temperature-dependent and the functional form of (2.1.5) can be somewhat modified.

This is one of very few mechanisms for inventing the kind of exponential in (2.1.5), in fact the only reasonable one I know. But the question is (1) "what defects" are present in the glass-forming liquid?, and (2) what are they defects in? What is the idealized state below T_0 of which the "defects" are modifications? This is the most open question.

We may suppose that at T_0, for a given p, one of two transitions might occur:

(1) This may be the actual liquid–solid transition in the sense that at constant V rather than p there may be a real liquid–solid critical line, or nearly so, at which the liquid is no longer even microscopically stable. If such a line exists, it must look on the V–T diagram very like the one pictured in the figure (2.1). At some temperature above T_m, the solid will become

absolutely unstable; at some T below T_m, the liquid also can no longer exist because the solid *at the same* V is lower in free energy. For the sake of this speculation one ascribes the first-order nature of the liquid–solid transition wholly to the volume effect, and when $F_{liq}(V, T) = F_{sol}(V, T)$, one assumes a true continuous second-order transition would take place: this is supposed to be T_0. Such a transition is envisaged in the two most popular theories of melting, the dislocation theory and the vacancy theory, when it is assumed in each case that melting of a solid occurs because of the appearance of a swarm of defects which, presumably, would occur even at constant V at sufficiently high T.

(2) A second possibility is that there really exists an "ideal" glass structure. It is presumably highly degenerate because of S_0, we assume, but once established every bond is satisfied reasonably well and flow of the structure cannot occur except by some kind of bond-breaking process. T_0 is then imagined to be the highest stable temperature of such an ideal random network, i.e., the temperature at which it first begins to have topological defects.

Given T_0, we may ask what defects may occur above it. If the underlying phase is crystalline, it is easy to describe defects (vacancies, interstitials, dislocations, disclinations), but not possible to see why the density of any of them should vary according to Vogel's law. Dislocations have a logarithmic energy per unit length, but the entropy per unit length is constant, not logarithmic. Disclinations are too massive, vacancies and interstitials not logarithmic.

If the underlying phase is a random network, even the question of what is a defect becomes obscure. We will encounter this problem again in the spin glass, but it is acute here. One ray of hope appears: in a random network, one can imagine a single broken bond (a "dangling bond") as a true topological defect. I leave it as an unsolvable exercise for the reader to guess the elastic theory of a dangling bond in a random network.

We are left finally with a satisfactory verbal, but an unsatisfactory physical, picture of the glass and the glass transition. We see the glass as a structure in which a density $n \sim \exp[-E_0/(T_g - T_0)]$ of some defect is frozen in at T_g, which defect may be a high density of dislocations in a fictional solid, or a high density of dangling bonds in an ideal random network, or a high density of holes in either. In any case, these objects can only move and annihilate with the help of each other, and when too few exist motion stops and we are left with a random but frozen state.*

* Note 2 added in proof.

2.2. "*Tunneling centers*" *and the linear specific heat*

The low-temperature properties of glass had been noticed to be a bit peculiar for many years – low-temperature physicists had noted that glass and other amorphous materials did not appear to be quite thermally stable even below 1°F, and O. K. Anderson and Bömmel [20] had long since observed that the Debye T^3 law predicted on the basis of acoustically measured elastic constants was lower than experiment. But the present state of knowledge began in 1971 with experiments on heat conductivity and specific heat by Zeller and Pohl [21]. These authors observed an apparently universal extra heat capacity in all kinds of glasses: polymers, fused quartz, GeO_2, and industrial glass, as well as a region of heat conductivity $\propto T^2$ which indicated strong resonant scattering centers.

The main term in the specific heat is of the characteristic metallic form

$$C = \gamma T ,$$

with γ somewhat less, in general, but of the order of that for metals: an effective T_F of 10–100 eV. And whatever states are causing this specific heat are also scattering phonons at a rate linear in T (since in the range of $\kappa \propto T^2$, the "phonon" specific heat $\propto T^2$ dominates, the mean free path must be $\propto T^{-1}$ since $\kappa \sim C_{sp} l v_s$).

A hypothesis to explain these observations was proposed independently by Phillips [22], and Anderson et al. [23]. We will follow the latter authors, and we will not work out all consequences in detail since that will undoubtedly be well covered by later lecturers. The fundamental physics, however, we will discuss in some detail, since this still remains open and interesting.

To see what these states might be, let us return to the kind of picture of the energy surface which we discussed previously. There are $3N$ coordinates for the multidimensional configuration space in which we must describe this surface. In such a configuration space, most points are almost infinitely distant from each other, since even a small motion of each of N atoms leads to a distance

$$S^2 = \sum_i^N (\delta r_i)^2 \propto N .$$

In this configuration space, the existence of a zero-point entropy tells us that there are a very large number,

$$M = (e^{S_0})^N \sim e^{N/2} \quad \text{or so} ,$$

of local energy minima whose energies differ by less than the order N – plus at least one, corresponding to crystalline order, which is far deeper but infinitely far from all these glass-like ones. Between these different minima there will be great distances, in general of order $\sqrt{N}$, and between such distant "valleys" we can expect that there will be correspondingly high "passes" or saddle-points of the energy surface, again of order N, i.e.

$$(\Delta E)_{\min i \to \min j} \sim N\,.$$

The whole point of the spin glass type of theory we will discuss later is that the presence of such high barriers can allow an absolute disconnection of different parts of configuration space to take place, and that this disconnection may take place suddenly at a critical temperature T_c, even though the pass- and valley-distribution of our complicated mountain range is totally random.

But in any case, at sufficiently low temperatures we can be sure that most of our minima are quite effectively cut off from each other. On the other hand, no argument that I can see requires that *all* of the alternative minima be unavailable. The suggestion of Phillips [22], and of Anderson, Halperin, and Varma [23] was that in fact there is a continuous distribution of *available* alternative minima, and that these cause the various phenomena mentioned above.

The situation is much more like a percolation theory than a localization theory. In the quantum-mechanical, finite dimensional localization theory there is no absolute disconnection and either all states must be isolated or all connected. But in percolation theory, even when percolation never occurs and all states are effectively isolated, there are still local clusters of connected states.

We hypothesize, then, that there are

$$e^{NS_1} = \text{no. of ``available'' alternative minima}$$

alternative states in the local cluster: e^{S_1} alternative configurations for each atom, which may amount to $\frac{1}{2}$ per mobile group such as an SiO_4 tetrahedron. Then the total number of minima is

$$e^{NS} = e^{NS_1}\, e^{NS_0}\,,$$

where S_0 is the rock-bottom zero-point entropy, assuming that all the e^{NS_1} states are available. A good way to think of the e^{NS_1} are that these are configurations where only a local group of atoms moves, while the e^{NS_0} require total rearrangement of a large part of the sample.

In fact, the Phillips–AHV theory requires that there be a large quantitative difference among availabilities of the e^{NS_1} states (see fig. 2.3), as well as of course a fluctuation in their minimum energies E_i. The energy difference $E_0 - E_i$ is a random variable which a priori is evenly distributed, at least on the scale of a few degrees K, and if thermally activated motion can take place, the system will occupy the lower. We ascribe the linear specific heat to the resulting constant density of available states E_i, of order 1 per few eV per atom. This will be due to those whose activation barriers are ($\sim$0.05–0.5 eV), quite low to allow motion in a finite time but not too low, because otherwise quantum-mechanical tunneling through the barrier will cause an energy splitting and a decrease of C at low temperatures.

The relatively small number of states with very low energy differences and low barriers $< \sim 0.05$ eV are responsible for the second phenomenon: the scattering centers. These are the "tunneling states": pairs of levels in which the tunneling matrix element T_i and the energy difference $E_i - E_0$ between the two minima are comparable. Such a pair of levels can give a strong resonant scattering of phonons of energy $E_i - E_0$. A striking confirmation of this process came when it was pointed out that normal ultrasonic measurements were made under conditions where these resonant tunneling states were saturated and "bottlenecked" by the high ultrasonic power levels usually used, and predicted that when lower levels were applied a very large increase in attenuation would be discovered. The success of this observation [24] conclusively disposes of any possibility that these phenomena are caused by "localized phonons" or any even approximately harmonic excitation, since it is very clear that these saturation phenomena are characteristic of a two-level system. This is probably also the cause of the specific heat discrepancy in the T^3 term, both because sound velocity is measured in the "bottlenecked" condition and because thermal phonons are very much more severely affected by the resonant scattering – which is proportional to ω/T – than are low-frequency ones. With the scattering will also appear a dispersion in velocity.

These "tunneling states" have been confirmed in very considerable experimental detail by the observation not just of "bottlenecking" and saturation but of several other non-linear phenomena associated with two-level systems such as spins: echoes, cross-saturation, etc. It is assumed that this double-minimum potential has an effective Hamiltonian

$$\begin{pmatrix} 0 & T \\ T & E_1(x) \end{pmatrix},$$

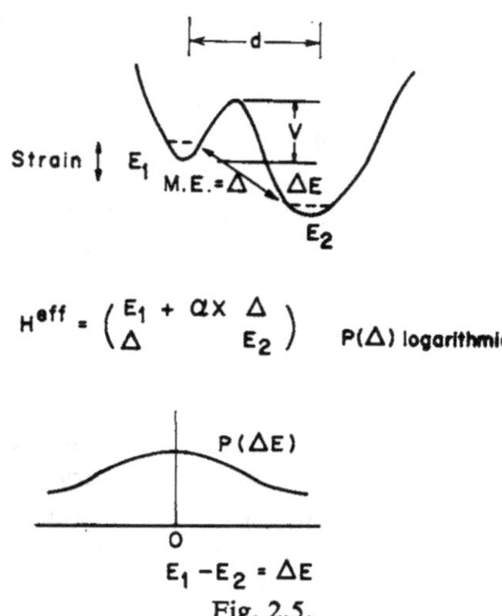

Fig. 2.5.

where T is the tunneling matrix element and $E_1(x)$ is the energy difference which is strain-dependent. The tunneling matrix element

$$T \simeq \hbar\omega_0 \, e^{-\lambda} \quad \text{with} \quad \lambda = \hbar^{-1} d\sqrt{2mV}$$

(see fig. 2.5 for definitions of d and V), and is probably relatively weakly dependent on strain compared to E_1. After diagonalization of H, the strain term can cause transitions when phonons are present, when T is not too small compared to E_1. The important "tunneling" levels are estimated to have barriers $V < 0.05$ eV, which is a relatively small fraction of all the double minima – it remains, to me, a bit of a surprise that so many of them exist in a wide variety of glasses.

I would like to go into this theory in a little more detail because there are some interesting physical questions and because there has been considerable experimental confusion and some experimental discrepancy with some of the results of these theories with regard to time-dependent specific heat.

In the first place, the reader should have been rather dubious when I allowed the system to find its way between e^{NS} alternative minima so easily. If we divide some reasonable volume of configuration space, say $(A_0)^N$, by the numbers of states e^{NS}, we have a volume per state of

$$V/\text{state} = (a_0 \, e^{-S})^N .$$

This seems very small – and yet we can show that the typical distance in configuration space between two such points is $\sqrt{N}a_0$, simply by observing that the volume of an N-dimensional sphere of radius r is

$$\frac{2}{N}\frac{\pi^{N/2}}{\Gamma(N/2)}r^N,$$

and the Γ-function decreases so strongly that r must grow as $\sqrt{N}$ to give a finite volume as $N \to \infty$.

Thus the success of the AHV–P theory is a puzzle in that is it very hard to understand why so many of the e^{NS_0} states are close to each other in configuration space. I would like to argue that it can be seen rather easily that the bonding constraints enforced by the electronic structure limit configuration space severely enough to allow the tunneling centers to occur.

The idea is that each of the alternative configurations can be thought of as the minimum energy atomic configuration belonging to a specific assignment of electrons to bonding and lone pair orbitals on the various atoms. (As we will see shortly, chemical pseudopotential theory allows us to assign electrons to local orbitals, not Bloch waves, if we like.) But the exclusion principle then tells us that we have only a finite number of degrees of freedom *per atom* since there are only 3 p orbitals per atom for us to assign the electrons to. Many of these assignments will be sterically hindered and not permitted energetically, so that we end up with e^{NS_0} alternatives. When we have a double-minimum AHV center, we may think of this as either an *atomic* barrier or an *electronic* transition with corresponding *atomic* motion. We will come to a deeper understanding of this situation in later chapters.

There has been considerable doubt thrown on the whole "double-minimum" theory by a couple of measurements of "fast-time" heat capacities which seem to disagree with the conclusions drawn from the ultrasonic measurements. It is clear that the minimum pairs with $V >$ 0.05 eV, say 0.05–0.5, will not tunnel very effectively and will communicate almost only by thermal activation. These will have a wide range of relaxation times, some of which will be in the range of time resolution of any specific heat measurement made. If we change the speed at which we measure C_{sp} by several orders of magnitude, the distribution of V's and E_1's deduced from ultrasonics suggests a fairly appreciable change in C_{sp}: that many of the levels will not communicate faster than seconds or minutes. Two such rapid measurements [25,26] did show some time dependence (contrary to strong statements of their authors) but did not agree with the numerical estimates from ultrasonic data. It is worth noting that much qualitative data exists on time-dependent specific heats in glasses in terms of the heat absorption near T_g mentioned earlier, and of "mysterious" heat leaks in low temperature apparatus; it is merely a matter of whether our estimates of the barrier height distribution are correct.

It is noteworthy that the phonon in glass turns out *not* to be a true low temperature elementary excitation or "collisionless" elementary excitation, since there are many more tunneling centers than there are phonons. The phonon can only be described as a hydrodynamic concept just as sound is in gases. In fact, even less so than in gases, since at sufficiently low excitations "phonons" become heavily damped.

I should point out for the sake of the record that this merely underlines the fact that we have no satisfactory theory of low-energy excitations in glass. One's intuitive assumption would have been that the phonon spectrum at least should consist of extended Goldstone-boson type modes with essentially Rayleigh k^4 scattering from the long-wavelength density fluctuations. I still, in fact, believe that in the absence of anharmonic centers the harmonic excitations of a random lattice should have that character, but doubts I will express later in the case of spin waves in spin glass have to be noted here also. It is remarkable that the Goldstone theorem (surely valid here, that the infinite-wavelength phonon is an eigenmode of zero frequency) does not seem to imply the corresponding branch of long-wavelength, low-energy excitations as in the true solid.

2.3. Electronic structure of glass: a non-problem?

I have discussed this problem in a preliminary way in the previous section. In this brief discussion I want only to reinforce our natural acceptance of the chemical point of view towards the structure of common insulating substances such as glass. The physicist, when confronted with a random polymer such as a protein or even a simple polymer such as polyvinyl chloride with random orientation, immediately tries to think in terms of a band structure and therefore immediately immerses himself in the deep problem of band structures of random systems. The chemist's first instinct is to think "bonds" and to see such a system as a relatively simple collection of localized bonds of various kinds. My concern here is to give the chemist a fundamental justification in terms of what I call the "*chemical pseudopotential*" theory.

Let us confine ourselves to an essentially one-electron theory – Hartree–Fock, or better a density functional or other theory which contains lowest order correlation corrections. We can then define a set of localized orbitals

ϕ_i belonging to specified units of the structure – atoms or bonds, specifically – by the equation

$$H\phi_i - \sum_{j \neq i}^{N} (\phi_j|(V - V_i|\phi_i)\phi_j = E_i\phi_i \,, \tag{2.3.1}$$

where V_i is that part of the potential energy primarily responsible for binding state i (either the potential of atom i or that of the few or two atoms it is bonded to), and the sum over i runs over all occupied states as defined by (2.3.1). It can be proven then that [27]

$$\sum_i E_i = \sum_\alpha E_\alpha \,,$$

where E_α are the energies of the occupied exact eigenfunctions of H, and E_i those of the solutions of (2.3.1). This requires that there be an energy gap, or at least an approximate one, between occupied and unoccupied states. The essence of the theory is that (2.3.1) cancels out most of the attractive potential energy of other atoms and has a bound state solution belonging to each atom or bond, and yet it is equivalent to the exact solutions of H. Thus as far as *insulating* glasses and amorphous materials are concerned, the separate occupied bond orbitals and atomic orbitals can be considered in isolation and the bonding state may be characterized locally, in spite of the complication of the actual band calculation. Only states actually in the gap, or excitations, need be treated using the full hamiltonian. This is why we essentially describe the electronic state proper as a "non-problem".

2.4. Metallic glasses

I have very little basic to say about the problem of metallic glass because I do not believe that any understanding of metallic glasses yet exists. There are a lot of data and speculations based on simply a random closest-packing model which fits well to the overall structure factor but poorly to one's instincts and particularly to EXAFS measurements by Hayes et al. [28]. EXAFS can measure the structure around the two types of atoms separately. It was found (in PtGe, I believe) that Ge is entirely surrounded by Pt neighbors but Pt by both. This suggests that Ge (the metalloid) gathers around itself a coordination polyhedron which (because of the high ratio of compositions) cannot be less than 4 and is probably 6–8 (something like a cube). This coordination group may be fairly rigid just because of the size of the metal atoms, and the sharing of metal atoms between coordination groups might then lead to a rigid structure; conversely, it could be that the metals prefer a certain angular arrangement of their coordinated metalloids.

What I can say is that various electronic "theories" purporting to explain the metallic glass do not as yet lend conviction.*

3. Electron motion in random systems; electronic centers and interactions

3.1. The Fermi glass

Landau's Fermi liquid theory – which some might say is not Landau's and not a theory, but it is clear that he phrased the idea far better than any predecessor and indicated very clearly what the eventual theory would be – is the absolutely essential basis for an understanding of the behavior of systems of fermions in free space or in regular lattices. Its essential content is not only to tell us that we may map the properties of the real, interacting Fermi particles on to those of a non-interacting gas, but also to show us in a precise way the limitations on this idea and what we may do to describe the effects of interactions. To use the phrase "Fermi glass' is going too far in claiming that we understand the properties of the interacting localized system or disordered system in terms of the as yet not totally understood properties of the non-interacting one. It is, however, a phrase worth using because it at least expresses what we must do – to imitate Landau [29], if that be possible – and second expresses what really is probably true, that the same exclusion principle restrictions which are responsible for the success of the Fermi liquid are of great help with the Fermi glass. But we must keep clearly in mind that there are very great differences, and these become especially acute when the appropriate one-electron theory has the mobility edge and the Fermi surface close to each other – of course, just the most interesting case.

We may, before plunging into the rather hairy mathematics, first enumerate the systems which exhibit localization of electrons and are, therefore, in some sense Fermi glasses, and discuss a little whether the term has any more resonances in the sense of whether there are thermal transitions from Fermi glass to Fermi (or classical) liquid.

All (or almost all) insulators are really Fermi glasses, in the sense that the one-electron theory would not really give an energy gap and hence one must rely on localization. Since the minimum metallic conductivity in three dimensions is e^2/ha where a is the inter-center spacing, which thus

* Note 3 added in proof.

enters only linearly, even for a purity of 10^{-9} this is still the relatively gigantic value of 1 $(\text{ohm-cm})^{-1}$. But in fact the systems one thinks of as Fermi glass are those in which the possibility of metallic conduction is at least thinkable and in which one wishes to study the metal–insulator transition. The classical one is the impurity band (see figs. 3.1 and 3.2). This has been exhaustively discussed and well summarized for instance in the book by Mott and Davis [30], so I need not detail it here. In all cases there appears to be no discontinuous transition as a function of n or T, so that it is an experimental fact that the Fermi glass and the Mott glass – as one might call the disordered magnetic insulator state corresponding to the Mott insulator – are identical physically – only an "Anderson transition" exists, and it is even more continuous than the glass transition. A number of other less well-studied impurity systems exist and were mentioned by Thouless. None violate the minimum metallic conductivity. The second now classical system is the MOSFET inversion layer [31,32], which David Thouless will discuss. This is a single layer of electrons or holes in the Schottky well on an Si or Ge surface. Its density is easily controlled, the

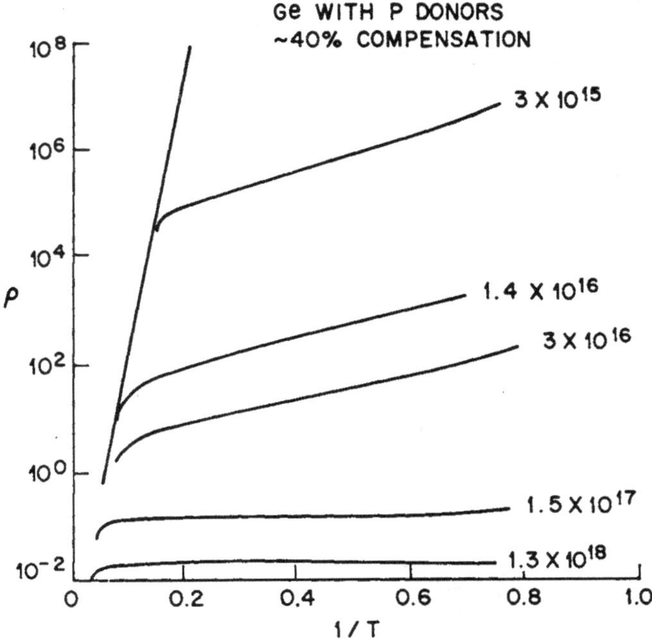

Fig. 3.1. Impurity conduction.

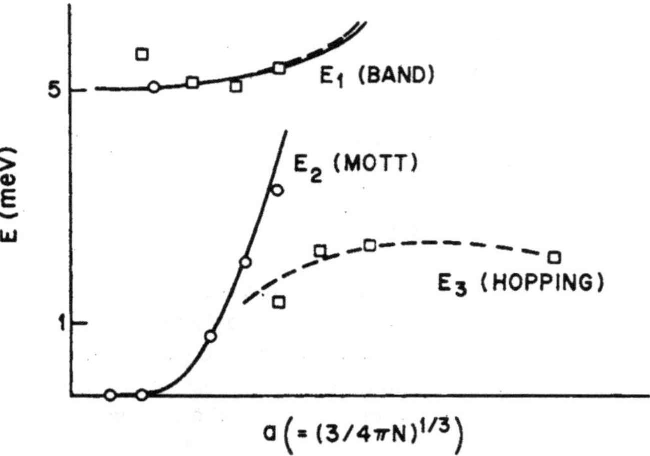

Fig. 3.2. Activation energies for impurity conductivity.

random potentials due to charges in the "0" layer less so, and it clearly exhibits an insulating, $\exp(T)^{-1/3}$ resistivity at small n, and again, a continuous transition from activated to metallic conductivity with confusing variations with the disorder sources; nonetheless, as in the previous case it clearly exhibits the minimum metallic conductivity in that the conductivity is never *below* that value at the transition.

Several indications suggest that the state is indeed glass-like in at least some instances, in that the "mobility edge" varies with concentration (as deduced from the activation energy for conduction, see fig. 3.3); and in that the "minimum metallic conductivity" is often too high, implying some kind of freezing into a localized state or "Mott glass" transition; and in that the Hall coefficient $R_H = 1/ne$ does not become activated at the conductivity transition. Adkins will discuss this situation in his seminar later.

A third canonical system is the thin metallic film with greater or less "islanding" leading to tunneling between metallic blobs (as is verified in the

loc ext

E_F E_C

$$\sigma \propto e^{-\left(\frac{E_C - E_F}{kT}\right)}$$

Fig. 3.3.

Dynes–Garno–Rowell [33] experiments by the observation of the bulk superconductivity T_C as an *increase* in resistivity due to gap formation). The observation of a conductivity transition dates back to Swann [34] in 1914 (see fig. 3.4) and it occurs in a wide variety of metals, and, with a great reliability, always at the minimum 2d conductance of 30,000 ohm^{-1} ($\pm 50\%$ or so). These measurements are particularly instructive on the question of percolation versus localization (or Mott–Fermi glass formation, at least), since these island systems perfectly exemplify the standard Cohen–Jortner [35] picture of critical percolation in a system of mixed metallic and insulating regions: in this picture one simply reduces the conductivity continuously to zero. As I remarked, the metallic regions in these films are sufficiently so to have bulk superconducting T_c's. Nonetheless, the number at which they go activated is

$$\sigma = \frac{e^2}{h} \pm 50\% \quad \text{or so} .$$

There seems to me to be no way to introduce the Mott number e^2/h (or in 3d, e^2/ha) into the critical percolation picture. Another way to express this point is to realize that conductivity at $T = 0$ is itself a rather mysterious process, which is necessarily of quantum nature: it contains h in its dimensional factor. There can be no "classical" theory of the metal–insulator transition since the metal itself is very basically a quantum phenomenon. For renormalization buffs, a way to say the same words is

Time of deposit (min)	Temperature (°C)					
	99.8	13.4	−180	13.4	99.8	14.8
50	Not measured	Not measured				
80	Not measured	Not measured	1200×10^6	281×10^6	224×10^6	286×10^6
110	24460	24390	28400	24400	24520	24465
130	3520	3499	3483	3500	3526	3506
160	2422	2407	2394	2408	2427	2414
190	1158	1147	1129	1145	1158	1150
220	706.6	699.7	687.3	698.5	706.4	700.5
250	620.3	613.8	601.4	613.8	620.4	614.4
460	272.4	268.5	262.1	268.7	272.2	268.8
960	148.1	147.2	142.4	146.5	146.3	142.8

Fig. 3.4. Electrical resistance of thin metallic films, 1914 data from Swann [34].

that for $T = 0$ conductivity, percolation crosses over to localization and not vice-versa.

There is a number of other similar systems of finely-divided metallic particles: cermets, smokes, metals co-deposited with rare gases. They behave more or less similarly to the films. An interesting and as yet unexplained observation is that the activated side often shows the Efros–Shlovskii [36] $\ln \sigma \propto T^{-1/2}$ law, not $T^{-1/d}$, which we will discuss later.

One seems never to see the actual "Anderson transition" in either amorphous semiconductors or amorphous semiconducting "chalcogenide" glasses. I will argue later that these are cases where the properties are determined by rather strong interactions – perhaps these would be the *best* characterized as "Fermi glasses" except that the electronic motions are also deeply involved in their nature as true, or "glassy" glasses, as Joffrin described them.

3.2. Repulsive interactions and magnetism

3.2.1. Many-body theory: quasiparticle states, anomalous diagrams and polarizations

I like to think of the "Fermi glass" proper as the situation in MOSFETs or impurity bands of weak electron–phonon interactions, where the electron–electron Coulomb interactions probably play the dominant role. This case is best approached from the ideal case in which we assume that we start from a very well localized picture – i.e. a situation in which the original "locator" perturbation theory for the one-electron theory is a very good approximation. The problem in this case has begun to be analyzed by Fleishman [37]. Even though the localized states are usually magnetic, it is instructive to first do as he does, ignore spin – which is a very tricky thing to do indeed, and will only be justified much later – and assume that the exclusion principle holds for all electrons. We also start at $T = 0$ so there are no real excitations present.

We now turn on interactions. These are of two kinds, electron–electron and electron–phonon, for each of which the story is slightly different. Let us do electron–electron first. Following Fleishman we make the argument from self-consistency. If the well-localized states are α, we have the basic self-energy diagrams as presented in fig. 3.5 (+ etc. of course. But as for the Landau theory the analysis of these suffices because these are the most troublesome).

First problem: we note that α'' need not equal α because there is no momentum conservation law. In other words, the interactions can, in

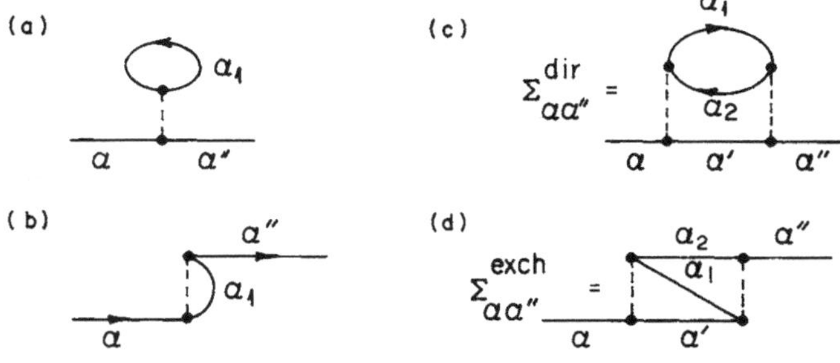

Fig. 3.5.

principle, change the localized states and the localization edge, not to mention that the Fermi level and hence Σ itself can change as the self-energies change. Let us make an assumption which we will then justify a posteriori: that for states at the final Fermi level, all contributions to the self-energy are real. This is already true in the Fermi liquid; it is even more true, we will find, in the Fermi glass – the point being that an excitation at the Fermi level cannot create real – as opposed to virtual – pair excitations $\bar{\alpha}_1\alpha_2$. If this is the case, we may write

$$G^0_{\alpha\alpha'}(E) = [(E - H_0 - \Sigma)]^{-1}_{\alpha\alpha'} \tag{3.2.1}$$

(which is Dyson's equation, or alternatively the definition of $\sum_{\alpha\alpha'}$ = sum of all diagrams which cannot be cut at a single line). But now since $\sum_{\alpha\alpha'}$ is real at $E = E_F = \mu$ the Fermi energy, a fortiori $E - H_0 - \Sigma$ is hermitian and one can diagonalize it. The result is a new set of localized states β at the Fermi level (in fact, when we come to introduce spin and phonons we will find that they are *more* localized):

$$(E - H_0 - \Sigma)_{\beta\beta'} = (G_0(\mu))_{\beta\beta'} = \delta_{\beta\beta'}/(\mu - E_\beta)\,. \tag{3.2.2}$$

The eigenstates ϕ_β are the appropriate quasiparticle states. E_β is not, except for states very close to μ, an actual pole of G; presumably the actual pole corresponding to E_β has a width and is complex, as we will shortly discuss, and the real eigenstates of H are complicated multiparticle excitations as in the Fermi liquid case.

(3.2.1)–(3.2.2) are the appropriate generalization to the localized case of the ideas of Kohn and Luttinger [38] (see also Balian and De Dominicis [39] and Bloch [40]) on "anomalous diagrams" in the Fermi liquid in a periodic solid, and those of Langer and Ambegaokar [41] on the single impurity. In both of those cases also, the one-electron states of H_0 are not

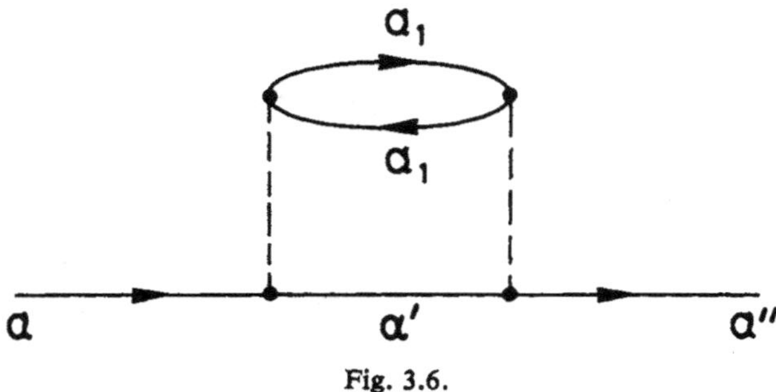

Fig. 3.6.

a correct basis for the interacting system (as also in the nucleus [42]) and we must redefine the basis and occupy different states, but then most of Fermi liquid theory remains. Incidentally, the occupation of the quasi-particle levels in (3.2.1) must be handled with great care including the "anomalous diagrams" of Kohn and Luttinger. (These are diagrams like fig. 3.6, which seem to vanish at $T = 0$ because of the statistical factors $f^+_{\alpha_1}, f^-_{\alpha_1}$, but involve vanishing energy denominators. If they are handled properly at finite T their inclusion has the effect of modifying the statistical factor by the self-energy.) Correspondingly, one must recognize that in (3.2.2), $\Sigma_{\beta\beta'}$ is a functional of the final result itself via the statistical factors.

(3.2.2) seems to prove a very important theorem: that there is no gap at the Fermi level due to the existence of the localization edge, at least until we get to the strong interaction case of sect. 3.3. Such a gap was, probably incorrectly, proposed by Efros and Shlovsky [36] for the Coulomb case. The basis for such a gap is the existence for localized states of a finite contribution like fig. 3.7, where the localized state statically self-interacts by polarizing the surrounding medium. These are negligible for extended states because the vertices are $O(1/\sqrt{N})$. In the phonon case they represent Franck–Condon displacements. The argument is that the particle, in hopping out into an unoccupied state β', leaves behind its polarization so there is a gap in G_1. Another expression is to suggest that the localized electron is always a self-trapped polaron.

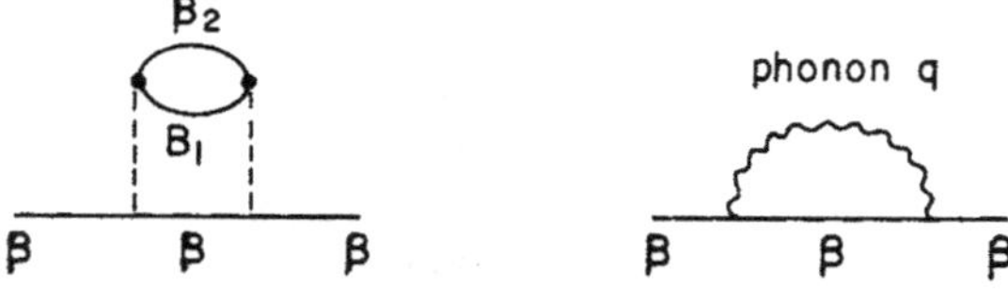

Fig. 3.7.

(3.2.2) tells us that the quasiparticle states including polarization effects are always the solution of some random potential problem which, at least for weak interactions, has a continuous density of states with (or without) a non-singular mobility edge in it. The localized quasiparticle states above E_F are just like those below and join continuously to them.

3.2.2. Imaginary parts and variable range hopping: localons

To keep to logical order, more or less, let us now justify a posteriori the reality of Σ at $E = \mu$, assuming that μ is in the localized region of E_β's. In fact, the statement can be made more strongly than in the Fermi liquid case: $\mathrm{Im}\,\Sigma(E)$ vanishes exponentially with T and with $|E - \mu|$.

First let us recapitulate Mott's famous $T^{-1/4}$ argument for the phonon case [43]. Let us consider first at $T = 0$ a state β, with $E_\beta - \mu$ position but very small, and ask for its probability for decaying into a state β' with emission of a phonon with

$$\hbar\omega_q = E_\beta - E_{\beta'} . \tag{3.2.3}$$

To find a state β' we must go out to a radius R such that

$$\tfrac{4}{3}\pi R^3 n(E)(E_\beta - E_{\beta'}) \simeq 1 \tag{3.2.4}$$

(neglecting the possibility of a nearby coincidence of E's, which the system avoids by the well-known argument and which in any case only gives us a strongly coupled local pair of states or "localon").

The matrix element for emission of a phonon surely depends among other things on the wave-function overlap

$$S_{\beta\beta'} = \int \phi_\beta \phi_{\beta'} V_{\mathrm{ph}}(r)\, \mathrm{d}^3 r \propto \exp[-\alpha(R_\beta - R_{\beta'})] , \tag{3.2.5}$$

where α is the inverse localization length, so that the rate of all processes $\beta \to \beta'$ goes like

$$\mathrm{Im}\,\Sigma_\beta = \mathrm{const} \times \int_0^{E_\beta} n_{\mathrm{ph}}(E_\beta - E_{\beta'}) \times \exp\left[-2\alpha\left(\frac{3}{4\pi(E_\beta - E_{\beta'})n(E)}\right)^{1/3}\right] \mathrm{d}E_{\beta'} ,$$

which is majorized by

$$\mathrm{Im}\,\Sigma_\beta < \mathrm{const} \times n_{\mathrm{ph}}(E_\beta - \mu) \exp\left[-2\alpha\left(\frac{3}{4\pi(E_\beta - \mu)}\right)^{1/3}\right] . \tag{3.2.6}$$

We see that the strongest term is the exponential which represents the famous "variable-range hopping". But in fact there are also energy-dependent terms like n_{ph} which alone make the dependence of $\text{Im}\,\Sigma_\beta$ on E_β at least as strong as in the usual Fermi liquid: the exponential term due to localization is extra.

Mott converts the energy-dependence of (3.2.6) into a temperature dependence by considering a finite temperature and optimizing the dependence (3.2.6) with respect to E_β – thus finding the largest $\text{Im}\,\Sigma$ for a given T. (All of these considerations, incidentally, require the condition

$$\alpha^3/n(E) \gg kT\,, \tag{3.2.7}$$

which is essentially that the width of the band of electronic energy levels be $\gg kT$.) In this case we multiply the rate by

$$n\left(\frac{E_\beta}{kT}\right)\left(1 - n\frac{E_{\beta'}}{kT}\right)n_{\text{ph}}\left(\frac{E_\beta - E_{\beta'}}{kT}\right),$$

at least one of which will fall off exponentially with E/kT when $E/kT \gg 1$. This leads to the optimization of

$$\exp\left(-\frac{E_\beta}{kT}\right)\exp\left(-\frac{A}{(E_\beta)^{1/3}}\right),$$

which gives the famous $T^{-1/4}$ law. Thus $\exp(-AT^{-1/4})$ is not only a limit on σ but also on $\text{Im}\,\Sigma$. But again we recognize that this is a far more rapid decrease to zero than in Fermi liquid theory (as was missed by Freedman and Hertz [44]).

Now let us consider the imaginary parts due to self-energy diagrams involving electron–hole pairs in the localized states. We will call these "localons"; decay with emission into the localon degrees of freedom was the problem specifically done by Fleishman et al. [45]. We have here, in some ways, precisely the same problem as before. Where the decay diagram looks like

(3.2.8)

in the phonon case, if we define a "localon" excitation

$$-0\text{-}0- \;=\; \cdots \bigcirc \cdots + \cdots \bigcirc \cdots \bigcirc \cdots + \cdots \tag{3.2.9}$$

i.e., a bubble sum in the localized states, it plays the same role as the phonon in (3.2.8). The only trouble is that we do not know whether the localon excitations themselves are localized or extended per se. They interact via a $1/R^3$ interaction. Each hole state β_1 has only a finite number of electron states β with which it overlaps, and these pairs are randomly arranged in frequency space and interact with each other at long range via their electric dipole–dipole matrix elements

$$V_{\beta\to\beta',\beta''\to\beta'''} = \mu_{\beta\beta'}\frac{1}{R^3}\left(1 - \frac{3\boldsymbol{RR}}{R^2}\right)\mu_{\beta''\beta'''}, \tag{3.2.10}$$

where

$$\mu_{\beta\to\beta'} = \int \phi_\beta(r)\boldsymbol{r}\phi_{\beta'}(r)\,\mathrm{d}^3r \tag{3.2.11}$$

(longer-range terms cancelling out by charge neutrality). In two dimensions the localon excitations will normally be localized in absence of phonons, but in three dimensions they are the unfortunate borderline case which is neither localized nor diffusing, but diffuses away by a peculiar variable-range law, according to my 1958 paper.

Experimental evidence on localon transport of course exists; they are responsible for the temperature-independent "$\omega^{0.8}$" ac (Pollak–Geballe [46]) conductivity observed in many localized systems, especially the impurity band. The conventional explanation of this term relies on the phonon-assisted hopping, essentially precisely the kind of calculation which gives the $T^{-1/4}$ law (and "$\omega^{0.8}$" which is really $\omega(\ln\omega)^4$, incidentally, where the $\ln\omega$ is essentially the radius of the localons and variable-range hopping is assumed). In diagrams one may write the standard calculation as that of fig. 3.8, i.e., of scattering of localon into phonon. It is, to me, by no means clear that the pure localon diagram (3.2.9) does not give the same or similar result. It is certainly amusing that the localon transport is temperature independent and thus could be a pure quantum transport process.

As for phonons only more so, Fleishman's calculation, even assuming extended localons, gives a very rapidly vanishing $\operatorname{Im}\Sigma$, of the form $T^2\exp(A/T)^{1/4}$ in terms of temperature. It is impossible to estimate whether the coefficient is reasonable but we rather suspect not, and that phonons are more efficient as they are in resistivity.

At high temperature, hopping always takes place to one of the nearest

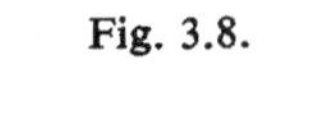

Fig. 3.8.

sites and one gets a more rapidly varying, approximately exponential, law from either phonons or localons, the former as calculated by Miller and Abrahams [47].

Having now explained the close connection of variable-range hopping calculations with conductivity, let us return briefly to the topic of gaps and edges due to interactions. First let us discuss the Coulomb case. The argument of Efros and Shlovskii was that each electron in a localized system is surrounded by a polarization cloud which essentially cancels its charge, and that in hopping away from the cloud to a distance R it must do so against a Coulomb energy barrier e^2/kR. By manipulations which are not clear to me, this leaves them maximizing

$$\exp\left(-2\alpha R - \frac{2}{\kappa RkT}\right),$$

which gives one a $T^{-1/2}$ law. As far as I can see, the sign is wrong – the electron has a bigger barrier to a more distant atom, not vice versa. This is, of course, in addition to the argument that it will probably not be difficult for the polarization cloud to move along with the particle, which is equivalent to saying that the self-energy has no thermal gap at E_F according to (3.2.2).

A more plausible (but still unproven) suggestion for $T^{-1/2}$ is the observation that it tends to occur in percolation-type systems in which localization occurs at barriers along percolation paths. Near p_c the percolation paths behave in an essentially one-dimensional way: they have a small number of branches per unit length and they visit a number of sites which does not grow very rapidly with pathlength

$$N \sim L^{1+\varepsilon(p-p_c)},$$

where $\varepsilon(p - p_c \rightarrow 0)$ is at least small. If we then assume localization exponential along these paths, we obtain $T^{-1/(2+\varepsilon)}$ as our law. This argument is not vulnerable to the objection of B. I. Halperin (private communication) that 1d transport proper does not obey $T^{-1/2}$, since above p_c the paths do branch considerably.

A case in which a kind of true gap is plausible is that of essentially self-trapped small polarons – i.e., the diagrams corresponding to static polarization are very large and lead to self-energies large compared to phonon energies (as is not possible according to Migdal's Theorem for the conventional metal). Then there would still be no thermal gap at the Fermi level but in order to hop into a neighboring localized self-trapped polaron state many phonon quanta would be needed. The electron then effectively acquires a large mass for tunneling and α becomes very large in effect. Such effects have been discussed by Ma [48] and may be responsible for some of the tunneling observations of J. J. Hauser on amorphous semiconductor tunneling barriers. In addition, one may, in this case, be totally unable to see the states optically. Mott [49] points out that it is possible also to have an energy barrier to self-trapping so that free electrons can have clean extended states and a sharp-edged gap optically, with longish lifetimes, but eventually drop into self-trapped localized states. Some such phenomenon may be occurring in quartz, especially for holes.

I have proposed [50] that it may be possible to have a true gap at the mobility edge due to self-trapping. I am not satisfied with the specific argument I gave but do not yet see why, because of the existence of the Franck–Condon self-trapping type of diagram, one cannot imagine Σ_β being a sharp function of $E_\beta - E_c$ so that some ranges of values could be excluded. There is no clear experimental evidence for this, although I am not sure about electrons in quartz, which have suspiciously high mobilities – i.e., there appears to be no localized or low-mobility tail of electron states. This effect is quite distinct from the non-existent gap at E_F or from the two-electron effects of the next section.

One final point about Coulomb interactions should be added here. They are seldom small: they more dilute the system, the bigger they are relative to hopping, since e^2/R falls only linearly with scale. When $E_c - E_F$ is not small, excitations above the mobility edge may easily have lost their simple single quasiparticle character, in that the motion of one electron changes the self-energies of all others by non-negligible amounts and can cause other localized levels to cross, etc. The true mobility edge as observed thermally can easily be a finite excitation energy for a state which either has quite lost its quasiparticle character or is a mixture of very many localons with a single quasiparticle. It is even possible that both hole and particle excitations share the same mobility edge. I think it a miracle that in so many cases activation to the mobility edge leads to conductivity with the minimum metallic mobility as the coefficient of the exponential, and do not see as any serious problem such questions as the failure to see the large Hall effect predicted on the purely one-electron picture [51]. We must await a proper interacting calculation of the Hall effect before drawing any strong conclusions.

3.2.3. Inclusion of spin

Almost all of the classical texts and discussions of localization almost ignore interactions, or even if they do not, they leave out the one degree of freedom which is clearly present in those very classical measurements on which the whole theory is based: spin. All of the impurity bands and many of the amorphous semiconductor samples showing classic variable-range hopping behavior are also Curie's law paramagnets. We have, in fact, no evidence that any Anderson transition takes place other than in paramagnetic systems, although both MOSFETs and islanded films are not easily measured.

My own feeling is that the classical Fermi glass behavior I have been describing relies almost always on the Coulomb self-interaction U between pairs of opposite spin particles in the same state being very strong, so that there is an effective exclusion principle which keeps two electrons out of the same state. If this is the case our "spinless" approximation is valid – we merely allow one rather than two electrons per state. But this has a large effect on localization, since it excludes half the sites from visitation by any given electron: Mott comes to the assistance of Anderson with a vengeance, and the spinny electron system is much more easily localized than the spinless one. In fact, for a half-filled band we have localization without disorder. Thus every Anderson transition is also a Mott, but not quite vice-versa.

There are then a large number of spin degrees of freedom, but the scale of interactions within them is assumed small, of order V^2/U, and is normally small compared to kT, much less to a typical $E_\beta - E_F$ in variable range hopping, or to the mobility edge. The possibility of spin-fluctuation assisted hopping exists but is small in such a regime. The interactions are antiferromagnetic and probably would give one a spin glass if the relevant range of T and H were accessible, which it appears not to be. Measurements of Romestain et al. [52] on CdS show the expected antiferromagnetic, small, Weiss θ. Mott has discussed the case of a mixture of singly and doubly occupied states, which I suppose may exist in the film case at least but I suspect not elsewhere. There is pretty good evidence of two Hubbard bands in Si donors, a "donor" D, and a "donor –" D^- band, in the latter of which the conductivity transition occurs for very low occupation.

As one approaches the mobility edge with E_F the question of whether U falls more or less rapidly than $E - E_c$ is germane. Unfortunately, by no means enough is known about the localized functions in this range. The experiments say that U stays up: the localized system stays paramagnetic, and, interestingly, spin diffusion starts up before conduction. On the extended side it is noteworthy that only weak spin fluctuation phenomena occur – measurements of Sasaki [53] suggesting a Kondo effect are not confirmed as yet elsewhere. The susceptibility in this range looks typical of a spin-fluctuation Fermi gas with a not particularly spectacular renormalization of χ until localization occurs, suggesting that almost localized resonances are not very predominant – again, complete absence of Cohen–Jortner [35] effects.

3.3. Attractive centers

3.3.1. Motivation and overall picture of electronic structure

The idea of attractive centers is by now so well-known that I really should not have to discuss generalities too deeply. Nonetheless I would like to tell you a bit of the history and motivation for the idea. The idea started from the realization that with regard to a great many insulating and semiconducting disordered materials, the standard picture built up until now and summarized in the Mott–Davis book just would not explain them satisfactorily. These were also those with the most characteristic disorderly properties: glasses, especially chalcogenide glasses, and amorphous semiconducting elements as then prepared.

They exhibited two sets of incompatible properties. The electrical properties indicated strongly that there was a high density of states in the gap, in that the Fermi level seemed to be pinned to the center of the gap very strongly, as one might expect from the simple Cohen–Fritzsche–Ovshinsky [54] model of overlapping bands (fig. 3.9). This was implied by insensitivity to doping, and by non-existence of any motion of the Fermi level due to strong electric fields or temperature – the resistivity is normally an excellent exponential function of temperature, becoming very high at low temperatures [30]. Later it was found that luminescence at half the gap

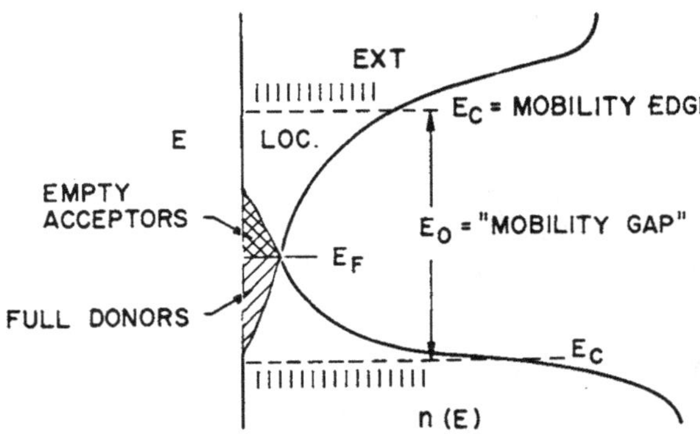

Fig. 3.9. Mott–CFO model for glasses.

frequency was relatively easily excited, such as could come from states near the Fermi level [55,56].

Another set of measurements showed clearly that there were no such states. As we have seen, a Fermi level in the band of localized states leads not just to Pauli but to Curie–Weiss paramagnetism, which is not there at all in the glasses [57,58]. At the time it was thought that even in a-Si and a-Ge paramagnetism was almost unrelated to the electrical properties; even now only in the "pure" material is a relation established, so what happens in all the others? There is a clear, well-defined one-electron band gap, optically or according to various spectroscopic techniques such as UPS. More disturbing, this band gap coincided within experimental error with the electrical gap (no polaron effects) and with the photoelectric one (no localized tails to speak of, see fig. 3.10) [30,59]. The only absorption in the gap seemed to come from the famous – and still mysterious – Urbach tails of form

$$\exp(-n\hbar\omega)/k(T + T_0)\,,$$

where neither n or T_0 are understood, but these tails are not at all special to this case and appear to be essentially a phonon effect.

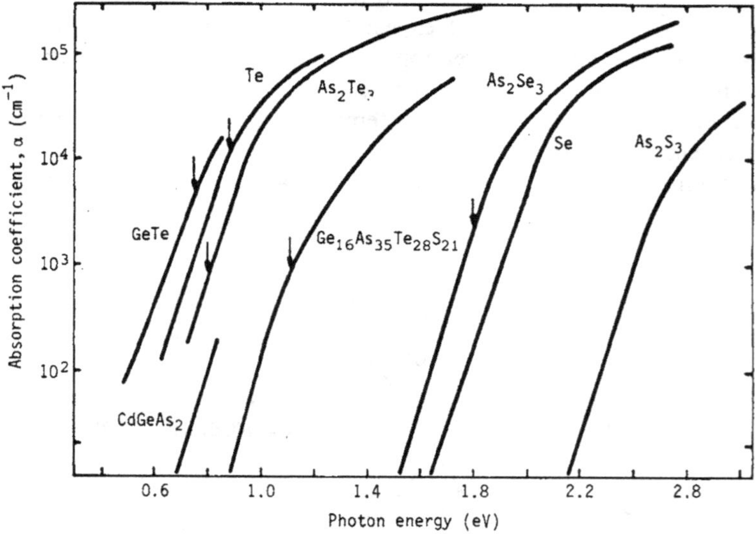

Fig. 3.10a. Exponential absorption edges in amorphous semiconductors at room temperature. The arrows mark the value of $2E$ for those materials in which the electrical conductivity has been observed to obey the relation $\sigma = C\exp(-E/kT)$ (from Mott and Davis [30]).

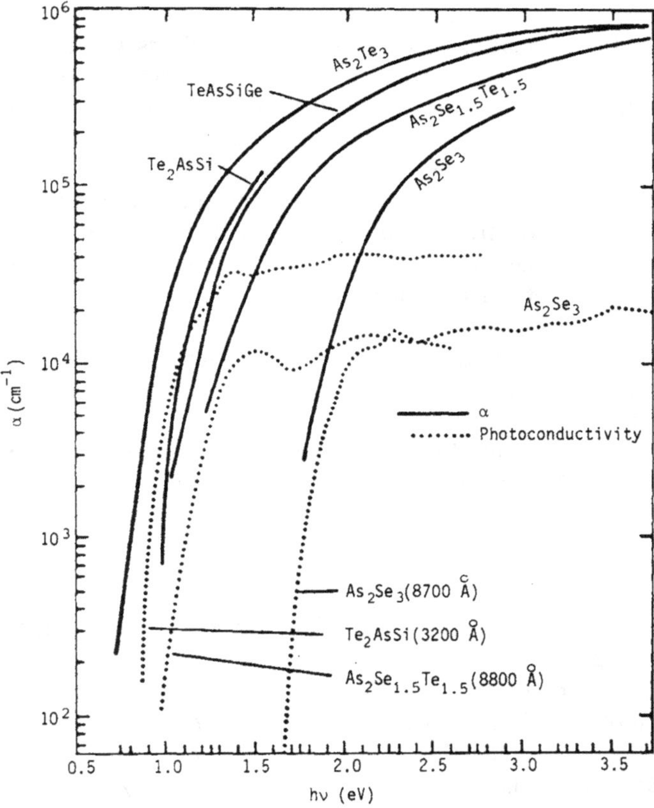

Fig. 3.10b. Plots of photoconductivity and optical absorption against photon energy in a few chalcogenide films at room temperature (from Rockstad [59]).

The solution to this dilemma was to propose that in these materials – and perhaps in general – the sign of the Mott–Hubbard U (so-called, actually due to Van Vleck) is negative rather than positive – i.e., the interaction of two electrons of opposite spin on the same site is attractive. There is an amusing parallel to the role of attractive interactions in superconductivity, where the Fermi level is determined by the pair excitations, not the single quasiparticles. In the attractive case one is rescued from the dilemma posed by eq. (3.2.2), the Fermi level and mobility edge can become the site of very peculiar many-electron effects, and it is easy to reconcile all the observations.

It is not peculiar for opposite spin electron attraction to take place – in fact it is the normal case, in spite of our instincts based on the Mott picture of metal–insulator transitions. Most materials we encounter contain mostly covalent bonds (in which electrons are paired) or closed-shell ions (ditto). Even most of the metals are superconducting, indicating a preference for attraction over repulsion. It is even true that probably more metal–insulator transitions are due to pairing and covalent bonding – like VO_2 – that are Mott transitions like V_2O_3 [60]. The model I set up for the semiconducting glasses [61,62], and in fact for at least some aspects of a much wider variety of materials including a-Si and real "glassy" glasses, assumes that many or most of the states are of this attractive character. I will first set up the model in a rather abstract sense, showing how it works mathematically, and then go more deeply into the chemistry and physics behind it.

The model envisages that all of the states of the electronic system of interest are to be identified with a set of localized sites with intrinsic one-electron energies E_i:

$$H_{00} = \sum_{i\sigma} E_i n_{i\sigma} . \tag{3.3.1}$$

Even from a one-electron point of view, and even in a liquid, as we have already emphasized there will be a tendency to form a covalent net in which those sites which are occupied by electrons are lowered in energy by atomic motions, those which are unoccupied are raised. This predisposes us to call them "bonds" though they may also be on certain *ions* or be *lone pair orbitals* on atoms from the V or greater columns of the periodic table. Because of the atomic motions, therefore, E_i will in general have a pseudogap, which for a very perfect net may even be nearly clear of states, and also may contain structure for certain special "defects" of the random net. We will include some specially strongly coupled parts of the electron–atom coupling explicitly, but the pseudogap is meant to help out by taking into account the favorable configuration of all of the rest of the random net. The resulting density of states of the E_i might look like the fig. 3.11a.

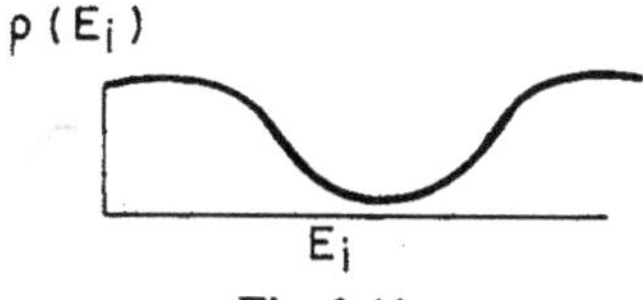

Fig. 3.11a.

As I said in sect. 2.3, one may always represent the relevant states by local orbitals, but one may *not* in general use a one-electron Hamiltonian without hopping integrals V_{ij} so we add V_{ij} to H_{00} to make H_0, the one-electron part of the hamiltonian.

$$H_0 = \sum_{i\sigma} E_i n_{i\sigma} + \sum_{ij\sigma} V_{ij} c^{\dagger}_{i\sigma} c_{j\sigma} \,. \tag{3.3.2}$$

The spectrum of H_0 is what the standard CFO picture from the texts looks like: we find that in general the gap will be narrowed by the V_{ij}, forming bands of extended states and a gap with tails, and possibly even with structure in it (fig. 3.11b).

I would like to emphasize – and in this I am perhaps to some extent a voice crying in the wilderness – that I cannot find plausible the idea that a sensible one-electron hamiltonian *even for a perfect random net* can have a perfect gap with no tails at all extending some way into it. Let us think a bit about the facts of the case. The energy of the amorphous state is 5–10% greater than that of the crystalline one. That means the average bond is distorted in such a way as to change E_0 by that much – which is in terms of the average gap of about 5 eV in Si, of order $\frac{1}{4}$ to $\frac{1}{2}$ an eV. Many bonds must be perfect; one expects the variance to be the same size, so that many bonds will be distorted (like 5–10%) by 1 eV or more. A potential fluctuation of 1 eV can easily bind a state in Si or Ge, even in the crystal, and in the amorphous bands with their flat, high density of states (which is implausible as a result of H_0 in any case, see fig. 3.12), this binding should

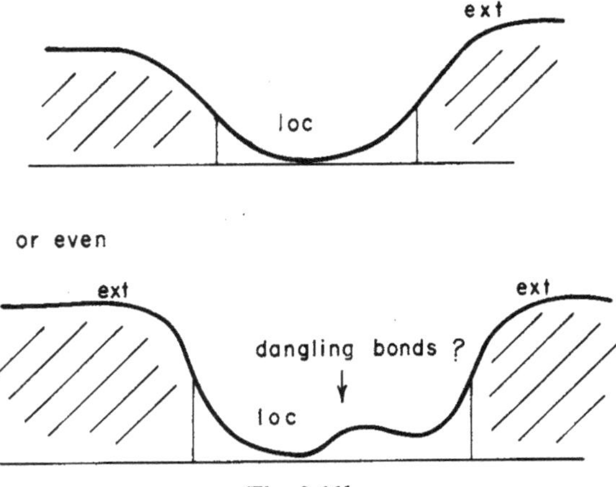

Fig. 3.11b.

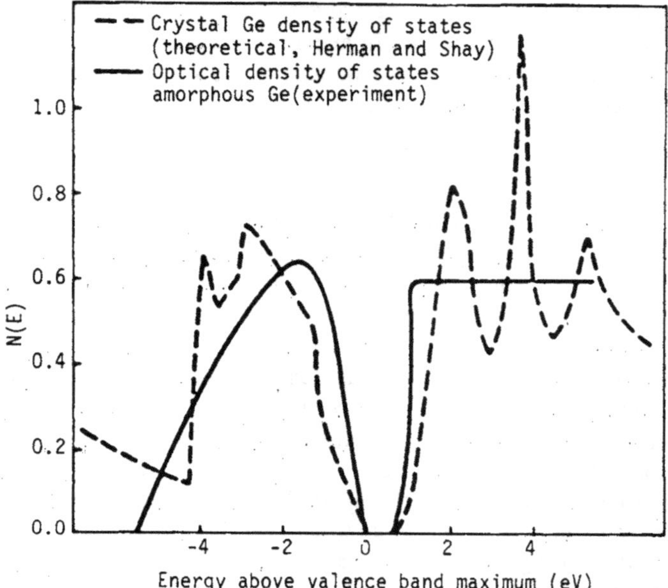

Fig. 3.12. Optical density of states for amorphous Ge as determined by photoemission, and calculated density of states for crystalline Ge (from Donovan and Spicer [63]).

be even stronger. Another way to put it is that with the bands markedly distorted internally, by eV amounts, what coincidence then allows the tails to be wiped out completely? From the first I have thought the remarkably sharp one-electron densities of states [63] to be one of the most mysterious of all the amorphous properties. In any case, what we can assume is that only in the case of rare, carefully prepared materials is there not a broad smearing of states into the gap, overlapping onto the Fermi level if the gap is narrow enough as in these various amorphous semiconducting materials.

Among all the Coulomb interactions, as we already discussed only one has a really major effect on properties: the Van Vleck–Mott–Hubbard interaction U between particles in the same local state:

$$H_{\mathrm{M-H}} = \sum_i U_i n_{i\uparrow} n_{i\downarrow} \,. \tag{3.3.3}$$

This is dominant in many situations – e.g., the impurity band, and in the dangling bond states of Si as Solomon has shown us. I believe that latter fact can be easily understood chemically as we shall see. Where U dominates, all localized states are Curie–Weiss magnetic, as I have said.

There is, however, another strong interaction, and as I have noted above, it is generally the stronger: the electron–phonon interaction, which tends to cause electrons to attract in pairs. For the sake of an oversimplified model, I am going to emphasize the aspect of this which is not familiar from textbooks on polarons, which are a relatively rare and esoteric phenomenon, but is universal and hence often unnoticed: the covalent bonding interaction. But in fact both can be described similarly except for the difference to which Mott [49] calls attention, which we do not discuss here. We assume that there is a configuration coordinate x_i which is strongly coupled to each local electronic state E_i. In most important cases, this represents mostly the length of one particular bond, which is short when the bonding electrons are there, long when they are absent; or conversely, if antibonding electrons are added in a given bond, it elongates because of the repulsion of the extra electrons for those already there. But it could also represent the radius of a coordination shell about an ion. The appropriate coupling term is then just $H^{\text{el-phon}} = \sum_i C_i n_i x_i$. Firstly, I make a very rough approximation which is useful because I do not want to get involved with phonon dynamics: the characteristic polaron approximation that the phonons are suitable for an Einstein model: $H_{\text{phon}} = \sum_i (P_i^2/2M) + \frac{1}{2}M\omega_{\text{ph}}^2 x_i^2$. The net electron–phonon hamiltonian is a model due to Holstein [64].

First, to fix ideas let me give a very simple example – one that will not be the most important in the gap but can sometimes occur. Consider an Si–Si covalent bond. If two electrons occupy the antibonding states, the atoms will wish to be far apart (fig. 3.13a). Under normal circumstances, as shown in the figure, this will be the higher energy state, while the low-energy state will be the covalent bond, and the 3-electron bond will be intermediate.

This will be the normal case; but imagine that the strains due to forming a random network without the assistance of hydrogen compelled the Si atoms to be held very far apart: it might well be that the ordering of energy levels went as in fig. 3.13b, with the doubly-charged pair of Si^- ions lowest, the 3-electron bond unstable, and the bonded configuration higher. In either case the odd-electron state is unstable. We note for future reference that the system would be *much* happier if two protons were to come along and covalently bond to each sp^3 orbit, or to one forming an H-bond with the other. The charge of the doubly negative center would have to be compensated by emptying 2 dangling bond orbitals, which costs energy.

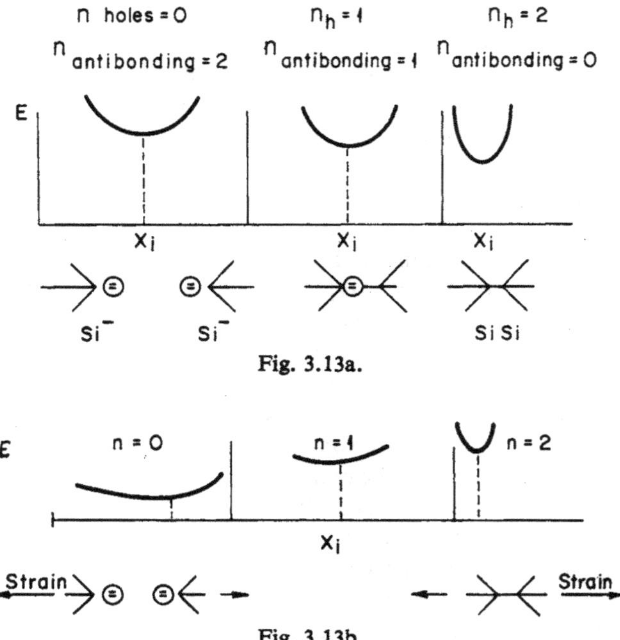

Fig. 3.13a.

Fig. 3.13b.

I have given this example because it is less familiar in the literature though probably common enough in nature; we will go into the important chemistry of the chalcogenide case later.

Now I would like to discuss what happens when we consider one of the localized states in the band gap of H_0. I am assuming that the localized states near the Fermi level are probably pretty well describable in terms of a single parent E_i. It is clear that such states are the least satisfactory bonds in the entire network, or non-bonding lone pair orbitals, or conversely the most nearly satisfactory antibonding states belonging to a bond which is heavily strained. Thus from now on, I will use i interchangeably for localized eigenstates or quasi-eigenstates and for bonds.

Consider the lattice to be stationary and find its potential minimum as a function of occupation of state i by taking $\partial E/\partial x_i = 0$. This gives three values:

$$\begin{aligned} E_0 &= 0\,, & n_i &= 0\,, \\ E_1 &= E_i - C_i^2/2m\omega_{\mathrm{ph}}^2\,, & n_i &= 1\,, \\ E_2 &= 2E_i - 2C_i^2/m\omega_{\mathrm{ph}}^2 + U_i\,, & n_i &= 2\,. \end{aligned} \qquad (3.3.4)$$

Fig. 3.14a.

If the Fermi level is at μ, the electrons for occupation are to be considered as coming from μ so in fact if we measure all energies from μ these are the true energies. Two essential cases occur:

$$(1) \quad E_0 + E_2 > 2E_1 \,,$$

in which case we have a *positive* effective U since the system prefers single occupation (fig. 3.14a), or

$$(2) \quad E_0 + E_2 < 2E_1 \,,$$

which is the *attractive* case (fig. 3.14b, see also fig. 3.16).

We may make this explicit by rewriting these "slow" energies as

$$E = E_i' n_i + U^{\text{eff}} n_{i\uparrow} n_{i\downarrow} \,, \tag{3.3.5}$$

with

$$E_i' = E_i - C_i^2/2m\omega_{\text{ph}}^2 \,, \qquad U^{\text{eff}} = U - C_i^2/m\omega_{\text{ph}}^2 \,. \tag{3.3.6}$$

It is, incidentally, also true that the V_{ij} renormalize, especially if $C_i^2/m\omega_{\text{ph}}^2 \gg \hbar\omega_{\text{ph}}$ as it often will be: the hopping $i \to j$ cannot take place then without either multi-phonon emission or tunneling through a phonon barrier: the latter modifies

$$V_j^{\text{eff}} \to V_{ij} \exp\left[-\left(\frac{C_i^2}{m\omega_{\text{ph}^2}}\right)\frac{1}{\hbar\omega_{\text{ph}}}\right].$$

This means that in the new representation the wave functions are ultra-localized like small polarons. The masses of small polarons are such that they localize very easily.

Let us now study the statistics of this simplest case, in which we may totally neglect V_{ij} and concentrate on the localized orbitals and their stationary states. In the repulsive case, $U_{\text{eff}} > 0$, nothing terribly interesting happens so long as all states are localized. The situation is identical with the typical Fermi glass with positive U. We imagine easily that the

Fig. 3.14b.

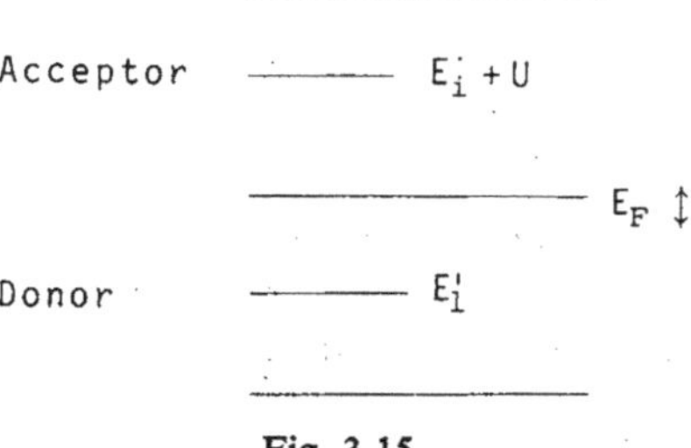

Fig. 3.15.

dangling bond states of a-Si are of this sort. Using that case to fix our ideas, $n_i = 1$ is a neutral state. If $E_i' < 0$, $E_i' + U > 0$ which is the common case for Si dangling bonds, we will have an occupancy of 1 at $T = 0$, and the state will act as either a donor of level E_i' or an acceptor of level $E_i' + U$, hence firmly trapping the Fermi level against doping when $U_{\rm eff}$ is small enough, as it seems to be in this case.

If E_i' is above the Fermi level, the center will be charged + and will accept an electron of the energy E_i'; while if $E_i' + U < 0$, it will be charged − and will donate an electron at $E_i' + U$. In either case the state farther from the Fermi level is suppressed and we have the conventional donor or acceptor.

Much more interesting is the case in which the electron–phonon coupling parameter C_i is large and we have

$$U^{\rm eff} = U - C_i^2/m\omega_{\rm ph}^2 < 0\,, \quad E = E_i' n_i - |U^{\rm eff}| n_{i\uparrow} n_{i\downarrow} \qquad (3.3.7,\, 8)$$

In this case one of the two extreme levels is always the lowest and will represent the normal state, and in the interesting case in which one of the energies is not deep within the band, the nearest state will have occupancy differing by two and not one (fig. 3.16).

We now try to show that the effect of such attractive centers is to introduce a gap around the Fermi level in the one-electron excitation spectrum – an amusing parallel is to the case of superconductivity, where there is no gap for excitations involving pairs, but all one-electron excitations have a gap. We assume a continuous density of such centers since they occur randomly in the system. It is perhaps best illustrated with the help of figures from my Autrans paper in the Journal de Physique Colloque

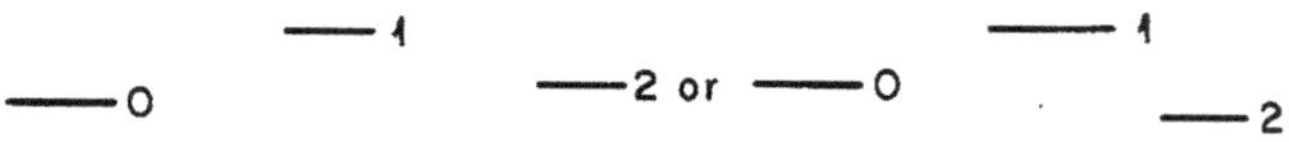

—— 1 —— 1

—— 0 —— 2 or —— 0 —— 2

Fig. 3.16.

[62]. First we consider what I called there the "slow" spectrum E_i', U^{eff} assuming the adiabatic adjustment (3.3.6). Any state with $2E_i' - |U^{\text{eff}}| < 0$ will by definition of E_F be occupied with two electrons otherwise totally empty. However, that means that (see fig. 3.17a)

$$E_i' < \tfrac{1}{2}|U^{\text{eff}}| \text{ occupied}, \qquad E_i' > \tfrac{1}{2}|U^{\text{eff}}| \text{ empty}.$$

However, the actual excitation energy of an electron pair is perfectly normal (see fig. 3.17b).

On the other hand, the one-electron excitation energy shows a gap. If i is empty, the energy to add one electron is still $E_1 = E_i'$; but if it is full, the energy to add a hole is

$$E_{\text{hole}} = -(E_i') + |U^{\text{eff}}| > \tfrac{1}{2}|U^{\text{eff}}|,$$

or per electron (see fig. 3.17c)

$$E_1 = -|U^{\text{eff}}| + E_i',$$

so that indeed there is a one-electron gap.

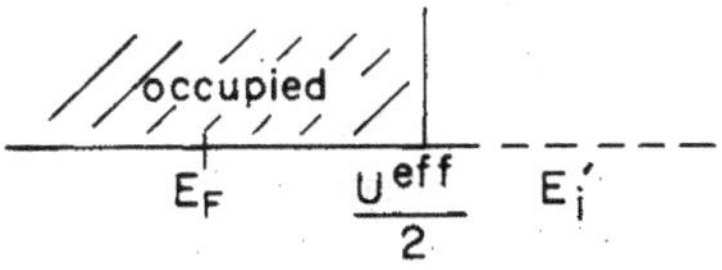

Fig. 3.17a. "Slow" site energies.

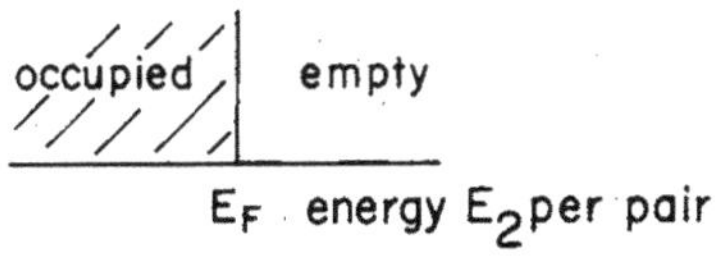

Fig. 3.17b. Energy per pair of electrons to switch from $n = 0$ to $n = 2$ or vice versa.

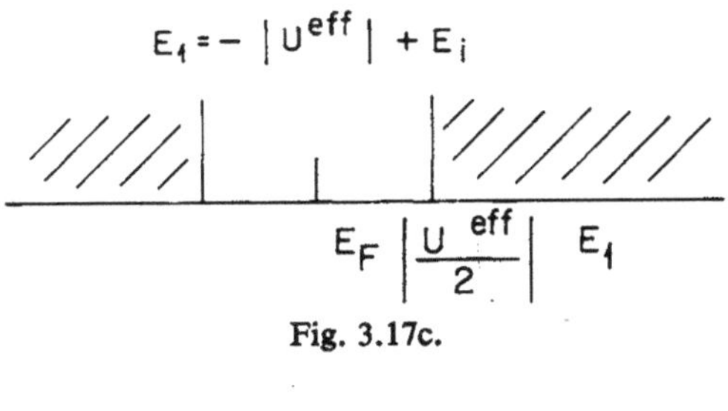

Fig. 3.17c.

Finally, I want to emphasize the important concept which I stated firmly in my opening lecture, and continue to emphasize that, Solomon or no, really everything is a distribution in a random system: a distribution which can often be surprisingly spiky but still spread. $|U^{eff}|$ is no exception – in any material it will have a more or less broad distribution, so that while this concept explains pinning without magnetism and without optical absorption very nicely and cleanly, it does not help with the sharpness of the band edges very much: we must *smear* U^{i} so that the spectrum of slow, localized one-electron states looks like fig. 3.17d. There is no reason that positive U and negative U centers cannot coexist, hence the band tails may even overlap E_F.

To comment: clearly E_F is nicely pinned if the density of two-electron states at E_F is high enough. In glasses it is at least 10^{19-20}; in some non-hydrogenated Si's probably also as big. These states seem, except for their pinning properties, extremely elusive experimentally; they seem always to be signalled only by the absence of something. One important property is that of relatively slow relaxation, even though their thermal occupancy can be quite high. Many semiconductor physicists still think of relaxation barriers as somehow related to the energies of states, but these are not. One might hope to locate them in relaxation effects.

This is as much of the idea as has been generally accepted and understood. It is considered likely that special chemical configurations – specifically dangling bonds – in chalcogenide glasses do provide a density of these two-electron gap states. My own feeling is that the concept is broader and extends to the entire spectrum in the following ways. In the first place, I apply it to the normal bonds of the random net just as much as to the special dangling bond situation envisaged by Street and Mott [65]. In other words, the idea is that the E_i spectrum encompasses the whole pseudogap region. There will in every random net be a certain number of particularly highly strained "normal" bonds, whose pair of electrons are

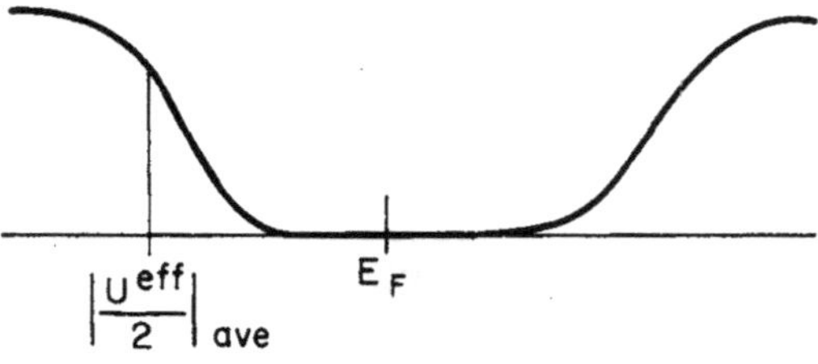

Fig. 3.17d.

unstable to the point of leaving entirely and joining two dangling bonds or making an extra bond somewhere else. I would agree that these highly strained bonds do not spread into the band gap just because of the negative U mechanism; their U^{eff} is greater than the gap, if anything, being of order the bond energy, and when one electron leaves, the other follows leaving simply a broken bond. This is part of the reason why band tails do not protrude from the conduction and valence band. A second is a bit more subtle and involves the interplay between states and dynamics. So far we have described the adiabatic motion of extra electrons assumed to occupy the tail states in localized fashion, and neglected V_{ij}. The one-electron states E_i in the bulk of the two bands will in general have a very large density of other states at the same energy for hopping, and so these states will delocalize easily and, as they delocalize, improve in energy by $\sim V_{ij}$ due to lowering their kinetic energy of localization. There will be a competition between the energy $-C_i^2/2m\omega^2$ and V_{ij} which the latter will usually win, so that the one-electron states in the band will often be not small polarons but free electron states (but not always; though, as for holes in SiO_2 glass, the small polarons may obey the Mott mechanism and require time to form).

States which are delocalized will thus be shifted *upwards* in energy (for both holes and electrons) by $C_i^2/2m\omega^2$ and *downwards* by about $\langle V_{ij} \rangle$; the result may well be that the localized, small-polaron-type states appear only as a rather weak tail below the mobility edge. That is, for fast optical states one has the situation in fig. 3.18a if the "slow" energy is E_i^1 and the "fast" energy is E_i^{opt}. And then with account of V_{ij}, the final optical density of states is given in fig. 3.18b. Again, by an equivalent of my earlier mechanism, the mobility edges appear as also optical edges. The full density of one-electron states will also have a more or less pronounced tail of localized, more or less self-trapped, one-electron states as well (fig. 3.18c). These localized tails will be magnetic and excited by optical radiation.

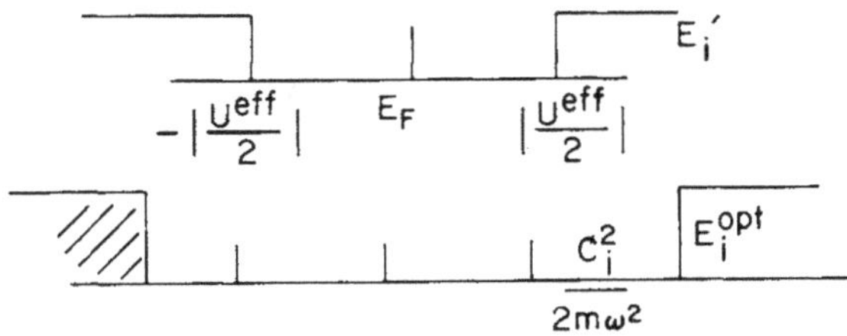

Fig. 3.18a.

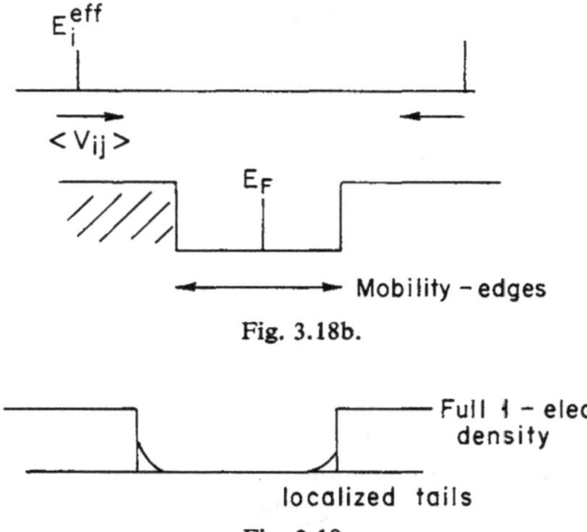

Fig. 3.18b.

Fig. 3.18c.

3.3.2. Chemistry of centers

Here I just want to summarize some of the suggestions which have so far appeared identifying particular behaviors with these states [65,66].

I want to emphasize the overlap between these ideas of chemical identification of centers and the very general point of view I have taken in the last section. Namely, one may think of the system as having relieved its *most* strained bonds by discarding them completely so that its random net contains only bonds which are less strained than a certain maximum amount. If carefully prepared, there will be relatively few and weakly strained bonds left. How completely this occurs is exemplified by Solomon's SiH_x, which seems to have its actual defects reduced down to the 10^{16} level. On the other hand, we must also realize that there is no restriction of the negative U idea to irregular systems. Two-electron centers undoubtedly exist as chemical impurities of regular lattices – I believe O in GaP is such a center.

In the chalcogenide glasses, one may identify three types of centers. The one which Mott–Street postulated in the literature is the dangling bond: a chain end of one Se chain, for instance, or a missing bond on As. This is really just like the Si dangling bond except its U is negative, as may be either roughly computed or based on the observations. The scheme takes advantage of the lone pair orbitals.

Consider for example an As dangling bond neighboring to an Se chain (see fig. 3.19). These are denoted as $D^{\pm}$. At any dangling bond there is

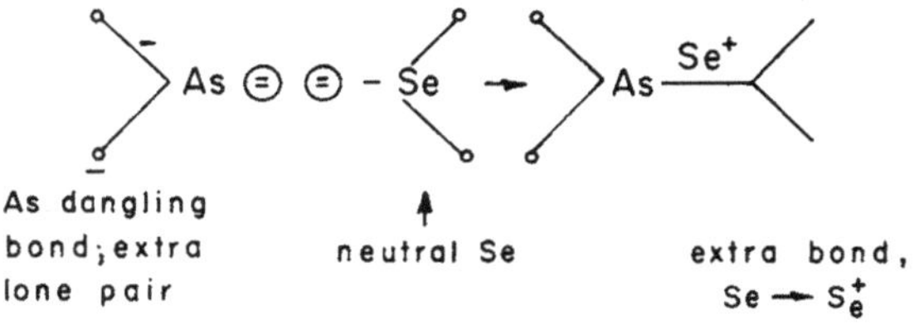

As ⊜ ⊜ - Se → As Se⁺

As dangling bond; extra lone pair

neutral Se

extra bond, Se → S_e^+

Fig. 3.19.

likely to be a nearby lone pair if there are chalcogenide atoms present, but of course the relative energies of D^+ and D^- ($=E_i$) vary over a wide range. These centers are topological defects and occur at the 10^{-4} level by electrical measurements. We also believe there are two other types of centers in those materials as follows: two lone pairs too close together as given in fig. 3.20a, which we call D^{++} centers

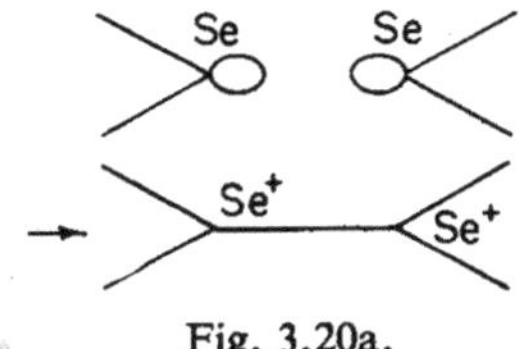

Fig. 3.20a.

and an excessively stretched bond as given in fig. 3.20b, called D^{--} centers.

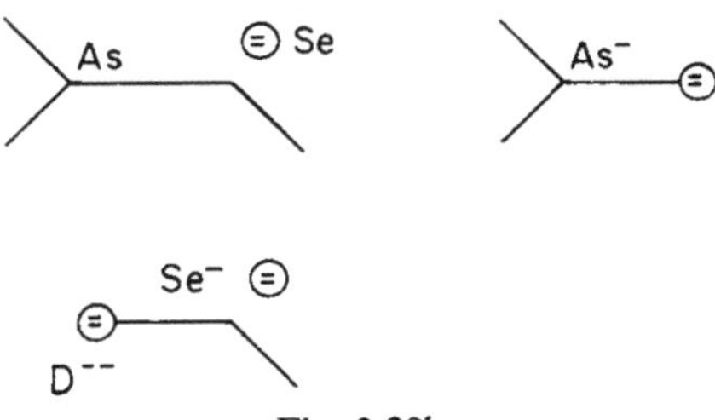

Fig. 3.20b.

Note that $D^{\pm}$ can surely be eliminated by doping with hydrogen, which will eat the dangling bond. D^{++} cannot at all; and D^{--} only partially – it will be converted to a dangling bond $D^{\pm}$ itself, probably, because there will be no room for two H's. The former is probably the reason that chalcogenides cannot be converted into dopable semiconductors with H.

In Si and other non-chalcogenide cases, where there are no lone pairs,

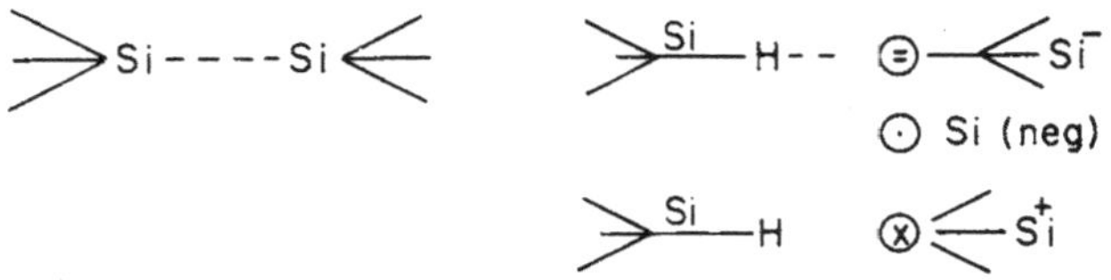

Fig. 3.21.

we have seen that the U of the dangling bond is positive, probably because there is no lone pair nearby for it to grab and bond to. D^{++} does not exist, D^{--} may. Hence H is capable of eliminating most centers. Fisch and Licciardello [67], however, have suggested a new center in hydrogenated Si, the center $D^{--} \rightarrow D^{\pm}$H–D which may be either magnetic or negative U or both depending on the degree to which a hydrogen bond may form, given the spacing of the two Si's (see fig. 3.21).

3.3.3. Other effects: optical, surfaces, initiation of switching, $1/f$, *etc.*

One of the characteristic properties of the two-electron center idea is that it is necessarily associated with a wide range of relaxation times, and also that the changes which take place due to electrical or optical excitation are in a very direct sense chemical bonding changes, not just electronic excitations. Many interesting and somewhat mysterious properties of amorphous (or even ordinary) semiconductors may find their explanation in these terms.

Ma [68] proposed a theory of the initiation of switching, a characteristic behavior of the chalcogenide glasses. We do not propose that the "on" state is describable by these simple methods. The principle is more or less as described in my Autrans paper [62]: that displacement of the two-electron centers to the less stable side can reduce the one-electron gap.

Many interesting photochemical and photoelectrical effects are found in some of these glasses. For instance GeS_2 shows the effect of photodissolution of Ag, and in other cases there are photochromic effects. The idea that displacement of a two-electron center or a set of two-electron centers is a genuine chemical bonding change is highly suggestive as a possible source of these effects, although no specific mechanisms have yet been proposed.

Varma and Pandey [69] and Harrison [70] as well as Anderson [62] have pointed out the role two-electron centers can play at the surface of a crystalline semiconductor in giving Fermi level pinning in precisely the way the pinning works in bulk glasses. Harrison points out that an unbalanced "Haneman model" even of a pure 111 surface can pin the Fermi level. The

Haneman model normally proposes equal numbers of D^+ and D^- bonds, as we have described them, for the odd bonds projecting from the 111 surface; but if we recognize that D^+/D^- is not necessarily unity for a charged surface, we immediately acquire a pinning method. Less perfect surfaces will have a more explicitly amorphous structure. But the principle that the surface is a copious locale of weak bonds and hence of two-electron centers must be a very general one. It is also significant that the pinning occurs only in the more covalent materials.

More speculative as yet is the suggestion that this wide range of relaxation times of two-electron centers and the relatively high occupation probabilities – the two dissociated completely from each other because the $n = 1$ state may be quite far from $n = 0, 2$ or very close – can be the source of the ubiquitous "$1/f$ noise" in semiconductors. This possibility has been explored by Ma in his thesis [48] (see also Anderson [62]), but unfortunately not conclusively as yet. Finally, Licciardello [71] has suggested modes of recombination via two-electron centers.

The key remark with which I end this section is that two-electron centers are not a strange, unique artifact, but extremely common, especially in technical materials which are not artificially purified and perfected. They have a great many unexpected properties, along with a tendency to remain hidden to the obvious probes. We can be sure that many familiar and some unfamiliar effects will turn out to result from them.

3.4. Resonant two-electron centers

In the course of a general study of the magnetic impurity model I pioneered (named after me [72] but basically due to Blandin and Friedel [73]), Haldane [74] pointed out that the "U" of that model is also not necessarily always positive, and drew a crude phase diagram for a "negative U Anderson Model".

The original model shows how a magnetic ion may retain its localized spin and magnetic moment in the presence of a metallic Fermi gas of electrons, by changing the bound state of the magnetic atomic shell into a resonance. Haldane showed that the negative U version of the same kind of center exists and has the same kind of properties as the gap states in amorphous semiconductors: a Fermi-level dependent "flip-flop" from $n = 0$ to $n = 2$.

The hamiltonian model used contains the same interaction terms

$$H_{\text{int}} = U n_{i\uparrow} n_{i\downarrow} + c_i x_i n_i + \hbar\omega_i (b_i^\dagger b_i + \tfrac{1}{2}) , \tag{3.4.1}$$

but the one-electron terms are those appropriate for a *resonance* not a localized or bound state:

$$H_{1-\mathrm{el}} = E_i n_i + \sum_{k\sigma} V_{ik}(c^*_{k\sigma}c_{i\sigma} + \mathrm{h.c.}) + \sum_k \varepsilon_k n_k\,, \tag{3.4.2}$$

where k runs over the states in a metallic band of extended states. Thus we have a local resonance in a metallic band coupled to a local phonon strongly enough to create a small polaron. Haldane was at first concerned with coupling to a screening model of the electron gas described as bosons, but this probably is not a terribly useful model, whereas the "polaron resonance" may play a considerable physical role in a number of systems.

As in the localized state case, we may at first treat the perfectly adiabatic situation with $\partial E/\partial x_i \equiv 0$. In this case, the lattice is treated as stationary but it readjusts depending on the average occupancy n_i of the resonance. Haldane shows that the approximate Hartree–Fock phase diagram of the resonance is as shown in fig. 3.22, where all the lines are first-order phase transitions in the sense that n jumps at these lines because of the phonon coupling, if that is treated adiabatically. In every case the state is a resonance rather than a sharp bound state, but the resonance may be magnetic or non-magnetic, and where it is non-magnetic may be essentially above or below the Fermi level. When magnetic there are two states which are automatically degenerate by spin symmetry, but the corresponding sym-

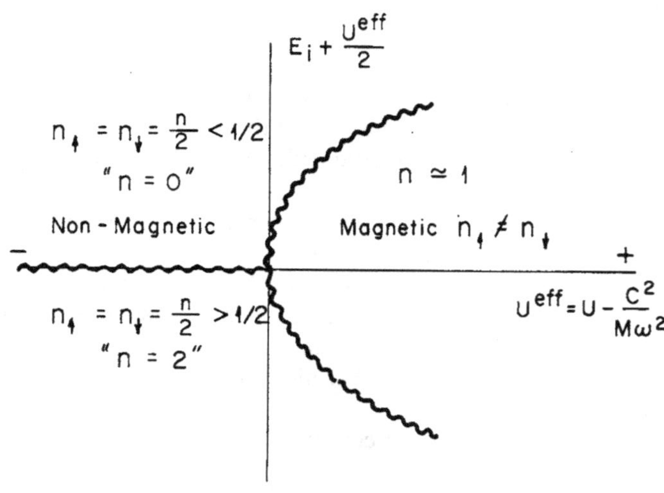

Fig. 3.22.

metry on the left side of the figures is of course non-existent and only at $E_d^{eff} = E_F$ is there a tendency to flip back and forth.

Nonetheless these centers may play a role in a number of problems. Haldane [74] has given a preliminary discussion of the properties of the right, $U > 0$ side of the figure in relation to mixed valence systems. Negative U centers may be relevant to other kinds of problems:

(1) Defect structures. By forming centers of this kind, systems can evade energetically unsatisfactory consequences of Luttinger's theorem: one may have local bonds formed as resonances using electrons from unstable parts of the Fermi surface. Such centers may even have Jahn–Teller-type degeneracies or near-degeneracies and hence be formed by entropy considerations at high temperatures.

(2) Such defects are likely to occur at surfaces, or in fact at any kind of macroscopic defect. Thus the idea of a trap with a possible slow relaxation time for changing its occupancy and scattering properties is not excluded by metallic properties. Such slow traps can be a source of $1/f$ noise in metals.

In general, the idea of the interaction of phonons with resonances is an almost unexplored one and some of the mysterious properties of more complex or impure metals – or even metallic glasses – suggest its further study.

4. Spin glasses – the "easy" case

4.1. Experimental facts and some general considerations

The experimental facts about the spin glass state have been beautifully summarized for you by Joffrin, so I will merely refer you to his lectures and to the few brief words in my notes for these. Particularly, I call attention to the possibility that gellation of polymers is a similar phenomenon but perhaps simpler. Experimentally, there appeare to be a phase transition in that a number of anomalous properties appear rather suddenly at a certain temperature, which is clearly characteristic of spin interactions which are "frustrated" in a certain very well-defined sense, which I have promised to talk about later, but which is roughly described as essentially *competing* with each other on a random basis, and preventing the establishment of any well-defined form of order. One must always keep in mind that in the experimental situation T_{SG} is usually 1–10^2 K, and, if we were to scale the glass transition down to that value, its apparent breadth of 10–20 K would be

$$\frac{10}{500} \times 20 = 0.4\,\mathrm{K}\,,$$

and the spin glass experimentalists are, unfortunately, either trained in a school which is not interested in that kind of precision or in one which is very precise but prejudiced against the existence of a phase transition. Thus the possibility of a continuous glass-like transition is not quite excluded, although none of the properties of the usual glass transition in terms of rate sensitivity, heat absorption, etc. are indicated (except see fig. 4.1 due to Tournier et al. [75] which indeed shows a continuous transition).

There are a number of different physical and mathematical points of view from which one may approach the spin glass transition, and most of them shed some light, some of them a great deal. But the light seems to come from different directions and to illuminate, not incompatible, but orthogonal, aspects of the total physics. Perhaps when two points of view meet in the middle we will have the problem solved.

One very old point of view which has been rather down-graded in the past, but in fact illuminates a number of aspects in quite an important way, is naive mean-field theory. I would like to start my discussion with that. Mean-field theory, in the case of almost all other phase transitions, is actually all that is necessary for physical understanding of the states at any distance at all from the phase transition, and the existence of a mean-field

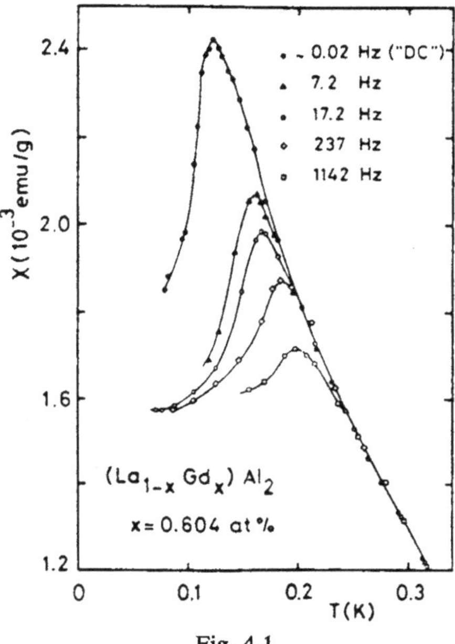

Fig. 4.1.

solution and the corresponding Gibbs–Landau free-energy functional from which to derive fluctuations is the essential starting point for the whole ridiculously ponderous apparatus of modern renormalization group theory. It is characteristic of most of the random phenomena we are studying here that this starting point does not exist and that we do not have a satisfactory way of starting our solutions. In localization, as Thouless said, no one has yet made a convincing connection to an appropriate statistics problem. It may well be that that appropriate problem is the spin glass problem; I am not sure.

Let us then start by returning to prehistory [76,77] and considering this mean-field equation for the spin glass:

$$\overline{\boldsymbol{S}_i} = L(\overline{\boldsymbol{H}_i}/kT)\,, \tag{4.1.1}$$

$$\overline{\boldsymbol{H}_i} = \sum_j J_{ij}\overline{\boldsymbol{S}_j} + h_i\,, \tag{4.1.2}$$

where L is the appropriate Langevin function for whatever type of spin we are using and h_i is an externally applied field on site i. For the moment

I would like to just discuss Ising spins because that is hard enough, so that

$$L(x) = \tanh x\,, \tag{4.1.3}$$

and the mean field equations are:

$$\overline{S_i} = \tanh\left[\frac{1}{kT}\left(\sum J_{ij}\overline{S_j} + h_i\right)\right]. \tag{4.1.4}$$

Now consider the linearization of (4.1.4) in that external field:

$$S_i = \mathrm{X}\left(\sum_j J_{ij}S_j + h_i\right). \tag{4.1.5}$$

This may be written

$$S_i = \sum_j \left(\frac{1}{1 - XJ}X\right)_{ij} h_j\,, \qquad \boldsymbol{S} = \sum_j \left(\frac{1}{kT - J}\right)\boldsymbol{h}\,, \tag{4.1.6}$$

where

$$(X)_{ij} = \frac{1}{kT}\delta_{ij}\,, \tag{4.1.7}$$

and J are linear operators in ij space.

Clearly we predict a transition and (4.1.6) diverges when

$$kT_c = J_{\lambda\,\max} \tag{4.1.8}$$

(where $J_{\lambda\,\max}$ is the largest eigenvalue of J), because at that point the "shattered susceptibility" (as we shall call it) diverges: some randomly oriented linear combination of spins polarizes. But the key point of Anderson [76] is that this is a false result if the largest eigenvalue of the random matrix J_{ij} is localized. So let us discuss the spectrum of J_{ij}, which we assume to be a typical random matrix of its appropriate dimensionality. (Note that I am here dismissing as almost irrelevant any special nature of J_{ij} such as whether or not it is Gaussian, ± 1, flat or even RKKY. What is essential is the nature of its positive eigenvalues.) The *general* random matrix (with $\mathrm{Tr}\,J = 0$ as it is) has a spectrum like fig. 4.2a, but in a few cases it can have the form of fig. 4.2b. These cases are: (1) Four or greater dimensions, if random perturbations are in some sense small enough that they cannot bind a state. (2) Very similar – the infinite random Gaussian matrix with elements in all positions [79]. (3) Of course, if J_{ij} is regular, either constant or periodic: $J_{ij} = f(r_{ij})$ on a periodic lattice. Finally, for <2 dimensions the entire spectrum is localized.

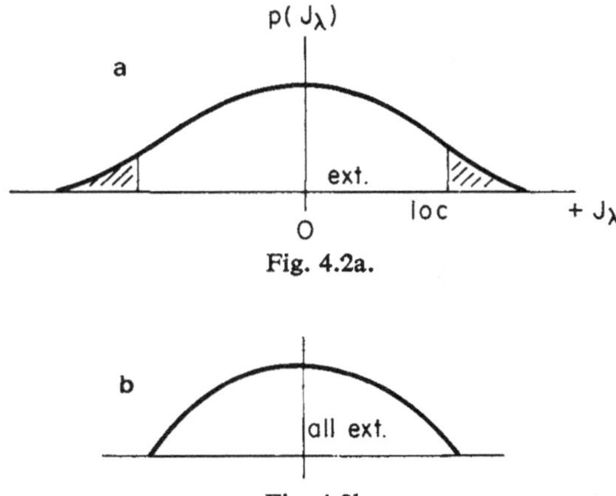

Fig. 4.2a.

Fig. 4.2b.

Now in case 4.2a the largest eigenvalue is automatically localized. But mean-field polarization into a localized state is nonsense, because a finite system cannot have a sharp phase transition. Consider the eigenvalue $J_{\lambda\,\max}$ with eigenfunction $\phi_\lambda(i)$. Long before we reach T_c as defined by (4.1.8) there will begin to be large fluctuations of the form $S_i = (\pm\,\text{const}) \times \phi_\lambda(i)$ which will cause the susceptibility to saturate down to all T instead of diverging:

$$X_\lambda \rightarrow \text{const} \quad \text{instead of} \quad X_\lambda \rightarrow \frac{1}{kT - J_{\lambda\,\max}} \rightarrow \infty\,. \tag{4.1.9}$$

The next eigenvalue can be expected to be similarly renormalized: we can think really of a kind of fluctuation renormalization of the entire matrix

$$J_{ij} \rightarrow J_{ij}(T)\,, \tag{4.1.10}$$

which we can see explicitly in operation later on in the TAP–Bethe theory. This renormalization takes place before any localized eigenvalue is actually reached. It is these localized eigenvalues which so convincingly mimic the "clusters" of the Néel interpretation of spin glass phenomenology above T_{SG}.

On the other hand, when T approaches any extended eigenvalue, there is no such automatic renormalization away of the resulting divergence. We know this to be the case in regular systems, and we see no firm reason to doubt it in random ones; in fact, there are counterexamples, in the TAP theory of the $Z \rightarrow \infty$ case for instance. But the question becomes: can the system renormalize away localized eigenvalues and work its way through the localized tail to the extended spectrum? If it does, the system will start to polarize in one of the extended eigenvalues. This eigenvalue represents some random polarization vector $S_i = \varepsilon\phi_{\lambda\,\max}(i)$.

There are three dimensionality ranges to consider. First, $d < 2$. We know that $d_c = 2$ is the critical dimensionality for localization so that there are no random matrices with *any* extended eigenvalues below $d = 2$. Thus, no divergence can ever occur and

$$\text{LCD} \geqslant 2 \text{ (Ising, SG)}\,.$$

Second, there is $d > 4$. As David Thouless told you, a 4d random matrix can become sufficiently weakly random that the mobility edge and the band edge coincide, and the density of states can rise from the mobility edge with a finite growth rate (or at worst like $\sqrt{E_c - E}$ in the $d = \infty$ case). Such a case could give a conventional critical point, and it must be this phenomenon which is signalled by the fact that many conventional methodologies show LCD = 4.

$$\text{LCD (more or less conventional behavior)} = 4\,.$$

But I emphasize that even so I have not proven that such a behavior exists; one awaits a satisfactory solution by renormalization group or other techniques.

Finally, the difficult and, unfortunately, physical cases of $d = 2, 3$. A possibility as rational as any other is that in these cases the matrix is renormalized in such a way that it eliminates localized tails but cannot eliminate the mobility edge itself: the spectrum at T_{SG} looks like fig. 4.3a, which means that it may polarize in any of a very large number of possible states sufficiently close to the edge, which could account for the simulation result that ground states never repeat upon raising above T_c and re-freezing. It is even possible that the spectrum of fig. 4.3 remains all the way

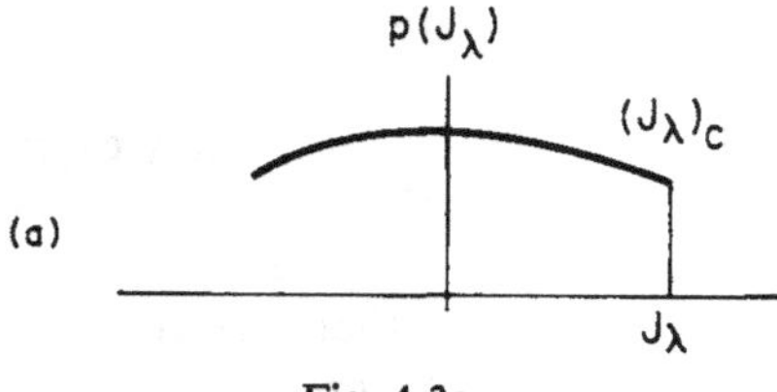

Fig. 4.3a.

where δS is the *fluctuating* part of S

$$= \frac{N}{T}(1 - q) + \frac{1}{T}\sum_{i \neq j}\langle S_i(t_1)S_j(t_2)\rangle ,$$

and assuming sufficiently random interactions the second term can be assumed to vanish; in any case, it seems to be regular and small.

If q in fact vanishes for sufficiently long times, we see that the system will have a Curie law susceptibility: a physically unlikely result which could even be correct if we keep in mind that X is defined for (a) infinitely *small* fields; (b) infinitely *long* times. Neither may be achievable physically. (4.2.4) is the key result for experimentalists, as realized by Mizoguchi et al. [83]. X is the physical measurement of choice – all others are derivative. It is regrettable that experimentalists find $X(H \to 0)$ so boring.

In an attempt to improve on this and devise a more sophisticated analysis for q, Sam Edwards brought into the problem the ubiquitous n-replica method which may now represent our best hope for analytical results on physical cases, if also our most serious trap for the unwary.

The n-replica method is a mathematical trick so hoary that its origins are lost in history. One wishes to find the average physical properties of a random system. These properties are safely described only by a free energy $F = -kT \ln Z$, not by the partition function,

$$Z = \operatorname*{Tr}_{S} \exp(-\beta H) ,$$

which is not extensive and may be biased by anomalous regions in the sample. But random averages of integer powers of Z are easy to take:

$$\langle Z^n\rangle = \int \{\mathrm{Tr}\exp[-\beta H(S)]\}^n P(H)\, \mathrm{d}H$$

$$= \prod_{\alpha=1}^{n} \operatorname*{Tr}_{S_\alpha} \exp[-\beta H(S_\alpha)]P(H)\, \mathrm{d}H , \tag{4.2.5}$$

so what is done is to try to analytically continue to very small n:

$$-\beta F = \lim_{n\to 0} \frac{\partial}{\partial n}\langle Z^n\rangle = -\beta\langle \ln Z\rangle . \tag{4.2.6}$$

Let me specifically do this for the spin glass hamiltonian

$$H = \sum_{\langle ij\rangle} J_{ij}S_iS_j , \tag{4.2.7}$$

where I choose a Gaussian distribution for the J's, relying on universality:

$$P(J_{ij}) = \frac{1}{(2\pi\langle J_{ij}^2\rangle)^{1/2}} \exp\left(\frac{J^2}{2\langle J_{ij}^2\rangle}\right) . \tag{4.2.8}$$

I ignore the possibility of a mean J, and concentrate on the spin glass problem; the crossover to ferromagnetism has been thoroughly discussed in the literature [84]. We have then

$$\begin{aligned}\langle Z^n\rangle &= \operatorname*{Tr}^n_{S_\alpha} \exp\left[\beta^2 \sum_{\langle ij\rangle} \left(\sum_\alpha S_{i\alpha}S_{j\alpha}\right)^2 \tfrac{1}{2}\langle J_{ij}^2\rangle\right] \\ &= \operatorname*{Tr}^n_{S_\alpha} \exp\left[\beta^2 \sum_{\langle ij\rangle} \tfrac{1}{2}\langle J_{ij}^2\rangle(n(S^2)^2) + \sum_{\alpha\neq\beta} S_{i\alpha}S_{j\alpha}S_{i\beta}S_{j\beta}\right] . \end{aligned} \tag{4.2.9}$$

Thus we have replaced a complicated and difficult, if direct, random problem with a regular one at the cost of (1) a rather complex form of interaction hamiltonian; (2) a very dubious and delicate mathematical extrapolation from finite integer n to $n \to 0$. It is not surprising at all that this extrapolation has already led us into deep, but not irrevocable, trouble even in exactly soluble cases; but what *is* clear is that any approximations in the finite n problem are very uncontrollable as $n \to 0$.

The philosophical question is an interesting one. $n \to 0$ does just what I warned you against, hiding the dirt under averages; but it is a sufficiently subtle kind of average that it may just come out with the right answer. Experience so far has shown that the dangers indeed outweigh the advantages.

The obvious first thing to do is produce a mean field theory of this problem, which no longer has any subtle problems with localized eigenvalues. We assume an order parameter which has some kind of comparability with our original q^i one:

$$\langle S_i^\alpha S_i^\beta\rangle = q^{\alpha\beta} . \tag{4.2.10}$$

We have then a mean field self-consistency problem: The effective hamiltonian for the variable $q_j^{\alpha\beta} = S_j^\alpha S_j^\beta$ is

$$H_{\text{eff}} = \beta^2 \sum_i \langle J_{ij}^2\rangle q_i^{\alpha\beta} q_j^{\alpha\beta} , \tag{4.2.11}$$

and hence in, for example, the Ising model, the mean field equation becomes (correct only to linear order in q because of the non-independence of the different q's):

$$q^{\alpha\beta} = \tanh \beta^2 q \sum_j J_{ij}^2 = \tanh \beta^2 q^{\alpha\beta} \bar{J}^2 , \tag{4.2.12}$$

thus defining $\bar{J}^2$ as the mean square J_{ij}^2. This first allows a solution for $q^{\alpha\beta}$ at $k^2T_c^2 = \bar{J}^2$ – interestingly, independent of n and thus the T_c as $n \to 0$.

The computation of the free energy for $q \neq 0$, while essentially an exercise in conventional mean field theory, is rather arithmetically complicated because of the correlations between the $q^{\alpha\beta}$'s which are caused by the fact that they are $n(n-1)/2$ variables made up from only n variables, which makes it complicated to estimate the entropy. The result I just repeat from Sherrington and Kirkpatrick's paper [85] where they calculated out in some detail the limiting case where there are Z neighbors, $Z \to \infty$ and $\langle J_{ij}^2 \rangle = \tilde{J}^2/Z$:

$$F = NkT\Big(-\underset{\substack{\uparrow \\ \text{4 for dividing into pairs}}}{\frac{\tilde{J}^2(1-q)^2}{(2kT)^2}} - (2\pi)^{-1/2} \int dz \exp(-z^2/2x) \times \ln\Big[2 \cosh \frac{\tilde{J}q^{1/2}z}{kT}\Big]\Big),$$

where by $\partial F/\partial q = 0$,

$$\lim_{n \to 0} q^{\alpha\beta} = q = 1 - (2\pi)^{-1/2} \int dz \exp(-z^2/2) \operatorname{sech}^2(Jq^{1/2}/kT)z .$$

The formulas are very similar if one wishes to do the case of finite Z as in the original, rather sloppy Edwards–Anderson paper, but this case is chosen because it is normally assumed that mean field theory is exact in the $Z \to \infty$ limit, and initially that is what S–K did assume.

Two special remarks need to be made about these calculations. One is that the basic solution is done by a saddlepoint integration on an auxiliary variable y which is coupled to q. In picking the saddlepoint, the natural assumption is made that $q^{\alpha\beta} \neq f(\alpha, \beta)$ is symmetric. However, as we shall soon see the results are physically incorrect and there is no reason for this choice except physical reasonableness, since there is no guarantee that a maximum for $n \geqslant 2$ is not the wrong saddlepoint as $n \to 0$. In fact, Thouless and a student [86] claim to have recently shown that the correct solution is surely unsymmetric with broken symmetry in $\alpha\beta$ space. This has led to a subculture of very complicated arithmetic which has not yet reached any conclusion [87].

Next, it is easy to show that (so long as, unfortunately, this symmetry is assumed correct) the mean spin $\langle S_i \rangle$ can be calculated as a derivative of an $n \to 0$ limit and we find that, as one might have hoped,

$$\lim_{n \to 0} \langle q^{\alpha\beta} \rangle = (\langle S_i \rangle)^2 = \langle q_i(t \to \infty) \rangle ,$$

so that $q^{\alpha\beta}$ is indeed in a sense a physical order parameter.

4.3. TAP

The reason we are sure that this extrapolation is wrong is that $S = -\partial F/\partial T < 0$ at low temperature as calculated using the S–K formulas. This is unphysical and impossible, and led Thouless, Palmer, and myself [88] to develop a different method for the same problem. It turned out also that the energy arrived at by S–K was visibly lower than the best computer simulations anyone was able to make on the ground state problem. It began to seem to us that we could understand some aspects of the S–K solution but that below T_c it was almost certainly not physical, presumably for the reason later identified by Thouless and recorded by Palmer and Van Hemmen [89] that so straightforward an $n \to 0$ extrapolation is not safe.

Our new solution has a great many physical implications for this problem so I would like to go into it in some detail.

It starts with the obvious remark that the form of the free energy when $q = 0$ is peculiar: it looks like

$$F = -\text{const}/T\,,$$

which is equal to one term in a high temperature statistical series for

$$F = -kT \ln \text{Tr}(e^{-\beta H}) = -\frac{1}{\beta} \ln \text{Tr}\left(1 - \beta H + \frac{\beta^2 H^2}{2!} \cdots\right)$$

$$= -\text{Tr}\frac{\beta \overline{H^2}}{2!} + \cdots . \tag{4.3.1}$$

All the other terms seem to have been dropped. When we carried out a diagram analysis of this series, we discovered the following facts:

(1) This high-temperature form is undoubtedly correct.

(2) T_c is also undoubtedly correct.

I will go through the diagram analysis just far enough to show you these facts, and then show you that our equation is so ridiculously obvious on physical grounds that it is almost impossible to dispute it. Then I will first motivate it in a second way which relates it to T_c and to the concept of frustration, and finally discuss the numerical and analytic techniques which led to a solution. I do this because I feel that this solution, special as it may be, is one of the few fixed points in spin glass theory and is worth really understanding. In an additional discussion we will consider generalization of TAP to finite Z and to vector spins, both of which lead to new physical results.

The high-temperature series for the Ising model is conveniently rewritten by using the identity:

$$\begin{aligned}\exp(-\beta J_{ij}S_iS_j) &= 2[\cosh \beta J_{ij}](1 - S_iS_j \tanh \beta J_{ij}) \\ &= 2 \cosh \beta J_{ij}(1 - S_iS_jT_{ij}) . \end{aligned} \tag{4.3.2}$$

We may then write

$$Z = \mathrm{Tr}\, e^{-\beta H} = 2^{NZ/2} \prod_{\langle ij \rangle} \cosh \beta J_{ij} \operatorname*{Tr}_{S_i} \prod_{ij} (1 - T_{ij}S_iS_j) . \tag{4.3.3}$$

Before taking the log and averaging we note that we may expand Z as a series of diagrams in which the lines represent T_{ij}'s, the vertices S_i's, and since $\mathrm{Tr}\, S_i^{2n+1} = 0$, any integer n, an even number of lines intersect at any vertex. The trick (4.3.2) has eliminated all repeated lines in favor of the universal term

$$\begin{aligned}F_{\text{pairs}} &= -kT \ln Z_{\text{pairs}} = -kT \prod_{ij} \ln \cosh \beta J_{ij} \\ &= \frac{-NZ}{2} \sum_{ij} \frac{J_{ij}^2}{kT} = -\frac{N}{2} \frac{\mathcal{J}^2}{kT}\end{aligned} \tag{4.3.4}$$

by the fact that $\beta^2 J_{ij}^2 = \beta^2 \mathcal{J}^2/Z \ll 1$.

When we come to consider the remainder of F,

$$F_{\text{loops}} = -kT \left\langle \ln \operatorname*{Tr}_{S_i} \prod_{\langle ij \rangle} (1 - T_{ij}S_iS_j) \right\rangle \tag{4.3.5}$$

we note that every term may be written as a connected diagram which contained, before the $\langle \ln \rangle$ operation, only single lines with even vertices (see fig. 4.4), but when we expand the ln out in powers and average, must contain only pairs of lines J_{ij} since $\langle J_{ij} \rangle = 0$. Hence the above diagrams lead to fig. 4.5.

Aside from combinatorial factors, we can calculate the magnitudes of these diagrams as follows:

(1) $\mathcal{J}^2/Z$ for each pair of lines.

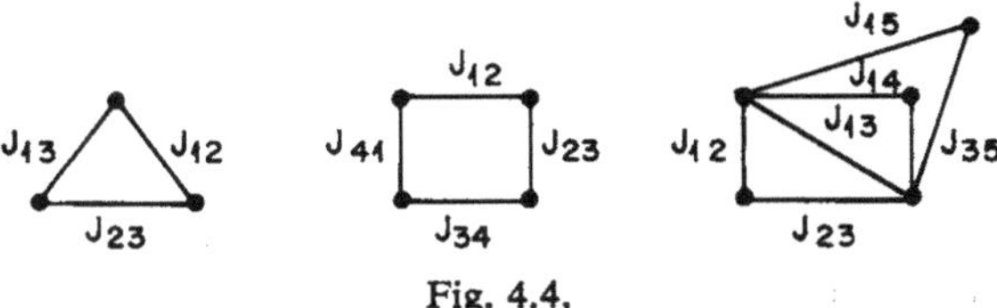

Fig. 4.4.

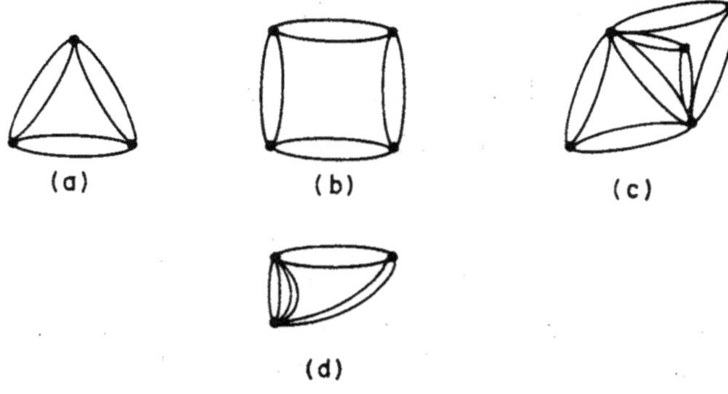

Fig. 4.5.

(2) N for one origin site in each diagram.
(3) Z for each other site.

Thus the largest such diagrams are all simple loops (a), (b) which are $O((N/Z)(\beta^2 J^2)^n)$ and all others have extra factors of $1/Z$, e.g., $c \sim (NZ^4 J^{14}/Z^7)$ $d \sim NZ^2 J^8/Z^4$, etc.

We see that all diagrams in F_{loops} are in principle negligible in the limit $Z \to \infty$, since they are of order $1/Z$ relative to F_{pair}. This is the reason for the peculiar form of the high T result. But it also leaves us at a loss, at first, as to why F_{pairs} is not valid at all T (which it clearly is not, since it exceeds obvious lower bounds on the energy and entropy as $T \to 0$; this alone is essentially a proof of a phase transition, since some analytic singularity is required to change F at some finite T). The reason turns out to be easy to see: that although the loop diagrams are formally of order $1/Z$, they *diverge* at a T_c and hence cannot be neglected *below* that temperature.

This divergence is obvious in the sum of the largest diagrams which are all the possible simple loops:

$$F_{\text{loops}} = \frac{N}{Z}\left(\tfrac{1}{2}\beta^4 J^4 + \tfrac{1}{3}(\beta^2 J^2)^3 + \tfrac{1}{4}(\beta^2 J^2)^4 + \cdots\right)$$

$$= \frac{N}{Z}\left[\ln\frac{1}{1-\beta^2 J^2} - \beta^2 J^2\right], \tag{4.3.6}$$

and this diverges at

$$kT_{\text{SG}} = J, \tag{4.3.7}$$

in agreement with (4.2.12). Below T_{SG} we must assume that some new state has intervened.

The choice of a new state is an obvious one: we must assume that each spin S_i has acquired a frozen-in average moment m_i, which will be a random variable to be determined from J_{ij} and the final field equations. Clearly

$$m_i^2 = q_i \quad \text{and} \quad \langle m_i^2 \rangle = q \tag{4.3.8}$$

is the original EA order parameter. But what we searched for and found was a set of equations giving the actual local moment arrangement m_i itself.

The formal mathematical principle is the following. We choose a soluble mean field hamiltonian

$$H_0 = -\sum_{\langle ij \rangle} J_{ij}(m_i m_j - m_i S_j - m_j S_i)\,, \tag{4.3.9}$$

and use the identity

$$\mathrm{Tr}\, e^{-\beta H} = (\mathrm{Tr}\, e^{-\beta H_0})\left[\frac{\mathrm{Tr}\, e^{-\beta(H-H_0)}(e^{-\beta H_0})}{\mathrm{Tr}\, e^{-\beta H_0}}\right],$$

which is to say

$$Z = (\mathrm{Tr}\, e^{-\beta H_0})\langle e^{-\beta(H-H_0)}\rangle_{H_0}\,. \tag{4.3.10}$$

In conventional mean field theory, the average of the fluctuation hamiltonian

$$(H - H_0)_{ij} = J_{ij}(S_i - m_i)(S_j - m_j) \tag{4.3.11}$$

is assumed to be zero and the $\langle\ \rangle$ in (4.3.10) unity; but in the present theory, we evaluate this term exactly instead, since even above T_{SG}, we have already discovered that the fluctuations lead to a finite effect. The essential assumption that is necessary to evaluate (4.3.10) is that the loop diagrams converge and hence, for the same reason as before, are negligible (of order $1/Z$).

This evaluation was originally done by a clever diagrammatic analysis by David Thouless. But actually there are at least two ways to do it directly which are simple and physical. In the TAP paper, we used the Bethe method but for reasons I will shortly discuss, I think that is misleading. A more direct physical way is as follows. Taking the log, we may write

$$F = F_0 + F_{\text{fluct}}\,,$$

$$F_0 = -kT \ln \mathrm{Tr}\, e^{-\beta H_0}\,, \qquad F_{\text{fluct}} = -kT \ln\langle\ \rangle_{H_0}\,,$$

H_0 is soluble: it is just spins in a mean field, and it leads to

$$F_0 = -\sum_{(ij)} J_{ij} m_i m_j + kT \sum_i \left[\frac{1+m_i}{2} \ln\left(\frac{1+m_i}{2}\right) + \frac{1-m_i}{2} \ln\left(\frac{1-m_i}{2}\right) \right]. \tag{4.3.12}$$

This is the mean field free energy and $\partial F_0/\partial m_i = 0$ leads to the conventional mean field equation.

If the loops are convergent, F_{fluct} is just the same physical effect as led to the high temperature result: the spins give a fluctuation mean field on each other

$$\overline{(\delta h_i)^2} = \sum_j J_{ij}^2 \overline{(\delta S_j)^2} ,$$

and this in turn gives a mean free energy

$$\begin{aligned} F_{\text{fluct}} &= -\tfrac{1}{2} \sum_i \chi_i (\delta h_i)^2 \\ &= -\frac{1}{2kT} \sum_{(ij)} J_{ij}^2 \overline{(\delta S_i)^2}\, \overline{\delta S_j^2} \\ &= -\frac{1}{2kT} \sum_{(ij)} (1 - m_i^2)(1 - m_j^2) J_{ij}^2 . \end{aligned} \tag{4.3.13}$$

All more complicated fluctuation terms are negligible because they involve higher powers of J_{ij}^2 or loop correlations, in either case bringing in powers of $1/Z$.

From this simple free energy, we may derive a new mean field equation via $\delta F/\delta m_i = 0$:

$$kT \tanh^{-1} m_i = h_i^{\text{eff}} = \sum_j J_{ij} m_j - \frac{m_i}{kT} \sum_j J_{ij}^2 (1 - m_j^2) . \tag{4.3.14}$$

This is the TAP "mean field" equation.

(4.3.14) gives the mean field values at which F is stationary, which is an elementary condition for stability of the solution. It clearly always may be solved by $m_i = 0$, but we know that below $kT_{SG} = \tilde{J}$, this is not an actual solution because the loop diagrams are divergent and positive: the system blocks itself and cannot reach this solution. We hope (and will show) that below T_{SG} there is at least one other solution with finite m_i (and thus smaller $\chi_i \propto 1 - m_i^2$) at which the loops are convergent.

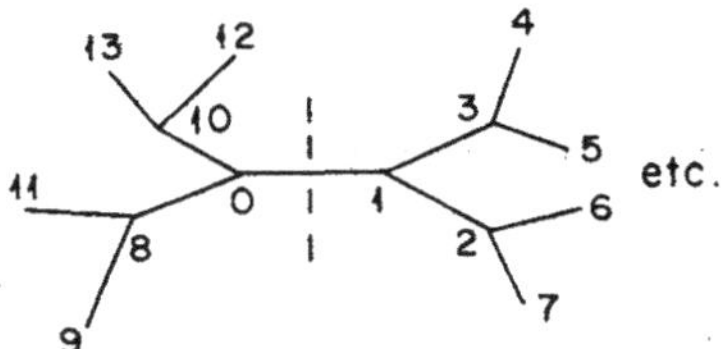

Fig. 4.6. Bethe lattice ($K = 2, Z = 3$).

As we showed in the original paper [88], a second way to derive F is to take advantage of the fact that in the absence of loops all diagrams can be drawn on a Cayley tree or "Bethe lattice". This is a lattice which branches indefinitely, as shown in fig. 4.6, and the Ising model on it is in principle exactly soluble by the so-called "Bethe" or cluster method carried out in lowest order. The argument is that if we consider a specific interaction J_{01} (see fig. 4.6), the only effect all of the parts of the lattice to the right can have on that to the left is by biasing the relative population of up and down states of 1, which in turns affects 0. Hence we can summarize all of the effect of 3, 4, 5, 6, 7, etc. by an effective field on 1, $h_1^{(0)}$, by which we mean the field acting on 1 as used in a cluster calculation in which site 0 and J_{01} is explicitly included.

The Bethe calculation is carried out explicitly in TAP. Consider first the cluster centered on δ_0, with hamiltonian

$$H^{(0)}_{\text{cluster}} = \sum_j J_{0j} S_0 S_j - h_j^{(0)} S_j \,. \tag{4.3.15}$$

It is easy to derive that

$$m_0 = \tan \beta \sum_j J_{0j} \tanh \beta h_j^{(0)} \tag{4.3.16}$$

(using $J_{0j} \ll 1$).

Now we also calculate the mean magnetization of a peripheral spin m_j, using (4.3.15) again and the smallness of J but not necessarily of h)

$$m_j = \tanh \beta h_j^{(0)} + m_0 \beta J_{0j} [1 - \tanh^2 \beta h_j^{(0)}] \,.$$

The manipulation in the original paper is now to consider the $h_j^{(0)}$ as auxiliary variables to be eliminated in favor of the m_j's, which can be done:

$$\tanh \beta h_j^0 \simeq m_j - m_0 \beta J_{0j} (1 - m_j^2) \,,$$

where we have neglected corrections of order $J^2 \sim 1/Z$. Inserting this into (4.3.16), we recover the field equation (4.3.14).

It is at first disturbing that the field equation follows from the Bethe method and the Cayley tree lattice. On a Cayley tree, there is no "frustration" at all: no competition between different paths, because no loop diagrams. But in fact we know this: (4.3.14) is correct *only if the loops converge*, and their divergence is the effect of frustration. What is the criterion for loop convergence? At first sight it seems unrelated to the above argument; but in fact we now show that it is identical in result.

The argument again follows from the Bethe cluster theory. I want first to show that the Bethe self-consistency condition is identical with the condition for convergence of the shattered susceptibility.

To do this we rearrange our thinking about the Bethe theory slightly. We imagine that from the Bethe self-consistency condition (4.3.14) we either raise T (lower β) keeping all m_i fixed, or raise the m_i keeping T fixed. In this case the equations are no longer self-consistent. Above T_{SG} there is no spontaneous m_i, and if T is too big, h_i does not produce a correspondingly large m_i, and all the order parameters q_j will decrease. That is to say that the staggered susceptibility is finite; even if we were to add a small δh_i in this situation, it would lead to no long-range effects. On the other hand, if we *lower* T or the m_i, we find that a spontaneous deviation of the n_i occurs without *any* δh_i to drive it; the shattered susceptibility is *infinite*!

Another way to express it is to study the propagation of order along the lattice, in the guise of the quantities $h_j^{(0)}$: can we express, say, $h_0^{(j)}$ in terms of the $h_j^{(0)}$ and hence study propagation of the ordering field? We do the following "gedanken" experiment: we introduce fields $h_i^{(N-1)}$ at the Nth generation of sites out from site 0 along the lattice (in fig. 4.6 1, 8, 10 are the first generation, 9, 11, 12, 13, 2, 3 the second, etc.) and ask what is, first, $m_0(h_i^{N-1})$, via asking how the fields themselves propagate along the lattice. It is very easy to show from (4.3.15) and (4.3.16) that in fact

$$h_0^{(j')} = \sum_{j \neq j'} J_{0j} \tanh \beta h_j^{(0)} \tag{4.3.17}$$

(again using the smallness of J). Let us make a small change $\delta h_j^{(0)}$, then (4.3.17) shows us that

$$\delta h_0^{(j')} = \sum_{j \neq j'} \chi_j J_{0j}\, \delta h_j^{(0)}\,, \tag{4.3.18}$$

where

$$\chi_j = \beta(1 - \tanh^2 \beta h_j^0) \simeq \beta(1 - m_j^2)\,.$$

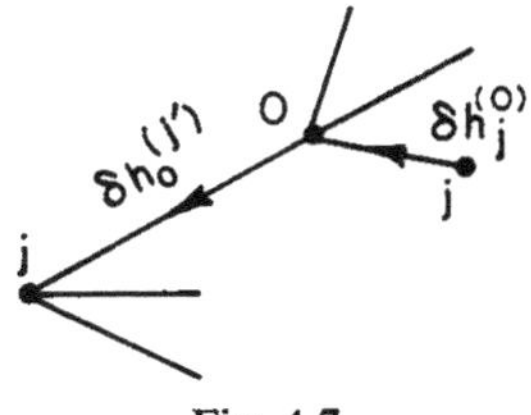

Fig. 4.7.

(4.3.18) means that the ratio of the variance of $\delta h^{j'}$ to that of δh^j is

$$\overline{(\delta h_0^{(j')})^2} = \left(\sum_{j \neq j'} J_{0j}^2 \chi_j^2\right) \overline{(\delta h_j^{(0)})^2} . \tag{4.3.19}$$

We may verify that the condition for T_c is precisely that

$$\beta^2 \tilde{J}^2 = \sum J_{0j}^2 \chi_j^2 = 1 :$$

$\overline{\delta h_j^2}$ does not increase. If

$$(\delta h_0^{(j')})^2 > (\delta h_j^{(0)})^2 , \tag{4.3.20}$$

that means that order grows exponentially in the lattice, because we will get a finite growth with each "generation" in the Cayley tree. This corresponds to a divergence of the "shattered susceptibility".

We now argue that this is also the condition for convergence of loop diagrams in any physical, finite (if large) dimensional lattice. If we try to embed a Cayley tree with $Z \gg 1$ in a lattice of dimensionality d, we find that it is reasonably possible for a few generations of neighbors; so the *local* structure of a Cayley tree and a physical large Z lattice of finite d are similar. But the long-range topological structure does not fit in any real sense even in the limit $d \to \infty$. This is because the non-repeating diagrams of length L in a d-dimensional lattice diffuse out to a distance $R \sim L^{1/2}$, containing a volume of lattice points

$$(n_L)_d \sim B_d L^{d/2} \quad (B_d \sim (d/2)!) ,$$

whereas the corresponding volume in the Cayley tree gives us

$$(n_L)_{\text{Cayley}} \sim Z^L ,$$

which is enormously larger as $L \to \infty$, if $L > d$. This means that in a real lattice there is a very high degree of multiple connectivity, and in particular that the Lth generation of the equivalent Cayley tree from site 0 will *always* have many other paths of length $l \ll L$ ($l \sim L^{1/2}$ in general) connecting to site 0 directly. Such paths mean that a mean spin m_0 on site 0 will always give us an effective polarization field

$$\delta h_{i,L}^{(f)} \sim \mathrm{e}^{-Al} m_0 \sim \mathrm{e}^{-BL^{1/2}} m_0$$

on the Lth generation of sites from site 0 which, if the order grows exponentially, will lead inevitably to a loop diagram divergence when the condition (4.3.20) is met, since this will in turn cause a polarization of site m_0 of exponential order in L.

The coincidence of the stability condition for convergence of loop diagrams, and the field equations ignoring loops, is thus proven. The convergence condition is the same as the condition that the "shattered susceptibility" χ_{ij} be just finite, which may be rewritten

$$|(\chi)_{ij}^{-1}| = \left|\frac{\partial^2 F}{\partial m_i\, \partial m_j}\right| = 0\,, \tag{4.3.21}$$

where $|\;|$ means determinant: i.e. the smallest eigenvalue just reaches zero at this point. The F which appears in (4.3.21) must be just the same F which determines the field equation, so we have:

$$\begin{aligned} &\frac{\partial F}{\partial m_i} \equiv 0 \qquad \text{(field equation)}\,, \\ &\left|\frac{\partial^2 F}{\partial m_i\, \partial m_j}\right| \equiv 0 \quad \text{(convergence)}\,. \end{aligned} \tag{4.3.22}$$

These two conditions occur at the same point (by the Bethe lattice argument) and hence we have a vitally important fact about the TAP result: the solution is a *saddle-point* of F, not a minimum (see fig. 4.8, from the original paper). This result was stated empirically but not proven in TAP. We feel that it is a key result which may be generalizable to the whole class of spin glasses, although we have not yet found a more general proof. It implies among other things that the entire temperature axis below T_c is a line of critical points, at which fluctuations will be anomalously large and responses anomalously slow. This at least is in agreement with observational fact.

Let us now, after this very general discussion, actually try to solve the TAP equations (4.3.14) as was done rather sketchily in the original paper. The first step is to note that the largest eigenvalue J_λ, and the distribution of eigenvalues $\rho(J_\lambda)$, of J_{ij} are known, although of course the corresponding eigenvectors are random.

$$(J_\lambda)_{\max} = 2\tilde{J}\,, \qquad \rho(J_\lambda) = \sqrt{(2\tilde{J})^2 - J_\lambda^2}\,. \tag{4.3.23}$$

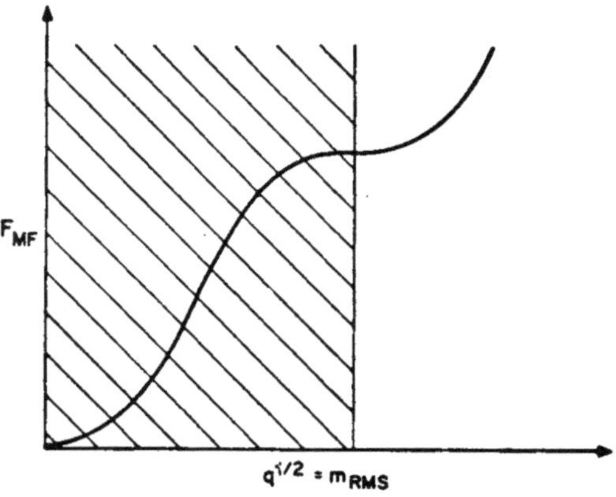

Fig. 4.8.

Denoting the normalized eigenvectors by $(i|\lambda)$ we have

$$\sum_j J_{ij}(j|\lambda) = J_\lambda(i|\lambda)\,,$$

and let us take the largest eigenvector as $(i|\lambda_m)$. Palmer (private communication) has proven that the eigenvectors are uncorrelated Gaussian variables, so we can neglect all correlations of odd powers and assume that the only non-vanishing averages are

$$\sum_j |(j|\lambda)|^2 = 1\,, \quad \sum_j (j|\lambda)^4 = 3/N\,, \quad \text{etc.} \tag{4.3.24}$$

The equation which must be solved may be linearized near T_c:

$$\sum_j J_{ij}m_j = \beta \tilde{J}^2(1 - \overline{m^2})m_i + T(m_i + \tfrac{1}{3}m_i^3 + \tfrac{1}{5}m_i^5 + \cdots)\,. \tag{4.3.25}$$

If we set $m_j = M(j|\lambda_m)$, $M \lll 1$, we get from the linear terms

$$(J_\lambda^m - \beta\tilde{J}^2 - T)M = 0\,, \qquad T\left(1 - 2\tilde{J}/T + \left(\frac{\tilde{J}}{T}\right)^2\right) = 0\,,$$

which is satisfied at $T_c = \tilde{J}$ but *only* by a *double zero*: the conventional "r" of the Ginzburg–Landau functional comes in at T_c with zero slope! This is another symptom of the peculiar nature of F and the fact that $m_i = 0$ is always a nominal minimum of F.

We now try to satisfy (4.3.24) with a solution

$$m_i = \sqrt{N}M(i|\lambda) + \delta m_i\,, \tag{4.3.26}$$

where

$$\sum_i \delta m_i(i|\lambda_m) = 0\,,$$

so that

$$\delta m_i = \sum_{\lambda \neq \lambda_m} (i|\lambda)\, \delta m_\lambda\,. \tag{4.3.27}$$

Substituting (4.3.26) in (4.3.25), we multiply through by $(i|\lambda_m)$ and sum over i, and use orthonormality as well as the Gaussian uncorrelated random nature of the $(i|\lambda)$ to obtain an equation valid to the necessary order (5th) in M

$$\begin{aligned} 2\bar{J}M &= \beta \bar{J}^2 M(1 - M^2) + T(M + M^3 + 3M^5) \\ &\quad N^{1/2} M^2 \sum_i (i|\lambda_m)^3\, \delta m_i\,. \end{aligned} \tag{4.3.28}$$

The last term on the right in (4.3.28) is *not* zero because δm_i, while orthogonal to $(i|\lambda_m)$, is not at all orthogonal to $(i|\lambda_m)^3$. Thus we must find an equation for δm_i, which we obtain by using (4.3.25) again, keeping the lowest order orthogonal terms

$$\begin{aligned} &\sum_\lambda \left(J_\lambda - \frac{\bar{J}^2(1 - M^2)}{T} - T \right) \delta m_\lambda (i|\lambda) \\ &\quad = \frac{T}{3} N^{3/2} M^3 \{(i|\lambda_m)^3\}_{\text{orthogonal piece}}\,. \end{aligned} \tag{4.3.29}$$

Now we must note that

$$(i|\lambda_m)\left[(i|\lambda_m)^3 - \frac{3}{N}(i|\lambda_m)\right] = 0\,,$$

because $(i|\lambda_m)$ is a Gaussian random variable, so the expression in brackets is the orthogonal part of $(i|\lambda_m)^3$ for use in (4.3.29). Next we define

$$f_\lambda = \sum_i (i|\lambda)\left[(i|\lambda_m)^3 - \frac{3}{N}(i|\lambda_m)\right], \tag{4.3.30}$$

and observe that by multiplying (4.3.29) by $(i|\lambda)$ and summing, we get

$$\delta m_\lambda \simeq -\frac{\frac{1}{3} T N^{3/2} M^3 f_\lambda}{2\bar{J} - J_\lambda}\,, \tag{4.3.31}$$

which determines δm_λ and, by (4.3.26), δm_i.

Finally we have

$$\sum_i (i|\lambda_m)^3 \, \delta m_i = \sum_\lambda \sum_i \left[(i|\lambda_m)^3 - \frac{3}{N}(i|\lambda_m)\right](i|\lambda) f_\lambda \frac{\frac{1}{3}TN^{3/2}M^3}{2\bar{J} - J_\lambda}$$

$$= \sum_\lambda f_\lambda^2 \frac{\frac{1}{3}TN^{3/2}M^3}{2\bar{J} - J_\lambda},$$

so that the last term of (4.3.27) is

$$\frac{T^2N^2M^5}{3} \sum_\lambda f_\lambda^2 \frac{1}{2\bar{J} - J_\lambda}.$$

This then leaves us only with the task of evaluating $\overline{f_\lambda^2}$ and doing the integral for $\overline{(2\bar{J} - J_\lambda)^{-1}}$. We do $\overline{f_\lambda^2}$, using simple Gaussian and uncorrelated assumptions, by squaring (4.3.30):

$$f_\lambda^2 = \sum_i (i|\lambda)^2 \left[(i|\lambda_m)^6 - \frac{6}{N}(i|\lambda_m)^4 + \frac{9}{N^2}(i|\lambda_m)^2\right] = \frac{6}{N^2},$$

while $\overline{(2\bar{J} - J_\lambda)^{-1}} = \bar{J}^{-1}$ so that the field equation gives eq. (21) of TAP:

$$\left(2\bar{J} - \frac{\bar{J}^2}{T} - T\right)M - \left(T - \frac{\bar{J}^2}{T}\right)M^3 + \left(\frac{2T^2}{\bar{J}} - 3T\right)M^5 = 0. \tag{4.3.32}$$

It is easily verified that the solution of (4.3.32) is to the only valid order a double zero:

$$1 - \left(\frac{T_c}{T}\right)^2 - \left(\frac{T_c^2}{T^2} - 1\right)M^2 + M^4 = 0,$$

$$\left(\left(\frac{T_c}{T} - 1\right) - M^2\right)^2 = 0.$$

Thus we confirm that $\partial^2 F/\partial(M^2)^2 = 0$: a saddle point in $M^2 = q_f$ as shown in fig. 4.8.

The manipulations necessary to arrive at a low temperature solution are also difficult, particularly since we have no external information like $\rho(J_\lambda)$ to guide us. Consideration of the low-temperature field equations, however, can give us a number of hints when we realize that the solution at $T = 0$ must satisfy

$$m_i^0 = \operatorname{sgn} h_i,$$

where

$$h_i = \sum_j J_{ij} m_j^{(0)} ,$$

and the overall energy has been minimized as well by rearrangement of the signs of m. We generated many such solutions numerically to study their properties. It became evident that stability of such solutions is not easily achieved, and in fact that the density of small fields h_i, $p(h_i)$, must necessarily be zero at $h_i = 0$. This is easily demonstrated, as is also that the slope must be linear.

Consider the subset of spins with $h_i < \delta h$, with δh a very small number. There are

$$\frac{n}{N} = \int_0^{\delta h} p(h)\,\mathrm{d}h \text{ of these .}$$

The energy due to all other spins may be assumed to essentially determine h, so that the energy lost relative to the remainder of the spins per spin by rearranging some of them will be δh.

On the other hand, all of these spins will be in interaction with each other with exchange integrals $\sim \tilde{J}/\sqrt{N}$. The variance of this interaction will be

$$\overline{(\Delta h)^2} = \frac{\tilde{J}^2}{N} n .$$

By rearranging only these spins, we can shift any of their effective fields by

$$\sim [\overline{(\Delta h)^2}]^{1/2} \sim \tilde{J} \left[\int_0^{\delta h} p(h)\,\mathrm{d}h \right]^{1/2} .$$

But this must not be bigger than δh or our original assumption was inconsistent. This requires that

$$p(h)\,\mathrm{d}h = \frac{h}{H_0^2}\,\mathrm{d}h , \tag{4.3.33}$$

so that there is a linear hole in the effective field distribution! (Quite contrary to all early speculations on the subject.) This hole was observed in the simulations.

The TAP equations plus the result that $\partial^2 F/\partial M^2 = 0$ allows us to determine the slope H_0 quantitatively. H_0 will increase, thus stiffening χ, until $\partial^2 F/\partial M^2$ is just zero, which is the boundary of the stability region. This we found as follows: given $p(h)$ near $h = 0$, i.e. H_0^2, it is possible to use the TAP equations to calculate the first correction to $\overline{m^2}$ at very low temperatures, which is of the form

$$\overline{m^2} = 1 - \alpha \frac{T^2}{\tilde{J}^2} \cdot \tag{4.3.34}$$

(Note that $\chi = (1 - m^2)/T$ thus becomes *linear* in T, in contrast to all previous theories of the Ising case.)

The equation determining m is

$$T \tanh^{-1} m_i = \left(h_i - \frac{m_i \tilde{J}^2}{T}(1 - \overline{m^2})\right). \tag{4.3.35}$$

It is possible to show that as $T \to 0$

$$h_i = h_i^0 - O(T^2)\,,$$

so that we may write

$$h_i = T \tanh^{-1} m_i + \alpha m_i T\,,$$

so that

$$1 - \frac{\alpha T^2}{\tilde{J}^2} = \int P(h)\, \mathrm{d}h\, m^2(h)\,,$$

which leads to, inserting $P(h \to 0) = h/H_0^2$,

$$\frac{\alpha}{\tilde{J}^2} = \frac{1}{H_0^2} \int_0^1 \mathrm{d}m\,(1 - m^2)(\tanh^{-1} m + \alpha m)\left(\frac{1}{1 - m^2} + \alpha\right), \tag{4.3.36}$$

which leads to the result quoted in the paper,

$$\frac{H_0^2}{\tilde{J}^2} = \tfrac{1}{4}\alpha + \frac{(2 \ln 2 + 1)}{3} + \frac{\ln 2}{\alpha}\,. \tag{4.3.37}$$

Note that indeed we find that H_0 cannot be less than $1.28\tilde{J}$ and lead to a finite χ, since the expression on the right has a minimum as a function of α. This is therefore the correct value of α (which determines χ as $T \to 0$) and H_0 (which determines $C_{sp}(T \to 0)$). The results for χ and for $M = \sqrt{q}$ are shown in fig. 4.9.

The trail which we have followed to these results is not a simple one, although we feel we justified all the steps along the way; and it is fascinating that the results disagree quite sharply with ,the replica theory. Some skepticism might be justified, but surely not the degree which has been shown in the literature (in that papers still seem to come out re-solving the S–K model with different results from ours). Thus it is perhaps fortunate

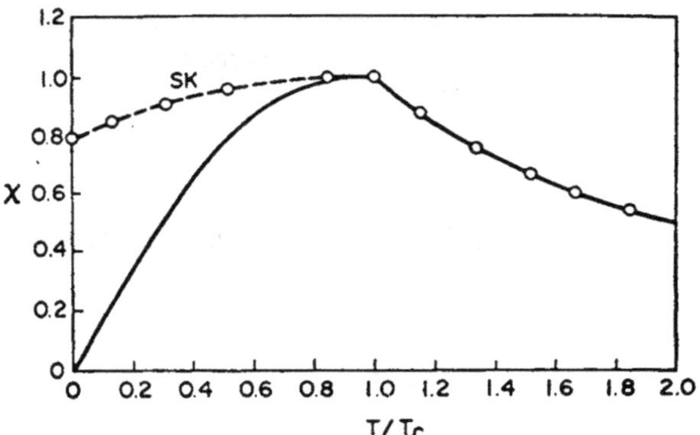

Fig. 4.9a.

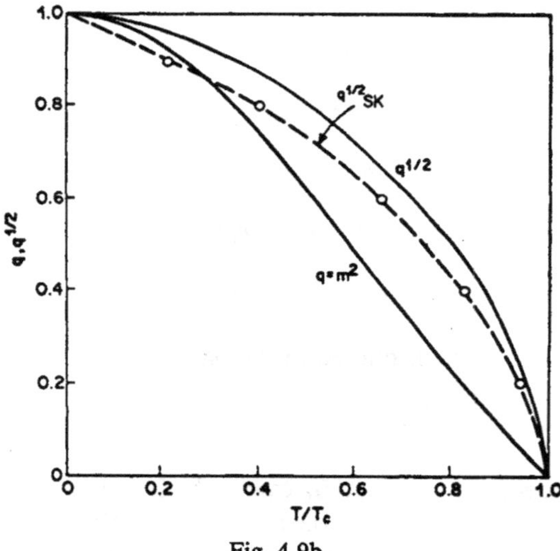

Fig. 4.9b.

that a careful computer simulation using $N = Z = 500$ quite unambiguously verifies the TAP results (see fig. 4.10, after Kirkpatrick and Sherrington [90]).

A minor industry has developed around trying to repair the replica method. Thouless and de Almeida [86] seem to show with some conviction that the problem is in the ambiguity of extrapolation to $n \to 0$; their contention is that the replica-symmetric solution is not the correct one to

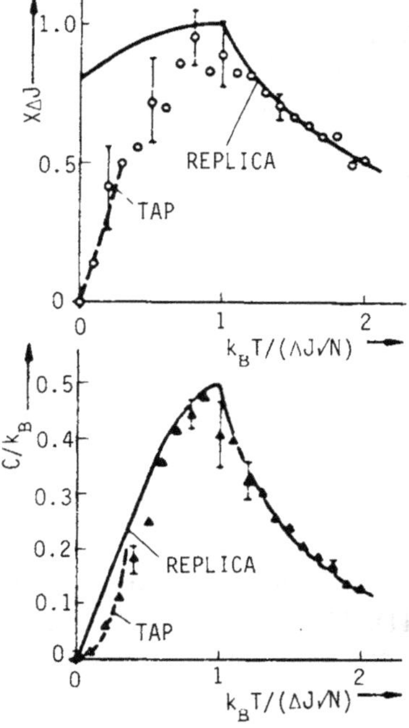

Fig. 4.10.

extrapolate. Bray and Moore [87] have carried this line further, and Palmer and a co-worker also reached a similar conclusion by a different route. I feel that, as can be seen from the Bray–Moore work, the repairs on the replica method may be very difficult, even if possible. More tangible and direct methods such as TAP may still be as useful. Bray and Moore do seem to be achieving very interesting and suggestive results.

Some additional results have been achieved by the generalization of TAP to other systems. Palmer and Pond (to be published) have discussed the case of vector spins. In this case the fluctuation free energy term takes a new form:

$$F_{\text{pairs}} = -\tfrac{1}{2}\sum_{ij}\chi_j(\delta H)_i^2 = -\tfrac{1}{2}\sum_{ij}J_{ij}^2(1 - m_i^2)\chi_j\,,$$

and for three-dimensional classical vectors of length l, for instance, the average χ is

$$\overline{\chi_j} = \frac{2m_j^2}{3h_j} + \frac{1}{3T}(1 - m_j^2)\,,$$

the first term being the susceptibility in the transverse direction to the field $\boldsymbol{h}_j$ (now a vector). This remains finite even as $m_j \to 1$ and makes the low-temperature effect stronger. The effective field acting on i (in the sense of the energy necessary to reverse S_i) then becomes

$$\begin{aligned} h_i^{\text{eff}} &= \sum_j J_{ij} m_j - \frac{\partial F_{\text{pairs}}}{\partial m_i} \\ &= \sum_j J_{ij} m_j - m_i \sum J_{ij}^2 \left(\frac{2m_j^2}{3h_j} + \frac{(1 - m_j^2)}{3T} \right). \end{aligned} \tag{4.3.38}$$

In the limit as $T \to 0$, $m_j \to 1$,

$$h_i^{\text{eff}} \to \sum_j J_{ij} m_j - \frac{2\tilde{J}^2}{3} \left\langle \frac{1}{h_j} \right\rangle. \tag{4.3.39}$$

Thus the state is not stable unless

$$h_i = \sum_j J_{ij} m_j > h^0 = \frac{2\tilde{J}^2}{3} \left\langle \frac{1}{h} \right\rangle. \tag{4.3.40}$$

That is, there is a *hole* in the field distribution: there are *no* values of $h < h^0$ (see fig. 4.11). This was verified by simulations. Palmer has estimated the size of h_0 and finds it quite appreciable – of order $0.5\tilde{J}$. The field distribution, then, becomes quite narrow – an interesting result physically.

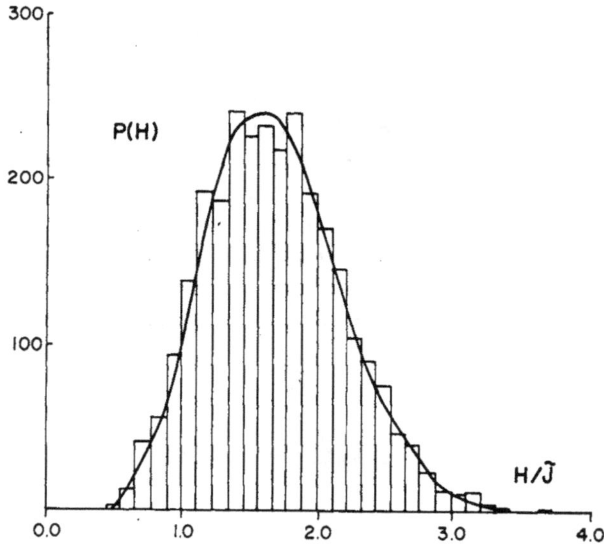

Fig. 4.11.

Nonetheless Palmer was able to show that the *specific heat C* is not determined by the field distribution in the usual way, but by the distribution of effective fields h_i^{eff}, and is quadratic in T as in the Ising case.

A second interesting generalization of TAP is to finite Z. For a physical model we picture a Cayley tree of finite Z embedded in a lattice of very high dimensionality d; the spin glass on a true Cayley tree is not a meaningful model because there would be no frustration and one may simply transform to a pure ferromagnet by flipping spin labels. But once one thinks in terms of a lattice with some, but very few, loops, exactly the same formal structure may be repeated and, using the Bethe method, one arrives at a set of field equations analogous to (4.3.14), and order propagation equations analogous to (4.3.17)–(4.3.19). These are very complicated in the general case. I have worked them out (with the help of C. M. Pond) in the linear region near T_c, linearized in m and h but, of course, not in J_{ij} (TAP is the opposite case, linearized in J but not in m and h).

The result for the field equation is rather interesting:

$$m_i = \sum_{j \text{ neighbor to } i} [m_j \sinh \beta J_{ij} \cosh \beta J_{ij} - m_i \sinh^2 \beta J_{ij}]\,, \tag{4.3.41}$$

$$m_i = \frac{\sum_{jni} (\sinh \beta J_{ij} \cosh \beta J_{ij}) m_j}{1 + \sum \sinh^2 \beta J_{ij}}\,.$$

Note that the coefficient of m_j is <1, and that a large $\sinh \beta J_{ij}$ reduces all neighboring coefficients; these correlations are apparently necessary to eliminate localized eigenvalues of the shattered susceptibility matrix defined by

$$(\chi^{-1})_{ij} = \beta[\delta_{ij}\left(1 + \sum \sinh^2 \beta J_{ij}\right) - \sinh \beta J_{ij} \cosh \beta J_{ij}]\,. \tag{4.3.42}$$

As far as we can see, χ^{-1} does not have localized zero eigenvalues, since it seems impossible for an eigenvector to have a maximum at any point.

For all $Z > 2$, there does seem to be a solution of (4.3.41) at some finite T (correctly, the linear chain gives no transition). We have not been able to show how far the finite Z case mimics the TAP theory at and below this transition, but as far as we can see, it is likely that it does.

The equation for propagation of the ordering field is just like (4.3.17) itself, essentially:

$$h_0^{(j')} = \sum_{j \neq j' \; j \text{ neighbor of } 0} h_j^{(0)} \tanh \beta J_{0j}\,, \tag{4.3.43}$$

where the tanh plays the role of βJ_{ij} in (4.3.14).

T_c is determined by the equation which gives the threshold for exponential growth:

$$\langle \ln(\tanh \beta_c J_{0j})^2 \rangle_{\text{ave}} = \ln \frac{1}{Z-1}.$$

This is quite different from what appears in the literature on the spin glass Bethe method [91].

With these preliminary results on the finite Bethe method, we conclude this rather long discussion of TAP. I think this method reveals very clearly the intrinsic features of the spin glass problem. It shows that the transition is basically a "*blocking*" transition: the spins, allowed independently to adjust to the random exchange energy, would achieve a much lower free energy. Unfortunately at T_{SG} this freedom suddenly is *blocked* by frustration effects: the spin recognizes that it is being given contradictory orders. The corresponding terms in the free energy are intrinsically *positive*: the mean energies due to long-range correlations along long loops are equally often positive and negative, leaving only fluctuation terms which are all positive. These terms still diverge, just as in a conventional broken symmetry signalled by divergence of the appropriate susceptibility; but this divergence is to be avoided, and the avoidance takes the form of the observed freezing of the polarization.

Correspondingly, at all $T < T_c$ the system retains the divergent shattered susceptibility, hence at all $T < T_c$ *even at infinite dimensionality*, the system remains at its critical point. This behavior is unique to the spin glass and is the reason why I have dealt with what is basically the mean field theory in such detail, and will treat all other techniques so relatively briefly. Where the mean field theory is already so aberrant, it is reasonable to treat any results obtained by conventional methods with skepticism.

4.4. Applications of the renormalization group and other "conventional" phase transition methods to the spin glass problem

Three more or less conventional theoretical methods for phase transitions have been used to attack the spin glass problem. These are: (1) renormalization group applied to the replica hamiltonian; (2) real-space renormalization; (3) high-temperature series methods. These methods give some solid results and a great deal of contradiction and confusion.

The first, and essentially the key, application of RNG to the replica hamiltonian was by Harris, Lubensky and Chen [92]. Both replica theory and TAP (and, in fact, physical intuition) would agree in assigning to the parameter $q(t)$ or $q_{\alpha\beta}$ certain intrinsic positivity conditions. In terms of replica theory, we have for integral powers p

$$\underset{(\alpha\beta\text{ etc.})}{\mathrm{Tr}}\ q^p = n(n-1)(n-2) \quad \cdots p \text{ times}\,, \tag{4.4.1}$$

since this is just

$$\sum_{\alpha\beta\gamma\ldots} S_i^\alpha (S_i^\beta)^2 (S_i^\gamma)^2 \cdots S_i^\alpha\,.$$

In particular, $\mathrm{Tr}\, q^3$ has a fixed value, proportional to n. Equally, if we think of q as a time-correlation $\langle S_i(t)S_i(0)\rangle$ this is clearly positive or zero. In order to do renormalization, we need to replace the replica hamiltonian plus correlation conditions on $q_{\alpha\beta}$ (see sect. 4.2) with a Ginsburg–Landau free energy functional, which must therefore have a term coupling to $\mathrm{Tr}\, q_{\alpha\beta}^3$. The existence of this q^3 vertex in the theory in addition to the conventional terms:

$$H_{\mathrm{eff}} = \mathrm{Tr}\{\gamma(T_c - T)q^2 + S(\nabla q)^2 - tq^3\} + w(\mathrm{Tr}\, q^2)^2 + \cdots, \tag{4.4.2}$$

introduces new orders of divergence into perturbation series, and new terms into the equations for exponents. In particular, we can add to any diagram in perturbation series two new vertices and one new line (fig. 4.12). If all lines and vertices are near $k = 0$ (and this is the source of the fluctuation divergences of critical point theory) the second diagram contains three new propagators k^2 in the denominator, and one extra d-dimensional integral over $k'' - k$. Thus for all dimensionalities $d \leqslant 6$, the new diagram is more divergent than the old, since it contains an extra factor of order

$$t^2 \int \frac{k^{d-1}\,\mathrm{d}k}{k^6}\,.$$

This means that perturbations from standard mean field theory must diverge for all $d \leqslant 6$, which implies that 6 is the upper critical dimensionality below which mean field theory is qualitatively wrong at T_c. One can then, by standard methods, expand in powers of $\varepsilon = 6 - d$, but this is three times less meaningful than the conventional $4 - \varepsilon$ expansion. In

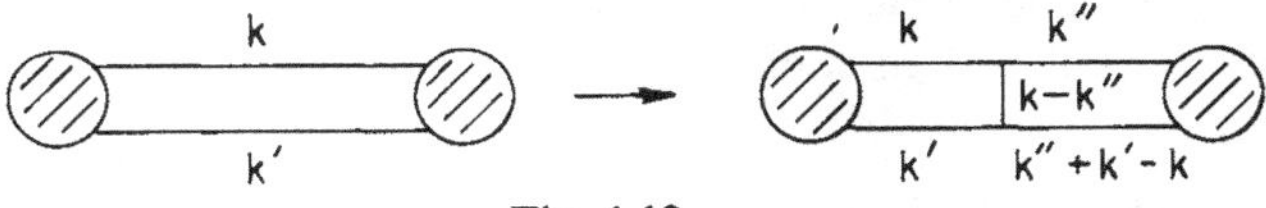

Fig. 4.12.

particular, a number of methods, some of which we will discuss, suggest that $d = 4$ is a *lower* critical dimensionality beyond which such expansions will not work, so that the physical case of $d = 3$ is unavailable, not just difficult.

It is important to realize that (4.4.2) is not to be handled in too literal a fashion, in that we must handle delicately values of q with the physically disallowed sign, by proper manipulation of r, t, and w. w must not become essentially negative, so that q^3 does not act like a Landau M^3 term and cause the transition to become first order. The rather delicate point that q must not be treated as a truly independent variable, but is just a surrogate for the real variables S_α^i, must be kept in mind at all times, and I am not sure this has been handled properly in all of the treatments of the replica method (though it surely was in Harris et al. [92]).

Below T_c the conventional replica method fails, as we have seen, and in particular TAP seems to leave us with a critical behavior at all T for $H = 0$. One naive approach is to assume that the q^3 "saddle-point" behavior of TAP is the appropriate mean field hamiltonian at all T, and that the stiffness r vanishes. Then $d = 2$ is surely a lower critical dimensionality even for the Ising model, by just the same argument as for vector spins. Bray and Moore [87] have made a serious attempt to tackle the modified $n \to 0$ method with replica asymmetry by these techniques, and seem to predict a lower critical dimensionality (below which there can be no true spin-glass order) of $d = 4$.

Real-space renormalization methods in this problem were pioneered by Young and Stinchcombe [93] and have also been used by Jayaprahesh, Chalupa, and coworkers [94] and Kinzel and Fischer [95]. The technique here is to combine the spins into "blocks" by one technique or another, replacing a lattice of spacing a by ones with spacing $2a$, $4a$, etc. One may use a given subset of the spins, integrating over the intermediate spins to obtain an effective coupling between the selected spins

$$\sum_{\substack{\text{subset}\\ ij \neq j', i'}} \exp\left(-\beta \sum_{ij} J_{ij} S_i S_j\right) \to \exp\left(-\sum_{i'j'} \beta J_{i'j'}^{\text{eff}} S_{i'} S_{j'}\right), \tag{4.4.3}$$

as done by Young and Stinchcombe. Alternatively, one introduces an artificial "block spin" representing in some sense the net spin of the block, by using the Kadanoff technique of introducing a weight-function involving the new variables S_i' in such a way as to keep the free energy invariant: multiply by

$$(1 + W(S_i, S_i')S_i') \exp(-\beta H(S_i)), \tag{4.4.4}$$

where

$$\operatorname*{Tr}_{S'_i} W(S_i, S'_i)S'_i = 0\,,$$

and do one's traces over S_i first to obtain an effective coupling between the S'_i. This method was used by Fischer [81] and by Jayaprahesh et al. [94]. The idea is that above T_c one expects J^{eff} to decrease indefinitely as we scale up, signaling that the effective interaction gets smaller and smaller and the fixed point is simply paramagnetic; while below T_c, J^{eff} should get bigger as we renormalize. Of course, all methods find that the distribution of J_{ij}'s changes, as well as that new kinds of interactions seem to enter; multispin or non-linear ones. The former can be handled by careful computation, the latter is universally neglected. Young and Stinchcombe, and Young [116], found that $d = 2$ did not exhibit a transition while $d = 3$ did, for Ising lattices; Jayaprahesh et al. [94] found that $d = 2$ did exhibit a transition but that extending the methods to thermodynamics, α was negatively large so the specific heat is not peaked, and also (incidentally) the effect of magnetic field is very strong as observed. Both are gratifying in view of the experimental data. But Fischer and Kinzel using two different methods found two different results, and make no comment as to which is better. One gives no transition at all in two or three dimensions, one gives one in all cases. The results then are quite inconclusive but at least the simpler and more direct ones do give transitions in the physical case.

There are three difficulties at least with this kind of method of which two are often cited. In the first place, the distribution of exchange interactions J_{ij} will be different from the original one, and one must guess or derive the actual fixed point distribution which results after enough stages of "decimation".

Second, the method generates interactions at each stage which are not in the simple spin glass hamiltonian: multiple spin and non-linear interactions. These are neglected, and this is an essential crudity of the technique in all cases, which means that it seldom works perfectly but is often a fair approximation.

Finally, and this is unique to the spin glass: it is not at all clear that the variable appropriate for describing a block remains a simple Ising or vector spin at every stage. This is most obvious for three-dimensional vectors: any block will have a best structure which is fundamentally three-dimensional and has three inequivalent axes, which can rotate in the space O3 of a triad of vectors. This contrasts with a spin which has only one axis and the space S2.

It is not at all clear that the decimation approximation retains this extra structure. Even for the Ising model, as blocks become bigger they will have internal degeneracies or near degeneracies (like the "frustrated plaquettes" of Toulouse [97] and Villain [98]), which are additional degrees of freedom not described by the assumed new hamiltonian. The internal entropy of the blocks can increase in some singular fashion as they become bigger. It is also a question how a line of critical points might appear, or algebraic order. It was noteworthy that Chalupa finds very slow growth of the interactions even below T_c; this may confirm the "sticking" property which leaves it critical for all $T < T_c$.

Fisch and Harris [99] have applied Padé approximant methods to the high-temperature series for the second field derivative of the susceptibility of a standard spin glass (Domb has shown that under fairly general conditions, the susceptibility itself, according to high T series, is not modified at all above T_{SG}). The principle is very simple: one calculates as many orders of diagrams at high T as one can manage (in this case quite a few) and then matches the result to the high temperature series of a form like $A/(T - T_c)^B$ + similar terms with at least three adjustable parameters in this case. They achieved excellent fits, and also rather strikingly found that at precisely four dimensions T_c began to move off the real axis, as B became large and negative, signaling a lower critical dimensionality. This seems excellent evidence that at least conventional ideas will not carry us below 4d. The fact that T_c remains finite at 4d, however, does not preclude the possibility of the kind of algebraic long-range order below 4d which occurs in the 2d xy and similar systems. Thus we have an extremely mysterious situation: experiment, simulations, and some real space renormalization calculations show at least some phase transition properties at 2d and 3d for some systems; the best theory shows an LCD property at four dimensions. My own conjecture is that below 4d we have a whole dimensionality range of anomalous behavior, with possibly true phase transitions but no true long-range order, and all temperatures below T_{SG} critical. I will discuss some reasons for this conjecture below.

4.5. *Other ideas and schemes: frustration, lower critical dimensionality methods, gauge, clusters and exponents, etc.*

4.5.1. *Frustration*

This is a term supposedly used by the present author to describe the essential nature of the interactions which lead to spin glass behavior, but later restricted and/or generalized by Toulouse and others to give a specific local measure of the degree of "glassiness" of a system. The general idea is that spin glass behavior requires *competition* among different interaction pathways, which has received the name "frustrated interactions". Several so-called "spin glass" models have been proposed which are not frustrated in this sense. In the first place, one may treat the unquenched, not quenched, case:

$$F = \ln\langle\exp(-\beta\sum J_{ij}S_iS_j)\rangle_{J_{ij}},$$

instead of

$$F = \langle\ln\exp(-\beta\Sigma J_{ij}S_iS_j)\rangle_{J_{ij}}.$$

This is simply a bond ordering problem and uninteresting from our point of view. Equally, one may discuss the "Mattis–Luttinger model" of *site* disorder [100,101]:

$$J_{ij} = J\varepsilon_i\varepsilon_j, \tag{4.5.1}$$

where $\varepsilon_i = \pm 1$ randomly on sites. Here one may define $S_i' = \varepsilon_i S_i$ and recover an ordered problem (sometimes in a random magnetic field, but that again is probably irrelevant). Sherrington [102] points out that (4.5.1) is non-trivial for *quantum* spins, at least dynamically. To define frustration generally, the present author [103] has proposed the following criterion. Imagine one's system cut in two along an arbitrary boundary, and all the exchange integrals across that boundary reduced to zero. Then we place each piece in its precise ground state, or in one of them. (They will in the vector case be free to rotate, in the Ising case to invert; and in either case we must assume the possibility of zero-point degeneracy of other kinds.)

If we now restore the bonds between the two *without changing the internal* states, there will be some change in the energy

$$\Delta E = \sum_{(\text{boundary})\,\text{AB}} J_{ij}\langle S_i\rangle_{\text{A}}\langle S_i\rangle_{\text{B}}. \tag{4.5.2}$$

If the system were ferromagnetic, we would have

$$\begin{aligned}\Delta E = J &\quad (\text{Area of boundary})\cos\theta_{\text{AB}}\\ &\propto\; (\text{Area of boundary})\end{aligned}$$

where θ_{AB} is the angle between the magnetizations S_{A}, S_{B} of our assumed state in the two fragments. This has extreme values and its variance $\overline{(\Delta E^2)}^{1/2}$ both *proportional to boundary areas.*

In the spin glass case, we will expect

$$\Delta E = \sum_{\text{boundary AB}} J_{ij}(\cos\theta_{ij}) \propto \sqrt{\text{Area}}, \tag{4.5.3}$$

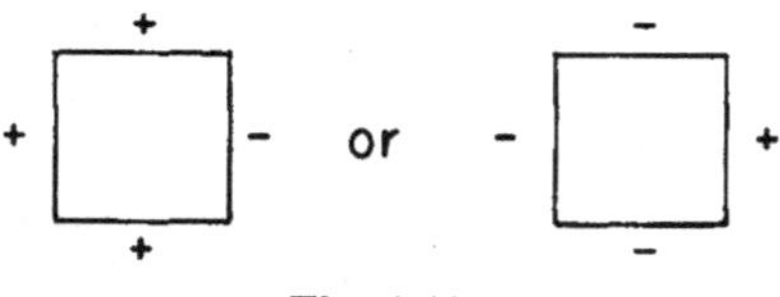

Fig. 4.13.

because it is certain that the J_{ij} and θ_{ij} are totally uncorrelated random variables. We take (4.5.3), or the equivalent

$$\overline{(\Delta E)^2} \propto \text{Area}\,, \tag{4.5.3'}$$

as the definition of a frustrated system. The ordering energy holding the separate pieces together is macroscopically less than for a true ordered system. This characteristic is surely the key to the fact that the spin glass is in a different "universality class" from all other transitions. Whether it has a deeper significance in defining the nature of the state is the important question which has so far been answered only speculatively.

Toulouse and Villain have centered their attention on a specific example of frustration: the so-called "frustrated plaquette" in which the interactions compete on the shortest possible path (fig. 4.13). Toulouse has proved a number of elegant results about degeneracies in square 2d lattices containing frustrated plaquettes, and defined a local density of "frustration" in these terms. He shows that a single frustrated plaquette causes extended defects in the ordering, and he and Villain visualize a theory of the spin glass as a system of a finite density of extended defects in a pseudo-regular system. So far, to my mind, these ideas are too specialized to the square lattice with constant $|J|$, and too speculative, to provide a general foundation for the theory without further work.

4.5.2. Lower critical dimensionality methods

There are two methods which are characteristically most effective with systems near lower critical dimensionality such as 2d ferromagnetic systems. Since we suspect that the physical spin glass is far closer to lower than upper ($d = 6$), these seem to be the methods of choice. They are (1) Migdal's version of the real space renormalization group; and (2) defect methods. A method related to Migdal's was suggested by myself and Pond [104]. Migdal's method has been shown by Mizoguchi et al. [83] and Kadanoff [105] to be equivalent to the simple idea of "bond moving". Near LCD, it is assumed, order breaks down even near absolute zero, so it is valid to assume that nearby spins are relatively slowly varying, i.e.,

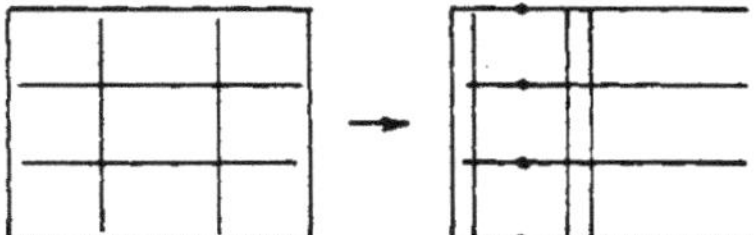

Fig. 4.14a.

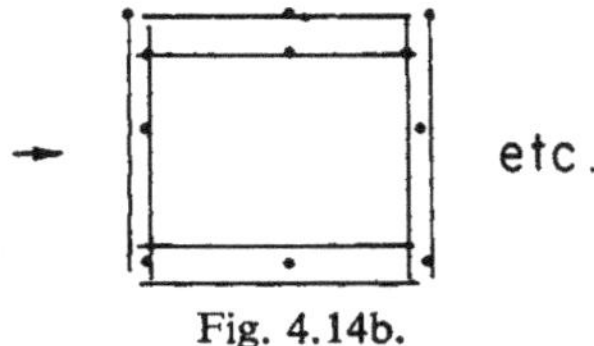

Fig. 4.14b.

there are strong short-range correlations. If so, a given bond contributes the same to the energy if we move it locally parallel to itself to coincide with another bond. We then move bonds as shown in fig. 4.14a, which leaves us with a series-parallel arrangement of double-strength and double-length bonds. We then do it the other way round, as shown in fig. 4.14b. So we have 2^{d-1} parallel bonds in series groups of 2. For ferromagnetic vector spins, it is easy to see that 2 parallel added bonds are twice as strong, 2 series are half as strong, so 2 is the critical dimensionality where there is no net effect.

In the spin glass, the parallel addition clearly must add like random numbers, i.e., the squares of the bonds must add

$$\begin{aligned} &(J^{\text{eff}})^2 = \sum J_{ij}^2 \\ &\overline{J_{\parallel}^2} \to 2^{d-1}\overline{J_{\parallel}^2} \quad \text{or} \quad J_{\parallel} \to 2^{(d-1)/2}\bar{J}\,. \end{aligned} \tag{4.5.4}$$

But when we come to add in series, we have several choices.

(1) We may trust to the replica method. In the replica method, the effective hamiltonian is

$$\sum_{\alpha,\beta}^{n} \sum_{ij} J_{ij}^2 S_i^\alpha S_i^\beta S_j^\beta S_j^\alpha \,,$$

containing the *square* of J_{ij}. In the replica method series bonds will add as squares. For multicomponent objects like vectors, series bonds add as inverses

$$\frac{1}{(J^{\text{eff}})^2} = \sum \frac{1}{J_{ij}^2}$$

and treating $q_i^{\alpha\beta}$ as a tensor, it is clear that again 2d comes out to be LCD in all cases.

On the other hand, one may easily treat this result very sceptically. Why should, in actual random systems, series bonds add as squares? Moving along such a bond, composed of, say, J_{12}, J_{23}, J_{34}, we may change the sign of S_2 if J_{12} is $-$ to make it $+$, which changes the sign of J_{23}; if that is now $-$, we change S_2, changing the sign of J_{34}; but now the sign of J_{34} determines the overall sign of J_{14}^{eff}, which is arbitrary in our model in any case. Thus the sign does not matter and series should add as they do for ferromagnets.

We are left, then, with a dilemma. The replica model, treated directly via Migdal, gives LCD always $= 2$. Direct Migdal on the Ising case says LCD < 2, and for vectors LCD $= 3$ (the ratio now being $2^{(d-1)/2-1} = 2^{(d-3)/2}$ as given in a slightly more careful argument by Anderson and Pond). I confess a prejudice for the latter result, as simply a matter of proprietary interest.

So far no successful defect theories have appeared in the literature. The energies of defects will be subtle to calculate, as pointed out by Anderson and Pond, since it is clear that $(\nabla q)^2$ may be too naive an expression for energy of distortion of the structure.

4.5.3. *Gauge theories*

Hertz [106] has attempted to describe the effect of random exchange integrals J_{ij} in terms of a ferromagnetic Landau–Ginzburg functional with gauge symmetry:

$$H_{\text{L-G}}^{\text{Hertz}} = A(T_c - T)M^2 + BM^4 + \xi^2[(\nabla - A(r))M]^2 ,$$

where A is now thought of as a quenched in, random function satisfying the appropriate conditions. The proper averaging process would then be the *quenched* average

$$F(A) = \ln \int \delta M \, e^{-\beta H} ,$$

$$F = \int \delta A \, F(A) \exp(-A^2/\overline{A^2}) ,$$

where the integrals are functional integrals over the fields M and A. But again it is easiest to treat A as a full fledged gauge field and do the unquenched averages – this seems to be the procedure of Dzyaloshinskii and Volovik [107], in a paper which does not convince this author that it treats the true spin glass, though it makes many suggestive speculative points. Hertz, handling his averages more carefully, seems to reproduce according to private communication the standard result that four is a lower critical dimensionality of some kind. So far the gauge approach is essentially an interesting speculation.

4.5.4. Clusters, exponents, and all that

The oldest theory of the spin glass is the "blocking" theory of Néel [108], still tenaciously advocated by Grenoble experimentalists and others of the European school, as well as Beck [109]. The concept is that for reasons which initially are mysterious, all of the alloys exhibiting spin glass and "mictomagnetic" behavior (low temperature weak remanence, hysteresis, field training, field sensitivity, slow susceptibility response) are composed of non-interacting clusters of spins which interact tightly among themselves and are hence each a random glob of spins of size N, moment $\sim\mu\sqrt{N}$, and random anisotropy also $\sim\sqrt{N}$, with N a large number and the number of clusters M very small (usually). NM is not necessarily related to the total number of spins. Since N is large, there is supposed to be a rather sharp temperature at which the anisotropy energy blocks further reorientation by the simple Arrhenius factor

$$\exp(-E_A/kT),$$

in the thermal relaxation process. The clusters themselves are what is called "superparamagnetic". They can reorient but have very large moments and at low temperatures are expected to saturate easily. This "theory" takes its extreme form with Beck, who seems willing to postulate any implausible sizes and degrees of uncoupling of his superparamagnetic clusters; some of the Grenoble work, on the other hand, is relatively straightforward and often nearly equivalent to a reinterpretation of the data without implausible microscopic assumptions.

The key point, of course, is that these ideas are *not* an alternative theory in any normal microscopic sense. No plausible microscopic assumption is advanced, rather the content of the "theory" is to say that some properties of all these systems are those that a collection of independent superparamagnetic clusters might have.

Naturally, the basic assumption is totally implausible for most well-made and well-characterized spin glass materials. There are no independent clusters of spins which fail to interact with the rest of the spins. For instance, the variance of the interaction energy, as we have noted already, rises as the area of the cluster, under certain assumptions.

There are also many experimental data which exclude the possibility of superparamagnetic clusters. For instance, just a short distance above T_{SG} the susceptibility becomes constant and non-saturating like a normal paramagnet, even though few degrees lower it can be non-linear on the scale of a few gauss. The kind of very large cluster which accounts for this is far too easily saturable.

Naturally there are a few tightly coupled groups of spins which are, for instance, near neighbors in the iron group noble metal alloys, and are frozen together at all low temperatures; but most of the interactions in or out of a cluster must not be too far from kT_{SG} in strength.

These tightly bound clusters have been proposed, undoubtedly correctly, by, for instance, Kirkpatrick et al. [90] as well as Soukoulis and Levin [110], as responsible for the fact that by the time one reaches just above T_{SG} in systems which are frustrated but with interactions of a dominant sign, the susceptibility has changed smoothly from a Curie–Weiss $C/(T + \theta)$ to a Curie law C/T: the residual *groups* of spins have interactions of almost purely random sign. This same kind of very limited cluster assumption is used successfully by Levin et al. to interpret neutron data.

Something like tightly coupled clusters do have a role to play in genuine microscopic theories in the understanding of data and results. In the case of a genuine transition, we realize that as we approach T_{SG} the correlation length ξ must rise to ∞, conventionally with some power α of $(T - T_{SG})^{-1}$:

$$\xi \propto (T - T_{SG})^{-\alpha} . \tag{4.5.5}$$

We may think of all the spins within a radius $\xi(T)$ as a "cluster": but now the cluster size is temperature-dependent: the spins within ξ move together, and have time-correlations just as long with each other as with themselves – in fact it may be assumed they are permanently locked, and surely reorient as a whole in time. This means that the entropy is only that of the clusters moving as a whole, so that the extra entropy S is proportional to the number of clusters:

$$S \propto \xi^{-3} \sim (T - T_c)^{3\alpha} . \tag{4.5.6}$$

This in turn gives a specific heat

$$C \sim \mathrm{d}S/\mathrm{d}T \sim (T - T_c)^{3\alpha - 1} . \tag{4.5.7}$$

Clearly if α is reasonably large this, which is the whole of the specific heat anomaly, is invisible, as is in fact observed. It is characteristic for α to be large near low critical dimensionality; in fact for the xy model, it is infinite (which means $\xi \propto \exp[A/(T - T_c)]$. This argument has been given by Chalupa [96] in more conventional "exponent" terms.

A very similar argument explains the strong field dependence. The "cluster" has a moment given by

$$\mu \sim N^{1/2}\mu_0 \sim \xi^{3/2}\mu_0 , \tag{4.5.8}$$

so that one will satisfy the susceptibility of the clusters at a field

$$\mu H \sim kT \quad \text{or} \quad \frac{H}{(T - T_{SG})^{3/2\alpha}} \sim \frac{kT}{\mu} . \tag{4.5.9}$$

At all T closer to T_{SG} no further effect of correlations will be felt, so the cusp will be smeared out on the scale

$$\Delta T/T_c \sim (\mu H/kT)^{2/3\alpha} . \tag{4.5.10}$$

Chalupa has estimated $\alpha \sim 5$ from this expression and the experimental data. This and the specific heat, far from being a difficulty for the spin glass transition theory are, to my mind, its strongest support.

Below T_{SG}, if we are correct in assuming that the behavior remains critical for all T, similar temperature-dependent "pseudoclusters" will surely exist. Explanations for the behavior in this regime are in a far less advanced state. One particularly interesting paper by Shastry and Shenoy [111] points out that the slow time relaxation of q implied by the TAP theory (since $F(q) \sim (q - q_0)^3$ gives an algebraic relaxation) will appear very strongly in a slow relaxation of magnetization m because of the coupling between q and m implied by the equation

$$\chi = (1 - q)\chi_0 ,$$

which implies a coupling term in the free energy

$$F \propto -qm^2 .$$

This slow relaxation is observed both in simulations and in experimental results. If we add to the TAP result the possibility that the order is only algebraic in nature at $d < 4$, and include exponents, it is clear that there is sufficient complexity available to match the complexity of the observations in this regime.

4.6. *Numerical simulation*

This subject has been well reviewed recently by Binder [112] and will be discussed in the Les Houches notes by Kirkpatrick at least in passing.

Therefore I will not try to go through it very completely, but merely hit a few highlights. I have referred repeatedly to simulation work in the previous sections and use that also as a reason for sketchiness.

Kirkpatrick observes that, with its relative the localization problem, the spin glass is one of the least amenable to Monte Carlo type simulation by machine calculation in terms of required machine time, even if there were no physical difficulties. But experiment and, now, even theory tells us that the physical system we try to simulate shows many anomalously long relaxations as well as rather subtle sensitivity to small perturbations. Thus it is gratifying that we have as much consistent data as we do.

I think there are two particularly important results of machine simulation work which both tell me the same rather important message: the hamiltonian we are using *is* correct, no peculiar clustering or electronic or long-range interactions are necessary, and the physics is in the models, not in some new microscopic effect.

The first of these is the Stauffer–Binder [113] simulation program as a whole. What is striking about all of their results is not how they confirm one or another theory, which as the results appear seems less and less to be the case, but how precisely they confirm the confusions and complications of the actual experiments. The Binder–Stauffer model is nearest neighbor exchange, usually Gaussian-random between, normally, Ising spins but also vector spins. This is a model which is very much schematized relative to real systems, which have long-range, very variable magnitude exchange, etc. In particular, the model certainly contains no hint of the Néel superparamagnetic clusters. All of the experimental observations which are supposed to indicate clusters are reproduced in the simulations: frequency-dependent χ below T_{SG}, field sensitivity, absence of specific heat singularity, remanence which decays anomalously with time, etc. The simulation results, in fact, give little firm support to any side on the various theoretical questions – what is LCD, whether there is no T_{SG} for vector (Heisenberg systems, etc.) since they are simply a crude version of the experimental results. But they do tell us that *all* the observed phenomenology is in the simple nearest neighbor model.

A similar message comes from the RKKY simulations of Walker and Walstedt [114]. These authors in particular show that the classical Heisenberg system contains low-frequency excitations which are spin wave-like in the sense of being extended spin procession modes, and which are of precisely the density even quantitatively to explain the observed linear specific heat of spin glasses as $T \to 0$. This very important result needs to be repeated on the simpler nearest-neighbor model with better statistics, because if true and general, it is extremely important to the whole theory of random systems. Again, in particular this result tells us that one of the most mysterious of all the experimental facts on spin glass is contained within the supposed model of random spacing and RKKY interactions, and probably in the even simpler nearest-neighbor model.

A third important result comes from both groups of simulations, as well as the work of Kirkpatrick and of Palmer, the former on general cases and both on the S–K model. This is the invariable fact of *non-uniqueness* of the "ground state". No spin glass model seems to repeat perfectly a lowest energy "equilibrium" state reached by adjustment from a different starting point. A reshuffle of the starting configuration invariably reshuffles the result. In the Walker–Walstedt work, there is local reproducibility but not overall; the same information has not been made available by Binder and Stauffer.

I interpret this result again in terms of the concept of the "mobility edge" and of the spectrum of extended states of χ_{ij}. If this spectrum has a very sharp edge, in any reasonable sized system there will be a very large number of possible states near the lowest one. Passing down through T_{SG},

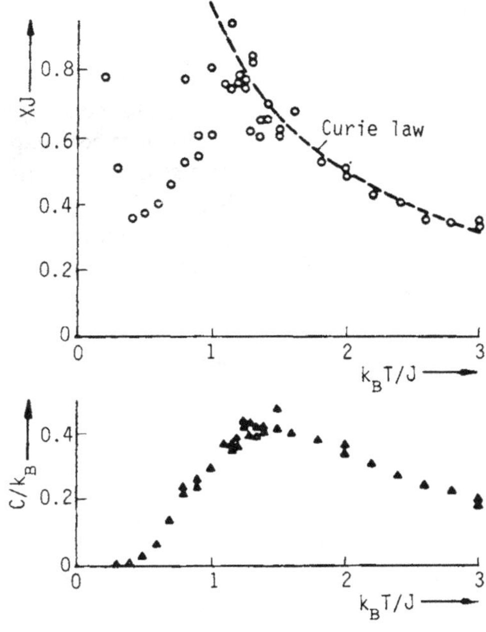

Fig. 4.15.

one will initially polarize in some arbitrary single one of these eigenstates, and thereafter m_i will evolve continuously from this starting state. It is not certain, but probable, that this evolution will not lead to a unique renormalized state, especially since we believe that the system sits permanently at a mobility edge with its accompanying high density of possible fluctuation states.

Further than these three indubitable facts, the simulations give support at will to almost any point of view. There is, perhaps, clearer evidence of a T_{SG} cusp in χ for Ising systems (see fig. 4.15), but $d = 2$ and $d = 3$ do not seem to differ. There is very clear evidence of no singularity in C_{sp}. Finally, the vector system remains unsettled. It seems possible that computer experiments are simply not refined enough for this problem. (One final note: Reed, Bray, and Moore [115] have purported to show by simulation that there is no stiffness energy, thus no domain boundary energy and can be no phase transition. This is not clear; given the lower order of boundary energy predicted, as e.g. by Anderson and Pond, the finite sample statistics of the Bray and Moore calculation seem too crude to pick up the expected boundary energy in the cases done.)

4.7. Conclusion and acknowledgements

In spite of 25 years of study, and 4 years since the first plausible proposals of a new kind of phase transition, the nature, and indeed the very existence, of this phase transition remain in doubt. It is clear both experimentally and theoretically that it has many features of dissimilarity from any other type of transition, and that the spin glass phase – which surely exists, even if it be basically a metastable one – is unique. The experimental data with its striking field dependence, and the sharp appearance of remanence, hysteresis, etc., at a particular temperature, are still the strongest argument for a true transition.

If I were to make a best guess at this point, I would propose two vital key points in the possible understanding of this transition. One is that it is probably a "blocking" transition in which the motivation for the change of state is not the opportunity to lower free energy but the prevention of susceptibility instabilities which would raise the energy. This probably leads to criticality of all $T < T_{SG}$. The second possible key is the relationship to the mobility edge discussed in the first section: that the statistics and dynamics of fluctuations will be controlled by the rather mysterious properties of extended states near a mobility edge. When we understand these, we can hope to approach the spin glass better. If either point is valid, orthodox phase transition theory will be of little help to us for the time being and thus the great bulk of the literature in the field is simply irrelevant.

My thoughts on this subject have been much clarified by discussions especially with Richard Palmer and Ronald Fisch. On the overall subjects of these lectures discussions with Don Licciardello, Scott Kirkpatrick, David Thouless, N. F. Mott, and J. C. Phillips have been of specific value.

"... Fondée en 1951, l'Ecole occupe un ensemble de chalets de montagne entourés de prés et de bois, dans les Alpes françaises près de Chamonix, face au Mont-Blanc. Sa situation offre des possibilités de promenade, de tourisme, d'esclade et de méditation..." (advertisements for the summer school of Les Houches).

Notes added in proof

1. (See page 165.) Chen [78] has been able to demonstrate fairly conventional glass transition behavior in several metallic glasses such as Pd–Cu–Si, Au–Ge–Si.

2. (See page 174.) The number of defects, as seen in fig. 2.4, is $< 10^{-10}$ at T_g since this is the exponential in eq. (2.1.1). This is far fewer than has been postulated for "dangling bonds" in chalcogenide glasses, and in my view adds to the mystery of their nature. I also call attention to the "entropy theory" of Adam and Gibbs [117] which is to a great extent a probably correct restatement of some of the above dilemmas.

3. (See page 182.) Recent experiments show that two-level centers are present in many metallic glasses, as studied by their relaxation effects on metallic electrons.

4. (See page 219.) A paper by Hertz, Fleischman and myself (to be published) further explores and explains these ideas.

References

[1] S. F. Edwards, Phil. Mag. 6 (1961) 617.
[2] J. L. Beeby, Proc. Roy. Soc. (London) A279 (1964) 82.
[3] N. F. Mott, Advan. Phys. 16 (1967) 49.
[4] L. Van Hove, Physica 21 (1955) 517.
[5] D. Turnbull and M. H. Cohen, J. Chem. Phys. 25 (1958) 1049.
[6] W. Kauzmann, Chem. Rev. 48 (1948) 219.
[7] A. J. Esteal, T. A. Wilder, R. K. Mohr and C. T. Moynihan, J. Am. Ceram. Soc. 60 (1977) 134.
[8] C. T. Moynihan, A. J. Esteal, M. A. DeBolt and J. Tucker, J. Am. Ceram. Soc. 59 (1976) 12, 16.
[9] F. Simon and F. Lange, Z. Physik 38 (1926) 227.
[10] G. E. Gibson and W. F. Giauque, J. Am. Chem. Soc. 45 (1923) 93.
[11] H. Vogel, Physik. Z. 22 (1921) 645.
[12] G. S. Fulcher, J. Am. Ceram. Soc. 8 (1925) 339.
[13] A. K. Doolittle, J. Appl. Phys. 22 (1951) 1471.
[14] M. L. Williams, R. F. Landel and J. D. Ferry, J. Am. Chem. Soc. 77 (1955) 3701.
[15] M. H. Cohen and D. Turnbull, J. Chem. Phys. 31 (1959) 1164.
[16] A. J. Barlow, J. Lamb and A. J. Matheson, Proc. Roy. Soc. (London) A292 (1966) 322.
[17] M. A. Phillips, A. J. Barlow and J. Lamb, Proc. Roy. Soc. (London) A329 (1972) 193.

[18] S. Glarum, J. Chem. Phys. 33 (1960) 639.
[19] J. M. Kosterlitz and D. J. Thouless, J. Phys. C6 (1973) 1181.
[20] O. K. Anderson and H. E. Bömmel, J. Am. Ceram. Soc. 38 (1955) 125.
[21] R. C. Zeller and R. O. Pohl, Phys. Rev. B4 (1971) 2029.
[22] W. A. Phillips, J. Low Temp. Phys. 7 (1972) 351.
[23] P. W. Anderson, B. I. Halperin and C. M. Varma, Phil. Mag. 25 (1972) 1.
[24] B. Golding, B. J. Graebner, B. I. Halperin and R. J. Schutz, Phys. Rev. Letters 30 (1973) 223.
[25] W. M. Goubau and R. H. Tait, Phys. Rev. Letters 34 (1975) 1220.
[26] R. B. Kummer, R. C. Dynes and V. Narayanamurti, Phys. Rev. Letters 40 (1978) 1187.
[27] J. D. Weeks, P. W. Anderson and A. G. H. Davidson, J. Chem. Phys. 38 (1973) 1388.
[28] T. M. Hayes, J. W. Allen, J. Tauc, B. C. Giessen and J. J. Hauser, Phys. Rev. Letters 40 (1978) 1282.
[29] L. D. Landau, Soviet Phys.-JETP 5 (1957) 101;
L. D. Landau, Zh. Eksperim. i Teor. Fiz. 32 (1957) 59.
[30] N. F. Mott and E. A. Davis, in: Electronic Processes in Non-Crystalline Materials (Clarendon, Oxford, 1971) (a) ch. 6; (b) fig. 7.31, p. 244; (c) fig. 7.24, p. 231.
[31] M. Pepper, S. Pollitt and C. J. Adkins, J. Phys. C7 (1974) 1273.
[32] M. Pepper, S. Pollitt, C. J. Adkins and R. A. Stradling, Critical Rev. Solid State Sci. (1975) 375.
[33] R. C. Dynes, J. P. Garno and J. M. Rowell, Phys. Rev. Letters 40 (1978) 479.
[34] W. F. G. Swann, Phil. Mag. 28 (1914) 467.
[35] M. H. Cohen and J. Jortner, Phys. Rev. Letters 30 (1973) 699.
[36] A. L. Efros and B. I. Shlovskii, J. Phys. C8 (1975) L49.
[37] L. Fleishman, to be published.
[38] W. Kohn and J. M. Luttinger, Phys. Rev. 118 (1960) 41.
[39] R. Balian and C. De Dominicis, Nucl. Phys. 16 (1960) 502;
R. Balian and C. De Dominicis, Compt. Rend. (Paris) 250 (1960) 3285, 4111;
R. Balian and C. De Dominicis, Ann. Phys. (NY) 62 (1971) 229.
[40] C. Bloch, Physica 26 (1960) 562.
[41] J. M. Langer and V. Ambegaokar, Phys. Rev. 164 (1967) 498.
[42] A. B. Migdal, Soviet Phys.—JETP 7 (1958) 996;
A. B. Migdal, Zh. Eksperim. i Teor. Fiz. 34 (1958) 1438.
[43] N. F. Mott, J. Non-Crystalline Solids 1 (1968) 1.
[44] R. Freedman and J. A. Hertz, Phys. Rev. B15 (1977) 2384.
[45] L. Fleishman, D. C. Licciardello and P. W. Anderson, Phys. Rev. Letters 40 (1978) 1340.
[46] M. Pollak and T. H. Geballe, Phys. Rev. 122 (1961) 1742.
[47] A. Miller and E. Abrahams, Phys. Rev. 120 (1960) 745.
[48] K. B. Ma, Thesis, Princeton (1976).
[49] N. F. Mott, in: Physics of SiO_2, ed. S. Pantelides (Pergamon, New York, 1978).
[50] P. W. Anderson, Nature Phys. Sci. 235 (1972) 163.
[51] C. J. Adkins, J. Phys. C11 (1978) 851;
C. J. Adkins, Abstract in this volume.

[52] R. Romestain, S. Geschwind and G. E. Devlin, Phys. Rev. Letters 35 (1975) 803; R. Romestain, S. Geschwind and G. E. Devlin, J. Physique Colloq. 37, C4 (1976) 313.
R. Romestain, S. Geschwind and G. E. Devlin, this volume.
[53] W. Sasaki, J. Physique Colloq. C4 (1976) 307.
[54] M. H. Cohen, H. Fritzsche, S. R. Ovshinsky, Phys. Rev. Letters 22 (1970) 1065.
[55] D. Engermann and R. Fischer, in: Proc. 5th Intern. Conf. on Amorphous Semiconductors, eds. J. Stuke and Brenig (Taylor and Francis, 1974) p. 111.
[56] R. A. Street, Solid State Commun. 24 (1977) 363.
[57] J. Tauc, F. J. Di Salvo, G. B. Peterson and D. L. Wood, in: Amorphous Magnetism I, eds. Hooper and De Graaf (Plenum, New York, 1973).
[58] F. J. Di Salvo, B. G. Bagley, J. Tauc and J. Waszczak, in: Proc. 5th Intern. Conf. on Amorphous Semiconductors, eds. J. Stuke and W. Brenig (Taylor and Francis, 1974).
[59] H. K. Rockstad, J. Non-Crystalline Solids 2 (1970) 192.
[60] D. B. McWhan, J. P. Remeika, T. M. Rice, W. F. Brinkman, J. P. Maita and A. Menth, Phys. Rev. Letters 27 (1971) 941; and subsequent papers.
[61] P. W. Anderson, Phys. Rev. Letters 34 (1975) 953.
[62] P. W. Anderson, J. Physique Colloq. 10, C4 (1976).
[63] T. M. Donovan and W. B. Spicer, Phys. Rev. Letters 21 (1968) 1532.
[64] T. Holstein, Ann. Phys. (NY) 8 (1959) 325.
[65] R. A. Street and N. F. Mott, Phys. Rev. Letters 35 (1975) 1293.
[66] D. L. Stein, D. C. Licciardello and K. B. Ma, Bull. Am. Phys. Soc. 23 (1978) 414.
[67] R. Fisch and D. C. Licciardello, Phys. Rev. Letters 41 (1978) 889.
[68] K. B. Ma, J. Non-Crystalline Solids 24 (1977) 345.
[69] C. M. Varma and K. Pandey, 1976.
[70] W. A. Harrison, Phys. Rev. Letters 37 (1976) 312.
[71] D. C. Licciardello, 1978.
[72] P. W. Anderson, Phys. Rev. 124 (1961) 41.
[73] A. Blandin and J. Friedel, J. Phys. Radium 20 (1959) 160.
[74] F. D. M. Haldane, Phys. Rev. B15 (1977) 281.
[75] R. Tournier et al., in: Proc. 15th Intern. Conf. on Low Temperature Physics – LT 15, to be published.
[76] P. W. Anderson, Mater. Res. Bull. 5 (1970) 549.
[77] R. W. Klein and R. Brout, Phys. Rev. 132 (1963) 124.
[78] M. S. Chen and M. Goldstein, J. Appl. Phys. 43 (1972) 1542; M. S. Chen, J. Non-Cryst. Solids 22 (1976) 135.
[79] E. P. Wigner, in: Proc. 4th Can. Math. Congr., Toronto 1959, p. 174 (from Mehta, Random Matrices (Academic Press, New York, 1967)).
[80] S. F. Edwards and P. W. Anderson, J. Phys. F7 (1975) 965.
[81] K. Fischer, Phys. Rev. Letters 34 (1975) 1438.
[82] C. Domb, J. Phys. A9 (1976) L17.
[83] T. Mizoguchi, T. R. McGuire, S. Kirkpatrick and R. J. Gambino, Phys. Rev. Letters 38 (1977) 89.
[84] D. Sherrington and B. Southern, J. Phys. F5 (1975) L49.
[85] D. Sherrington and S. Kirkpatrick, Phys. Rev. Letters 35 (1975) 1792.
[86] J. R. O. de Almeida and D. J. Thouless, J. Phys. A11 (1978) 983.
[87] A. J. Bray and M. A. Moore, Phys. Rev. Letters 41 (1978) 1068;
A. J. Bray and M. A. Moore, J. Phys. C, to be published.

[88] D. J. Thouless, P. W. Anderson and R. Palmer, Phil. Mag. 35 (1977) 593.
[89] R. G. Palmer and L. van Hemmen, to be published.
[90] S. Kirkpatrick and D. Sherrington, Phys. Rev. B18 (1978).
[91] S. Katsura, J. Phys. C9 (1976) L619.
[92] A. B. Harris, J. I. Lubensky and J. H. Chen, Phys. Rev. Letters 35 (1976) 1792.
[93] A. P. Young and R. B. Stinchcombe, J. Phys. C9 (1976) 4419.
[94] C. Jayaprahesh, J. Chalupa and M. Wortis, Phys. Rev. B15 (1977) 1495.
[95] W. Kinzel and K. H. Fischer, J. Phys. C11 (1978) 2775.
[96] J. Chalupa, Solid State Commun. 24 (1977) 429.
[97] G. Toulouse, Commun. Phys. 2 (1977) 115.
[98] J. Villain, J. Phys. C10 (1977) 1717;
J. Villain, this volume.
[99] R. Fisch and A. B. Harris, Phys. Rev. Letters 38 (1977) 785.
[100] D. C. Mattis, Phys. Letters A56 (1976) 421.
[101] J. M. Luttinger, Rev. Letters 37 (1976) 778.
[102] D. Sherrington, this volume.
[103] P. W. Anderson, 1978.
[104] P. W. Anderson and C. M. Pond, Phys. Rev. Letters 40 (1978) 903.
[105] L. P. Kadanoff, Ann. Phys. (NY) 100 (1976) 359.
[106] J. A. Hertz, Phys. Rev. B18 (1978).
[107] I. E. Dzyaloshinskii and G. E. Volovik, preprint (1978).
[108] L. Néel, J. Phys. Soc. Japan 17 (1962) Suppl. B1, 676;
L. Néel, Ann Géophysique 5 (1949) 99.
[109] P. A. Beck, Met. Trans. 2 (1971) 2015; plus other works.
[110] C. M. Soukoulis and K. Levin, Phys. Rev. Letters 39 (1978) 581.
[111] B. S. Shastry and S. R. Shenoy, preprint (1978).
[112] K. Binder, in: Festkörperprobleme (Advances in Solid State Physics), Vol. 17, ed. J. Treusch (Vieweg, Braunschweig, 1977) pp. 55.
[113] D. Stauffer and K. Binder, Z. Physik B, to be published.
[114] L. R. Walker and R. E. Walstedt, Phys. Rev. Letters 38 (1977) 514.
[115] P. Reed, A. J. Bray and A. M. Moore, J. Phys. C11 (1978) L139.
[116] A. P. Young, in: Amorphous Magnetism II, eds. R. A. Levy and R. Hasegawa (Plenum, New York, 1977).
[117] G. Adam and J. H. Gibbs, J. Chem. Phys. 43 (1965) 139.

COURSE 4

ELEMENTARY ALGEBRAIC TOPOLOGY RELATED TO THE THEORY OF DEFECTS AND TEXTURES

Valentin POÉNARU

Département de Mathématiques, Faculté des Sciences d'Orsay, Université Paris XI, 91405 Orsay, France

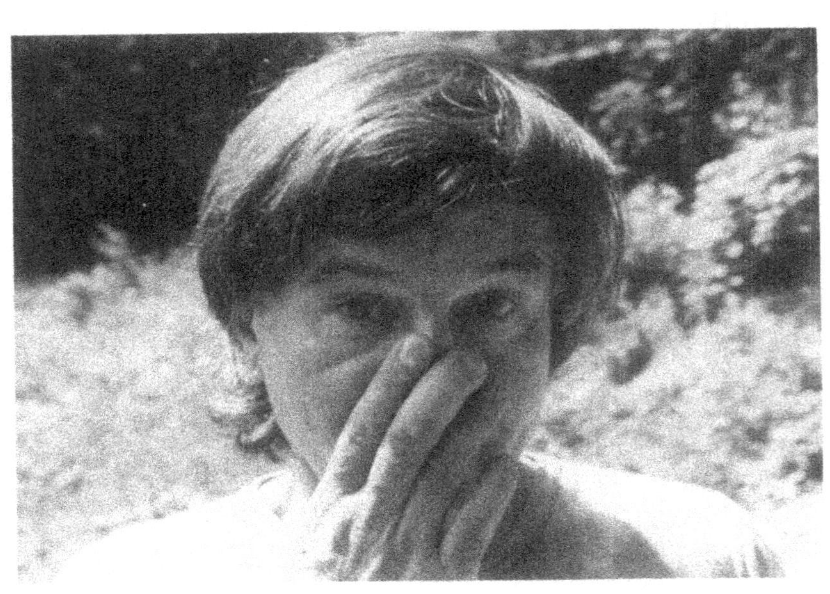

Contents

R. Balian et al., eds.
Les Houches, Session XXXI, 1978 – La matière mal condensée/Ill-condensed matter

Introduction

This set of lectures is an introduction to algebraic topology, in particular to homotopy theory, in view of applications to the physics of ordered media (in particular of their defects). The parts of topology presented here are very standard and at the end of this course we give a list of some textbooks which cover the same topics with more details. The present text will rather try to give an intuitive idea of how things work; proofs will be often sketchy and just replaced by drawings and heuristic arguments. In earlier days, textbooks on quantum mechanics and textbooks on Hilbert space were often indistinguishable. I think times are ripe now for topology and ordered media to be blended together in a somehow similar fashion.

An *ordered medium* will be, roughly speaking, a physical space M, where to a point $x \in M$, an *order parameter* $\Phi(x)$ will be attached. The order parameter will vary *continuously* with x and take values in a space V, called the *manifold of internal states*, characteristic for the order in question. The point is that Φ is not everywhere defined. There will be a set $\Sigma \subset M$ where Φ is not defined. The points of Σ are called *defects*, and, in general, there will be *topological obstructions* to extend Φ *continuously* to Σ on parts of Σ. This kind of model goes back to Landau. Later on, a different kind of topological distortion, called *texture*, will be also described.

Since all this sounds pretty abstract, some examples will be necessary. We follow the presentation given in ref. [8] of the general references at the end of this course.

(1) *Spins*. The physical space is euclidean 3-space $\mathbf{R}^3$ or a part of $\mathbf{R}^3$. (Unless otherwise stated, this will be the case in all the next examples too.) The order parameter is a *unit vector*,

$$\Phi(x) = \frac{V(x)}{\|V(x)\|},$$

pointing in any direction. This is for instance the case of an isotopic ferromagnet. Here V is the 2-sphere S^2.

Suppose that there is some ball $B \subset \mathbf{R}^3$ such that on the boundary ∂B the order parameter is everywhere defined and points towards the *outside*.

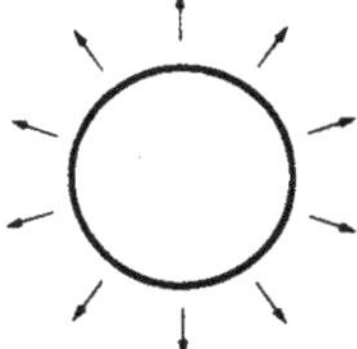

It is intuitively clear (and can be rigorously proved) that in this situation *there have to be defects inside B*. This is maybe one of the simplest instances of "applications of topology to the theory of defects". Of course, at this level one might argue that plain geometrical intuition is really all that is needed. The point is that in similar instances, with more complicated V's, rather subtle arguments belonging to the field of algebraic topology, will be necessary in order to describe and understand *what* is actually going on.

(2) *Nematic liquid crystals.* This is a medium consisting of long molecules with the symmetry of ellipsoids of revolution. The order parameter is a *direction* (the direction of the axis of revolution), or if one prefers, a "vector without an arrowhead". The space of directions in $\mathbf{R}^3$ is called the *real projective plane* (RP^2) and this is our V. It can be also described as a sphere S^2 with antipodal points identified. As we shall see later, here already the topological theory of defects becomes rather subtle.

(3) *Crystalline solids.* We consider three independent vectors V_1, V_2, V_3, and the Bravais lattice

$$\mathscr{L} = \{z_1 V_1 + z_2 V_2 + z_3 V_3\},$$

where $z_i \in Z = \{\text{the integers}\}$. In a perfect crystal there is a density function which is periodic: $\rho_0(x + V) = \rho_0(x)$, where $V \in \mathscr{L}$. In reality there will be a vector valued function $W(x)$ (varying very slowly with respect to $\mathscr{L}$) such that the actual density function is

$$\rho(x) = \rho_0(x + W(x)).$$

Note that if $r \in \mathscr{L}$, passing from $W(x)$ to $W(x) + r$ does not change anything. Hence it is natural to do the following; first, identify two points $x', x'' \in \mathbf{R}^3$ if $x' - x'' \in \mathscr{L}$. The "quotient-space" is the three-dimensional torus $T^3 = S^1 \times S^1 \times S^1$. Call $\Phi(x)$ the value of $W(x)$ *in* T^3. This is our order-parameter function, and $V = T^3$.

(4) *Dipole-locked-A-phase of* He_3. (For an introduction to the physics of superfluid helium-3, one can read the beautiful article by Mermin and Lee in Scientific American, Dec. 1976.)

The order parameter is a pair of distinguished orthonormal axes, arbitrarily oriented. Here V can be identified to SO(3).

(5) *A more detailed discussion of planar spins*. First an abstract definition: If $X \underset{f_1}{\overset{f_0}{\rightrightarrows}} V$ are two continuous maps, we will say that f_0 and f_1 are *homotopic* if one can deform continuously f_0 into f_1. More precisely, this means that there exists a continuous map $F: X \times [0, 1] \to V$ such that $F(x, 0) = f_0(x)$, $F(x, 1) = f_1(x)$. F will be called a *homotopy* (connecting f_0 to f_1). Think of $f_t(x) = F(x, t)$, for fixed t, as being another map $X \to V$. When t goes from 0 to 1, "f_t varies continuously from f_0 to f_1". We shall write $f_0 \sim f_1$ if f_0 and f_1 are homotopic. This is clearly an equivalence relation.

Consider now a *planar* isotropic ferromagnet, or more generally a *planar spin*. Let $p \in \mathbf{R}^2$ be an isolated defect point of this ordered medium, and let S be a small circle around p. The vectors $\Phi(x)$ where x runs around S give us a continuous map from the circle S to the unit circle $V = S^1$. Call this map $\varphi_p: S \to S^1$. Then one can talk about the number of times φ_p *winds* around S^1. This *winding number* $w(\varphi_p)$ is an arbitrary integer and it turns out that it is a complete invariant for the homotopy class of φ_p (which means $\varphi'_p \sim \varphi''_p \leftrightarrow w(\varphi'_p) = w(\varphi''_p)$). Also $w(\varphi_p)$ does not depend on the particular choice of S, but only on p (and the particular spin in question). Here are some examples:

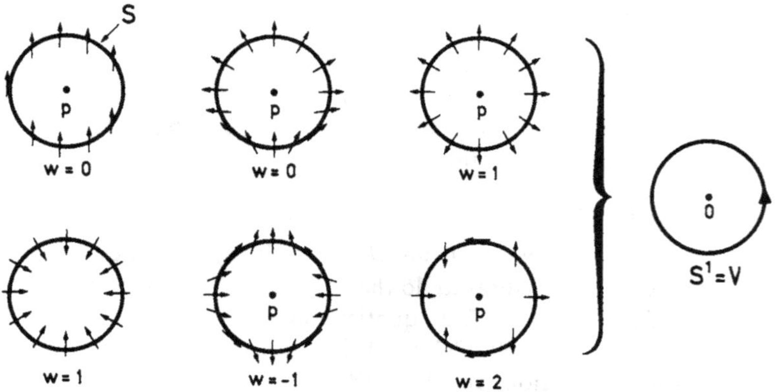

Here are some (easy) facts:

(1) One can extend the order parameter, by a very small perturbation, so that it is also defined at p iff $w = 0$.

(2) Two defects with the same w "look alike" (if one allows continuous deformations).

(3) When two punctual defects are slowly approaching each other until finally they *coalesce*, the numbers w *add up*.

At this point, these facts may, or may not seem intuitively clear. They (and similar but much more complicated facts) will become (hopefully) transparent in the next sections.

1. Review of some basic concepts (spaces, groups, ...)

1.1. Discrete groups

A discrete group G is just a group endowed with the "discrete topology", or if one prefers, without any topology at all. We will usually write such a group multiplicatively (and sometimes, when we know for sure that it is abelian, additively). As a basic *convention*, I will denote the unit element of a group G, but if G is reduced to its unit element, it may also be denoted by 1. This is the "multiplicative notation". As long as one works with abelian groups only, 1 will be replaced by 0 ("additive notation"). If G is *finite*, one can completely describe it by giving (explicitly) its table of multiplication.

But most of the groups we will have to deal with in these lectures will be infinite. The convenient way to describe them will be the following. A set (possibly infinite) of elements $x_1, x_2, \ldots \in G$ will be given, such that any element in G can be represented as a product of x_i's at various powers (these powers being arbitrary integers). $x_1, x_2, \ldots$ will be called *generators* and so any element in G can be represented (but *not* necessarily in a unique fashion) as a finite *word* written with the letters $x_1, x_1^{-1}, x_2, x_2^{-1}, \ldots$. Of course, some words in x may happen to be equal to $1 \in G$. Consider the *free group F* generated by the letters $x_1, x_1^{-1}, \ldots$ (no relation except the obvious ones). The natural map $F \xrightarrow{\varphi} G$ is a surjective homomorphism and

G is completely determined if we specify the *invariant* subgroup $\mathbf{R} = \operatorname{Ker}(\varphi) \subset F$.

This can be done by specifying a set of elements $W_1, W_2, \ldots \in \mathbf{R}$ such that $W_1^{\pm 1}, W_2^{\pm 1}, \ldots$ and *their conjugates* (in F) generate $\mathbf{R}$. The "words" $W_1, W_2, \ldots$ (written in the "alphabet" $x_1, x_1^{-1}, x_2, x_2^{-1}, \ldots$) are called a "*set of relations*" for G. So G is completely described by giving a set of generators and a set of relations. This is called a "presentation of G". Such a presentation is not at all unique and, moreover, can be highly redundant.

1.2. Lie groups and coset spaces

Let G be a Lie group and let $H \subset G$ be a closed subgroup (which will be automatically a Lie group). I recall that the *coset space* G/H is the set of *left* cosets gH endowed with the natural topology coming from G. The map $G \to G/H$ which associates gH to g is continuous and surjective. The group G acts *transitively* on G/H by the rule $g_1 \cdot gH = (g_1 g)H$ and the *isotropy group* of the coset H is the subgroup $H \subset G$ itself.

There is a converse to this description. Suppose G acts on some space X, which means that a continuous map $G \times X \to X$ taking $(g, x) \mapsto g \cdot x \in X$ with the obvious properties $g_1 \cdot (g_2 x) = (g_1 g_2)x$, etc. is given. This breaks X into disjoined subsets of the form Gx called *orbits*. If X is a single orbit, the action is called *transitive*.

Suppose this is the case, and chose a point $x_0 \in X$. Let $H \subset G$ be the subgroup of those elements $h \in G$ such that $h \cdot x_0 = x_0$ (the "isotropy group of x_0"). Clearly this group is closed. There is an obvious continuous map. $G/H \xrightarrow{\varphi} X$ sending gH to $g \cdot x_0$. The map φ is a continuous bijection but φ^{-1} need not be continuous. However, in many instances, for example when X itself is a *compact space*, φ^{-1} is continuous, and hence (in these cases, which are the only ones considered in these lectures) φ is a homeomorphism $G/H \xrightarrow[\approx]{} X$. This homeomorphism carries the natural action of G on G/H into the given action of G in X.

Examples. SO(3) acts transitively on the unit sphere $S^2 \subset \mathbf{R}^3$, and the isotropy group of $(0, 0, 1) \in S^2$ is the set of rotations in the plane $\mathbf{R}^2 = \{x, y\}$, denoted by SO(2). So

$$S^2 = \mathrm{SO}(3)/\mathrm{SO}(2) .$$

Let $K \subset \mathrm{SO}(3)$ be the subgroup generated by SO(2) and by $(x, y, z) \xrightarrow{T} (x, -y, z)$. It is easy to see that

$$\mathrm{RP}^2 = \mathrm{SO}(3)/K .$$

Most of the "manifolds of internal states" V considered in these lectures will be of type G/H. More exactly, our physical systems will have two phases: a phase of high energy characterized by a symmetry G, and a less symmetric phase of low energy, characterized by a subgroup $H \subset G$ ("breaking of symmetry"). Points in $M - \Sigma$ will be in the phase of low energy, and the order parameter will take its values in $G/H = V$ (one should keep in mind, here, that a fast rotating object acquires extra symmetries...).

Defects will be in the high energy phase; so they will be costly, energetically speaking. This means that they will spontaneously pass to the low energy phase, whenever this will be topologically possible.

The example of biaxial nematics (see de Gennes: *Liquid Crystals* (Oxford, 1974)). These are, for the time being, hypothetical materials. They are liquid crystals where the molecule has as symmetry group a *finite* subgroup $H \subset \mathrm{SO}(3)$ (for example an ellipsoid with three unequal axes). Here V is naturally $\mathrm{SO}(3)/H$, and topology has a lot of interesting things to say about it.

I end this section with a remark which will be useful later. Let $G_0 \subset G$ be the connected component of the identity element. It is easy to show that G_0 is a closed invariant subgroup. The quotient group G/G_0 is a *discrete group* which will be denoted by $\pi_0 G$.

1.3. "Spaces"

The notion "space" which will be used, relatively loosely, in these notes always means "a topological space". The reader who might find this notion forbidding can as well think of a "space" as being some piece of the euclidean n-space, for some (possibly high) n. Very often on such a space an equivalence relation will be given, and the "quotient space" will be considered. The simplest such instance is when a piece Y of the space X is crushed to a single point. For example, let

$$X = (\text{the unit } n\text{-disk}) = \{x_1^2 + \cdots + x_n^2 \leqslant 1\}$$

and

$$Y = (\text{the unit } (n-1) \text{ sphere}) = \{x_1^2 + \cdots + x_n^2 = 1\}\,.$$

Crushing $Y \subset X$ to a point produces a new space which is *homeomorphic* to the unit n-sphere S^n. The reader should try to work this out as an exercise.

Often "spaces" have many different, not obviously equivalent descriptions. Here is an example. Think of RP^2 as being the set of directions in $\mathbf{R}^3$. This is clearly equivalent to the set of pairs of antipodal points on the unit sphere. All of these pairs are already represented by the points of the "northern hemisphere", with the ambiguity that on the equator itself each pair continues to be represented twice. So, RP^2 is a 2-disk D^2 with any pair of antipodal points *in the boundary*, glued together.

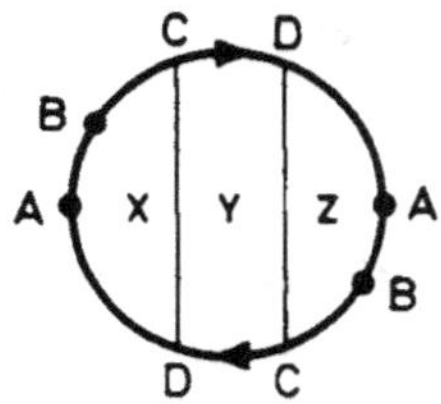

Cut this disk into three parts X, Y, Z as in the figure above. Then follow our recipe for glueing points together, first *separately* in Y and $X + Z$. The result in Y is a "Möbius band":

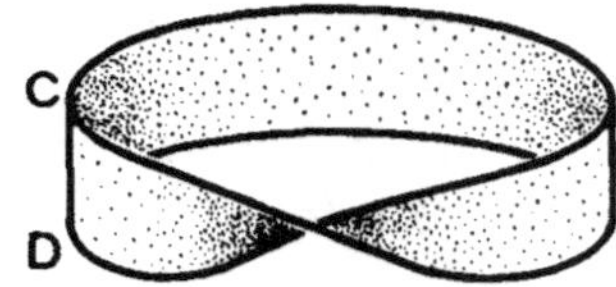

It is an easy exercise to see that if one takes X and Z, and glues $A \in X$ to $A \in Z$, $B \in X$ to $B \in Z$, etc., one gets (topologically speaking) another 2-disk:

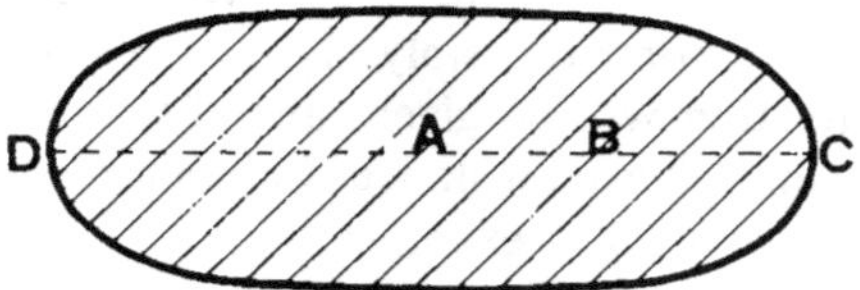

By glueing the boundary of this 2-disk to the boundary of the Möbius band (which is a unique "circle") one gets back our RP^2.

Exercise

More generally; describe $\mathbf{RP}^n$ = (the set of directions in $\mathbf{R}^{n+1}$) as the result of glueing an n-disk to $\mathbf{RP}^{n-1}$.

1.4. SO(3) *and* SU(2)

I will recall some basic facts about SO(3) = (the group of rotation with determinant 1 in $\mathbf{R}^3$). Some algebra will be needed now.

Let Q be the algebra of *quaternions*. These are linear combinations with real coefficients: $a_0e_0 + a_1e_1 + a_2e_2 + a_3e_3$, ($a_j \in R$) where the algebra generators e_0, e_1, e_2, e_3 multiply by the following rules:

$$e_0e_j = e_je_0 = e_j\,, \qquad e_1^2 = e_2^2 = e_3^2 = -e_0\,,$$

$$e_ie_j = -e_je_i \quad (i, j \geqslant 1)\,, \qquad e_1e_2 = e_3\,, \qquad e_2e_3 = e_1\,, \qquad e_3e_1 = e_2\,.$$

The algebra thus defined is associative but *not* commutative. We define the *conjugate element by*

$$\overline{\sum a_je_j} = a_0e_0 - a_1e_1 - a_2e_2 - a_3e_3\,.$$

This is an involution which is an anti-automorphism. We define the *norm* by:

$$N\left(\sum a_je_j\right) = \sum a_j^2\,.$$

Clearly $N(q_1q_2) = N(q_1)N(q_2)$ and, if $q \neq 0$, $N(q)^{-1}\bar{q}$ is an inverse for q. (So Q is actually a non-commutative *field*.)

The elements $q \in Q$ such that $N(q) = 1$ will be called *spinors*. They form a group under multiplication, denoted (sometimes) by Sp(1) and clearly Sp(1) is homeomorphic to S^3. So S^3 is the underlying space of a Lie group, like S^1.

THEOREM. Let $H \subset$ Sp(1) be the invariant subgroup consisting of the two elements e_0 and $-e_0$. Then

$$\mathrm{Sp}(1)/H = \mathrm{SO}(3)\,.$$

Proof. Let $\mathbf{R}^3 = \{a_1e_1 + a_2e_2 + a_3e_3\} \subset Q$. It is easy to check that if $q \in \mathrm{Sp}(1)$ then $q\mathbf{R}^3q^{-1} \subset \mathbf{R}^3$. In this way one defines a homomorphism $\mathrm{Sp}(1) \xrightarrow{\varphi} \mathrm{GL}(3, \mathbf{R})$. It is not hard to show that φ is actually a *surjective* homomorphism $\mathrm{Sp}(1) \xrightarrow{\varphi} \mathrm{SO}(3)$, and that $\mathrm{Ker}(\varphi) = H$. In other words $\mathrm{SO}(3) = \mathrm{Sp}(1)/H$ (quotient-group).

COROLLARY. SO(3) is homeomorphic to $\mathbf{RP}^3$.

Exercise

Prove this corollary directly, geometrically, without any algebra. (A possible *hint*: Any non-trivial rotation is described by an *axis* and an *angle* of rotation. We think of this angle as being written in the form $\pi + \theta$ where $-\pi < \theta < \pi$. The axis is a point in RP^2, or rather, a pair of antipodal points in S^2.

Of course, given an axis and an angle one has also to specify in which direction (left or right) to turn. If this has been done for *one* rotation, there is a continuous parametrization of all non-trivial rotations, as pairs (x, θ) where $x \in S^2$, $\theta \in (-\pi, \pi)$, with the ambiguity that (x, θ) and $(-x, -\theta)$ represent the *same* rotation.

Think now of S^2 as being the equator of S^3 and θ the "latitude" on S^3, identify north and south poles with the trivial rotation, etc.)

Let us come back to algebra. Q contains a subfield isomorphic to $C = \{\text{the complex numbers}\}$, namely the set of all elements $a_0e_0 + a_1e_1$. So it can be considered as a C-vector space (of complex dimension 2). We will write

$$q(a + ib) = q(ae_0 + ie_1)\,.$$

In the basis (e_0, e_2) left-multiplication by e_0, e_1, e_2, e_3 corresponds for instance to the following ("Pauli") matrices

$$\begin{pmatrix}1 & 0\\0 & 1\end{pmatrix}, \quad \begin{pmatrix}i & 0\\0 & -i\end{pmatrix}, \quad \begin{pmatrix}0 & -1\\1 & 0\end{pmatrix}, \quad \begin{pmatrix}0 & -i\\-i & 0\end{pmatrix}.$$

Using this kind of interpretation, the group Sp(1) can be identified to SU(2). We will not use this fact in these lectures.

2. The fundamental group

2.1. The problem

Let V be a space, which for simplicity will be assumed to be *connected*. We have seen already that the following *problem* is relevant for the study of ordered media: if f, g are two continuous maps from the circle S^1 into V, when are f and g *homotopic*? Although this problem is a purely topological one, it will turn out (surprisingly) that the answer will take us into the realm of group theory. To the space V we will attach a certain group, called the *fundamental group of* V (noted $\pi_1(V)$); f and g will be interpreted in a natural, well-defined way, as *conjugacy classes* $[f]$ and $[g]$ of $\pi_2 V$, and it will turn out that f and g are homotopic iff $[f]$ and $[g]$ are the same conjugacy class.

2.2. The group $\pi_1(V, x_0)$

Let us choose a "*base point*" $x_0 \in V$, which for the moment will be fixed, and consider the set of all continuous paths in V, starting and ending at x_0, parametrized by the unit time-interval. So, we consider *continuous* maps $f: I \to V$ (where $I = [0, 1]$) such that $f(0) = f(1) = x_0$. We will denote by $\Omega(V, x_0)$ the set of all such closed *loops*, "based" at x_0.

By definition f_0 and f_1 (belonging to $\Omega(V, x_0)$) are *homotopic* if one can join them by a continuous path $f_u (u \in [0, 1])$ in $\Omega(V, x_0)$, in other words if there exists a continuous map

$$\underset{\substack{\uparrow \\ \text{variable } t \\ \text{(time)}}}{F : I} \times \underset{\substack{\uparrow \\ \text{variable } u \\ \text{(parameter)}}}{I} \to V$$

such that $F(t, 0) = f_0(t)$, $F(t, 1) = f_1(t)$, $F(0, u) = F(1, u) = x_0$. We represent the source of F which is a square, in fig. 1.

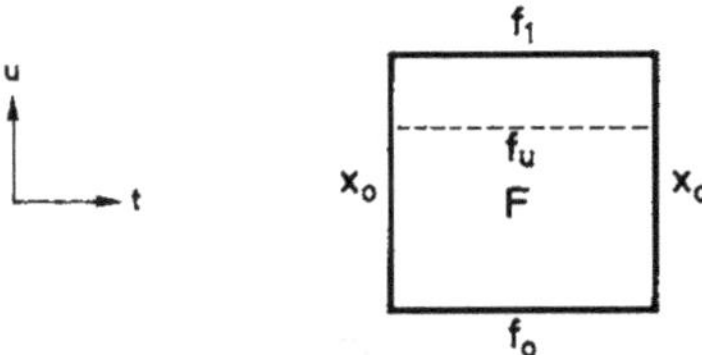

Fig. 1. The source of F. The letters tell us how F is defined on various points of the square.

If f_0 and f_1 are homotopic, we will write $f_0 \sim f_1$, and clearly this is an equivalence relation.

If $f, g \in \Omega(V_0, x_0)$ I will define their *product* $f \cdot g$ by the formula:

$$(f \cdot g)(t) = \begin{cases} f(2t) & \text{if } 0 \leqslant t \leqslant \frac{1}{2} \\ g(2t-1) & \text{if } \frac{1}{2} \leqslant t \leqslant 1 \end{cases}.$$

Note that this definition makes sense, since $f(1) = g(0) = x_0$. Intuitively speaking $f \cdot g$ is the same thing as going first along path f (with double speed) and then along path g (with double speed). This makes sense because f and g are actually loops based at the same point x_0. If we did not double the speed we would get a map from $[0, 2]$ to V. In other words doubling the speed means just reducing the scale from $[0, 2]$ to $[0, 1]$, like in the figure below:

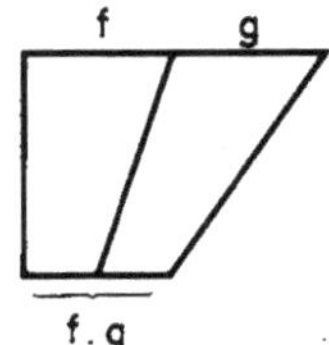

LEMMA 1. If $f_0 \sim f_1$, $g_0 \sim g_1$, then $f_0 \cdot g_0 \sim f_1 \cdot g_1$.

Proof. The homotopy in between f_0 and f_1 (respectively g_0 and g_1) is a square in which a map F (respectively G) is defined. We can glue together these homotopies and then reduce the scale to get a homotopy in between $f_0 \cdot g_0$ and $f_1 \cdot g_1$ (fig. 2).

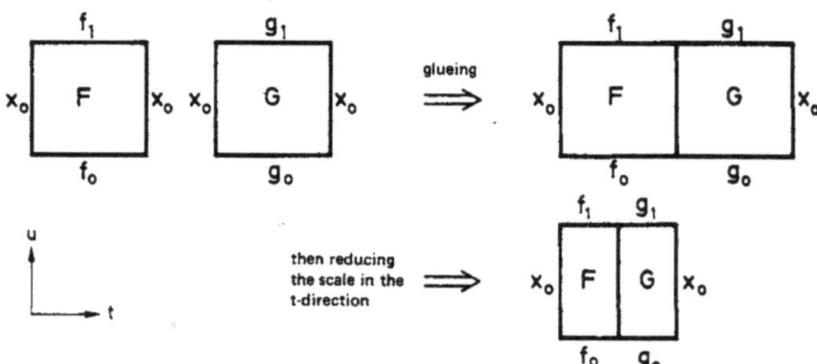

Fig. 2.

Let us call $\pi_1(V, x_0)$ the set of homotopy classes of loops based at x_0. If $f \in \Omega(V, x_0)$, its homotopy class will be denoted by $[f]$. We can *define* a product-operation in $\pi_1(V, x)$ by

$$[f] \cdot [g] \underset{\text{def}}{=} [f \cdot g],$$

and lemma 1 shows that this definition makes sense.

THEOREM. $\pi_1(V, x_0)$ (with the product law we have just defined) is a *group*. The unit element of this group (denoted by 1) is the homotopy class of the constant loop, and the inverse element of $[f]$ is the homotopy class of the loop $t \mapsto f(1-t)$ (i.e. f with the time reversed).

Sketch of proof

(1) *Associativity*. One has to show that $(f \cdot g) \cdot h \sim f \cdot (g \cdot h)$, for $f, g, h \in \Omega(V, x_0)$. This is schematized by the square in fig. 3. This square is mapped into V in such a way that it realizes a homotopy in between $(f \cdot g) \cdot h$ and $f \cdot (g \cdot h)$. One "combs" the square by a family of straight lines joining top to bottom and in each line the value in V is constant.

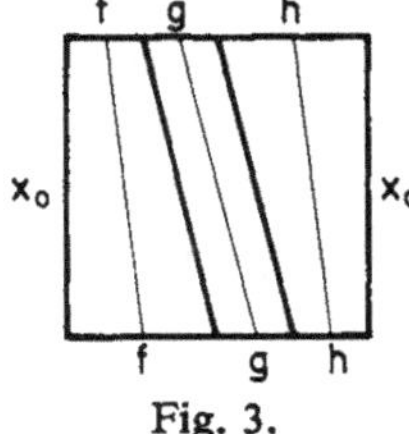

Fig. 3.

(2) *Unit element*. Let C: $[0, 1] \to x_0$ be the constant loop. One has to show that $C \cdot f \sim f$. This is schematically shown in fig. 4.

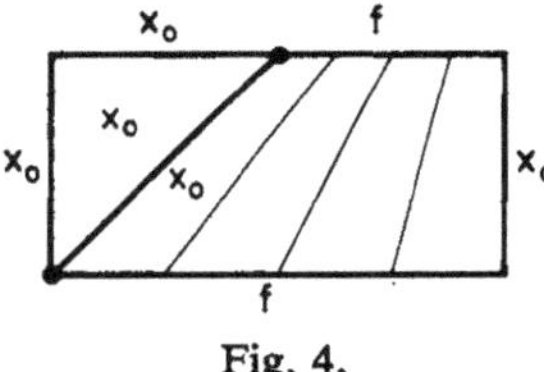

Fig. 4.

(3) *Inverse elements*. Let f^{-1}: $[0, 1] \to V$ be the loop $f^{-1}(C) = +(1 - t)$. One has to show that $f \cdot f^{-1} \sim C$. This is shown in fig. 5.

The explanation of these sequences of drawings is the following: let $f|_u$ be the path $f|[0, u]$, which means f restricted to $[0, u] \subset [0, 1]$. One

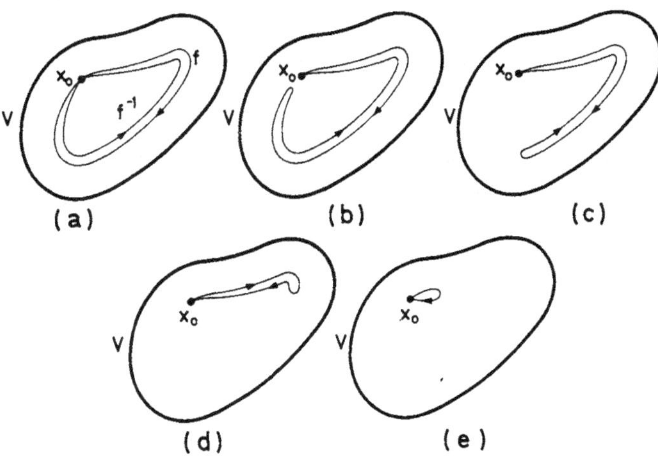

Fig. 5.

remarks that $f|_u \cdot (f|_u)^{-1}$ is a *closed* loop based at x_0. When u varies from 0 to 1 this defines a homotopy in between C and $f \cdot f^{-1}$.

If $F: (V, x_0) \to (V', x'_0)$ is a continuous map such that $F(x_0) = x'_0$, any loop of V based at x_0 is mapped by F into a loop of V' band at x'_0. This induces a *homomorphism* of groups

$$F_*: \pi_1(V, x_0) \to \pi_1(V', x'_0) .$$

The following properties are easily checked:

(1) If

$$(V, x_0) \xrightarrow{F} (V', x'_0) \xrightarrow{F'} (V'', x''_0)$$

then $(F'F_*) = F'_* F_*$.

(2) If

$$(V, x_0) \underset{F_1}{\overset{F_0}{\rightrightarrows}} (V', x'_0) .$$

are homotopic, by a homotopy keeping $x_0 \mapsto x_0$, then $(F_0)_* = (F_1)_*$.

A map $X \xrightarrow{f} Y$ is called a *homotopy equivalence* if there exists another map $Y \xrightarrow{g} X$ such that $X \xrightarrow{f} Y \xrightarrow{g} X$ is homotopic to the identity map of X and $Y \xrightarrow{g} X \xrightarrow{f} Y$ is homotopic to the identity map of Y. In other words X and Y are homotopy-equivalent if one can deform one into the other.

A homotopy equivalence induces an isomorphism at the level of π_1.

2.3. *Comparison of $\pi_1(V, x_0)$ with $\pi_1(V, y_0)$*

What happens to $\pi_1(V, x_0)$ when we change the base point x_0 to another point $y_0 \in V$?

Here the assumption that V is connected will be essential. We will join x_0 to y_0 by a continuous path $\beta: [0, 1] \to V$ ($\beta(0) = x_0, \beta(1) = y_0$). If $f \in \Omega(V, y_0)$ then $\beta \cdot f \cdot \beta^{-1} \in \Omega(V, x_0)$ (see fig. 6).

One checks easily that if f and g are homotopic loops in $\Omega(V, y_0)$, then $\beta \cdot f \cdot \beta^{-1} \sim \beta \cdot g \cdot \beta^{-1}$. So there is a well-defined map $\beta_*: \pi_1(V, y_0) \to$

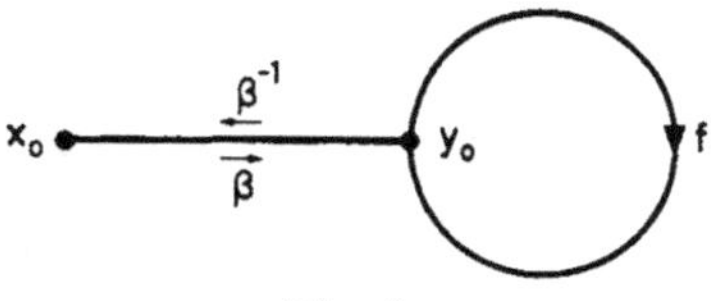

Fig. 6.

$\pi_1(V, x_0)$ which takes $[f]$ into $[\beta \cdot f \cdot \beta^{-1}]$. By the same argument which was used to prove that $\pi_1(V, x_0)$ possessed inverse elements, one checks that

$$\beta \cdot f \cdot \beta^{-1} \cdot \beta \cdot g \cdot \beta^{-1} \sim \beta \cdot f \cdot g \cdot \beta^{-1},$$

but this means that β_* *is a group homomorphism.*

THEOREM. β_* is an isomorphism.

Proof. Clearly $(\beta^{-1})_*$ is an inverse of β_*. So $\pi_1(V, x_0)$ and $\pi_1(V, y_0)$ are isomorphic. But a different choice of β can give a different isomorphism. Anyway (as long as V is connected) as an *abstract group* $\pi_1(V, x_0)$ does not depend on the choice of base point, and will be denoted by $\pi_1 V$. This is the *fundamental group* of V. The following properties can be easily checked:

(1) If $\beta_0, \beta_1 : [0, 1] \to V$ are paths joining x_0 to y_0, which are homotopic, by a homotopy which keeps their endpoints fixed, then

$$(\beta_0)_* = (\beta_1)_* .$$

(2) The whole construction makes sense for the case $x_0 = y_0$, $\beta \in \Omega(V, x_0)$. In this case $[\beta] \in \pi_1(V, x_0)$ and

$$\pi_1(V_1, x_0) \xrightarrow{\beta_*} \pi_1(V_1, x_0)$$

is the *inner automorphism*

$$[f] \mapsto [\beta][f][\beta]^{-1} .$$

(3) The isomorphisms $(\beta)_*$ are independent of β (for all x_0, y_0, β) iff $\pi_1 V$ is abelian. In that case it makes sense to add loops based at *different* points.

(4) Let $f \in \Omega(V, x_0)$. I can also think of f as being a map from $S^1 \to V$, since $[0, 1]$ with the two endpoints identified is just $S^1 \cdot S^1$ is, of course, the boundary of the 2-disk D^2. Then $[f] = 1 \leftrightarrow f$ extends to a continuous map $D^2 \to V$.

(5) Let $f, g \in \Omega(V, x_0)$, and consider f and g as maps from $S^1 \to V$. These maps are homotopic (without any condition on x_0) iff $[f]$ and $[g]$ are *conjugate* in $\pi_1(V, x_0)$.

2.4. *Topological groups: first examples*

Let G be a (connected) topological group. We shall prove the following:

THEOREM. "$\pi_1 G$ is abelian".

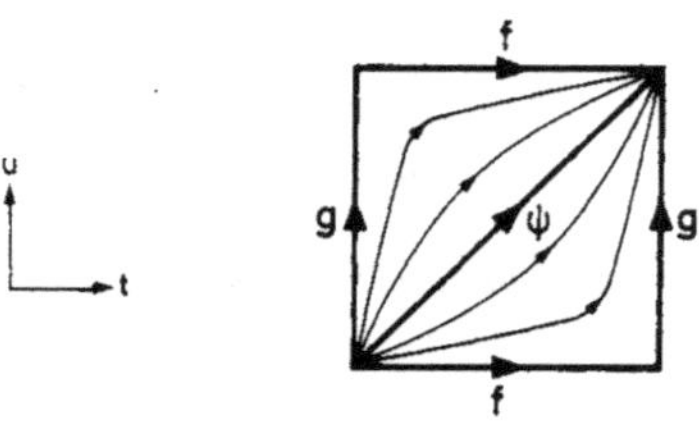

Fig. 7.

Proof. Let $1 \in G$ be the unit element, let $f, g \in \Omega(G, 1)$, and denote by $\otimes$ the multiplication in G. We consider

$$F: [0, 1] \times [0, 1] \to G$$

defined by $F(t, u) = f(t) \otimes g(u)$ (see fig. 7).
Let also ψ: $[0, 1] \to G$ be the map: $\psi(t) = f(t) \otimes g(t)$.
Looking at fig. 7 one reads

$$g \cdot f \sim \psi \sim f \cdot g$$

which implies what we were looking for. But moreover this says that the multiplication in $\pi_1 G$ can be defined *directly*, using the multiplication in G.

Note. G itself is *not* necessarily abelian.

Examples.

(1) S^1 itself is a topological group so $\pi_1 S^1$ is abelian. It can be shown that if f, g: $S^1 \to S^1$ are two continuous maps, then f and g are homotopic iff the number of times they wind around the target is the same. It is "intuitively clear" that the winding number induces an isomorphism $\pi_1 S^1 = Z$ (where Z is the additive group of the integers). Actually *proving* this fact is not completely trivial. A direct elementary argument can be found in Massey's book *Algebraic topology, an introduction*, pp. 68–74. A more conceptual proof follows the theory of covering spaces sketched in sect. 4.

(2) A connected space V will be called *simply-connected* (or 1-connected) if $\pi_1 V = 0$. A space V is called *contractible* if the following two maps of V into itself are homotopic: $f(x) = x, f_0(x) = x_0$. It is easy to see that a contractible space is simply connected. Examples of contractible spaces: $\mathbf{R}^n$, any convex set. (For $\mathbf{R}^n$ choose $x_0 = 0$, and the homotopy connecting the identity map and the constant map is $F(t, x) = tx$.)

(3) Let $n \geqslant 2$. I claim that S^n is *simply connected.*

Proof. Remark first that if $P \in S^n$, then $S^n - P = \mathbf{R}^n$ (homeomorphism). So if P is given, any continuous map $f\colon S^1 \to S^n$ such that $f(S^1) \not\ni P$ is null-homotopic. Of course $f(S^1)$ might cover the whole of S^n (Peano) but any f can be approximated as closely as we wish by a continuous map $g\colon S^1 \to S^n$ whose image consists of a finite collection of geodesic arcs. So $g(S^1)$ does not cover S^n, hence it is null-homotopic. But since g is very close to f it is not hard to see that they are homotopic.

2.5. Various computations

The process of attaching to a space V its fundamental group $\pi_1 V$ (and to the continuous map $V' \xrightarrow{f} V''$ the group homomorphism $\pi_1 V' \xrightarrow{f_*} \pi_1 V''$) is, in a certain sense, a dictionary translating geometry (or rather topology) into algebra. The present section gives more details about this kind of dictionary.

Product spaces. The following holds:

THEOREM. $\pi_1(V' \times V'') = \pi_1 V' \times \pi_1 V''$.

The proof is easy. One starts with the remark that a loop in $(V' \times V'', x_0' \times x_0'')$ is exactly a pair of loops of (V', x_0') and (V'', x_0''). We leave the details to the reader.

Example. Consider the two-dimensional torus $T^2 = S^1 \times S^1$

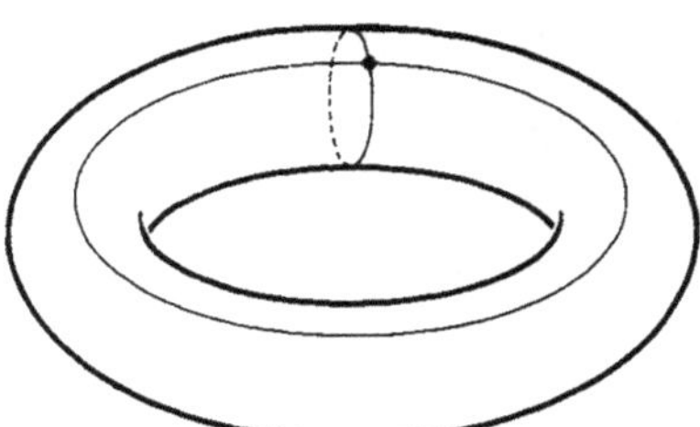

One has $\pi_1 T^2 = Z \times Z$, and, more generally:

$$\pi_1(\underbrace{S^1 \times \cdots \times S^1}_{k\text{-times}}) = \underbrace{Z \times \cdots \times Z}_{k\text{-times}}$$

So far all the π_1's we have seen were commutative groups. This is very far from being the general situation. The next paragraph will be an illustration of how *non-commutative* fundamental groups occur.

Glueing together spaces (*or groups*). Assume $V = V_1 \cup V_2$. One will think of V as being obtained by *glueing* together V_1 and V_2 along $V_1 \cap V_2$.

There is a corresponding notion for *groups* called the *amalgamated product*. Consider the groups G_1, G_2, H and the homomorphisms

$$H \begin{array}{l} \xrightarrow{f_1} G_1 \\ \xrightarrow[f_2]{} G_2 \end{array}$$

We will construct a group $G_1 \underset{H}{*} G_2$ ("glueing G_1 and G_2 along H") as follows. Consider *presentations* for G_1, G_2, H:

	Generators	Relations
G_1	$x_1, x_2, \ldots$	$P_1(x), P_2(x), \ldots$
G_2	$y_1, y_2, \ldots$	$Q_1(y), Q_2(y), \ldots$
H	$z_1, z_2, \ldots$	$R_1(z), R_2(z), \ldots$

The homomorphisms f_1, f_2 are completely determined by expressing $f_1(z_i), f_2(z_i)$ as *words* in x or y:

$$f_1(z_i) = A_i(x)\,, \qquad f_2(z_i) = B_i(y)\,.$$

The group $G_1 \underset{H}{*} G_2$ is defined by the following presentation: Generators: $x_1, x_2, \ldots, y_1, y_2, \ldots, z_1, z_2, \ldots$, relations: all the $P_i(x)$, $Q_j(y)$, $R_k(z)$, *and*

$$z_i^{-1}A_i(x)\,, \qquad z_i^{-1}B_i(y)\,.$$

(*Note*. This presentation is redundant; the reader should find another presentation using only x's and y's.)

Examples.

(1) If $H = \{1\}$, then $G_1 \underset{\{1\}}{*} G_2$ (denoted just by $G_1 * G_2$) is obtained by putting together the presentations of G_1 and G_2. An element of this group is of the form:

$$\cdots g'g''h'h''\cdots$$

where $\ldots, g', h', \ldots \in G_1$, and $\ldots, g'', h'', \ldots \in G_2$. When all the g's and h's are $\neq 1$ such an expression *cannot* be simplified. Contrast this with $G_1 \times G_2$ whose elements are just *pairs* (g', g'').

(2) $\{1\} \underset{H}{*} G$ is obtained from G by adding extra relations. Remark that the amalgamated product is the most natural notion of "glueing groups".

Consider connected spaces $V = V_1 \cup V_2$ and the natural inclusions:

$$x_0' \in V_1 \cap V_2 \begin{array}{l} \overset{i_1}{\hookrightarrow} V_1 \\ \underset{i_2}{\hookrightarrow} V_2 \end{array} .$$

This induces, at the level of fundamental groups, some homomorphisms $(i_1)_*, (i_2)_*$ which are not necessarily injective. ($V_1 \cap V_2$ is also assumed to be connected.)

$$\pi_1(V_1 \cap V_2, x_0) \begin{array}{l} \xrightarrow{(i_1)_*} \pi_1(V_1, x_0) \\ \xrightarrow[(i_2)_*]{} \pi_1(V_2, x_0) \end{array} .$$

THEOREM (Van Kampen).

$$\pi_1(V_1 \cup V_2) = \pi_1 V_1 \underset{\pi_1(V_1 \cap V_2)}{*} \pi_1 V_2 .$$

We will not give a proof. But the intuitive content is clear: our dictionary transforms the operation of glueing spaces, into glueing groups.

Examples.

(1) *Wedges*. Let (V_1, x_1), (V_2, x_2) be two spaces *with* base points. We can decide that $x_1 = x_2$ and glue V_1 and V_2 accordingly. The result is called the *wedge* of V_1 and V_2, and denoted by $V_1 \vee V_2$.

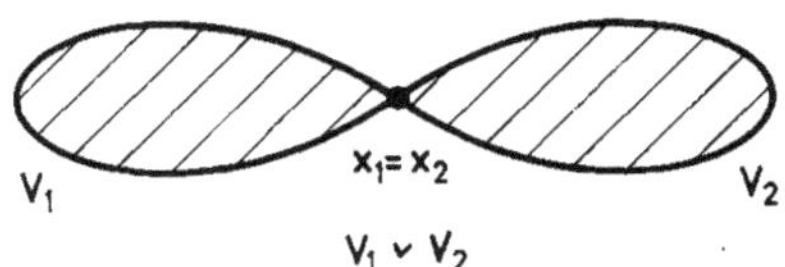

Then

$$\pi_1(V_1 \vee V_2) = \pi_1 V_1 * \pi_1 V_2 .$$

In particular, for a wedge (or "bouquet") of circles one finds:

$$\pi_1(\underbrace{S^1 \vee \cdots \vee S^1}_{k\text{-times}}) = \underbrace{Z * \cdots * Z}_{k\text{-times}}$$

(2) *Attaching "cells" (or balls)*. Let V be a connected space, D^n the n-ball, $S^{n-1} = \partial D^n$ its boundary, and $f\colon S^{n-1} \to V$ a continuous map. One can define a new space $V \underset{f}{\cup} D^n$ obtained by *glueing D^n to V along f.* Assume first that $n = 1$. Then $V \cup D^1$ can be deformed into $V \vee S^1$,

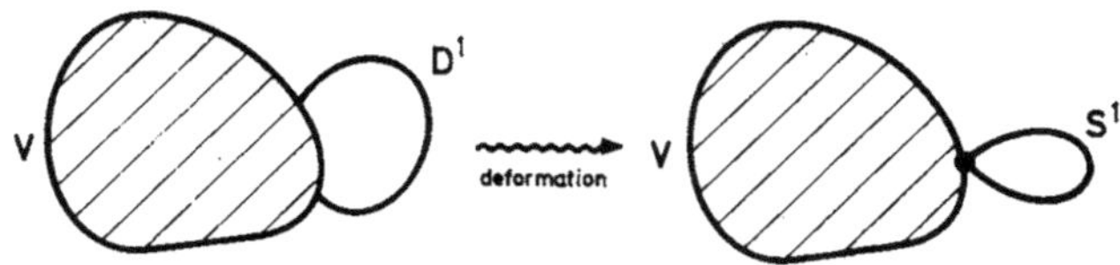

so

$$\pi_1(V \cup D^1) = \pi_1 V * Z\,.$$

If $n = 2$, then $\pi_1 S^1 = Z$ and the natural diagram

$$S_1 \begin{array}{l} \xrightarrow{\;f\;} V \\ \hookrightarrow D^2 \end{array}$$

gives rise to:

$$1 \in \mathbf{Z} \begin{array}{l} \xrightarrow{\;f_*\;} \pi_1 V \ni f_*(1) \\ \longrightarrow 1 \end{array}$$

By (a slight generalization of) the Van Kampen theorem

$$\pi_1(V \underset{f}{\cup} D^2) = \{\pi_1 V \text{ with the extra relation } f_*(\overset{\substack{\text{generator}\\ \text{of } Z\\ \downarrow}}{1}) = \underset{\substack{\uparrow\\ \text{unit element}\\ \text{of } \pi_1 V}}{1}\}$$

So $\pi_1(V \underset{f}{\cup} D^2)$ is obtained from $\pi_1 V$ by *killing* one element (and hence the invariant subgroup generated by this element; $\pi_1(V \underset{f}{\cup} D^2)$ is the corresponding quotient group).

If $n > 2$ it is possible to show that the operation of passing from V to $V \underset{f}{\cup} D^n$ does *not* alter the fundamental group. Now it turns out that by starting with a single point and successively attaching cells, we can obtain most of the usual spaces. And for all these we have actually described a theoretical procedure for computing π_1.

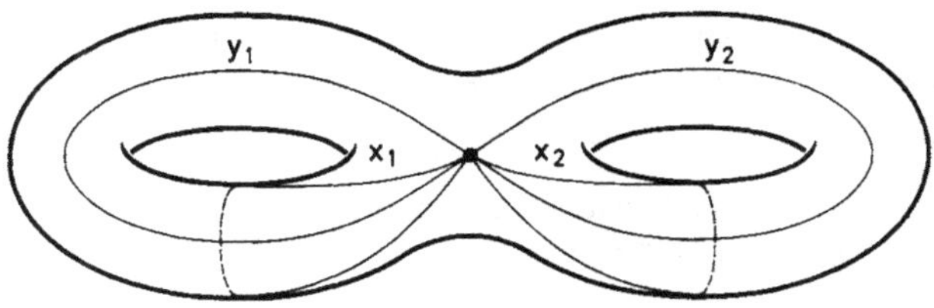

Fig. 8.

(3) We will illustrate this in a special case. Consider the torus with two holes, and cut it along the four curves in fig. 8. The result is topologically a two disk. By reading things backwards, one gets the following description of the torus with two holes: start with a bouquet of 4 circles, labelled x_1, y_1, x_2, y_2:

Consider an octogon (which is topologically just a 2-disk), with sides labelled like in fig. 9.

Attach this octogon to the bouquet of 4 circles *according to the labelling*. The result is the torus with 2 holes! Hence π_1 (of the torus with 2 holes) has the following presentation: generators: x_1, y_1, x_2, y_2, and a single relation: $x_1 y_1 x_1 y_1^{-1} x_2 y_2 x_2^{-1} y_2^{-1} = 1$.

Note. The same works for the case of the torus with p holes. If $p = 1$ this

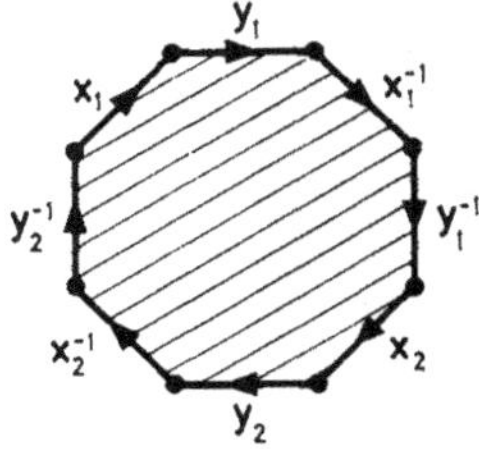

Fig. 9.

is what we already know since the group generated by x, y, with the single relation $xyx^{-1}y^{-1} = 1$ is exactly $Z + Z$. In the general case, one has:

$$x_1y_1x_1^{-1}y_1^{-1}x_2y_2x_2^{-1}y_2^{-1}\cdot\ldots\cdot x_py_px_p^{-1}y_p^{-1} = 1\,.$$

In a similar vein RP^2 can be described by attaching a single disk to a single circle x, by the following recipe.

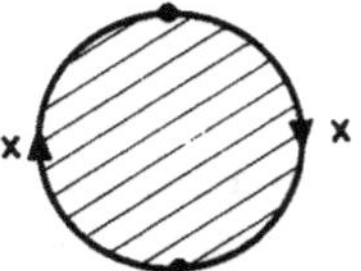

Hence $(\pi_1 \mathrm{RP}^2)$ has one generator x with the single relation $x^2 = 1$. This is a commutative group isomorphic to the integers-modulo-2: $Z/2Z$.

Final remark. Any group G can be described in terms of a presentation by k generators and l relations, which, in turn can be mimicked by considering a wedge of k circles to which one attaches l disks of dimension 2. In this way we construct a two-dimensional space V such that $\pi_1 V = G$.

3. Higher homotopy groups; homotopy and defects

3.1. Definition of $\pi_n V$

Again let V be some space and choose a base point $x_0 \in V$. Let I^n be the n-cube, with coordinates $(t_1, \ldots, t_n)$ where $0 \leqslant t_i \leqslant 1$. Let ∂I^n be the boundary of I^n. We consider the set $\Omega^n(V, x_0)$ consisting of all continuous maps $f\colon I^n \to V$, such that $f(\partial I^n) = x_0$. If $f, g \in \Omega^n(V, x_0)$ we will write $f \sim g$ (and say that f and g are "*homotopic*"), if f and g are homotopic by a homotopy preserving the condition $\partial I^n \mapsto x_0$. We define $f \cdot g \in \Omega^n(V, x_0)$, by generalizing the operation of composition for closed loops:

$$f\cdot g(t_1, \ldots, t_n) \underset{\text{by def}}{=} \begin{cases} f(2t_1, t_2, \ldots, t_n) & \text{if } \ 0 \leqslant t_1 \leqslant \frac{1}{2} \\ g(2t_1 - 1, t_2, \ldots, t_n) & \text{if } \ \frac{1}{2} \leqslant t_1 \leqslant 1\,. \end{cases}$$

Exactly as in the case $n = 1$, for arbitrary n, the set of homotopy classes is thus endowed with a group law, and the corresponding group is denoted by $\pi_n(V, x_0)$. So far there is nothing really new with respect to π_1. But now comes the following:

THEOREM. If $n \geqslant 2$, the group $\pi_n(V, x_0)$ is *abelian*. (Accordingly its composition law will be written additively.)

Proof. I will give the argument for $n = 2$, but it will be clear that it is general. An element $[f] \in \pi_2(V, x_0)$ is the homotopy class of a map $f: I^2 \to V$ sending ∂I^2 to x_0. We represent it symbolically by the following square:

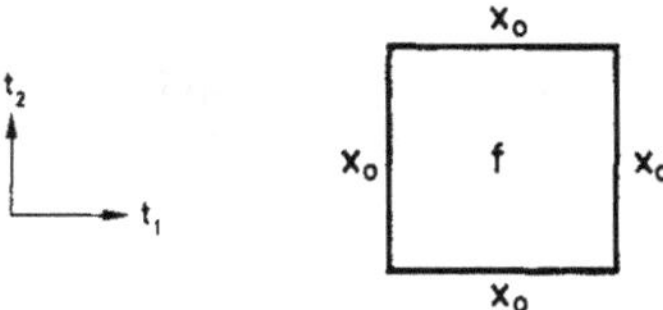

I claim that $f \cdot g \sim g \cdot f$ (where g is another such map). Note that according to our rules of the game $f \cdot g$ is represented by the following square:

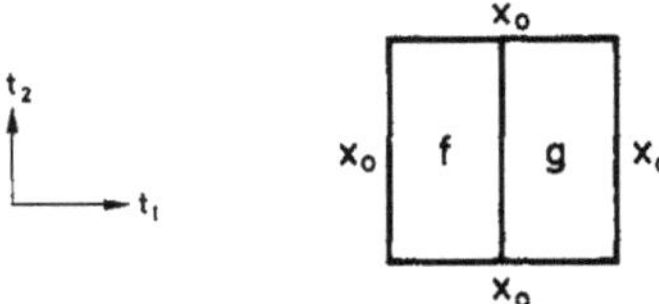

By a homotopy, keeping $\partial I^2 \mapsto x_0$ one can change this map so that it is constant except in two small islands, looking like f and g respectively.

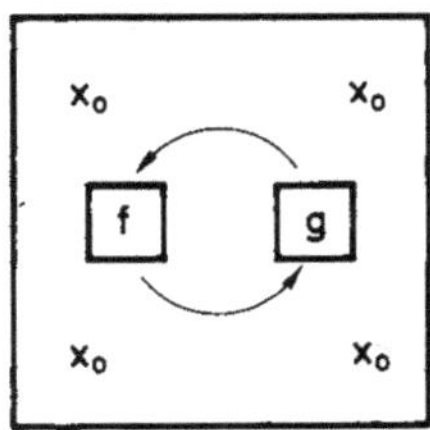

Again by homotopy (keeping $\partial I^2 \mapsto x_0$), the position of the two islands can be interchanged, like the arrows indicate. This leads to:

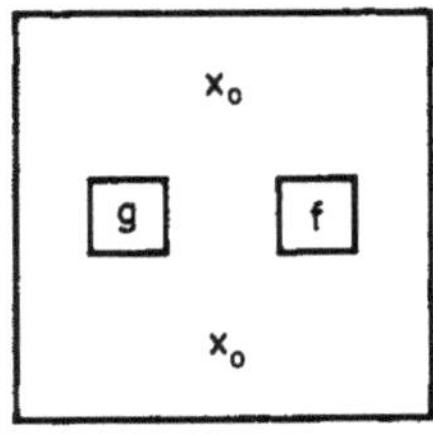

and reversing the whole argument, this is homotopic to $g \cdot f$.

There is another, more intuitive way to conceive $\pi_n(V, x_0)$. Think of I^n with the boundary crushed to a point, $I^n/\partial I^n$, as being the n-sphere S^n with a distinguished point $* \in S^n$ (image of ∂I^n).

This is an S^n given once for all, endowed with a "singular" coordinate system, namely $t_1, t_2, \ldots, t_n$. (Actually all that interests us is $*$ and the "orientation".) Choose once for all an equator S^{n-1} passing through $*$. Crushing the equator to a point defines once for all, a (God-given)-map

$$S^n \xrightarrow[\Phi]{} S^n \vee S^n$$

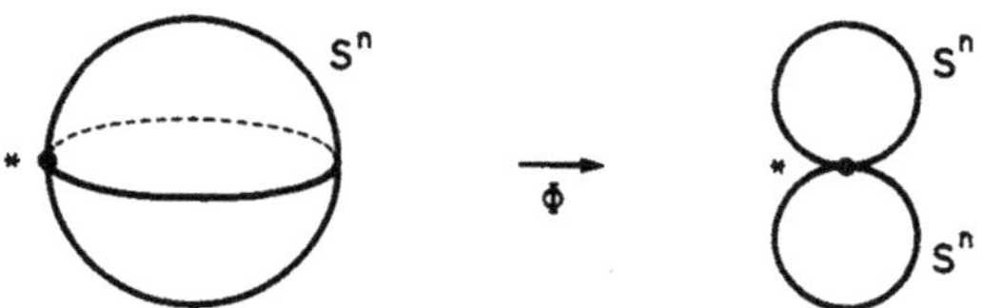

Another (God-given) map is the symmetry with respect to the equator

$$(S^n, *) \xrightarrow[\Psi]{} (S^n, *) \,.$$

Now, $f \in \Omega^n(V, x_0)$ can be thought of as a map $f\colon (S^n, *) \to (V, x_0)$. In this interpretation $f + g$ and f^{-1} are given by the following two diagrams:

$$S^n \xrightarrow[\Phi]{} S^n \vee S^n \quad V \qquad \text{and} \qquad S^n \xrightarrow[\Psi]{} S^n \xrightarrow{f} V \,.$$

(Note that f and g glue together to a unique, well defined, continuous map $S^n \vee S^n \xrightarrow[f \vee g]{} V$.)

3.2. The "action" of $\pi_1(V, x_0)$ on $\pi_n(V, x_0)$

Suppose now that V is connected. How can we compare $\pi_n(V, x_0)$ and $\pi_n(V, y_0)$? Again we will generalize the considerations from the preceding chapter. For simplicity's sake let us take $n = 2$ and consider a continuous path β: $[0, 1] \to V$ such that $\beta(0) = x_0$, $\beta(1) = y_0$. Let $f \in \Omega^2(V, y_0)$. I will define $\beta \cdot f \in \Omega^2(V, x_0)$ as follows:

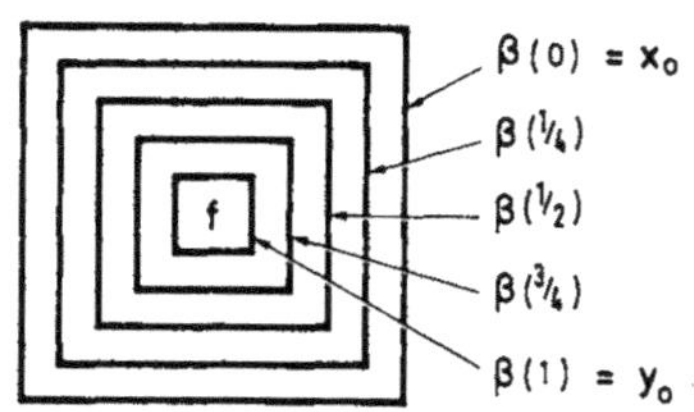

Inside the unit square I take once for all the square of size $\frac{1}{2}$ (and center $(\frac{1}{2}, \frac{1}{2})$) and define on it f (reduced to scale). Then on boundaries of concentric bigger squares I put the values of β as in the figure above. This clearly defines an element in $\Omega^2(V, x_0)$ whose homotopy class depends only on the homotopy class of f (and on the homotopy class of β, by homotopies keeping the endpoints fixed). One defines in this way a map

$$\beta_*: \pi_n(V, y_0) \to \pi_n(V, x_0),$$

which is actually an isomorphism of groups ($(\beta^{-1})_*$ is the inverse of $(\beta)_*$).

Hence, as long as V is connected, as an *abstract* group π_2 does not depend on the base point. This abstract group is denoted by $\pi_2 V$.

But we can also consider the case $x_0 = y_0$. In this way we define an *action* $\pi_1(V, x_0) \times \pi_n(V, x_0) \to \pi_n(V, x_0)$ denoted by $(\beta, f) \mapsto \beta \cdot f$. For fixed β this is an isomorphism of $\pi_n(V, x_0)$. Moreover $\beta \cdot (\gamma \cdot f) = (\beta \cdot \gamma) \cdot f$, etc.

If for all $\beta \in \pi_1(V, x_0)$ one has $\beta \cdot f = f$ we say that V is *n-simple*.

Note. This is not necessarily the case even if $n \geq 2$ and π_1 is abelian. If $n = 1$ then $\beta \cdot f = \beta f \beta^{-1}$.

Here is a more intuitive definition of $\beta_*: \pi_n(V, y_0) \to \pi_n(V, x_0)$. Consider the space $I \vee S^n$ where $I = [0, 1]$ is glued to $* \in S^n$. So this is a sphere with a stick attached. Consider on S^n an equator S^{n-1} as far away as possible from $*$, and

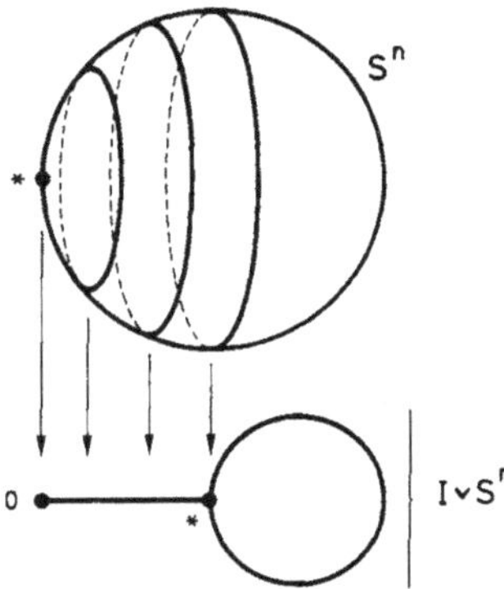

a family of concentric $n - 1$ spheres running from $*$ to S^{n-1} as in the figure above. By crushing each of the members of the family to one point, one gets a map (defined once for all)

$$S^n \underset{x}{\to} I \vee S^n$$

(taking $* \mapsto 0$, $S^{n-1} \mapsto *$). Our $\beta \cdot f$ is then given by the following diagram

$$S^n \xrightarrow{x} I \vee S^n \quad V \qquad (\beta: I \to V,\ f: S^n \to V)$$

One should think of $\beta \cdot f$ as being obtained by pushing $f(S^n)$ with one's finger along β^{-1}, as in this figure:

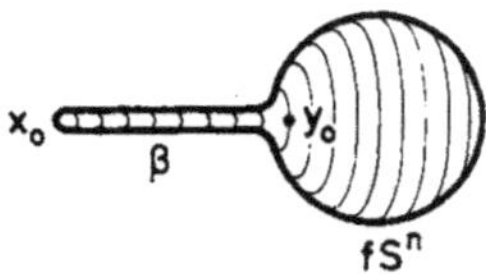

Exercises.

(1) A topological group is n-simple.

(2) Let $f, g: (S^n, *) \to (V, x_0)$. Then: (a) f and g are homotopic, keeping $* \mapsto x_0$, iff $[f] = [g] \in \pi_n(V, x_0)$; (b) f and g are homotopic, iff there exists $\beta \in \pi^1(V, x_0)$ such that $\beta \cdot [f] = [g]$.

(3) If V is contractible then: $\pi_n(V, x_0) = 0$; for all n.

(4) If $p < n$ then $\pi_p(S^n, x_0) = 0$; if $p > 1$, $\pi_p(S^1) = 0$.

(5) $\pi_n(X \times Y) = \pi_n X \times \pi_n Y$. Hence for $n \geqslant 2$, π_n of the p-dimensional torus is 0.

3.3. Homotopy groups and defects

Let M^3 be the (three-dimensional) physical space, $\Sigma \subset M^3$ the set of defects and $\phi: M^3 - \Sigma \to V$ the order parameter. It will be convenient to think of Σ as being built in the following fashion. In M^3 we consider a finite set of points $\Sigma^0 \subset M^3$. Then one attaches 1-cells (i.e. arcs) to Σ^0 (as in sect. 2), and gets a larger one-dimensional space $\Sigma^1 \subset M^3$. Then one attaches 2-cells (i.e. 2-disks) to Σ^1 and one gets an (even) larger two-dimensional space Σ^2. One way to think of Σ is to consider that Σ is such a Σ^2 (call this the *two-dimensional model for* Σ).

Such a model is not unreasonable, because:

(i) For energetic reasons it is natural to assume that Σ does not really contain any solid three-dimensional chunk of M^3.

(ii) Once (i) is granted, if Σ is not in some sense or other "wild", it can always be described as above. It will be convenient to keep in mind the sequence of nested spaces:

$$\Sigma^0 \subset \Sigma^1 \subset \Sigma^2 = \Sigma .$$

We will call the 2-cells *walls*, the 1-cells to which no 2-cells are attached *defect lines* and points on Σ^0 to which nothing is attached *punctual defects.*

It is also convenient, sometimes at least, to thicken a bit Σ into a very thin three-dimensional object ("three-dimensional submanifold of M^3") staying very close to Σ and "looking like Σ". *We call this the three-dimensional model for* Σ. Note that several topologically distinct two-dimensional models can lead to the *same* three-dimensional model (up to topological equivalence).

To be even more precise, in the three-dimensional model we will think of Φ as being already defined on the *boundary* of the three-dimensional set of defects.

Now think of a wall $D^2 \times [-\varepsilon, +\varepsilon]$ in the three-dimensional model, take $x \in \operatorname{int} D^2$ and consider $p = x \times \varepsilon$, $q = x \times (-\varepsilon)$

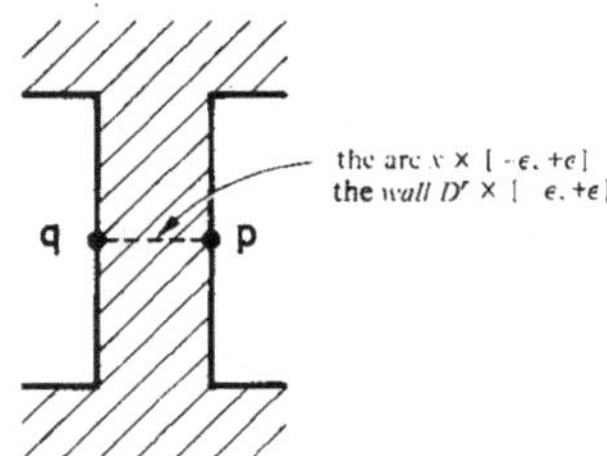

If $\Phi(p)$, $\Phi(q)$ are in the *same* connected component of V, then we can extend Φ continuously over the arc $x \times [-\varepsilon, +\varepsilon]$. So we have made a hole through the wall, after which by continuity we can make the whole wall disappear as in the following figures.

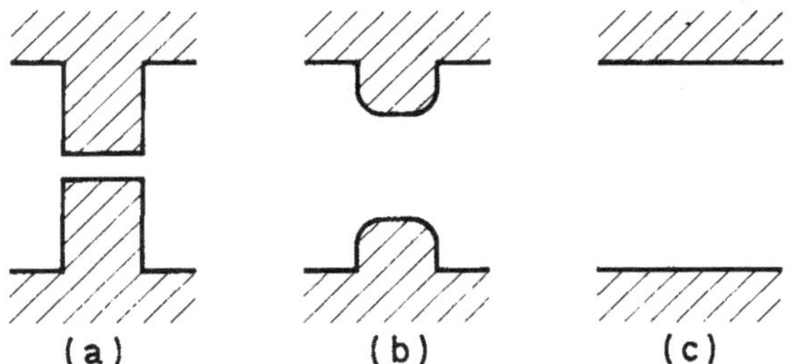

The pierced wall disappears

Hence, *if V is connected (which will be written symbolically $\pi_0 V = 0$), there are no (topologically stable) wall-defects.*

Similarly, consider a defect line, in the three-dimensional model and

the closed loop γ going around it, as in the figure shown above. Since ϕ is defined on γ we have a continuous map

$$\gamma = S^1 \xrightarrow{\Phi/\gamma} V$$

which defines an element in $\pi_1 V$, up to conjugacy. If this element is $0 \in \pi_1 V$ one can extend Φ over the disk Δ and cut the defect line. In particular, *if* $\pi_1 V = 0$ *there are no (topologically stable) defect lines.*

Similarly, a punctual defect, in the three-dimensional model is a 3-ball. Φ restricted to the boundary is an element in $\pi_2 V$, well defined up to multiplication by an element in $\pi_1 V$. Again, we can wipe out the punctual defect if the corresponding element is $0 \in \pi_2 V$.

This is clearly a very general kind of discussion. To summarize it, think of an "n-dimensional physical space" M^n and of a set of defects $\Sigma \subset M^n$ (with a model $\Sigma^0 \subset \Sigma^1 \subset \cdots \subset \Sigma^k = \Sigma$, k being the "dimension of the set of defects" and Σ^i being obtained from Σ^{i-1} by attaching i-cells). Our discussion can be summarized as follows (Kleman–Toulouse): *Suppose* $\pi_0 V = \pi_1 V = \cdots = \pi_{d-1} V = 0$ and $\pi_d V \neq 0$. *Then, if* Σ *is topologically stable, we have*

$$k \leqslant n - d - 1\,.$$

(In fact, generally speaking $k = n - d - 1$.)

So the existence of (stable) defects, in a certain range of dimensions, has to do with the non-vanishing of certain homotopy groups of the manifold of internal states V. Moreover, specific walls, lines, points (in the set of defect) are related to more or less well-defined elements is specific homotopy groups of V. So the "classification" of defects is, at least in a very rough first approximation, connected with the homotopy theory of V.

There is another side to the story: the "dynamics of defects", namely coalescence of punctual defects, crossing of defects lines, etc. This turns out to be connected to more specific questions involving the algebra of homotopy groups. Here is a set of references for these kinds of questions:

[a] V. Poénaru and G. Toulouse, The Crossing of Defects in Ordered Media and the Topology of 3-Manifolds, J. Physique (1977).

[b] Volovik and Mineev, Study of Singularities in Ordered Systems by Homotopic Topological Methods, Zh. Exp. Teor. Fiz. (1977).

[c] M. Kléman, Relationship between Burgers Circuit, Volterra Process and Homotopy Groups, J. Physique (1977).
[d] V. Poénaru and G. Toulouse, Topological Solitons and Graded Lie Algebras, J. Math. Phys., to appear.

3.4. How to compute $\pi_i(G/H)$ for $i = 1, 2$

Much of the relevant information concerning homotopy groups for the manifolds of internal states considered in these lectures is contained in the following theorem:

THEOREM. Let G be a *simply-connected* Lie group ($\pi_0 G = \pi_1 G = 0$) and let $H \subset G$ be a closed (Lie) subgroup.

There are "canonical" isomorphisms

$$\pi_1(G/H) \xrightarrow{\alpha} H/H_0$$

and

$$\pi_2(G/H) \xrightarrow{\beta} \pi_1 H_0 .$$

Via these isomorphisms, the action

$$\pi_1(G/H) \times \pi_2(G/H) \to \pi_2(G/H)$$

can be described as follows:

Let $h \in H$, and consider the map $H_0 \xrightarrow{h} H_0$ defined by $h(h_0) = hh_0h^{-1}$. It is easy to see that the induced isomorphism $\pi_1(H_0) \xrightarrow{h_*} \pi_1(H_0)$ depends only on the image of h in H/H_0 (here $\pi_1(H_0) = \pi_1(H_0, 1)$). Then, if $x \in \pi_1(G/H)$ and $y \in \pi_2(G/H)$ one has:

$$\beta(x, y) = \alpha(x)_*(\beta(y)) .$$

The proof of this theorem will be given in the next section (as a corollary of some more general results.) A more elementary and direct approach can be found in Mermin's notes.

However, here are some examples:

(1) *Superfluid* He_3 (*dipole-locked* A *phase*). Here, as we have seen $V = SO(3)$, and according to what has been said in sect. 1,

$$SO(3) = Sp(1)/H .$$

Sp(1) is homeomorphic to S^3 (hence $\pi_0 Sp(1) = \pi_1 Sp(1) = \pi_2 Sp(1) = 0$)

and H is the (discrete group) consisting of two elements, denoted by $Z/2Z$. So

$$\pi_1 \mathrm{SO}(3) = Z/2Z \quad \text{and} \quad \pi_2 \mathrm{SO}(3) = 0 .$$

(*Note:* according to a very general theorem of Cartan for any (compact) Lie group G, $\pi_2 G = 0$.) Of course, also $\pi_0 \mathrm{SO}(3) = 0$. So in this case we have only lines of defect. Moreover there is only one kind of defect lines. The structure of $\pi_1 V$ also suggests that two defect lines meeting together always annihilate each other. In a simply connected M^3 single defect lines can disappear (since $\pi_2 V = 0$).

(2) *Three-dimensional spins.* Here:

$$V = S^2 = \mathrm{SO}(3)/\mathrm{SO}(2) .$$

But we can also describe V as $\mathrm{Sp}(1)/\varphi^{-1}\mathrm{SO}(2)$, where $\varphi^{-1}\mathrm{SO}(2)$ is the lift of SO(2) to the group of spinors. It is not hard to see (using, for example, the description of Sp(1) given in the next chapter), that $\varphi^{-1}\mathrm{SO}(2)$ is (abstractly speaking) another copy of SO(2), covering the original one twice.

This leads to the following result:

$$\pi_0 S^2 = \pi_1 S^2 = 0 , \qquad \pi_2 S^2 = Z .$$

So, there are only punctual defects, each characterized by an integer. The defect is stable iff this integer is $\neq 0$. The law of combining three punctual defects is just adding the corresponding numbers.

(3) *Crystalline solids.* Here $V = S^1 \times S^1 \times S^1$ and with or without any reference to the theorem above $\pi_1 V = Z + Z + Z$ (and all the other $\pi_i V$'s are 0). $\pi_1 V$ is canonically isomorphic to the group of integral translations in $\mathbf{R}^3$, hence to a Bravais lattice. There are only defect lines and when one goes around the defect line one finds an element in $\pi_1 V$ (conjugacy is trivial in an abelian group). This element can be interpreted as a vector in our Bravais lattice.

This is the so-called *Burgers vector* from chemical crystallography.

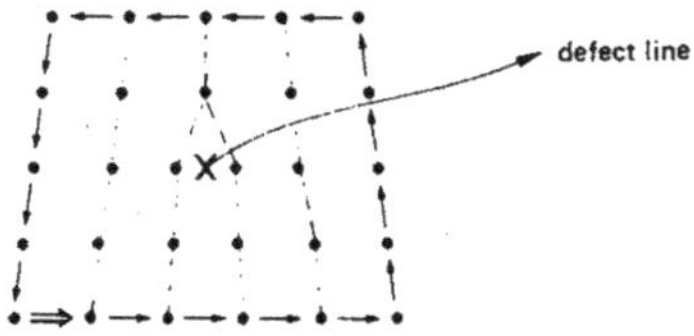

The sum of all arrows adds up to ⇒ which is the Burgers vector.

(4) *Nematic liquid crystals.* Here $V = \mathrm{RP}^2$, and according to sect. 1,

$$\mathrm{RP}^2 = \mathrm{SO}(3)/K$$

where K is the subgroup generated by SO(2) and $T: (x, y, z) \to (x, -y, -z)$. So $K_0 = \mathrm{SO}(2)$ and $K|K_0 = Z|2Z$. Notice that the operation $\mathrm{SO}(Z) \xrightarrow{T_*} \mathrm{SO}(Z)$ which consists of conjugating by T is the sign-reversing automorphism of SO(2). (An easy argument is to think of a two-dimensional rotation as being multiplication by $e^{i\theta}$ in the *complex* plane $\mathbf{C} = \{x + iy\}$. Then multiplication by T is complex conjugation and $\overline{e^{i\theta}\bar{u}} = e^{-i\theta}u$.)

But in order to apply our *theorem* we think of RP^2 as being

$$\mathrm{RP}^2 = \mathrm{Sp}(1)/\varphi^{-1}K .$$

Again it is not hard to see (using for instance the description of Sp(1) given in the next chapter) that $\varphi^{-1}K$ has two connected components and that $(\varphi^{-1}K)_0$ is, abstractly speaking SO(2), and covers K_0 twice. Let $\tilde{T}$ be a lift of T to the group of spinors. Since the diagram

$$\begin{array}{ccc} (\varphi^{-1}K)_0 & \xrightarrow{(\tilde{T})_*} & (\varphi^{-1}K)_0 \\ \downarrow \varphi & & \downarrow \varphi \\ K_0 & \xrightarrow{T_*} & K_0 \end{array}$$

is clearly commutative, $(\tilde{T})_*$ is necessarily the sign-reversing automorphism of $(\varphi^{-1}K)_0$. So:

$$\pi_0\mathrm{RP}^2 = 0\,, \qquad \pi_1\mathrm{RP}^2 = Z/2Z\,, \qquad \pi_2\mathrm{RP}^2 = Z\,,$$

but: *the (unique) non-trivial element of* $\pi_1\mathrm{RP}^2$ *acts on* $\pi_2(\mathrm{RP}^2) = Z$ *by changing the signs* (of the corresponding *integer*). We will come back to this in the next chapter; anyway here there are both punctual *and* linear defects. As we will see punctual defects combine in a rather subtle way.

(5) *Biaxial nematics.* Here $V = \mathrm{SO}(3)/H$ where H is finite. We use the description:

$$V = \mathrm{Sp}(1)/\varphi^{-1}H\,,$$

which tells us that:

$$\pi_0 V = 0\,, \qquad \pi_1 V = \varphi^{-1}H\,, \qquad \pi_2 V = 0\,.$$

Hence there are only linear defects.

4. A quick trip through the basic techniques of homotopy theory

This section will indicate the kind of ingredients which go into the proofs of the various statements made in the last paragraph of the previous section. More details can be found in any of the many standard text books about homotopy theory. Towards the end of the chapter we discuss a number of (*topological*) "*confinement mechanisms*".

4.1. Relative homotopy groups and exact sequences

We consider the standard unit n-cube I^n, with coordinates $t_1, \ldots, t_n$, where $0 \leqslant t_i \leqslant 1$. For the moment we will assume that $n > 1$. We will think of the unit $n - 1$ cube I^{n-1} as being $I^{n-1} = I^n \cap (t_n = 0)$ and we denote by J^{n-1} that part of ∂I^n which is *not* in I^{n-1}.

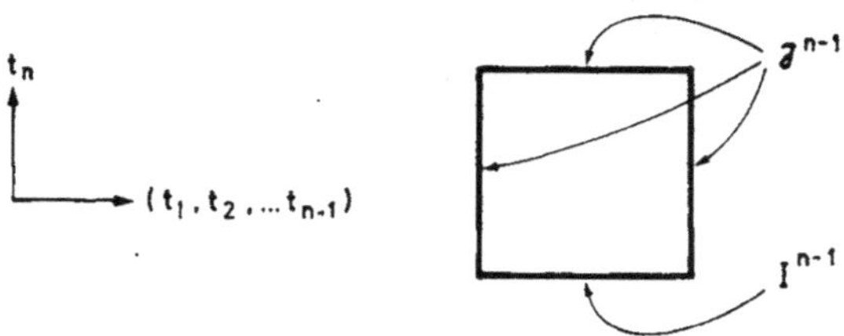

So $\partial I^n = I^{n-1} \cup J^{n-1}$ and $I^{n-1} \cap J^{n-1} = \partial I^{n-1}$. Let V be some space, $W \subset V$ a "subspace" and choose a base point $x_0 \in W \subset V$. We consider the set $\Omega^n(V, W, x_0)$ of all continuous maps $f = I^n \to V$ having the properties: (a) $f(J^{n-1}) = x_0$; (b) $f(I^{n-1}) \subset W$.

If $f, g \in \Omega^n(V, W, x_0)$ we write $f \sim g$, and say that "f and g are homotopic", if f and g can be connected by a homotopy of maps $I^n \to V$ satisfying (a) and (b).

The set of homotopy classes is endowed with a composition law, defined exactly as for $\Omega^n(V, x_0)$ (same formula!) and it turns out to be a group (same proof again). We denote this group by $\pi_n(V, W, x_0)$. It is abelian if $n > 2$. Note that $\pi_n(V, x_0, x_0) = \pi_n(V, x_0)$ and that any map $f: I^n \to W$ such that $f(J^{n-1}) = x_0$ represents the unit element in $\pi_n(V, W, x_0)$.

This last fact is easy to understand if we give another, more intuitive description of $\pi_n(V, W, x_0)$. Consider the n-disk D^n with a base point $* \in S^{n-1} = \partial D^n$. Then the various elements of $\pi_n(V, W, x_0)$ are exactly represented by homotopy classes (in the appropriate context) of maps $D^n \xrightarrow{\varphi} V$ such that $\varphi(*) = x_0$ (condition (a)); $\varphi(\partial D^n) \subset W$ (condition (b)). This is really like an n-dimensional possibly open bag in V, with its mouth or border-rim in W, and some specially chosen points (chosen once for all) fixed at x_0. If the whole bag lies already in W one can slowly retract it to x_0 (staying inside $W \cdots$).

There is an obvious map

$$\pi_n(V, x_0) \xrightarrow{\beta_n} \pi_n(V, W, x_0)$$

which is a group-homomorphism (think of an element in $\pi_n(V, x_0)$ as being a closed bag).

But, also, restricting $f: I^n \to V$ to $I^{n-1} \subset I^n$ gives us another group-homomorphism:

$$\pi_n(V, W, x_0) \xrightarrow{\partial_n} \pi_{n-1}(W, x_0)$$

(keep only the border-rim of the bag...).

There is a third group homomorphism which we will have to consider. Think of the natural inclusion map

$$(W, x_0) \xrightarrow{i} (V, x_0) .$$

This induces a (*not* necessarily injective) group homomorphism.

$$\pi_n(W, x_0) \xrightarrow{i_n} \pi_n(V, x_0) .$$

For simplicity's sake we omit to write x_0 from now on.

Before we continue with homotopy theory, we need some more algebra.

A sequence of group homomorphisms is called *exact*, if

$$G \xrightarrow{\alpha} H \xrightarrow{\beta} K , \quad \operatorname{Im} \alpha = \operatorname{Ker} \beta .$$

Exercises.

(1) The sequence $1 \to G \xrightarrow{\alpha} H$ is exact iff α is injective.

(2) The sequence $H \xrightarrow{\beta} K \to 1$ is exact iff β is surjective.

(3) The sequence $1 \to G \to H \to K \to 1$ is exact iff G is an (invariant) subgroup of H and $K = H/G$.

A long sequence of group homomorphism is called *exact* if any partial subsequence of three terms is exact.

THEOREM. The long sequence

$$\pi_{n+1}(V, W) \to \pi_n(W) \xrightarrow[i_n]{} \pi_n(V) \xrightarrow[\beta_n]{} \pi_n(V, W) \xrightarrow[\partial_n]{} \pi_{n-1} W \to \cdots$$

is exact.

We will not give the complete proof, which is not hard, anyway. Consider, for example, $[f] \in \pi_{n-1} W$, where $f: (S^{n-1} *) \to (W, n_0)$. If $i_{n-1}[f] = 0$ this means that f extends to a map $D^n \xrightarrow{f} V'$. But $[f] \in \pi_n(V, W)$ and

$\partial_n[f] = [f]$. So we have proved that Ker $i_{n-1} \subset \text{Im}\, \partial_n$. Five similar little arguments complete the proof.

So far, our exact sequence has stopped at $\pi_1(V, x_0)$. In order to continue it, we need to somehow extend our framework. A *pointed set* is, by definition, a set S together with a chosen point $x \in S$. If S happens to be a group, the chosen point will always be the unit element. We can define exact sequences for maps of pointed sets, respecting the chosen points.

I will consider the *pointed* set $\pi_0(V)$ consisting of all the connected components of V, the chosen element being the component which contains x_0. I will also consider the set of homotopy classes of maps $f: I \to V$, such that $f(0) = x_0$, $f(1) \in W$. This is a pointed set $\pi_1(V, W)$ whose chosen point is the class of maps $(I, 0) \to (W, x_0)$. It is not hard to see that our long sequence stays exact if we continue it as follows:

$$\to \pi_1 W \to \pi_1 V \to \pi_1(V, W) \to \pi_0 W \to \pi_0 V .$$

Here $\pi_1(V, W) \to \pi_0 W$ is the map induced by $f \to f(1)$.

4.2. *Fibrations and the covering homotopy theorem*

A fibration is a continuous map

$$\begin{array}{c} E \\ \big\downarrow p \\ B \end{array}$$

with certain special properties:

(1) B is connected and p is surjective.

(2) There is a space F called the "*fiber*", such that *locally* E looks like $B \times F$. Precisely, this means the following: For any $x \in B$, there is an open neighborhood of x in B, call it U, and a homeomorphism $\psi: p^{-1}(U) \to U \times F$ such that the following diagram is commutative:

$$\begin{array}{ccc} p^{-1} U & \xrightarrow{\psi} & U \times F \\ & {\scriptstyle p}\searrow \quad \swarrow & \\ & U & \end{array}$$

E is called the *total* space and B is called the *base* space. If $E = B \times F$ and p is the natural projection $B \times F \to B$ we will say that the fibration is *trivial*. (A general fibration is only *locally trivial*.)

Examples.

(1) The map $R \xrightarrow{p} S^1 = \{$the complex numbers of modulus 1$\}$ given by $x \to e^{ix}$. This is a special case of what is called a *covering space*, which

means a fibration such that the total space is connected and the fiber discrete.

(2) The map $S^1 \to S^1$ given by $e^{ix} \to e^{inx}$ where n is some fixed integer. This is also a covering space.

Actually examples (1) and (2) describe *all* covering spaces where the basis is S^1.

(3) Let G be a Lie group and $H \subset G$ a closed subgroup. *The canonical map* $G \xrightarrow{p} G/H$ *is a fibration of fiber* H.

(The idea of the proof is the following. Suppose first that there exists a continuous map $G \xleftarrow{s} G/H$ such that for all $x \in G/H$ one has $ps(x) = x$. Such a map is called a *cross-section* (and it does not always exist). If such an s exists any $z \in G$ can be written *uniquely* as $z = p(z) \cdot h$ with $h \in H$ and hence $G = G/H \times H$.

There is a theorem which says that for any $x \in G/H$ there is a *local* cross-section, defined in a small neighborhood of x. We will not prove this in general, but it is easy to check in the various special cases considered in these lectures.

Now, from the existence of the *local* cross-sections one can derive, without much trouble, that $G \xrightarrow{p} G/H$ is a fibration (i.e. that it is *locally* like a product).)

THE COVERING HOMOTOPY THEOREM. Let $E \xrightarrow{p} B$ be a fibration, K a (let us say compact) space and f, F maps as below:

$$\begin{array}{ccc} K \times 0 & \xrightarrow{F} & E \\ \downarrow & & \downarrow p \\ K \times [0,1] & \xrightarrow[f]{} & B \end{array}$$

We assume this diagram to be commutative. Then there is a map

$$K \times [0,1] \xrightarrow{\Phi} E$$

such that the following larger diagram is also commutative:

$$\begin{array}{ccc} K \times 0 & \xrightarrow{F} & E \\ \downarrow & \overset{\Phi}{\nearrow} & \downarrow p \\ K \times [0,1] & \xrightarrow[f]{} & B \end{array}$$

The statement asks for some explanations. What it means, in plain English, is the following: Suppose a map $F: K \to E$ is given together with a homotopy $K \times [0,1] \xrightarrow{f} B$ which starts with $p_0 F$. Then there is a homotopy $K \times [0,1] \to E$, starting with F, which projects down to f ("covers f").

In the special case when K is a point this means that any path in B can be lifted to E, with prescribed initial point. The general case is a parametrized version (with parameter space K) of this "path lifting property".

We will sketch the proof in the case when K itself is an internal, $K = I$. The general case is not very different. Note first that the theorem is immediate if the fibration is trivial. Second, divide the square $I \times [0, 1]$ into smaller squares $S_1, S_2, \ldots$ as below:

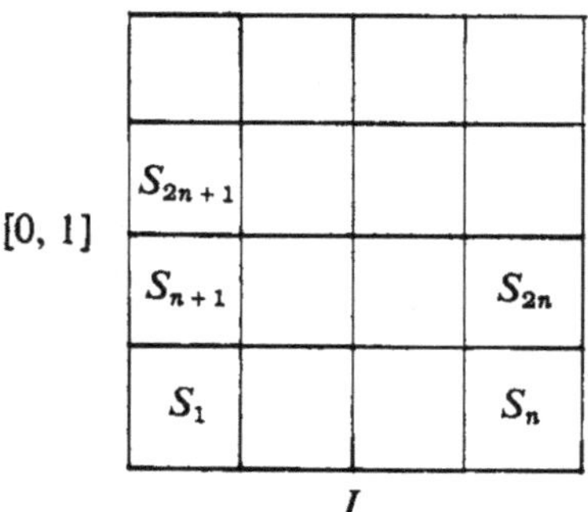

The squares S_i are supposed to be sufficiently small, so that our fibration, when restricted to $f(S_i)$ is trivial. Then one can extend f successively over $S_1, S_2, \ldots$. Note that a similar argument works when K is an n-cube, and this is really all we need in these lectures.

4.3. *The exact homotopy sequence of a fibration*

We consider a fibration $E \xrightarrow{p} B$. We will choose a base point $x_0 \in B$, and by thinking of $p^{-1}(x_0)$ as a copy of F we have an inclusion $F \underset{i}{\subset} E$. Choose also a base point for E, $y_0 \in F$.

THEOREM. There is a canonical isomorphism

$$\pi_n(E, F, y_0) \xrightarrow{P} \pi_n(B, x_0) .$$

Proof. Projecting down gives an obvious arrow P. I will show how to construct an inverse. We start by choosing once for all a homotopy

$$\underset{u}{\overset{\uparrow}{I^{n-1}}} \times \underset{t}{\overset{\uparrow}{I}} \xrightarrow{\Psi} I^n ,$$

such that:

(1) $\Psi_t(\partial I^{n-1})$ is always the identity map of ∂I^{n-1} into itself;
(2) $\Psi_0(I^{n-1}) = J^{n-1}$;
(3) Ψ_1 is the identity map $I^{n-1} \to I^{n-1}$.

Such a homotopy can be visualized in the following way:

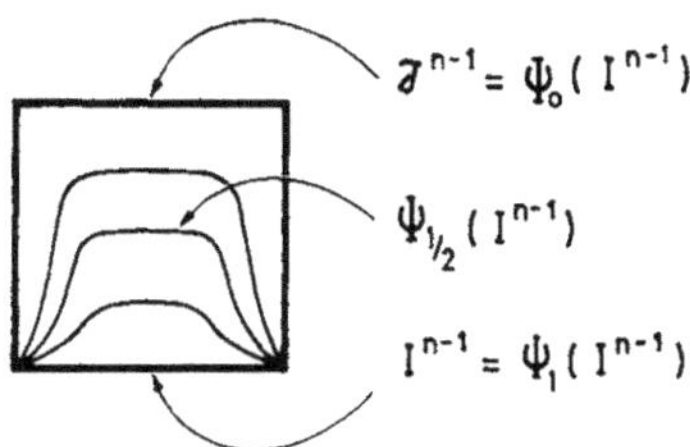

Consider now $g: (I^n, \partial I^n) \to (B, x_0)$, $[g] \in \pi_n(B, x_0)$, and the constant map $F: I^{n-1} \to y_0 \in E$. If we apply the covering homotopy theorem to the diagram

$$\begin{array}{ccc} I^{n-1} \times 0 & \xrightarrow{F} & E \\ \downarrow & \overset{\Phi}{\nearrow} & \downarrow p \\ I^{n-1} \times I & \xrightarrow{g \cdot \Psi} & B \end{array}$$

we obtain a map $\Phi \in \Omega^n(E, F, y_0)$. Another application of the covering homotopy theorem shows that $[\Phi] \in \pi_n(E, F, y_0)$ depends only on $[g] \in \pi_n(B, x_0)$. It is not hard to show that $[g] \mapsto [\Phi]$ is the inverse (group-homomorphism).

COROLLARY. If

$$\begin{array}{ccc} F & \hookrightarrow & E \\ & & \downarrow p \\ & & B \end{array}$$

is a fibration, then the following long sequence is exact:

$$\cdots \to \pi_n F \xrightarrow{i_*} \pi_n E \xrightarrow{p_*} \pi_n B \xrightarrow{\partial \cdot p^{-1}} \pi_{n-1} F \to \cdots .$$

Applications.

(1) Let $E \xrightarrow{p} B$ be a covering space. Then $\pi_n F = 0$ if $n > 0$ and one deduces that:

(a) $\pi_n B = \pi_n E$ if $n > 1$ (the isomorphism is p_*).

(b) At the level of π_1:

$$\pi_1 E \xrightarrow{p_*} \pi_1 B \, ,$$

is injective. It can be shown that any subgroup of $\pi_1 B$ appears in this way (in a more or less unique fashion). [A special case of (a) is $\pi_n S^1 = 0$, for $n > 1$.]

(2) Let G be a compact Lie group with $\pi_0 G = \pi_1 G = 0$. Let $H \subset G$ be a closed subgroup, and $H_0 \subset H$ the connected component of the identity. Inside our exact sequence we find

$$\pi_2 G \to \pi_2(G/H) \to \pi_1 H \to \pi_1 G .$$

Now $\pi_1 G = 0$ by hypothesis and $\pi_2 G = 0$ by a general theorem of Cartan. (This is anyway proved directly in the cases we have been studying in these lectures.) So one has an isomorphism

$$\pi_2(G/H) \xrightarrow{\beta} \pi_1 H_0 .$$

(3) Under the same assumptions, we also have an exact sequence

$$\pi_1 G \to \pi_1(G/H) \to \pi_0 H \to \pi_0 G .$$

A priori this is just an exact sequence in the context of *pointed sets*. But it so happens that $\pi_0 H_0, \pi_0 G$ are again groups, and $\pi_0 H \to \pi_0 G$ a group-homomorphism. What the middle map is, from the standpoint of group theory will be explained below.

Here again $\pi_1 G = \pi_2 G = 0$ and clearly $\pi_0 H = H/H_0$. (Remember that $H_0 \subset H$ is an invariant subgroup, so H/H_0 is actually a group, not just a pointed set.) This provides us with a bijective map $\chi\colon \pi_1(G/H) \to H/H_0$. It is not hard to see that $\chi(ab) = \chi(b)\chi(a)$ so that the map $a \to (\chi(a))^{-1}$ (which I will call $\alpha(a)$) is actually an isomorphism of groups:

$$\pi_1(G/H) \xrightarrow{\alpha} H/H_0 .$$

(4) We will analyze now the action of $\pi_1(G/H)$ in $\pi_2(G/H)$.

We will denote by $\otimes$ the multiplication in G and by $\cdot$ the action of π_1 on π_2. Let $h \in H$ and denote by H_1 the coset $hH_0 \in H/H_0$. Let also $f\colon [0, 1] \to G$ be a continuous path such that

$$f(0) = h , \quad \text{and}$$

$$f(1) = \{\text{the unit element of } G\} .$$

The homotopy class of the *closed loop* $p(f)$ is an (arbitrary) element $x \in \pi_1(G/H)$ and our conventions are such that $\alpha(x) = H_1$. Let also $(D^2, \partial D^2, *) \xrightarrow{\varphi} (G, H, 1)$ (where $*$ is a point of ∂D^2, chosen once for all) be a continuous map and denote the restriction of φ to ∂D^2 by $S^2 \xrightarrow{\psi} H$. If we consider:

$$y = [p\varphi] \in \pi_2(G/H), \qquad z = [\psi] \in \pi_1 H = \pi_1 H_0,$$

then again, $\beta(y) = z$.

Let $X = (S^1 \times [0, 1]) \cup D^2$ where $S^1 \times 1$ is glued to $S^2 = \partial D$ (X is really another copy of D^2). On X we define a continuous map $\Phi: X \to G$ as follows:

$$\Phi | D^2 = \varphi, \text{ and, if } \theta \in S^1,\ t \in [0, 1];$$

$$\Phi(\theta, t) = f(t) \otimes \psi(\theta).$$

This Φ is really a map

$$(D^2, \partial D^2, *) \xrightarrow{\Phi} (G, H_1, h).$$

Note that $\Phi | \partial D^2 = h \otimes \psi$, and since H_0 is an invariant subgroup of H, $h \otimes \psi \otimes h^{-1}$ is a map of S^1 into H_0.

It is easy to see that $\Phi \otimes h^{-1}$ is a map $(D^2, \partial D^2, *) \to (G, H, 1)$ such that $[p(\Phi \otimes h^{-1})] = x \cdot y \in \pi_2(G/H)$.

Conjugation by h defines a map

$$h: H_0 \to H_0,$$

and clearly $\beta(x \cdot y) = h_*(z)$.

This completes the proof of the theorem from sect. 3.4.

4.4. *More about spinors and covering spaces*

Recall that $\mathrm{Sp}(1) \xrightarrow{\varphi} \mathrm{SO}(3)$ is a covering space, and that

$$\pi_1 \mathrm{Sp}(1) = 0, \qquad \pi_1 \mathrm{SO}(3) = Z/2Z.$$

Call I (respectively $\tilde{I}$) the identity element of SO(3) (Sp(1)).

Consider the set of all paths f: $[0, 1] \to \mathrm{SO}(3)$ starting at I, and consider two such paths as being equivalent if they have some endpoints and are homotopic by a homotopy keeping *both* endpoints fixed. The space of equivalence classes, with the natural topology will be denoted by $\widetilde{\mathrm{SO}(3)}$. Sending f to $f(1)$ gives a continuous map $\widetilde{\mathrm{SO}(3)} \xrightarrow{p} \mathrm{SO}(3)$.

There is a natural map

$$\widetilde{\mathrm{SO}(3)} \xrightarrow{L} \mathrm{Sp}(1)$$

defined as follows. If $[f] \in \mathrm{SO}(3)$ then I can lift the path f to Sp(1) starting with $\tilde{I}$. This lifting is actually unique and will be called $\tilde{f}$. It is easy to see that $\tilde{f}(1) \in \mathrm{Sp}(1)$ depends only on $[f]$ (all these facts use, of course, the covering homotopy theorem and the information that Sp(1) is simply-connected). We can set $J([f]) = \tilde{f}(1)$. I claim that *J is a homeomorphism inbetween* $\widetilde{\mathrm{SO}}(3)$ *and* Sp(1). The inverse map is the following: if $x \in \mathrm{Sp}(1)$ there is always a path g_x joining $\tilde{I}$ to x in Sp(1), and since Sp(1) is simply connected this path is unique, up to a homotopy keeping the ends fixed. The map $x \to [\varphi g_x]$ is the inverse of J. All this is a new (and pleasant) description of Sp(1).

Exercises.

(1) Let $y \in \mathrm{SO}(3)$ be a rotation of angle $\pi + \alpha$ around the axis A. In the space of rotations around A consider the following two paths:

$$t \to t(\pi + \alpha), \qquad t \to t(-\pi + \alpha), \qquad (t \in [0, 1]).$$

These are exactly the two spinors living above our y.

(2) In $\widetilde{\mathrm{SO}(3)}$ one can multiply two paths in an obvious way, using the composition law of SO(3). This gives a Lie group which is the same as Sp(1), via J.

The passage from SO(3) to SO(3) is a special case of a very general construction. Let V be a connected space, "locally well-behaved" (which in the present context means "locally connected" and "locally simply-connected") and let $x_0 \in V$. One can consider the space of paths in V starting at x_0, and build $\tilde{V}$ out of V in the same way as $\widetilde{\mathrm{SO}(3)}$ was built out of SO(3). It can be shown that $\tilde{V}$ is always simply connected and that $\tilde{V} \xrightarrow{p} V$ is a covering space. This is called the *universal* covering space of V. It is the unique covering space of V whose total space is simply-connected. All the other covering spaces of V are quotients of $\tilde{V}$.

It can be shown that $\pi_1 V$ is isomorphic to the group of homeomorphisms $\tilde{V} \to \tilde{V}$ which commute with the projection p. A special case: The exponential map $R \to S^1$ $(t \to e^{it})$ is the universal covering space of S^1, and this shows that $\pi_1 S^1 = Z$.

Exercise.

We know already that $\pi_1(\mathrm{RP}^2) = Z/2Z$. Show that the following loop in the space of directions is the generator of this group.

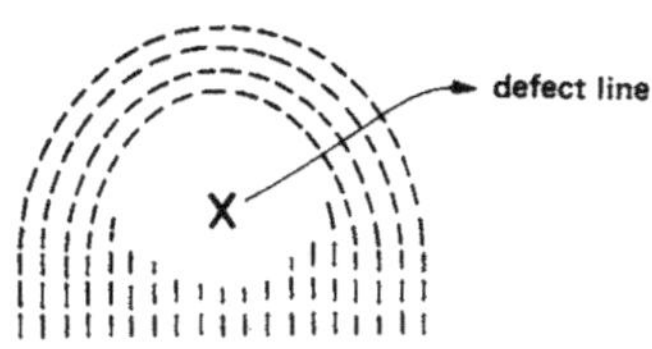

Note: Hence, in a *nematic liquid crystal*, if we see a configuration like this around a defect line, that defect line is (topologically) stable. Punctual defects in a nematic liquid crystal are classified by $\pi_2(RP^2) = Z$ and as we know that the action of the generator of $\pi_1(RP^2)$ on $\pi_2(RP^2)$ is non-trivial (change of sign); this means that two punctual defects coming close to each other on one side or the other, of such a defect line, will or *will not* cancel each other out. This remark is due to Volovik and Mineev [1] (end of sect. 4.5), (see also [2]).

4.5. *The Hopf fibration, $\pi_3 S^2$ and "textures"*

Think of S^3 as being the set of pairs of complex numbers (z_1, z_2) such that $|z_1|^2 + |z_2|^2 = 1$ and think of S^2 as being the *complex projective line*; this means that S^2 is the set of pairs (z_1, z_2) such that $(0, 0)$ is excluded and $(\lambda z_1 \lambda z_2)$ is by definition the same point as (z_1, z_2). The map $S^3 \ni (z_1, z_2) \xrightarrow{H} (z_1, z_2) \in S^2$ is a fibration, called the "Hopf fibration".

Exercise.

Prove that H is actually a *non-trivial* fibration of fiber S^1.

The exact homotopy sequence

$$\rightarrow \underbrace{\pi_3 S^1}_{0} \rightarrow \pi_3 S^3 \rightarrow \pi_3 S^2 \rightarrow \underbrace{\pi_2 S^1}_{0}$$

tells us that $\pi_3 S^2 = \pi_3 S^3 = Z$. Vector fields in $\mathbf{R}^3$ which are everywhere non-zero and constant close to ∞ are the same thing as maps $S^3 \rightarrow S^2$, and hence they are classified by $\pi_3 S^2 = Z$. Here one has the example of a "texture", an ordered medium where the order parameter is everywhere defined, and the regions where it is not constant play the role of defects. See [3] for more details.

[1] Volovik and Mineev, Study of Singularities of Ordered Systems by Homotopic Topology Methods, JETP (1977).
[2] V. Poénaru and G. Toulouse, The Crossing of Defects in Ordered Media and the Topology of 3-Manifolds, J. Physique (1977).
[3] Bouligand, Derrida, V. Poénaru, Y. Pomeau, and G. Toulouse, Distortions with Double Topological Character: The Case of Cholesterics (to appear).

4.6. *Topological "confinement mechanisms"; double distortions in the cholesterical orders*

This paragraph is basically an exposition of ref. [3] (end of sect. 4.5). It will (hopefully) make explicit, and clearer, some points which are only implicitly hinted at in ref. [3].

Some background might be necessary in order to appreciate the peculiar type of "*topological atom*" described in ref. [3]. This background is contained in refs. [a] and [d] (end of sect. 3.3); it can be briefly summarized as follows. The sequence of homotopy groups of a (connected) manifold of internal states V:

$$\pi_1 V, \pi_2 V, \pi_3 V, \ldots,$$

can be organized as a *graded Lie algebra*, by introducing the so-called *Whitehead product*

$$\pi_p V \times \pi_q V \xrightarrow[{[\,,\,]}]{} \pi_{p+q-1} V .$$

(Very roughly speaking this product is defined as follows: we consider S^p, S^q with base points $*_p$, $*_q$ and given orientation. Inside $S^p \times S^q$ we consider $S^p \vee S^q = (S^p \times *_q) \vee (*_p \times S^q)$. We can slightly thicken $S^p \vee S^q$ into a $(p + q)$-dimensional manifold N, which can be crushed (collapsed) onto $S^p \vee S^q$ by a map $r: N \to S^p \vee S^q$. One should notice that the boundary of N is S^{p+q-1}; and this sphere will be oriented like $-N$.

If

$$f: (S^p, *_p) \to (V, x_0), \qquad g: (S^q, *_q) \to (V, x_0)$$

are given, then the composite map

$$S^{p+q-1} \xrightarrow{r} S^p \vee S^q \xrightarrow{f \vee g} V$$

is in the homotopy class of the Whitehead product of $[f]$ with $[g]$. See ref. [d] of sect. 3.3 for more details.)

This operation is a kind of commutator, in the sense that it has the formal properties of a graded Lie algebra which is also called a "super-algebra" in the context of *supersymmetry*.

To be more explicit, if

$$\alpha \in \pi_p V, \qquad \beta \in \pi_q V, \qquad \gamma \in \pi_r V,$$

then one has:

$$[\alpha, \beta] = (-1)^{pq}[\beta, \alpha] \quad \text{(anticommutativity)},$$

and

$$(-1)^{qr}[\alpha, [\beta, \gamma]] + (-1)^{rp}[\beta, [\gamma, \alpha]] + (-1)^{pq}[\gamma, [\alpha, \beta]] = 0 .$$

If $p = q = 1$ and $\alpha \in \pi_1 V$, $\beta \in \pi_1 V$, then the Whitehead product $[\alpha, \beta]$ is just the usual commutator $\alpha\beta\alpha^{-1}\beta^{-1}$. If $\alpha \in \pi_1 V$ and $\beta \in \pi_p V \; (p > 1)$, then:

$$[\beta, \alpha] = (-1)^p[\alpha, \beta] = \beta \underset{\uparrow}{-} \underset{\uparrow}{\alpha} \cdot \beta$$

the group operation of π_p — the action of π_1 on π_p

If V itself is a topological group, then all the Whitehead products are 0. In particular, the kind of "confinement" we describe in the first part of this paragraph will not exist if the manifold of internal states is a Lie group.

We consider now a "physical space" M, of dimension $n = p + q + 1$, and an "ordered medium":

$$\underset{\substack{\uparrow \\ \text{defects}}}{M - \Sigma} \xrightarrow[\substack{\uparrow \\ \text{order parameter}}]{\Phi} \underset{\substack{\uparrow \\ \text{manifold of} \\ \text{internal states}}}{V}$$

We consider "defect cells" L^p, $N^q \in \Sigma$ of dimensions p and q respectively, and also the "gedanken-process" of *crossing* these defect cells (without entanglement) as in the series of drawings below. (In these drawings $p = q = 1$, and dim $M = 3$.)

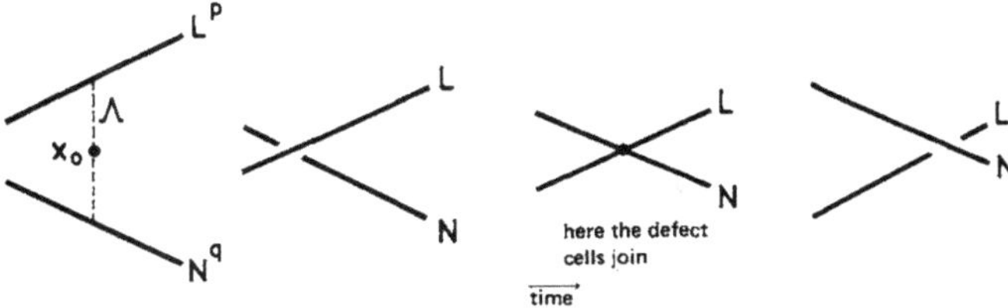

The question is whether this process is *physically possible*, without a very large input of outside energy which would bring all the surrounding physical space into the high energy phase of Σ. Technically this will be supposed to mean that we are allowed to move defects around (without changing the topology), add *defect lines*, or cut defect cells of any dimension, as in sect. 3.3 (whenever this *is topologically possible*).

Notice that L and N come close to each other along a line Λ, whose middle point will be denoted by x_0. By surrounding L (respectively N) with a q-dimensional (p-dimensional) sphere, as in sect. 3.3, and by pushing the base points of those spheres into x_0, along Λ (as in sect. 3.2) one gets two elements (well-defined up to their sign)

$$\alpha_q \in \pi_q(M - \Sigma, x_0)\,, \qquad \beta_p \in \pi_p(M - \Sigma, x_0)\,.$$

The figure below gives the idea of what is happening (when $p = q = 1$).

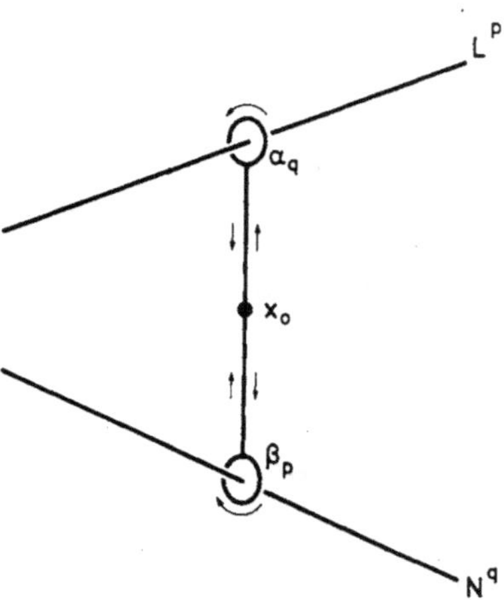

By computing the order parameter along α and β one gets:

$$\Phi_* \alpha_q \in M_q V\,, \qquad \Phi_* \beta_p \in \pi_p V\,.$$

In ref. [d] (end of sect. 3.3) it is shown that *the crossing of L and N is possible if and only if the elements $\Phi_*\alpha_q$, $\Phi_*\beta_p$* ***commute****, in the sense that their Whitehead product vanishes*

$$[\Phi_*\alpha_q, \Phi_*\beta_p] = 0 \in \pi_{p+q-1} V\,.$$

For linear defects in a three-dimensional physical space this is a matter of how commutative the group $\pi_1 V$ is (see ref. [a] (end of sect. 3.3) for a detailed discussion of this case). This kind of "confinement mechanism" coming from non-commutativity should be observable in biaxial nematics (if and when they are synthesized) but has been, most likely, already observed in cholesterics. There is also a four-dimensional very appealing speculation: in a four-dimensional physical space a *loop-like defect* ($p = 1$) and a *bag-like defect* ($q = 2$) which are linked (this is, of course, possible in dimension 4) cannot be separated (without a tremendous cost of energy) if

$$[\underset{\substack{\uparrow \\ \pi_2}}{\Phi_*\alpha}, \underset{\substack{\uparrow \\ \pi_1}}{\Phi_*\beta}] = \Phi_*\alpha - \Phi_*\beta \cdot \Phi_*\alpha \neq 0 \in \pi_2 V\,.$$

This should look somehow like this drawing

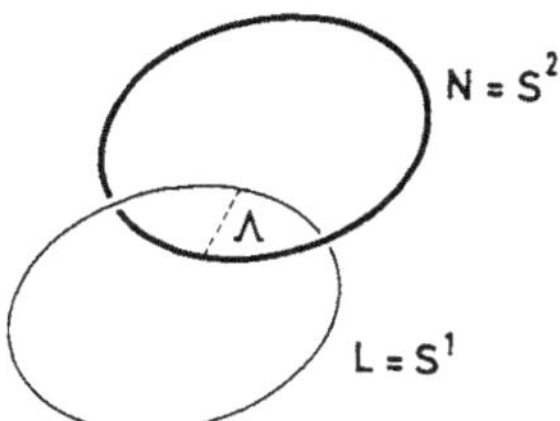

After this brief discussion of "*confinement mechanisms*" *related to non-commutativity* we change (for a while) the topic, and discuss *cholesterics*. In the undistorted cholesterics one has a *local* frame of three directions: direction 1 (the molecular direction), direction 3 (the cholesteric direction), direction 2 (the binormal). (The reader can find more details about the cholesteric order in de Gennes's book: *Liquid Crystals* (Oxford, 1974) or in Kleman's more recent book: *Points, lignes, parois*, vol. I, 1977.) Let D_2 be the group of 180°-rotations around the three directions of such a local frame. According to Volovik and Mineev, the manifold of internal states for cholesterics, V, can be identified to $SO(3)/D_2$ (ref. [1], sect. 4.5). There is a certain subtlety involved in this discussion, since the ordered phase symmetry group contains also translation elements. Anyway, $\pi_1 V$ is the group consisting of the 8 spinors living over D_2:

The elements of $\pi_1 V$

$$I\,,\quad J\,,\quad e_1\,,\quad -e_1\,,\quad e_2\,,\quad -e_2\,,\quad e_3\,,\quad -e_3\,.$$

The combination rules

$$e_1e_2 = -e_2e_1 = e_3, \ldots,$$
$$e_i^2 = e_1e_2e_3 = J\,,\qquad J^2 = I\,,\qquad Je_i = e_iJ = -e_i\,.$$

(This is sometimes called the "quaternion group" and it should also appear in the case of biaxial nematics.)

The only non-trivial element in the center of this group is J. So, according to the previous theory there should be obstructions for crossing $\pm e_i$-defect lines (and this seems to be the case indeed), while there should be no obstructions at all for the crossing of J-defect lines.

Now, there is one special fact about cholesterics: for energetical reasons, there are no discontinuities for the molecular director alone, so there is a "secondary order parameter" *everywhere well-defined* (hence a *texture*):

$$\{\text{physical space}\} \xrightarrow{\varphi} RP^2\,.$$

Yves Bouligand (J. Physique 35 (1974) 959) has extensively studied the cholesteric order and here is a brief summary of some of his findings:

– He developed an experimental technique, with which when examined in suitable polarized light, a distorted cholesteric exhibits black lines, which give the locus of points where the molecular director is vertical. So the black lines are $\varphi^{-1}(p)$ where

$$p = \{\text{the "vertical direction"}\} \in \mathrm{RP}^2 .$$

– He was able to establish that along these black lines live defect lines of type J of the whole cholesteric order.

– He found pairs of black lines linked as in the figure below,

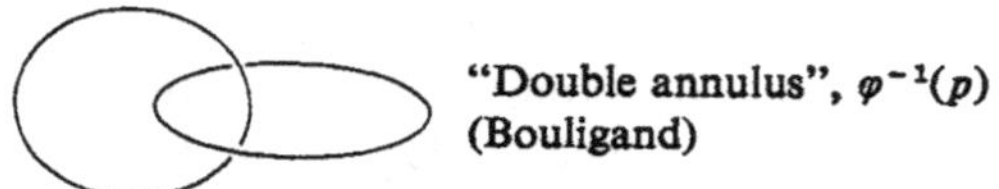

"Double annulus", $\varphi^{-1}(p)$ (Bouligand)

which *do not separate*. Since J is in the center of $\pi_1 V$ (which means that J commutes with everything else), our previous "non-commutative confinement mechanism" does not account for the existence of this new kind of "topological atom". In order to explain the existence of the "double annulus" a different "topological confinement mechanism is needed". This is what comes next.

In the preceding paragraph, we showed that $\pi_3 S^2 = Z$. An explicit isomorphism can be described as follows. (For more details, we refer to Milnor's elegant little book: *Topology from the Differentiable Viewpoint.*)

Let $f\colon S^3 \to S^2$ be a let us say smooth map. S^3 has a given orientation and S^2 a provisional orientation. For a "*general point*" $x \in S^2, f^{-1}(x)$ is a (finite) collection of disjoint single closed loops in S^3. The provisional orientation of S^2 induces a transverse orientation of $f^{-1}(x)$ which together with the chosen orientation of S^3 orients $f^{-1}(x)$. If $y \neq x$ is another general point, we can compute the *linking number*:

$$\mathrm{Lk}(f^{-1}(x), f^{-1}(y)) \in Z .$$

This number measures how many times $f^{-1}(x)$ goes around $f^{-1}(y)$. It also measures the amount of work done by the field generated by a unit electric current along $f^{-1}(x)$ on a unit magnetic charge moving with unit constant speed around $f^{-1}(y)$. Notice that Lk is invariant for change of the provisional orientation of S^2 or for $x \leftrightarrow y$. It can be proved that it establishes our isomorphism $\pi_3 S^2 = Z$.

Since S^2 is a covering space of RP^2 one also has $\pi_3(\mathrm{RP}^2) = Z$. Also, since S^3 is simply connected, any continuous map $\varphi\colon S^3 \to \mathrm{RP}^2$ has a lift

$\tilde{\varphi}: S^3 \to S^2$ (this means that one can put arrows to the directors, in a consistent continuous fashion). If $p \in \mathrm{RP}^2$ is the director corresponding to $(x, -x) \in S^2$, one has a disjoint decomposition $\varphi^{-1}(p) = \tilde{\varphi}^{-1}(x) \cup \tilde{\varphi}^{-1}(-x)$.

If p is "general" assigning to $\varphi: S^3 \to \mathrm{RP}^2$ the integer $\mathrm{Lk}(\tilde{\varphi}^{-1}(x), \tilde{\varphi}^{-1}(-x)) \in Z$ establishes the isomorphism $\pi_3(\mathrm{RP}^2) = Z$. Notice that here we used only *one* point in RP^2, in order to make the computation. In this way one gets an easy and elegant way of computing, let us say, the homotopy class of a field of directors

$$R^3 \xrightarrow{\Phi} \mathrm{RP}^2 .$$

which is constant in the vicinity of ∞; such a field can, of course, be thought of as a map $S^3 \to \mathrm{RP}^2$.

Because of the fact that the local frame of the cholesteric order is moving in space, even in the undistorted situation we cannot assume constant boundary conditions for the molecular director:

$$\{\text{physical space}\} \xrightarrow{\varphi} \mathrm{RP}^2 .$$

So, we will assume instead that our physical space is a box D^3 (homeomorphic to a 3-cell), that for the vertical director p, the set $\varphi^{-1}(p) \subset D^3$ is just a Bouligand "double anneau", not touching the boundary ∂D^3, and that $\varphi^1|\partial D^3$ is some (non-constant) continuous map $\psi: \partial D^3 \to \mathrm{RP}^2$ which will be supposed fixed. (This ψ is our boundary condition.) Since D^3 is simply connected, the field of molecular directors $\varphi: D^3 \to \mathrm{RP}^2$ can be lifted to a vector (spin) field $\tilde{\varphi}: D^3 \to S^2$. Let us say that x and $-x$ are the two vertical vectors corresponding to p. Bouligand has analyzed and described the experimental data with so much detail and precision, that one can actually perform the passage from φ (directors) to $\tilde{\varphi}$ (vectors = directors with consistent arrows) *explicitly*.

One finds that *one of the components of the Bouligand "double anneau" is* $\tilde{\varphi}^{-1}(x)$ *and the other one is* $\tilde{\varphi}^{-1}(-x)$. (The arrows on the two components point in *opposite* directions.)

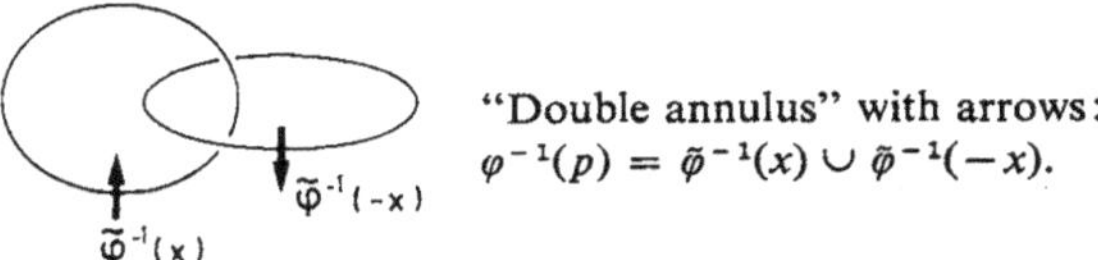

"Double annulus" with arrows: $\varphi^{-1}(p) = \tilde{\varphi}^{-1}(x) \cup \tilde{\varphi}^{-1}(-x)$.

In ref. [3] (end of sect. 4.5) it is shown that *the following process is impossible because it hits upon a non-vanishing element of* $\pi_3(\mathrm{RP}^2)$: *It is **not** possible to deform our cholesteric in such a way that the boundary conditions*

$\psi: \partial D^3 \to \mathrm{RP}^2$ stay fixed, but at the end of the deformation, for the new field of molecular directors $\varphi_1: D^3 \to \mathrm{RP}^2$, the set $\varphi_1^{-1}(p)$ consists of exactly two unlinked components.

The argument runs as follows: Let us assume that there is a homotopy

$$\varphi_t: D^3 \to \mathrm{RP}^2$$

($t \in [0, 1]$) such that $\varphi_0 = \varphi$, $\varphi_t|\partial D^3 = \psi$, and such that $\varphi_1^{-1}(p)$ consists of two *unlinked* circles like in the figure below.

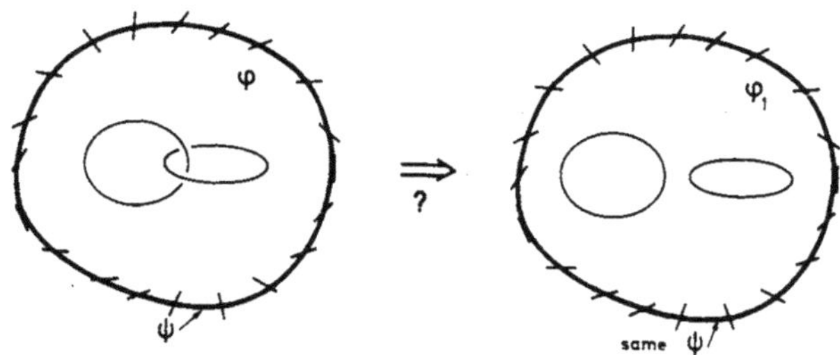

The sources of φ and φ_1 are really the *same* box D^3 in the physical space. But let us think, abstractly, of D^3 as being split into *two* boxes D_0, D_1 such that $\partial D_0 = \partial D_1$ and that φ is defined on D_0 and φ_1 is defined on D_1

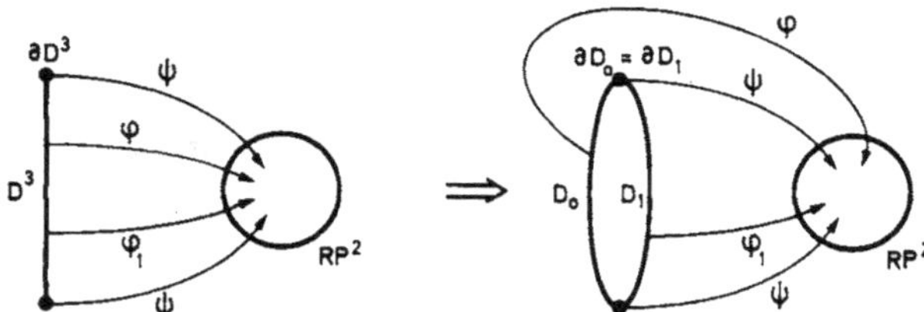

Clearly $D_0 \cup D_1 = S^3$ and φ, φ_1 combine together into a well-defined continuous map

$$S^3 \xrightarrow[F = F_{\varphi,\varphi_1}]{} \mathrm{RP}^2 .$$

It is very easy to show that the desired homotopy φ_t exists, if and only if $[F] = 0 \in \pi_3(\mathrm{RP}^2)$.

But $[F]$ is really the integer:

$$[F] = \underbrace{\mathrm{Lk}(\tilde{\varphi}^{-1}(x), \tilde{\varphi}^{-1}(-x))}_{\text{this is } \pm 1} + \underbrace{\mathrm{Lk}(\tilde{\varphi}_1^{-1}(x), \tilde{\varphi}_1^{-1}(-x)}_{\text{this is, anyway, } 0} .$$

So $[F] \neq 0 \in \pi_3(\mathrm{RP}^2)$ and our deformation does not exist. The kind of discontinuous, three-dimensional jump from one part of RP^3 to another, which is implied here, cannot be achieved continuously.

Remarks. The same kind of argument shows that *textures* $D^3 \xrightarrow{\varphi} V$, *with given boundary conditions, are classified by* $\pi_3 V$. This is a kind of *local phenomenon*, like the discussion in sect. 3.3 or like the way in which homotopy groups appear in gauge theories.

In more *global phenomena* (involving defects, textures, gauge fields. . .) homology theory (as opposed to homotopy groups) also becomes relevant. A small sketchy example of what this kind of comment is supposed to mean, is provided in the next section.

5. A seminar on the defect cycle and Volterra processes

This section presupposes knowledge of *homology theory* (see e.g. [4] or [5] (end of this section)) and of *surgery on defects*, as explained in [2,3]. The reader is also supposed to be familiar with the *Volterra process* (see e.g. [1]).

5.1. Introduction; the defect cycle

It seems pretty safe to say that by now one understands how homotopy groups occur "naturally" in the study of defects. The use of homotopy groups corresponds to a somehow *local* study of defects (like for instance finding out whether it is possible to cut a specific defect line). This talk will present another, very tentative, *global* approach, where one says something about the whole set of defects, via homology theory. Whether or not this is actually physically meaningful I do not know, but at least it should help those people who know about defects and homotopy, to understand what homology is all about. Anyway it seems to me that once defects are considered, the "defects cycle" described below, is there, and hence should be looked at.

I will consider an ordered medium whose manifold of internal states V, has the following properties:

(1) $\pi_0 V = \pi_2 V = 0$.

(2) $\pi_1 V$ is abelian. So $\pi_1 V = H_1 V$, and for simplicity's sake, this group will be denoted by π.

The physical space will be an orientable three-manifold M with a chosen orientation. So we know what a "right screw" in our physical space means. I will assume that M is triangulated and that the set of defects is a one-dimensional subcomplex $\Sigma \subset M$. The order parameter is defined, as always, outside this Σ:

$$M - \Sigma \xrightarrow{\Phi} V .$$

I will consider (simplicial) p-chains of M with coefficients in π; which means formal sums of the type $\sum_i g_i \otimes \sigma_i$ where σ_i is an oriented p-simplex of M and $g_i \in \pi$. The set of p-chains is denoted by $C_p(M, \pi)$.

Let $\tau_1, \tau_2, \ldots, \tau_h$ be the set of all 1-simplexes of Σ, and choose arbitrarily, for each τ_i an orientation. Around each τ_i one takes a small close loop $\gamma_i \subset M - \Sigma$ oriented in such a way that the following is true: consider a small disk S_i, if boundary γ_i, tranversal to τ_i, and oriented like γ_i. Then: (orientation of τ_i) $\times$ (orientation of S_i) = (orientation of M), where this formula is understood to mean that a "right screw" which turns like γ_i moves like τ_i.

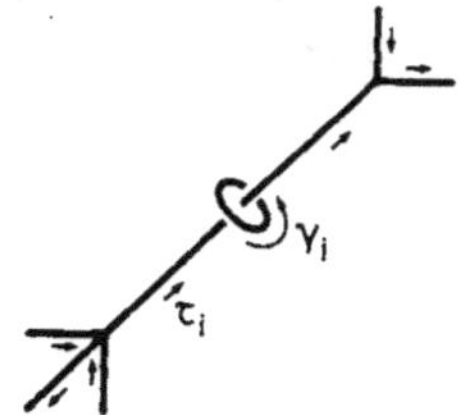

$\Phi(\gamma_i)$ is a 1-cycle of V, whose homotopy class will be denoted by $[\Phi(\gamma_i)] \in H_1 V$.

I will consider the following 1-chain of M with coefficients π:

$$\Delta = \sum_i [\Phi\gamma_i] \otimes \tau_i \in C_1(M, \pi) .$$

Note that Δ does *not* depend on the chosen orientations of the τ_i's. Note also that $\Delta = 0 \in C_1(M, \pi)$, if and only if one can wipe out the set of defects by (negative) surgery.

PROPOSITION 1. Δ is a *cycle* (this means that $\partial\Delta = 0$, where ∂ is the boundary-homomorphism $C_1(M, \pi) \to C_0(M, \pi)$); this cycle will be called the *defect cycle*.

Proof. Let x be a vertex of the defect set and let $\tau_1, \tau_2, \ldots, \tau_e$ be the defect lines touching x. Without any loss of generality we can assume that all the orientations of $\tau_1, \tau_2, \ldots, \tau_e$ point away from x and that all the circles $\gamma_1, \ldots, \gamma_e$ are contained in a small sphere S of center x.

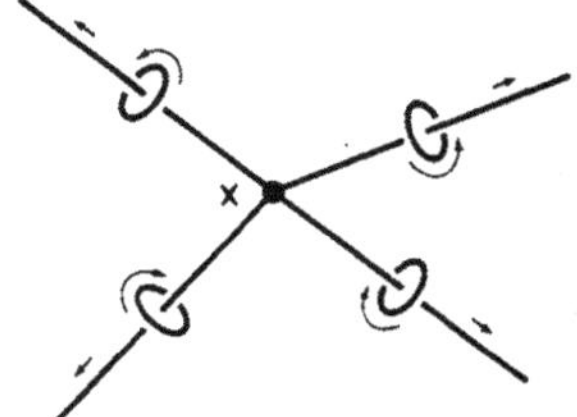

Then $\gamma_1 + \gamma_2 + \cdots + \gamma_e$ is a cycle of $C_1(M - \Sigma, Z)$ which is homologous to 0 (in the integral homology), so that:

$$[\Phi\gamma_1] + \cdots + [\Phi\gamma_e] = 0 \in H_1 V .$$

Hence the coefficient of x in the 0-dimensional chain of M with coefficients π, $\partial\Delta$ is 0, and since the same argument works for an arbitrary x, one finds that

$$\partial\Delta = 0 \in C_0(M, \pi) .$$

Note. This is really nothing else than the "règle des nœuds" (the Kirchhoff law for defects) known already to Friedel and Grandjean a very long time ago.

5.2. The Volterra process

We will reformulate here, in our context, a classical construction which is described in detail in Kléman's book [1]. For convenience of the reader we give the following very elementary lemma:

LEMMA 2. Let S be a surface with connected boundary, $f: S \to V$ a given map, $\xi \in H_1 V$ a given homology class. There exists a map

$$F = F(f, \xi): S \times [0, 1] \to V$$

such that:

(1) $F|S \times 0 = F|S \times 1 = f$.

(2) If $x \in S$, I consider the map F_x: $[0, 1] \to V$ given by $F|x \times [0, 1]$. Note that $F_x(0) = F_x(1)$ so that F_x defines a 1-cycle of V (oriented like

$[0, 1])$. Then $[F_x] = \xi \in H_1V$. Moreover, such a map F is unique up to a homotopy fixed on $S \times 0 \cup S \times 1$.

Proof. Choose a point $x_0 \in \operatorname{int} S$ and a set $K \subset S$ such that:

(1) $x_0 \in K$.

(2) K is obtained from x_0 by attaching 1-cells $I_1, I_2, \ldots, I_g$.

(3) S retracts in K.

Notice that $S \times [0, 1]$ retracts in

$$X = (S \times 0) \cup (K \times [0, 1]) \cup (S \times 1),$$

so we need only to consider extensions to X. Also X is obtained from $(S \times 0) \cup (S \times 1)$ by adding a 1-cell $x_0 \times [0, 1]$ and g two-dimensional cells $I_1 \times I, \ldots, I_g \times I$.

Construct first $F|x_0 \times [0, 1]$, so that condition (2) is fulfilled for x_0. Then F is already constructed in $\partial(I_i \times I)$. Notice that:

$$[F(\partial(I_i \times I))] = 0 \in H_1V = \pi_1V.$$

So one can extend our F from

$$(S \times 0) \cup (x_0 \times [0, 1]) \cup (S \times 1)$$

to all of X. Uniqueness follows from the assumption that $\pi_2V = 0$.

Consider now our ordered medium $M - \Sigma \xrightarrow{\Phi} V$. It will be convenient to thicken Σ to a regular neighborhood $N(\Sigma)$ so that Φ is defined in $M - \operatorname{int} N(\Sigma)$. In order to define a Volterra process one has to give the following items:

(1) An orientable surface S, with only one boundary component ∂S and a proper embedding $(S, \partial S) \subset (M - N(\Sigma), \partial N(\Sigma))$.

(2) A transversal orientation for S.

(3) A homology class $\xi \in H_1V$.

With these data one can define a new ordered medium $M - \Sigma \xrightarrow{\Phi} V$, as follows. First one replaces S by S $\times$ [0, 1], as in the figure below.

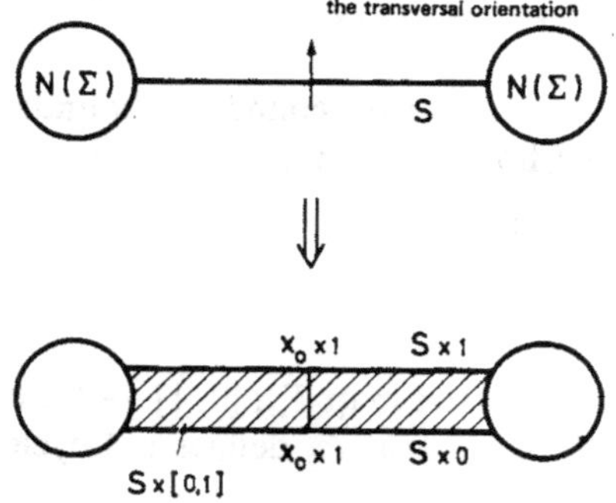

One keeps the definition of Φ in the complement of int $N(\Sigma) \cup (S \times (0, 1))$, in such a way that $\Phi|S \times 0 = \Phi|S \times 1 = \Phi|S$.

One redefines the order parameter in $S \times (0, 1)$ as being $F(\Phi|S, \xi)$ (lemma 2). This construction is the Volterra process, in our given context.

We consider now the homology class of the defect cycle:

$$[\Delta] \in H_1(M, \pi) .$$

THEOREM. The set of defects can be wiped out by surgery (which means: creating or cutting defect lines when possible, or wiping out isolated defect points, which is always possible since $\pi_2 V = 0$) and Volterra processes if and only if

$$[\Delta] = 0 \in H_1(M, \pi) .$$

Proof. I will show that if $[\Delta] = 0$ one can wipe out Σ, and leave the converse to the reader. The assumption $[\Delta] = 0$ means that there exists a simplicial 2-chain:

$$\Gamma = g_1 \otimes \omega_1 + \cdots + g_N \otimes \omega_N \in C_2(M, \pi) ,$$

where $g_i \in \pi$, and ω_i is an oriented 2-simplex, such that $\partial S = \Delta$. By applying positive surgery (which enlarges Σ) one can assume without loss of generality that the boundaries of the triangles ω_i lie in Σ.

The proof of the theorem follows now from:

LEMMA 3. Suppose that

$$\Delta = \sum_1^k \psi_i \otimes \tau_i \quad (\psi_i \in \pi) ,$$

and consider an oriented 2-simplex ω of M such that $\partial\omega = \tau_1 + \tau_2 + \tau_3$. (We assume here that we arrange the orientations of τ_1, τ_2, τ_3 so that the signs in this formula are correct.)

We consider the Volterra process defined by the following conditions:

(1) $S = \omega$;

(2) (the transversal orientation) $\times$ (the orientation of ω) $=$ (the orientation of M);

(3) Some $\xi \in H_1 V$ is given.

Let us denote by Δ' the defect cycle for the ordered medium which is obtained in this way. Then:

$$\Delta' = \Delta + \partial(\xi \otimes \omega) = (\psi_1 + \xi) \otimes \tau_1 + (\psi_2 + \xi) \otimes \tau_2 + (\psi_3 + \xi) \otimes \tau_3 + \sum_{i=4}^k \psi_i \otimes \tau_i .$$

The proof of this lemma is an easy geometrical exercise left to the reader. Now, in order to wipe out the defect Σ, we perform the Volterra processes corresponding to $(\omega_1, -g_1), \ldots, (\omega_N, -g_N)$, and since $\Delta - \partial S = 0$, one obtains a new ordered medium, for which the defect cycle is 0. Then one can kill everything by surgery.

Remarks.

(1) Dennis Sullivan suggests that there should be a *non*-commutative version of this little theory.

(2) What we really do here is an exercise in "obstruction theory" (see [4]). The Poincaré dual of the "defect cycle" is the classical "obstruction cocycle". Chern classes and Pontryagin classes, which appear in gauge theories, are another aspect of this theory.

(3) This whole theory can be viewed as a kind of "conservation law". One can always pass from one defect cycle to another (by surgery and Volterra processes) as long as they are in the same homology class: $[\Delta] = [\Delta'] \in H_1(M, \pi)$, and during such processes the homology class $[\Delta] \in H_1(M, \pi)$ *does not change*. If M is contractible, then $[\Delta] = 0$, so non-trivial effects can be exhibited only if the physical space itself has some non-trivial topology.

[1] M. Kléman, Points, lignes, parois, dans les fluides anisotropes et les solides cristallins, vol. 1 (Editions de Physique, 1977).

[2] V. Poénaru and G. Toulouse, The Crossing of Defects in Ordered Media, and the Topology of 3-Manifolds, J. Physique (1977).

[3] V. Poénaru and G. Toulouse, Topological Solitons and Graded Lie Algebras (to appear in J. Math. Phys.).

[4] E. Spanier, Algebraic Topology (McGraw-Hill, New York).

[5] V. Poénaru, Initiation à la topologie algébrique (lecture notes written by F. Axel, A. Gervois, R. Seznec, A. Voros, D.Ph. T., Saclay, 1978).

Acknowledgements

I wish to thank Lauris Balian for her typing, Françoise Axel for filling in the formulas and producing the drawings at the school, Paul Ginsparg for proofreading sect. 4.6, and Dennis Sullivan for suggesting to me the idea of the defect cycle.

General references

(*A*) *Mathematics.* Here are some suggestions for further reading:

[1] Massey, Algebraic Topology, an Introduction (Harcourt-Brace, New York).
[2] E. Spanier, Algebraic Topology (McGraw-Hill, New York).
[3] Steenrod, The Topology of Fibre Bundles (Princeton University Press, Princeton, MA).
[4] Milnor, Topology from the Differentiable Viewpoint (University Press of Virginia).
[5] H. Cartan, Topologie algébrique, Séminaire ENS 1948–1949.
[6] Gramain, Topologie des surfaces (P.U.F., Paris).
[7] Godbillon, Topologie algébrique (Hermann, Paris).

(*B*) *Applications to ordered media*

[8] Mermin, The Topological Theory of Defects in Ordered Media, preprint, Cornell University.

The literature on the subject is already quite vast. A *good bibliography*, up to September 1977 can be found in:

[9] L. Michel, Topological Classification of Symmetry Defects in Ordered Media, preprint IHES.

More specific references appear in the above text (see in particular sects. 3.3, 4.5, and 5.2).

COURSE 5

MODELS OF DISORDERED MATERIALS

Scott KIRKPATRICK

IBM Thomas J. Watson Research Center,
Yorktown Heights, New York 10598, USA

Contents

R. Balian et al., eds.
Les Houches, Session XXXI, 1978 – La matière mal condensée/Ill-condensed matter

1. Introduction, philosophy, and sundry disclaimers

These lectures will focus on the properties of simple models of interesting disordered materials, and the means by which they may be calculated. To impose some order on the unwieldy menagerie of models and methods which have attracted interest, I will treat the percolation problem by itself first, in some detail, then discuss various kinds of random magnetic systems, and inhomogeneous conductors, and finally consider systems with classical frustration of quantum mechanical interference. Simple concrete models which might capture the properties of real glasses or polymers are still lacking or quite rudimentary, so I will not be able to say anything about either of these interesting topics.

With a few exceptions, I will describe models which employ regular lattices, short-ranged interactions, and statistical variables without long-ranged correlations. There are enormous practical advantages to doing so, and very little physics is lost. For example, consider the problem of determining the macroscopic room-temperature conductance of a "cermet" (ceramic–metallic composite) such as those described in the review of Abeles et al. [1], or a very thin evaporated metallic film [2,3], which consists of overlapping blobs or islands of metal on an insulating substrate. One can define a spatially varying conductivity, $\sigma(R)$, to relate local currents to local electric fields. It would be possible, though tedious, to map out the actual spatial arrangement of conducting and insulating material for a small region of a sample and solve numerically for the potential $\varphi(R)$ and current distributions. To do so one introduces a regular square or cubic grid of points, R_i and solves for the potential distribution satisfying the current conservation law, $0 = \nabla\cdot\sigma(R)\nabla\varphi(R)$ at the grid points, using finite differences. The resulting equations:

$$\sum_{j\text{ inside sample}} \sigma_{ij}(\varphi_j - \varphi_i) = S_i\,, \tag{1.1}$$

where $\sigma_{ij} \equiv \sigma[\frac{1}{2}(R_i + R_j)]$, $\varphi_i = \varphi(R_i)$ and S_i is zero except possibly at the sample surface, are Kirchhoff's laws for a simple cubic or square network of conductances, σ_{ij}. As a first approximation one takes a grid just coarse enough so that the σ_{ij} are statistically independent of one another. This yields a bond percolation model in which the fraction of conducting bonds is equal to the volume fraction of metal in the material.

The threshold concentrations obtained for bond percolation on square and cubic lattices ($x_c^{(b)} = \frac{1}{2}$(2D), $\approx\frac{1}{4}$(3D)) are in agreement with the threshold volume fractions of many real composites. Even in special cases, like some cermets, where the thresholds differ, we may expect that critical behavior, such as the vanishing of the conductivity, will depend only upon dimensionality, and thus be correctly obtained in this first approximation.

Even when microscopic structural considerations affect the threshold of the actual material, it is often possible to calculate x_c by approximate arguments which make no reference to any underlying lattices (see Shante and Kirkpatrick [4]). Thus in bond percolation the threshold is found to occur when the average number of neighbors which can be reached from a given site reaches a critical value, n_c:

$$n_c \approx d/(d-1)\,, \tag{1.2}$$

in d dimensions.

For site percolation models one can construct [5] an appropriate allowed fraction which at threshold takes a value of $\simeq 0.17$ in 3D and roughly $\frac{1}{2}$ in 2D.

Kirchhoff's laws (1.1) are identical in form to the set of equations which determine the equilibrium orientations of spins in a ferromagnet close to the ordered state at zero temperature. One can show by microscopic calculation (see, e.g., Kirkpatrick [6,7] that the macroscopic conductance, σ, of a resistor network is equal to the exchange stiffness, A, of a network of classical spins with its exchange interactions J_{ij} numerically equal to the σ_{ij}. From macroscopic considerations this is an obvious result.

In the electric case the energy dissipation per unit volume is given by $\frac{1}{2}\sigma(\nabla\varphi)^2$ while the fluctuation energy density in the magnetic case when the mean local orientation of the magnetization deviates through an angle $\theta(R)$ from the ordered state is $\frac{1}{2}A|\nabla\theta|^2$. In the first case, one applies a potential difference Φ across a sufficiently large sample, calculates the local φ_i's, and determines the energy density by summing local contributions $\frac{1}{2}\sigma_{ij}(\varphi_i - \varphi_j)^2$ in order to determine σ. In the second case one twists the spins on one edge of a sample by an angle Θ, keeping the spins on the opposite edge fixed, sums the actual local strain energies $\frac{1}{2}J_{ij}(\theta_i - \theta_j)^2$, and identifies A from the result.

Thus equivalence between σ and A is independent of model and, in particular, makes no reference to details of the lattice. It is useful, since in ferromagnets A governs both magnetic domain phenomena and spin wave excitations at long wave lengths. (For a development of this point of view, along with the necessary extensions to treat ferri- and antiferro-magnets, see Harris and Kirkpatrick [8].)

The types of disorder I shall consider separate naturally into *dilution* – situations in which the strength of some local property, such as conductivity, varies from point to point, and *opposition* – where different regions may exhibit tendencies which conflict with or frustrate each other.

Dilution causes many effects which are properly the concern of what Phil Anderson has called the "old school" of disorder physics – modifications of material properties which one may with sufficient effort understand through the construction of an appropriate effective medium. It also causes novel effect – the percolation threshold.

Percolation thresholds are fast becoming "old school" as well. Exact mappings into fairly conventional second-order phase transitions can be constructed for many models. Extremely accurate computer simulations have now been performed for some (but not all) properties affected by a threshold. Still I shall spend the greater part of my time discussing dilution phenomena because, first, they are various and ubiquitous and, second, they quite dominate the experiments which are performed to isolate effects of the second sort of disorder. Thus a quantitative understanding of the consequences associated with weak fluctuations or percolation thresholds is a prerequisite in designing or interpreting experiments on localization, spin glasses, or the glass transition.

Simplified geometric pictures of what happens at a percolation threshold are extremely useful in understanding transport properties and magnetic phenomena. They are also helpful in extending results obtained by simple model calculations to more complicated realistic systems or more complex properties. I shall describe three such pictures, in which the infinite cluster close to the percolation threshold is viewed as made up of "channels" (fig. 1a), as a regular network (fig. 1b), or as a "texture" (fig. 1c). The idea of the first picture is that percolation reduces the volume of material (and hence the cross-section) available for transport. The second picture (due originally to de Gennes [9], Stauffer [10], and Skal and Shklovskii [11]) adds the notions that the channels must intersect and that the characteristic length scale of their intersections should be given by the percolation coherence length. The third picture [12] incorporates effects of large numbers of cross links in the network with lengths less than ξ. I shall argue that the first picture is hopelessly wrong and the second reasonable but slightly in error, while the third is the correct way to think about the experimentally observed thresholds in 2 or 3 dimensions. Simplified

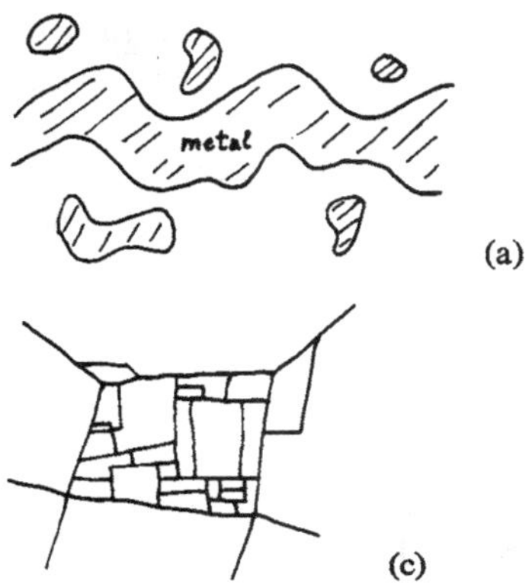

Fig. 1. Three simplified pictures of a transport process occurring just above the percolation threshold, arranged (a)–(c) in order of increasing sophistication.

pictures such as those of fig. 1 are playing a role in models of polymer kinetics. It remains to be seen whether geometrical arguments will be useful in the problems of localization or glasses.

There are two theoretical tools for dealing with model disordered systems which have developed sufficiently to merit reviewing in the course of these lectures. Real-space renormalization involves, like the momentum-space renormalization amply described in Wilson and Kogut [13] and Ma [14], analysis of transformations of a dynamical system of N variable to one of N' variables, where $N' < N$. In the momentum space approach the dynamical variables are Fourier coefficients of the order parameter and the interactions are coefficients of an appropriate Ginzburg–Landau functional. The connection between these couplings and the parameters of the actual material is obscure. Thus the principal appeal of rescaling transformations expressed in real space, where the interactions are, at least initially, those of the actual hamiltonian, is the possibility they provide of calculating properties of systems far from critical points, where microscopic details become important. For disordered systems it will be necessary to smooth out through rescaling both thermal fluctuations and the quenched disorder. The nature of the disorder terms is usually more obvious in real space. The drawbacks and restrictions of the real space methods will become apparent in due course.

Real-space renormalization for translationally-invariant magnetic systems has a large literature. Useful reviews are those of Wilson [15] and Kadanoff [16], to which I will refer heavily. One technical point deserves mention at this stage. Rescaling in momentum space usually consists of integrating out a small shell of large momentum degrees of freedom and rescaling the magnitudes of the remaining components so that they may correspond to excitations of, e.g., blocks of coupled spins. One may perform such transformations to new block spin variables in real space as well, but an appealing short cut is simply to perform partial traces, letting the new dynamical variables be a subset of the old ones. This trick, known as "decimation", makes computations on complicated models tractable, and will be used in all the calculations which I describe. Unfortunately, the evidence, which I will discuss in the later lectures, is that decimation is not adequate for problems like spin glasses or localization, and instead block transformations may have to be developed.

Computer simulations or Monte Carlo give an accuracy which is competitive with real space rescaling calculations (and the old-fashioned power series methods which have been applied to disorder problems – see especially the review of Essam [17]). Convergence of direct simulations is a relatively straightforward question of obtaining results for a sufficiently large sample, while convergence of RG and series calculations is often uncertain. Scaling arguments can be used to extract information even from simulations on relatively small samples by making use of the size dependence of the results.

The most valuable aspect of computer simulations is that, like real experiments, they may surprise us occasionally. There is also a wealth of microscopic information available in simulations which would not be experimentally available but can be used to develop simplifying pictures such as the ones introduced above. Finally, I shall discuss "experimental details" of simulations and try to convey some idea of the factors that must be considered in designing a successful computer experiment.

The most important "experimental detail" of a computer simulation is the complexity of the basic algorithms employed, i.e., whether the computing effort scales with the size, N, of the problem (number of sites, bonds, spins, etc.) as N^α or perhaps as $\exp(N)$. This sort of scaling usually determines whether a sufficiently large sample can be studied to obtain useful results. The improvements made by reducing α, or replacing a bad ($\exp(N)$) algorithm by a good (polynomial (N)) one far outweigh the gains available through streamlining computer programs or finding time on faster computers.

Anyone interested in carrying out simulations should be aware of the series of books by Knuth [18], which give a lucid and exhaustive treatment of the complexity of the elements of any algorithm – sorting, searching, data structures, etc. Most of what is known at present about the inherent difficulty of and best strategies for common problems in graph theory, linear algebra, and optimization is described in the book by Aho et al. [19], as well as in several other books with similar titles which have recently appeared. Most of the problems one simulates involve short-ranged interactions, and the matrices expressing these interactions are therefore sparse, having very few non-zero elements. To conserve computer storage as well as for algorithmic efficiency it is highly desirable to preserve this sparseness as much as possible while manipulating such matrices. There are several useful discussions of such techniques in a conference proceedings edited by Rose and Willoughby [20]. Since that time, significant progress has been made in finding eigensystems of sparse matrices, but this work has appeared only in applied maths journals. One reference is Edwards et al. [21].

A central goal of modern research on computational complexity is the identification of problems for which one can prove the non-existence of a good algorithm. For one large class of combinatorial questions, the NP-Complete problems [22], it is presently believed that an effort proportional to $\exp(N)$ is required to guarantee a solution. Finding the ground state energy of a spin glass is known to be in this class (see Ref. [22], p. 282). For many of these inherently hard problems there exist heuristic procedures which run in polynomial time and either (a) get the right answer in all but a small fraction of cases, or (b) come within a known error of the exact answer in all cases. For spin glasses, however, all situations of interest to the physicist appear to be "worst cases" of the general problem.

2. The pure percolation problem

The theory of the percolation problem has been given extensive reviews by Shante and Kirkpatrick [4] and by Essam [17]. These two papers added considerable data and some discussion of applications to an earlier review of Frisch and Hammersley [23], which is also quite readable, but contained few ideas which could not have been anticipated in 1963. The connections to the Potts model and to dilute Ising magnets at zero temperature were known in 1972, but had not been exploited. Reviews of percolation theory by Stauffer, by Essam, and by Pfeuty and Guyon are presently in preparation or available in preprint form.

Since that time, renormalization group ideas have had a radical effect on analytic calculations of percolation threshold properties as well as on the design and interpretation of computer experiments. An upper critical dimension (6D) has been established for percolation thresholds [24], and the role of dimensionality in percolation has been somewhat clarified. The finite time and space available here prevent my giving a full review of the current state of the percolation problem. Instead I will focus on the real-space RG methods which have been developed for percolation, using them in the analysis of computer simulations and also to generate simple generalizable models of the effects which accompany a threshold.

2.1. *Formalism and exact transformations*

Consider first a 1D chain, whose nearest neighbor bonds are present with probability p. Most of the properties of this model can be written down by inspection. For example, the probability $g(r - r')$ that two sites, separated by n bonds, are members of the same connected component, is just p^n. This sort of correlation has the exponential form $g(r - r') = \exp(-|r - r'|/\xi)$ where the correlation length $\xi = a/(\ln p)$, for lattice constant a. When p approaches 1, $\ln p \sim 1 - p \equiv q$, so ξ diverges with a simple pole. The feature that shows this model is trivial is that $g(r - r')$ has the same form when $|r - r'| < \xi$. Correlations near percolation thresholds in higher dimensionalities usually have a power law form for distances less than some coherence length, then decay exponentially beyond.

This model also provides a simple example of space rescaling. If we lump the bonds into groups of b consecutive bonds, the probability that a lump provides a connection is p^b. Since $p^b < P$ repeated rescaling of the problem quickly produces a picture in which only a few, isolated bonds are present. The system has been driven to a homogeneous fixed point analogous to the high temperature, non-interacting fixed point of conventional magnetic systems.

The other trivial percolation model may seem rather artificial, but it proves useful. Consider two points which are linked by a large number of bonds which act in parallel. Each bond has probability p of being present. If these bonds are lumped in groups of b the probability that there is *not* a connection must be raised to the power b. Thus the lumped probability $p' = 1 - (1 - p)^b$, and p' always exceeds p. Rescaling this model repeatedly drives the concentration to 1, a second homogeneous fixed point analogous to the low temperature ordered fixed point of magnetic systems.

Simple models such as these are readily studied using only the calculus of independent probabilities, but in actual 2 and 3D percolation problems conditional probabilities and other correlations appear. A rigorous formal-

Fig. 2. Hypothetical star–triangle transformation for bond percolation.

ism, obtained by establishing an equivalence to a Potts model, is useful to treat more complicated cases. Instead of using the full-blown expressions derived first by Kasteleyn and Fortuin [25] and discussed in Tom Lubensky's lectures, I will use the form (for bond percolation):

$$Z = \lim_{s\to 1} \operatorname{Tr}_{\xi_i} \prod_{\langle ij\rangle} (q + p\delta_{\xi_i\xi_j}), \tag{2.1}$$

where the ξ_i are s-state Potts variables, the $\langle ij\rangle$ are the possible bonds between neighbors, and $\delta_{\xi_i\xi_j}$ is a Kronecker delta. The idea behind (2.1) is that its expansion yields a desired cluster series, but we can exhibit its properties more simply by checking that it exhibits the correct transformation properties for the two trivial models. Lumping 2 bonds in a chain is accomplished by multiplying the two terms of (2.1) and tracing out the intermediate variable:

$$\begin{aligned}
&\lim_{s\to 1} \operatorname{Tr}_{\xi_1} (q + p\delta_{\xi_0\xi_1})(q + p\delta_{\xi_1\xi_2}) \\
&\quad = q^2 \lim_{s\to 1} \operatorname{Tr}_{\xi_1} 1 + pq \operatorname{Tr}_{\xi_1}(\delta_{\xi_0\xi_1} + \delta_{\xi_1\xi_2}) + p^2\delta_{\xi_0\xi_2} \\
&\quad = (1 - p^2) + p^2\delta_{\xi_0\xi_2}\,.
\end{aligned} \tag{2.2}$$

By extension of (2.2) b times, $p' = p^b$ as desired. Likewise expanding the product $(q + p\delta_{\xi_1\xi_2})^b$ and using the fact that $(\delta_{\xi_1\xi_2})^n = \delta_{\xi_1\xi_2}$ gives $q^b + (1 - q^b)\delta_{\xi_1\xi_2}$.

This machinery becomes useful as soon as we consider 2D problems. For example, the star–triangle transformation sketched in fig. 2, when applied to Ising models [26] yields a reduction of a large class of 2D lattices into the soluble honeycomb or triangular lattices. Multiplying out the three bonds of the triangle we find

$$\text{“triangle”} = q^3 + pq^2 \sum_{\langle ij\rangle} \delta_{\xi_i\xi_j} + (3p^2q + p^3)\delta_{\xi_1\xi_2\xi_3}, \tag{2.3}$$

while the star, after performing the trace over the central site, is

$$\text{“star”} = q'^3 + 3p'q'^2 + q'p'^2 \sum_{\langle ij\rangle} \delta_{\xi_1\xi_2} + p'^3\delta_{\xi_1\xi_2\xi_3}\,. \tag{2.4}$$

Equating coefficients of the two-site terms gives

$$p' = 1 - p\,. \tag{2.5}$$

Equating either of the other pairs of coefficients and using (2.5) gives

$$q^3 = p^3 + 3p^2q\,, \tag{2.6}$$

which is the condition for p_c on the triangular lattice (see Thouless' lectures). Thus only at this value of p is the star–triangle transformation simple. Otherwise we will have to add some three spin correlation on one or the other side of fig. 2 to obtain an identity. The lower symmetry of the Potts interactions, evidenced by the failure of the star–triangle transformation, is the apparent reason why there are few exact results in percolation theory. The full power of this machinery becomes evident when the hamiltonian is modified to generate a site percolation model [27] or various kinds of polymer statistics (Hilhorst [28] and Lubensky, these lectures).

Calculations with real space renormalization groups will almost always involve a matching procedure. One maps the system into a region of concentration, temperature, etc., where its correlations are easy to compute, matches these to the transformed values of the original correlations, and then solves for the bare correlation functions. Mapping from an arbitrary set of initial parameters to some one point where the answer is known is only possible if the length scale can be varied continuously. The momentum space renormalization group is couched in terms of continuous fields, and naturally leads to the desired expressions for a differential change of scale. The same can be done in the real space renormalization group if we introduce a continuation from integer rescaling factors, b, to arbitrary values. To prove the validity of this procedure, I will apply it to the 1D case first:

Let $b = \mathrm{e}^{\delta}$, where δ is small so that b is close to 1. Then $b \approx 1 + \delta$ and the recursion $p' = p^b$ becomes

$$p' \sim pp^{\delta} = p\,\mathrm{e}^{\delta \ln p} \approx p(1 + \delta \ln p)\,. \tag{2.7}$$

As a natural notation for a differential renormalization we separate $\mathscr{R}(b)$ into the identity operator and δ times its infinitesimal generator $\mathscr{L}$. Thus, formally

$$\mathscr{R}[p] = p + \delta\mathscr{L}[p]\,, \tag{2.8}$$

and for the 1D recursion relation (2.7)

$$\mathscr{L}_{1\mathrm{D}}[p] = p \ln p\,. \tag{2.9}$$

$\mathscr{L}$ can readily be integrated to recapture $\mathscr{R}$ for arbitrary values of b. Since

$$\mathrm{d}p/\mathrm{d}(\ln b) = p \ln p\,, \tag{2.10}$$

one integrates

$$\int_p^{p'(b)} \frac{\mathrm{d}p}{p \ln p} = \int_0^b \mathrm{d}\delta\,, \qquad \ln(\ln p'/\ln p) = b\,, \tag{2.11}$$

exponentiates both sides twice and recovers $p' = p^b$.

Finally, we consider the model with many bonds in parallel. Lumping b bonds in parallel when $b = \mathrm{e}^\delta \approx 1 + \delta$ gives $p' = p - \delta(1 - p)\ln(1 - p)$, so

$$\mathscr{L}_{\parallel}[p] = -(1 - p)\ln(1 - p)\,. \tag{2.12}$$

2.2. Real-space RG's for 2 or more dimensions

I will describe two transformations of the bond percolation problem on cubic lattices in an arbitrary number of dimensions. The first is a discrete transformation. It was introduced by Reynolds et al. [29], and recently applied to the conduction problem, as I will discuss later, by Bernasconi [30]. The basic idea is sketched in fig. 3a: replace a cell of 2^d sites and the associated bonds by one site and d bonds by requiring each of the new bonds to express the possibility that opposite sides of the cell are connected in that direction. This construction in 2D requires that we combine probabilities in the "Wheatstone bridge" arrangement indicated in fig. 3b.

By a simple enumeration one finds the transformed probability,

$$p' = p^5 - 5p^4q + 8p^3q^2 + 2p^2q^3\,. \tag{2.13}$$

The Wheatstone bridge is self-dual: the probability of a connection when the bonds are present with probability p is equal to the probability of a block when the bonds are present with probability $1 - p$, as is easily seen in the course of performing the enumeration leading to (2.13). Therefore (2.13) exhibits a fixed point at $p = \frac{1}{2}$.

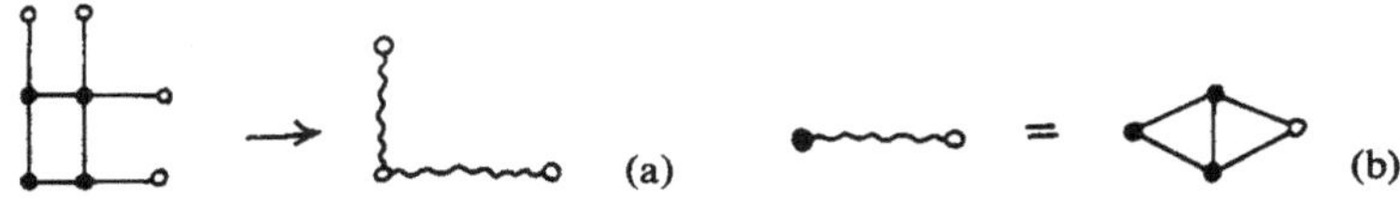

Fig. 3. Real space rescaling for the 2D bond percolation problem. The effective bonds produced in (a) are defined precisely by the construction (b).

The $p = \frac{1}{2}$ fixed point is unstable: rescaling the model with p initially $>\frac{1}{2}$ $(<\frac{1}{2})$ drives the rescaled probability to 1 (or to 0). Close to the two stable homogeneous fixed points the behavior of the transformation is rather like what we have already encountered in the two trivial cases. Expanding (2.13), one finds

$$p'(p) \sim 2p^2, \qquad p \leqslant 1, \tag{2.14a}$$

$$\sim 1 - 2q^2, \qquad p \sim 1, \tag{2.14b}$$

$$\sim \tfrac{1}{2} + \tfrac{13}{8}(p - \tfrac{1}{2}), \quad p \sim \tfrac{1}{2}. \tag{2.14c}$$

The effect of repeated applications of the transformation (2.13) is shown in fig. 4. Taken literally, fig. 4 is a prediction of the increasing sharpness of the percolation threshold which might be measured on finite samples of increasing size. It also shows that on any scale there will be ranges of concentrations over which the system appears uniform, with the characteristics of either the $p = 1$ or $p = 0$ fixed point, plus an intermediate

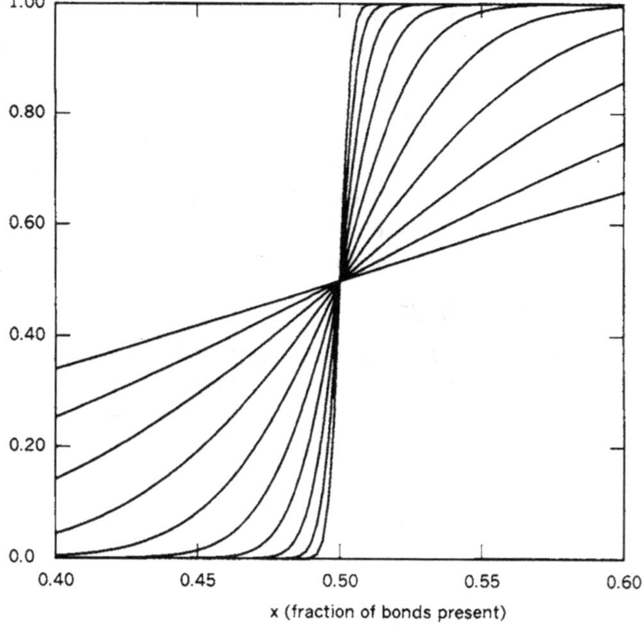

Fig. 4. Result of repeated iteration of the transformation in fig. 3. The lines predict the fraction of samples which will be connected for a series of sample sizes increasing by factors of 2 from 2 × 2 to 1024 × 1024 sites.

range in which fluctuations in connectivity will be felt. The second interpretation makes possible a simple calculation of correlation functions and permits, in particular, identification of the correlation length, $\xi(p)$, as that scale on which, for a particular concentration, the system first appears homogeneous. Using (2.14c) we express the result of rescaling all lengths by ξ as

$$p'(p, \xi) \sim \tfrac{1}{2} + (\tfrac{13}{8})^{\ln_2 \xi}(p - \tfrac{1}{2}) = \tfrac{1}{2} + \xi^{\ln_2(13/8)}(p - \tfrac{1}{2}) . \tag{2.15}$$

For the system to appear homogeneous, p' must equal 1, or 0 so ξ is determined as

$$\xi(p) = (\text{const}) \times (p - \tfrac{1}{2})^{-1/\ln_2(13/8)} , \tag{2.16}$$

and ν is identified as $[\ln_2(\frac{13}{8})]^{-1} = 1.43$. This is close to the best numerical estimate, $\nu = 1.365$, which we describe below. For arguments that ν_p is exactly given by $\ln(3^{1/2})/\ln(\frac{3}{2}) = 1.355$, see Klein et al. [137].

To go beyond this and calculate correlation functions we need to know the length scaling of some operator which couples to the property of interest. For example an external field, h_i, in the $s \to 1$ Potts model couples to the probability that site i is a member of the infinite cluster (for the precise definition, see Thouless' lectures). Suppose we carry out the rescaling of h for p close to p_i and find that $h'(p) = 2^{\lambda_h} h(p_i)$. To evaluate the percolation probability, $P(p)$, we simply rescale by ξ, at which point the percolation probability in the transformed system is unity and the transformed field,

$$h' = 2^{\lambda_H \ln_2 \xi} h = \xi^{\lambda_H} h , \tag{2.17}$$

the percolation probability is now extracted by taking a logarithmic derivative of the rescaled "Z", and normalizing the result to the rescaled volume, ξ^d

$$\begin{aligned} P(p) &\sim \xi^{-d} \frac{\partial}{\partial h_i} \ln Z' = \xi^{\lambda_h - d} \frac{\partial}{\partial h_i'} \ln Z' = \xi^{\lambda_h - d} , \\ &\propto (p - p_0)^{\nu(d - \lambda_h)} , \end{aligned} \tag{2.18}$$

so we identify the order parameter exponent β as

$$\beta = (d - \lambda_h)\nu . \tag{2.19}$$

For this transformation, (2.19) gives $\beta = 0.19$, which is close to the accepted value of 0.14, and extrapolation to cells of 4 or 5 sites on a side gives agreement to two decimal places [31].

Correlations such as the probability, $G(r_i - r_j)$, that sites i and j are both in the infinite cluster, can also be obtained by rescaling

$$\frac{\partial}{\partial h_i}\frac{\partial}{\partial h_j}\ln \text{“}Z\text{”}\,,$$

until sites i and j are nearest neighbors. The result is

$$\begin{aligned} G(r_i - r_j) &\sim |r_i - r_j|^{2\lambda_h - 2d} G'(1) p'(p, |r_i - r_j|) \\ &= |r_i - r_j|^{-2\beta/\nu} p'(p, |r_i - r_j|)\,. \end{aligned} \tag{2.20}$$

Since p' is just a number close to p_i we see that $G(r)$ falls off algebraically for $r \leqslant \xi(p)$. If we scale over distances greater than $\xi(p)$, it is no longer correct to use λ_h as calculated near p_c. Instead, $\lambda_h^{(p')}$ must approach d as $p' \to 1$. The algebraic factor will saturate at roughly $\xi(p)^{-2\beta/\nu} = (p - p_c)^{2\beta} = P(p)^2$, while $p'(p, |r_i - r_j|)$ approaches 1. Using (2.14b) we find $p'(p, |r_i, r_j|) \sim 1 - (q^*)^{|r_i - r_j|/\xi(p)}$, where q^* is the point at which we match on to the asymptotic form, in which correlations decay exponentially.

Notice that the customary identification of ξ from the exponentially decreasing part of the correlation function $G(r_i - r_j)$ leaves some ambiguity in the coefficient of $(p - p_c)^{-\nu}$. It seems more straightforward to identify ξ as the averaging length, or scale on which matching is possible. This corresponds to piecewise-linearizing $p'(p)$, as sketched in fig. 5, before rescaling until $p'(\xi) = 1$.

A differential real space transformation of wide applicability was invented by Migdal [32], then rederived and given a certain amount of justification by Kadanoff [16]. Since the application of the Migdal–Kadanoff transformation to all sorts of dilution phenomena is reported in

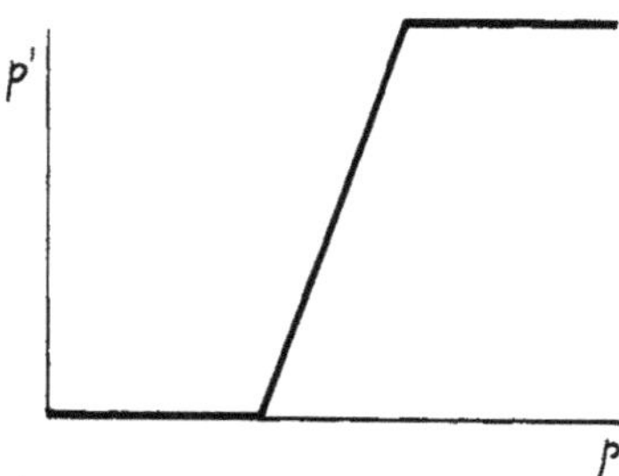

Fig. 5. Piecewise-linear approximation to $p'(p)$, which distinguishes two homogeneous regimes and one critical regime.

Kirkpatrick [33], I will treat in these notes only those features which differ significantly from conventional transformations.

The central trick to the transformation is that of modifying the lattice by moving interactions in such a way that decimation is facilitated. Shifting interaction terms can be shown to cause negligible errors close to the homogeneous fixed points [16]. The transformation interpolates between the two trivial fixed points. It may not be quantitatively accurate at the unstable fixed point in between, but should give sensible qualitative predictions over the whole range of parameters. Experience with applications in magnetism (see especially, José et al. [34]) supports this hope.

To rescale one coordinate by b we first move the interactions which are perpendicular to that axis, grouping them b bonds at a time, separated by 1D chains of b bonds. The chains can be replaced by single effective bonds, using the first of our exact transformations. The effect of lumping b bonds is given by the second exact transformation. After this has been repeated for each of the d axes, each set of bonds has been subjected to one 1D transformation, and $(d - 1)$ parallel transformations. The result will depend upon the sequence of the transformations, but taking $b \to 1$ and introducing infinitesimal generators will in principle fix this:

$$\mathscr{R} = \mathscr{R}_{\parallel}^{d-1}\mathscr{R}_{1D} \to \mathbf{1} + \delta[(d - 1)\mathscr{L}_{\parallel} + \mathscr{L}_{1D}] . \tag{2.21}$$

From an earlier discussion, we have $\mathscr{L}_{\parallel}$ and $\mathscr{L}_{1D}$, and can write

$$\mathscr{L}[p] = p \ln p - (d - 1)(1 - p) \ln(1 - p) . \tag{2.22}$$

The vanishing of $\mathscr{L}$ identifies a fixed point. For $d = 2$ these again occur at 0, $\frac{1}{2}$, and 1. The exponent ν is obtained by linearizing the differential recursion relation at the unstable fixed point, since if

$$\mathrm{d}p/\mathrm{d}\delta = A(p - p_c) + O((p - p_c)^2) , \tag{2.23}$$

we can integrate until $p \approx 1$:

$$A^{-1} \int_p^1 \frac{\mathrm{d}(p - p_c)}{p - p_c} = \int_0^{\delta^*} \mathrm{d}\delta \Rightarrow A^{-1} \ln\left[\frac{1 - p_c}{p - p_c}\right] = \delta^* ,$$

and exponentiate both sides to obtain ($\xi(p) = \mathrm{e}^{\delta^*}$):

$$\xi(p) = (1 - p_c)^{A^{-1}}(p - p_c)^{-A^{-1}} . \tag{2.24}$$

For the transformation (2.22)

$$\frac{1}{\nu} = d + (1 - p_c)^{-1} \ln p_c = d + \frac{d - 1}{p_c} \ln(1 - p_c) , \tag{2.25}$$

which for $d = 2$ gives $\nu = 1.63$, about 20% too high. The scheme is best at low dimensions and gives clearly wrong results for large d. In particular one finds $p_c \to e^{-(d-1)}$, which violates the bound $p_c \geqslant 1/(2d - 1)$, and $\nu(d) \to 1$, instead of $\frac{1}{2}$. Therefore we will use it only for $d = 2$ or 3.

2.3. Computer studies

At this point it is desirable to test the predictions of the rescaling theories against the simplest experiments available, computer simulations. At least in two dimensions and to some extent in 3D as well one can have the computer prepare pictures of portions of actual infinite clusters, examine those pictures, and see if the concepts and quantities I have been discussing are the appropriate ones, or have the claimed behavior. I have published such a set of pictures in the ETOPIM conference proceedings [80], so I will not repeat them here. One can see evidence of $\xi(p)$ directly in such pictures. From its role as an averaging length, it must be some multiple of the linear dimension of the largest missing or disconnected regions in the picture. These missing regions are evidently diverging in size as the percolation threshold is reached, so the pictures give support for a large length scale associated with fluctuations in connectivity. One can also ask questions about the effects of structure on smaller scales in the pictures, but I will take up this topic later in the context of magnetism or conduction near threshold.

Most computer studies of percolation thresholds have aimed at quantitative information. Dean and Bird [35,36] investigated cluster statistics as a function of p for site percolation on many common 2D and 3D lattices. Although recent workers have been able to treat samples much larger than Dean and Bird's, their extensive tabulations remain valuable for questions of detail and magnitude. Several recent studies (Reynolds et al. [31], Roussenq et al. [37], and my own work to be reported here) have focussed on the most basic prediction of the real space rescalings: the sample size dependence of the apparent percolation threshold. The work of Roussenq et al. was anticipated by several papers of the Leningrad group. See, especially, Levinshtein et al. [138].

In each of these studies a population of sample cells of a given size is constructed and the computer program determines for each cell whether there is a complete path across it joining a selected pair of faces. The resulting observations are expressed, for a given lattice type, as a function $p'(p, L)$, the probability of connectedness over a scale L when the bare probability is p. My results for bond percolation on 2D square and 3D

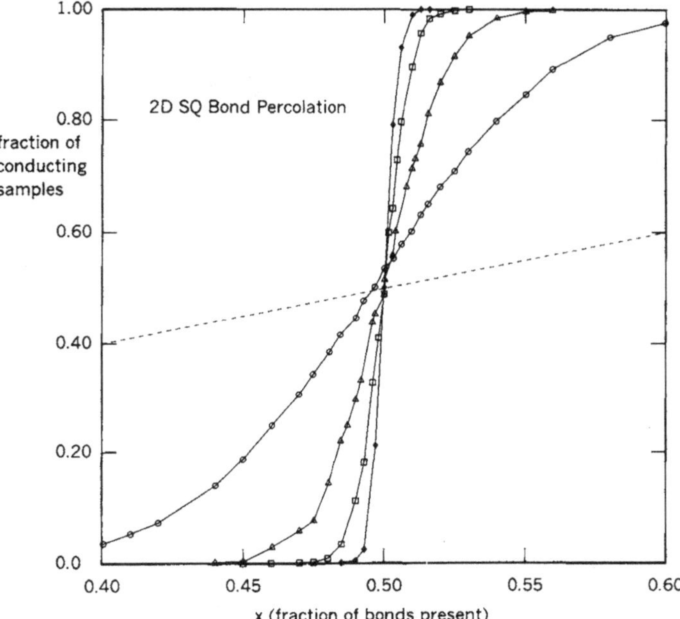

Fig. 6. Fraction of samples which are connected, $p'(p)$, for 2D bond percolation on a square lattice. The dashed line indicates $p' - p$. Data points are for samples of 16 × 16 sites (circles – 10,000 cases averaged per point), 64 × 64 (triangles – 1000 cases each), 200 × 200 (squares – 1–3000 cases each), and 512 × 512 (diamonds – 600–1000 cases each).

cubic lattices, on sample sizes ranging from small to rather large, are shown in figs. 6 and 7. This method determines the percolation threshold to a very high accuracy. From fig. 7 one obtains p_c(3D SC bond perc.) = 0.2495 ± 0.0005 by two extrapolation methods. The points $p'(p, L) = \frac{1}{2}$ tend to decrease with increasing L, while the "fixed points" p^*, of the scale transformation by L, $p'(p^*, L) = p^*$, tend to increase with L. Extrapolations of the two quantities agree to within the quoted error.

[A digression: Is either $p_c = 2^{-(d-1)}$ or $= 1/[2(d-1)]$ an exact result for bond percolation on hypercubic lattices? The first cannot be, since for $d \geqslant 4$, $\frac{1}{2}^{(d-1)} < 1/(2d-1)$, but the second is tempting. However, one expects for d sufficiently large that $p_c \sim 1/(2d-1)$, the Bethe lattice result, so the apparently simple result in 3D may be coincidental. In unpublished calculations similar to the 2D and 3D studies in figs. 6, 7, 8, I find $p_c = 0.1435 \pm 0.001$ for bond percolation on the 4D hypercubic lattice, not at all consistent with $[2(d-1)]^{-1} = \frac{1}{6}$, but close to the limiting result,

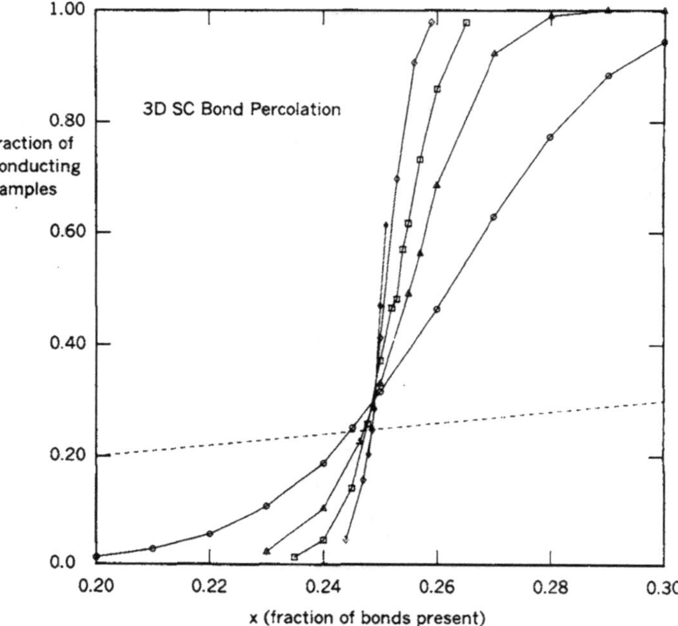

Fig. 7. Fraction of samples which were connected, as in fig. 6 but for 3D simple cubic lattices. Data points represent 10 × 10 × 10 samples (circles – 20,000 cases), 20 × 20 × 20 (triangles – 5000 cases), 30 × 30 × 30 (squares – 2000 cases), 50 × 50 × 50 (diamonds – 1000 cases), and 80 × 80 × 80 (solid diamonds – 500 cases each).

$p_c \sim 1/(2d-1) = \frac{1}{7} = 0.1429$. Results for ν_p were less accurate, but consistent with $\nu_p = 0.66 \pm 0.02$.]

The concept of regarding $p'(p, L)$ as a rescaled probability in the original problem has been developed particularly clearly by Reynolds et al. [31]. It leads naturally to a very accurate determination of ν if one identifies

$$\mathrm{d}p'(p, L)/\mathrm{d}p = L^{1/\nu}, \tag{2.26}$$

just as in (2.16). A log–log plot of the observed slopes versus L, in fig. 8, shows that (2.26) is satisfied rather accurately. The exponents obtained are $\nu = 1.365 \pm 0.015$ (2D), $= 0.845 \pm 0.015$ (3D).

Finally we can obtain estimates of $\xi(p)$ free of ambiguities about leading coefficients by approximating each curve in figs. 6 and 7 by three straight lines as in fig. 5 and identifying the intersections of the sloping lines with

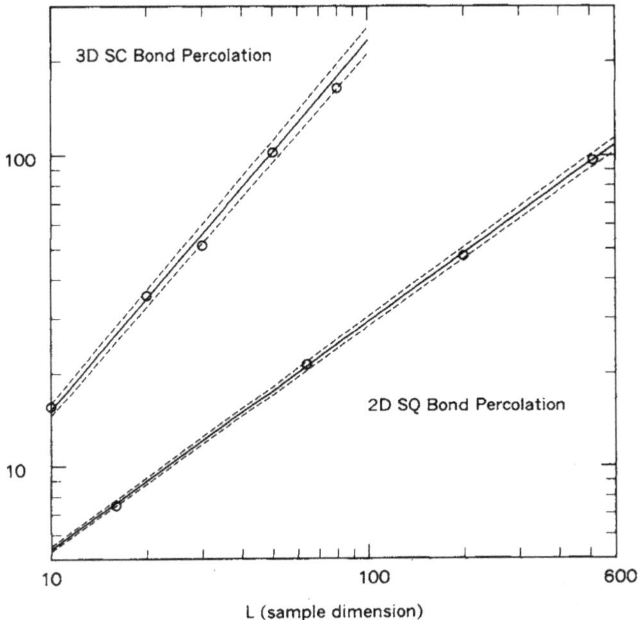

Fig. 8. Slope of $p'(p)$ near p_c, for 2D and 3D bond percolation, using data from figs. 6 and 7. The solid lines give the prediction of (2.26) for $\nu = 1.365$ (2D) and 0.845 (3D). The dashed lines represent the result of values of ν varying by ± 0.015 from these figures.

the $p' = 0$ or 1 axes as defining the values of p for which $\xi = L$. These will be useful later in fleshing out simple pictures which express the consequences of a semi-microscopic length scale on transport or other threshold properties.

To calculate the percolation probability we must first define it for a finite sample. The simplest definition is:

$$P(p) \equiv \frac{\text{number of sites (or bonds) in largest cluster}}{\text{number of sites (or bonds) in sample}} . \tag{2.27}$$

Figures 9a and 9b show the typical appearance of $P(p)$ defined in this way. $P(p) \approx p$ for $p \gtrsim 0.40$ (the dashed line), but for $p \lesssim 0.40$ it quickly falls to zero. The transition is rounded by effects of the finite sample size (fig. 9b) but we can employ this rounding to extrapolate to the behavior which would be seen in the limit $L \to \infty$, just as was done in the interpretation of figs. 6 and 7.

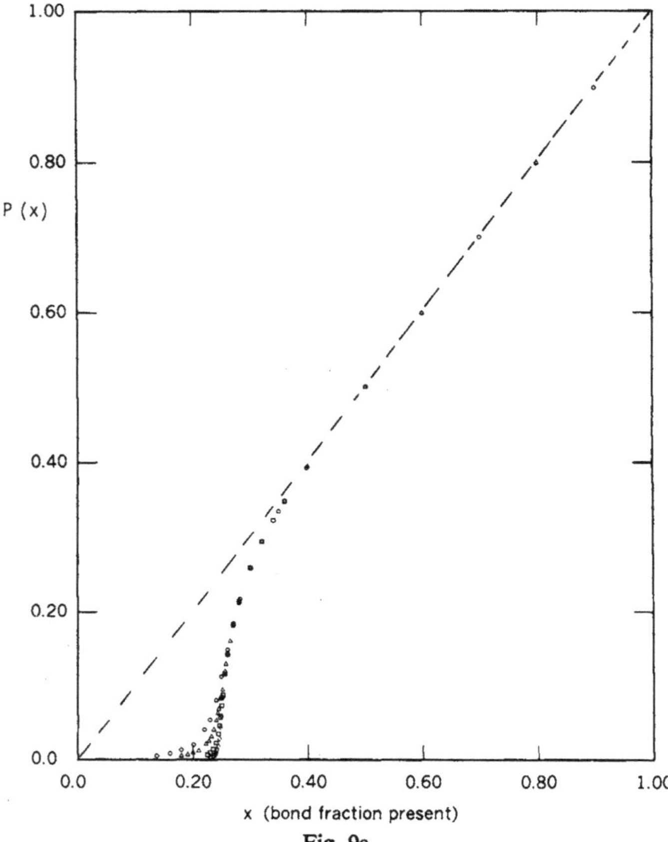

Fig. 9a.

Rounding occurs when the ratio $L/\xi \lessdot 1$. If we could increase the sample size as $p \gtrdot p_c$ to keep this ratio constant at some value >1, the result would approach a sharp threshold with exponent β. The scaling form

$$P(p, L) \sim L^{-\beta/\nu} f(L^{1/\nu}(p - p_c)), \tag{2.28}$$

provides the desired critical behavior when L/ξ is kept constant, and can be used to map the data in fig. 9b onto a single universal curve (fig. 9c). This procedure is less accurate than the length scaling analysis used in fig. 8. Nonetheless, Sur et al. [38] have been able to determine $\beta = 0.41 \pm 0.01$ for 3D site percolation by these finite-size scaling arguments. A better universal curve can probably be obtained with the data in figs. 9 by includ-

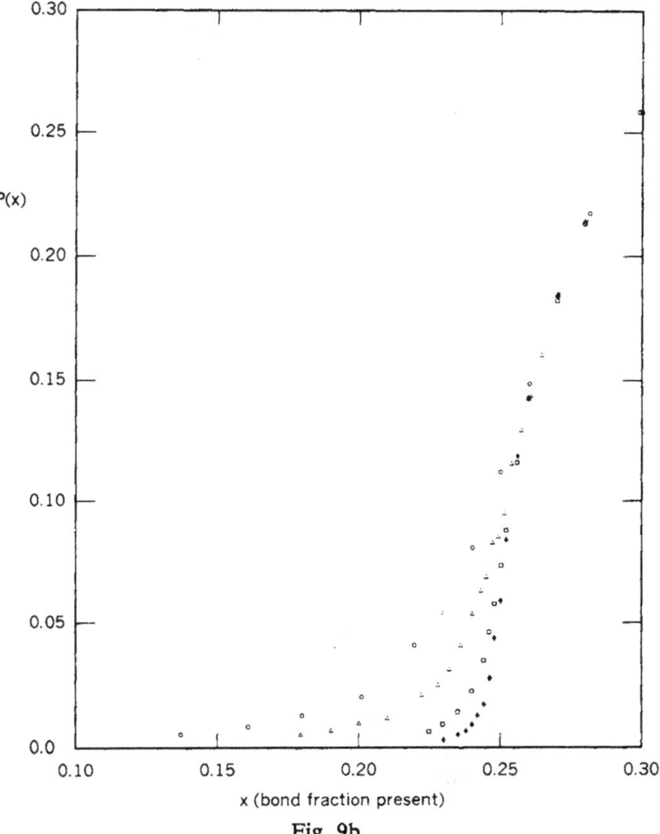

Fig. 9b.

ing shifts in the apparent position of the threshold. Reynolds et al. [31] have given a careful analysis of this, and find agreement with

$$(p_c(L) - p_c) \propto L^{1/\nu} \tag{2.29}$$

when $p_c(L)$ is determined as in figs. 6 and 7.

2.4. *Algorithms for percolation*

To find the largest cluster in a sample random lattice with bonds or sites randomly present one must at least examine every site or bond to see if it is

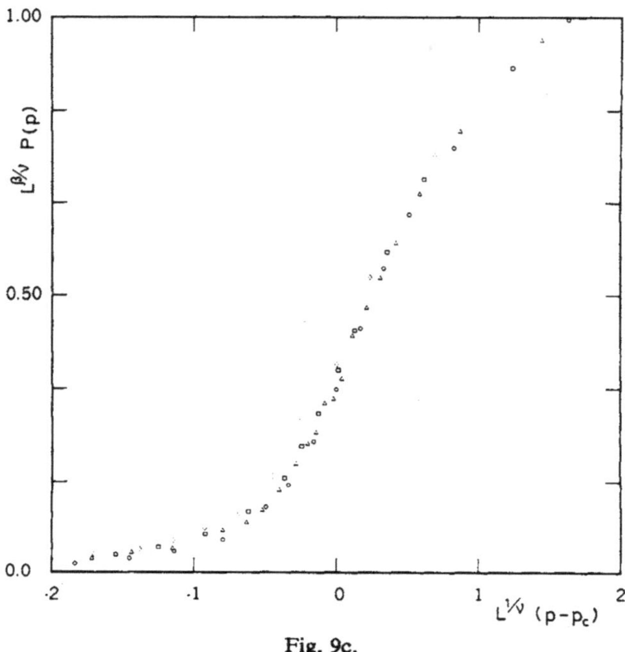

Fig. 9c.

Fig. 9. Percolation probability data, for the 3D bond problem on simple cubic lattices of size 10 × 10 × 10 (circles), 16 × 16 × 16 (triangles), 30 × 30 × 30 (squares), and 50 × 50 × 50 sites (diamonds). (a) shows the data on a large scale, (b) the same data on an expanded scale close to p_c, and (c) the universal function obtained by using the finite-sized scaling form of (2.28).

present, so an amount of computational work proportional to N is inescapable. It is also sufficient, since the following algorithm for site percolation meets that condition. One puts in the computer an array of flags representing the sample. A flag set "on" represents a site which is present. Search this array in sequence for a site which is present. Place the location of that site in the first location of a list and set its flag to "off". Now examine all neighbors of the first site in the list. If any are present, add their locations to the end of the list and set their flags "off". Continue by examining the neighbors of the second element, the third, etc., until the list is exhausted. This completes the enumeration of one cluster, so its size or any other information needed may be stored. Continue the search of the sample for a new starting site and examine the next cluster. When the search of the sample is completed, all clusters will have been enumerated.

Each site will have been examined once, as will all neighbors of any site which is present. If Z is the number of neighbors of each site, the computational effort required will be $\leqslant (Z + 1)N$. Storage is required for N flags and a list of addresses which need only accommodate the largest possible cluster. This will not exceed pN sites.

Some trading off of time for reduced storage is possible within an order N algorithm. Hoshen and Kopelman [39], for example, describe an algorithm in which only two $(d - 1)$-dimensional layers of the sample are created and stored at any time. In order to piece together the parts of clusters which are first encountered separately but are later found to be connected, a data structure of cluster labels must be built, and this incurs some extra overhead. Reynolds et al. [31] report good results with an extension of this method. A very similar algorithm has been developed by Stoll, and used in a wide variety of percolation studies [141].

To calculate the $p'(p, L)$ data shown in figs. 6 and 7, however, it was not necessary to enumerate all clusters exhaustively, only to find a path across the sample. The shortest such path usually contains of order $N^{1/d}$ sites, so with luck one need inspect only $O(N^{1/d})$ sites to determine that a connected sample is connected, while all clusters touching one face must be enumerated to determine that a disconnected sample is disconnected.

The algorithm just described is called a "breadth-first search", since all new members of the cluster associated with a given site are located before proceeding to the next site in the list. The alternate search strategy, "depth-first search", provides a way of discovering paths across the sample as early as possible without incurring much extra overhead if it proves necessary to search the whole cluster. First a search order is assigned to the various nearest neighbor directions, with the desired search direction taken first. Then as the cluster is searched two lists are constructed; one containing locations of active sites currently being searched for neighbors, and the second containing, for each site, the number corresponding to the nearest neighbor direction last explored. As soon as the ordered search of neighbors of a site in the list is successful the location of the neighbor is added to the next location in the address list, its flag is set off, and its position in the search sequence is noted. The enumeration continues with the new site, starting with the most desirable search direction. If no neighbor of a site is found the program backtracks along the list until a site with unexamined neighbors is encountered, and continues from that site. If the opposite side of the sample is reached, the sample must be connected and enumeration can terminate. If backtracking brings us to the beginning of the list the cluster is exhausted and a new cluster may be searched.

Although two lists are maintained, storage requirements for the depth-first search are less than for breadth-first search, since only a small part of the cluster is ever active at one time. Performance comparisons are harder to draw, since they depend on the fraction of samples which are connected. For $50 \times 50 \times 50$ site samples and the range of concentrations shown in fig. 7, however, the depth-first search averaged 6 times faster than a breadth-first search which also halted as soon as a continuous path was found. Using an IBM model 370/168 computer I was able to search, on average, one such 50^3 bond percolation sample per second.

3. Dilute magnetic systems

This has been an active area of research for roughly as long as there has been a statistical mechanical theory of magnetism. For applications in technology, one desires quantitative predictions of the effects of dilution or alloying. The literature of such calculations, usually carried out in one of the many effective field theories of magnetism, is vast and not very useful to a theorist since little attempt is made to understand the validity of most approximations. Only quite recently have reliable calculations of the simplest properties (critical temperatures) of the simplest model alloys (dilute 2D models) become available [40,41].

An apparently unlimited variety of intermetallic amorphous alloys can be made by rapid quenching or vapor deposition, and most of these are magnetic. There are also at least two families of ideal dilute magnets for which extensive and accurate experimental data are available. These are insulating crystalline alloys for which the exchange interactions, mostly due to super-exchange, have been determined. The compounds $KMn_{1-x}R_xF_3$ and $K_2Mn_{1-x}R_xF_4$ have been studied by De Jong and co-workers [42, 43]. The first offers a simple cubic array of magnetic sites with nearest neighbor exchange, the second a set of well-separated planar square arrays. Inelastic neutron scattering data of Birgeneau et al. [139] is also available for two samples from this class of compounds. The rare-earth iron garnets have similar exchange interactions, but more complicated structures. They can be prepared with substitutions on any of three distinct magnetic sublattices, and considerable data has accumulated on these materials because of their utility as magnetic bubble memories (see Harris and co-workers [8,44] for references).

3.1. Real-space RG

The problem presented by dilute magnets is that of simultaneously treating thermal fluctuations and quenched-in-disorder. We previously described interactions as having merely a probability, p, of being present, a zero-temperature point of view. Now, the strength, K, of an interaction is also of interest. The fraction of bonds with $K \neq 0$ is p, and this is renormalized under scale changes as before. The rescaling of the interactions, however, depends upon p. Schematically, a scale change, discrete or infinitesimal, induces:

$$K \to K'(K, p) \tag{3.1a}$$

$$p \to p'(p) \tag{3.1b}$$

which gives a general picture like that sketched in fig. 10.

The behavior near the two interesting fixed points has been calculated directly for the dilute-bond 2D Ising model [41]. For $p \sim 1$,

$$T_c(1)^{-1}\, dT_c(p)/dp = 1.329\,, \tag{3.2}$$

while close to p_c,

$$\exp[-2J/kT_c(p)] = 2\ln 2(p - p_c) + O(p - p_c)^2\,. \tag{3.3}$$

The steep rise of the curve in fig. 10 from $T_c(p_c) = 0$ is a consequence of the finite energy of excitations from the Ising ground state, which makes

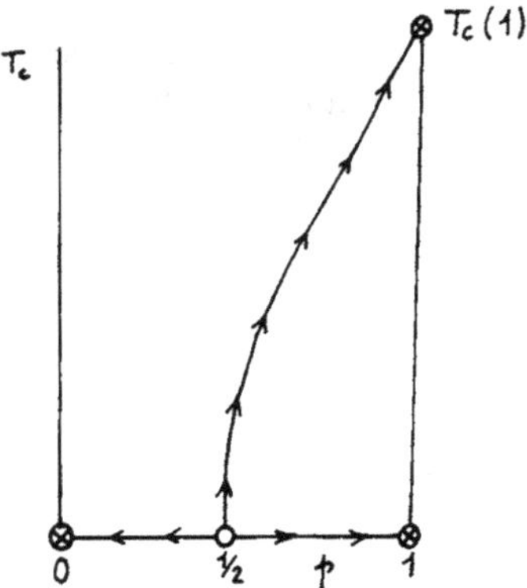

Fig. 10. Phase diagram and flows under rescaling for a dilute magnet.

$e^{-2J/kT}$ the appropriate "temperature" variable. Systems with a continuous excitation spectrum should show smoother power-law thresholds, with $T_c(p) \propto (p - p_c)^\varphi$.

Lubensky [45] has studied magnetic systems with small fluctuations in the exchange interactions and concluded that such fluctuations were irrelevant. Pure system behavior was obtained over sufficiently large scales as long as the specific heat exponent, $\alpha < 0$. (I will discuss one derivation of this criterion shortly.) Since of the common magnetic models, only the 3D Ising model is thought to have $\alpha > 0$, this result supports the general idea of relating critical phenomena in abstract models to those of real crystals, with defects, etc. However, the distribution of interactions present at an arbitrary point in fig. 10 will certainly not be narrow. Rescaling will introduce statistical correlations between nearby interactions. It is also not clear how much rescaling is needed to make a dilute alloy look like a pure system, or what contributions are made to the free energy by fluctuations on these smaller scales.

As an experiment to get a qualitative idea of these effects, I applied decimation (following Wilson [15], with a few new wrinkles of my own) to some large (1024×1024 sites) random 2D Ising models, and studied the distribution of interactions obtained when the systems had been reduced to only 4×4 or 8×8 sites. Some typical results, for systems with 5% missing bonds at three temperatures close to $T_c(p)$, are shown in fig. 11. Without a caption, one cannot tell what sort of bare model produced fig. 11; after a few renormalizations, I found, they all look qualitatively alike. The average interaction strengths flowed quickly to values outside, inside, or on the pure system critical curve. Thus one could easily see that one case in fig. 11 is below T_c (0.95), a second is above it, and the third is very close. Values of $T_c(p)$ obtained by this sort of test using the mean interaction strength and relatively few rescalings are in excellent agreement with the curve sketched in fig. 10 (the ends of which are exact).

But the distributions in fig. 11 remain broad long after their averaged behavior is clear. To do a useful calculation, therefore, one has to follow the "swarm of bees" shown in fig. 11 until it decides at which fixed point to collect. Stinchcombe and Watson [46] have provided a formalism which permits this. It is necessary to specify the distribution $g(J_{ij})$ of any type of interaction, singling out the non-zero interactions by:

$$g(J_{ij}) = (1 - p)\delta(J'_{ij}) + p\tilde{g}(J_{ij}) . \tag{3.4}$$

Rescaling now sends $p \to p'$ as before and g into a new distribution g'. For

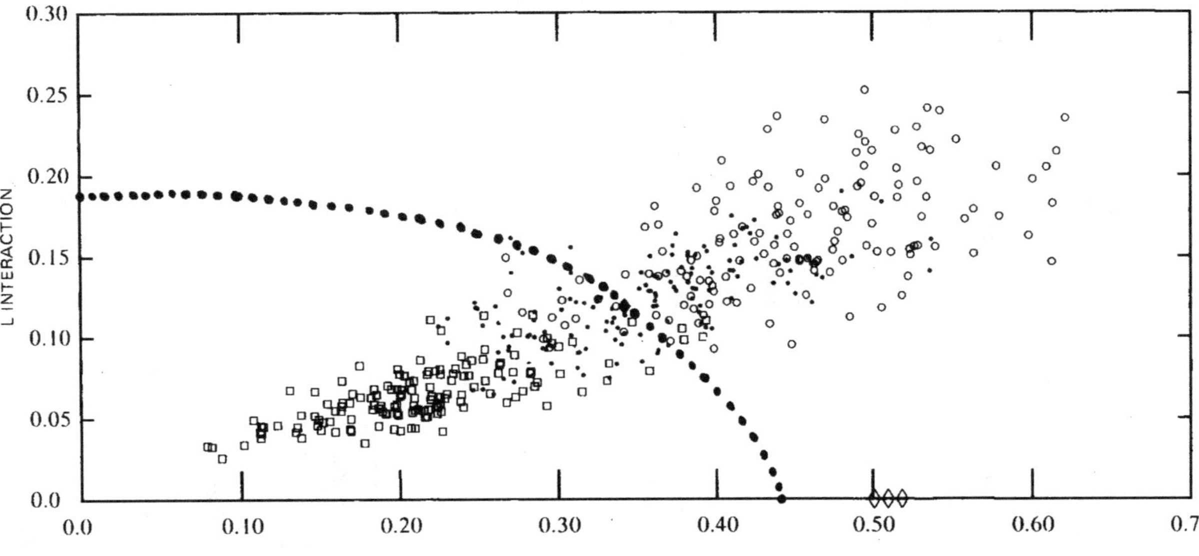

Fig. 11. Scatter plot of the effective interactions remaining after eight rescalings by factors of 2 when one starts at temperatures just below, just above and at the critical surface in a dilute 2D square Ising magnet (bond dilution, $p = 0.95$). The horizontal axis indicates the nearest-neighbor interaction, the vertical axis the next-neighbor interaction, and the dotted line the critical surface for the pure system. Strengths of the bonds initially present in the three cases are indicated by the hollow diamonds.

a given p, one finds fixed distributions, g^*, such that

$$\tilde{g}^*(J_{ij}) \to \lambda \tilde{g}^*(\lambda J_{ij}), \quad (3.5)$$

in which the interactions are reduced by a factor of λ in the recursion without any change in the shape of the distribution. Eq. (3.5) gives a complete description of the evolution of the system when p is stationary (i.e., $p = p_c$, $T > 0$).

Also we can think of the line in fig. 10 from the percolation threshold to the magnetic critical points as a line of fixed distributions – as the system evolves down this line with increasing scale, the distribution g^* will narrow, finally reaching a delta function at $p' = 1$. The product of factors λ along the trajectory gives the reduction of T_c from its value for $p = 1$. It is possible in principle, but apparently not yet in practice, to include spatial correlations in this description. A second convenient approximation is the replacement of the exact g^* by some simple approximate form, such as a few delta functions.

The Migdal recursion relation, applied in its differential form within this formalism is extremely successful in treating the dilute Ising magnet. For the 2D SQ lattice, the pure system fixed point and percolation threshold are exact, and the slopes of $T_c(p)$ and $\exp(-2J/hT_c(p))$ at the two ends of the line are within 3–4% of their exact values. Details of the calculations can be found in Kirkpatrick [33], and Jayaprakash et al. [47].

Using a discrete ($b = 2$) Migdal recursion, Jayaprakash et al. have also calculated the susceptibility and specific heat as functions of temperature. They find rather striking corrections resulting from the disorder. The critical region rapidly gets narrower with decreasing concentration of bonds, and is too small to see for $p \leqslant 0.85$. What remains are broad maxima in both quantities, at temperatures higher than the true T_c. These maxima reflect the freezing in of short ranged correlations, effects which one can think of as occurring along the trajectory the system follows until it reaches the appropriate pure-system fixed point. The result is an appealing one – in particular it explains why high precision specific heat measurements done in disordered magnets at temperatures close to the maximum may not reveal any critical phenomena. Unfortunately, recent Monte Carlo calculations by Zobin [48] suggest that the smearing in Ising systems is not quite so strong. In this data the critical region remains apparent until $p \leqslant 0.60$. Since the Migdal procedure gives a poor value of α ($\alpha \simeq 0.7$ and a cusp in the specific heart rather than $\alpha = 0$ and a logarithmic singularity), so Jayaprakash's calculation may actually give a closer description of dilute Heisenberg systems which also have $\alpha < 0$. (See the recent Monte Carlo results by Klenin [142] for the dilute 3D Heisenberg ferromagnet.)

3.2. When do random magnets exhibit a critical temperature?

Besides obscuring the critical phenomena, disorder may also modify it by causing a new fixed point to appear when $\alpha > 0$ in the pure system. There is no evidence as yet that the predicted changes, which are small, actually occur, and no compelling physical picture of how the physics of the random fixed point might differ from that of the pure system, but the argument for the role of the sign of α is a plausible one, due originally to Harris [40]. He questioned whether a random alloy would show a single sharp T_c or some sort of distribution of local critical temperatures, and introduced the following construction to resolve the issue:

Divide the macroscopic material up into cells of linear dimension L. The concentrations of the different cells will fluctuate over a range proportional to $L^{-d/2}$. If the cells exhibit a range of critical temperatures these will presumably also have a spread proportional to $L^{-d/2}$. In order to speak of the cells with nearly average concentrations as near critical and still only weakly interacting with each other we must choose $L \geqslant \xi(T) \propto |T_c - T|^{-\nu}$. For this choice of L, the range of potential local T_c's, δT_c,

$$\delta T_c \propto |T_c - T|^{+d\nu/2} . \tag{3.6}$$

As long as $\delta T_c \ll |T_c - T|$, the cells with unusual concentrations will not significantly smear out the behavior of the more typical cells. This will occur for sufficiently small $|T - T_c|$ as long as

$$\tfrac{1}{2}d\nu - 1 > 0 . \tag{3.7}$$

using the scaling relation $d\nu = 2 - \alpha$, in (3.7) gives

$$\tfrac{1}{2}\alpha < 0 , \tag{3.8}$$

as the condition for a sharp unique T_c to be observable.

This construction determines an "averaging length" for the alloy with $\alpha < 0$, and it diverges in the way one expects. For the 3D case $\alpha > 0$, we must modify the construction slightly, taking a larger length L to accommodate the longer range of correlations in the cells closer to criticality than average. The important cells to treat properly are the ones which first percolate. Their local T_c's will deviate from the mean by a constant number of standard deviations. Thus we now require

$$L \geqslant \xi(|T - T_c - AL^{-d/2}|) \propto |T - T_c - AL^{-d/2}|^{-\nu} , \tag{3.9}$$

where A is a small constant. One can solve (3.9) self consistently for all T. The nature of the solution can be seen transforming (3.9) into (for $T > T_c$):

$$L^{-1/\nu} + AL^{-d/2} = T - T_c . \tag{3.10}$$

Far from T_c, our length increases as $|T - T_c|^{-\nu}$, as before, but close to T_c the increase is more rapid,

$$\zeta \propto (T - T_c)^{-2/d} . \tag{3.11}$$

The increase in the effective value of ν has the effect of restoring $\alpha = 0$. Khmel'nitskii [49] has obtained a slightly stronger result by a $4 - \varepsilon$ expansion. He finds a raised value of ν and $\alpha_{\text{eff}} < 0$.

There is another multi-cell construction which can give useful insights into the stability of various ground state configurations of random magnets. It has been used by Harris and Kirkpatrick [8] to discuss the susceptibility of dilute antiferromagnets, by Imry and Ma [50] to treat the effect of random external fields and by Alben et al. [51] to study the effects of random anisotropy. Since the last topic is relevant to recent experiments on spin glasses, I will review the argument here.

The model is

$$H = -\sum_{\langle ij \rangle} J_{ij} S_i S_j - D \sum_i (\hat{n}_i \cdot S_i)^2 . \tag{3.12}$$

To an otherwise uniform ferromagnetic system is added a single-ion anisotropy term which aligns S_i preferentially along a randomly distributed anisotropy axis $\hat{n}_i$. To determine whether the anisotropy term will modify the properties of the ordered state of the model we construct an approximate ground state.

First divide the system into cells of side L. Consider the spins in each cell to be roughly parallel and align them along the overall anisotropy axis of the cell (found by averaging the $\hat{n}_i$). This gives an anisotropy energy averaging $\propto DL^{-d/2}$ per unit volume. Adjacent cells will not be parallel. When the differences are allowed to relax smoothly from cell to cell the resulting cost in exchange energy density will be $\propto AL^{-2}$. For $d < 4$ the energy is minimized by the choice.

$$L_{\text{stable}} \propto (A/D)^{2/(4-d)} . \tag{3.13}$$

This identifies a Heisenberg–Ising cross-over. Where viewed on scales $\ll L_{\text{stable}}$ the spins are seen pointing in all directions and subject to slow fluctuations. When viewed on scales $> L_{\text{stable}}$, the cells act like two-state Ising spins, with the appropriate incoherent thermal fluctuations. The question of whether the Ising-like cell spins will order into a spin glass or ferromagnet is still being debated at present, and will be discussed below. The case $d = 4$ can be treated by extending the construction. For those cells where $\langle \hat{n}_i \rangle_{\text{cell}}$ is large we proceed as before. When it is small, we must break up the cell into 2^d subcells and calculate $\langle \hat{n}_i \rangle_{\text{cell}}$ for these. If $\langle \hat{n}_i \rangle_{\text{cell}}$ is a standard deviation or more away from zero in a subcell an anisotropy axis can be assigned, if not, the subdivision continues. In this way, extra logarithmic factors in the anisotropy energy are gained, which stabilize the random anisotropic ground state in 4D. (The construction is described in ref. [8].)

4. Dilute or highly inhomogeneous conductors

4.1. Crossover picture of threshold properties

First I will talk about all kinds of percolation thresholds simultaneously. Later in this section I consider features and techniques specific to the conduction problem. Figure 12 illustrates the very general observations of Straley [52] and Efros and Shklovskii [134] that all percolation thresholds have two sides and a middle. The vertical axis in fig. 12 is not labelled. It could be the Curie temperature of an alloy, conductivity, or several other quantities. As $p \to p_c$ from above, the conductivity of a mixture of conducting (of weight p) and perfectly insulating $(1 - p)$ material will drop to zero with the indicated power law, $\sigma(p) \propto (p - p_c)^t$. If the material

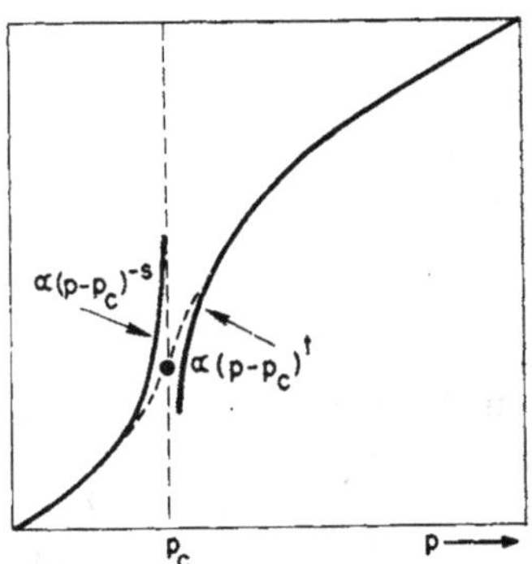

Fig. 12. Schematic behavior of properties such as conductivity, magnetic transition temperature, etc., near the percolation threshold in a mixture.

were a mixture of conducting $(1 - p)$ material and superconducting (p) regions then as $p \to p_c$ we would expect a singular increase in the conductivity. Following Straley's conventions we describe this as $\sigma(p < p_c) \propto (p - p_c)^{-s}$.

The situation below p_c in a metal–insulator composite when the insulator is a poor but measurable conductor is similar to that in the superconductor–metal mixture. The conductivity increases as p_c is approached from below, but the singularity is cut off by the finite conductivity of the metallic material. At p_c, σ takes on some compromise value, indicated by a large dot in fig. 12. This will depend on both σ_1, the conductivity of the insulating material, and σ_2, the conductivity of the metallic regions. A plausible form for the compromise value is:

$$\sigma(p_c) \simeq \sigma_1^u \sigma_2^{1-u} . \tag{4.1}$$

The discussion extends without change to the case when the system contains a mixture of weakly and strongly ferromagnetic interactions, so that fig. 12 would represent $T_c(p)$. Another application of this picture is conduction at finite frequency when the insulating material acts as a capacitative shunt, and the vertical axis in fig. 12 will represent the magnitude of the complex admittance of the material – the measured ac conductivity. If we take $\sigma_1 \equiv i\omega C$, $\sigma_2 \equiv \sigma_0$, (4.1) predicts $\sigma(p_c) \sim \omega^u$. This sort of frequency-dependent conductivity is commonly observed in poor conductors. (For a discussion of frequency-dependent conduction and the related complex dielectric constant in composites see Bergman [53,54] or Dubrov et al. [132].)

The curve in fig. 12 has some resemblances to the smoothed out singularity seen when magnetization is plotted versus decreasing temperature for a ferromagnet in an applied field. Using this analogy Straley [52,55,56] developed a scaling formula to unify properties observed just above, at, and just below p_c. The principal consequence of the scaling construction is the relation:

$$u = t/(s + t) , \tag{4.2}$$

connecting the middle to the two sides of the transition.

Unfortunately, the analogy between fig. 12 and a ferromagnet in an external field is imprecise. Microscopic calculations [57,58] have recently shown that the simple scaling picture and (4.2) are correct for $T_c(p)$ in an alloy, but not for a mixture of two differing conductances. The reason for the failure of the analogy is that neither $T_c(p)$ nor $\sigma(p)$ is an order parameter. $T_c(p)$ is an energy density, while σ in the context of magnetism is a stiffness coefficient at zero temperature, hence a parameter in a correlation function.

A simple argument differentiates the two cases. In discussing $T_c(p)$, the order of two limits, $p \to p_c$ and $T \to 0$, is important. Take $T \to 0$ first, and one has the percolation problem. Only by letting $p \to p_c$ while $T \neq 0$ can one study the magnetic transition. Thus there is a (p, T) plane of possible directions from which to approach the percolation critical point. A scaling function with one crossover exponent can express this sensitivity to the order of two limits, just as such a scaling function describes the sensitivity of the magnetization to the order of limits $h_{ext} \to 0$, $T \to T_c$. To describe conductivity, however, we must take three limits in prescribed order. First one takes $T \to 0$ to consider only the elastic or linearized spin fluctuations, then $q \to 0$ to extract the coefficient of $\chi(q)$ which is the stiffness, then finally we may let $p \to p_c$. (The susceptibility at finite q and the exchange stiffness at finite T are also interesting, but different, properties of an alloy.) There is a volume of possible directions from which to approach P. Stephen [57] in fact finds that there are two crossover exponents for the conduction problem rather than one as originally proposed. What is lost is the possibility of explaining s in terms of t and the other static exponents. Equation (4.2) remains valid.

4.2. *Effective medium theory*

Often the experimentally accessible properties of an inhomogeneous conductor do not depend sensitively on the critical effects near the percolation threshold [135]. When a material has a local conductivity which varies smoothly over a wide range, as is the case for many glasses and highly disordered semiconductors, one does not observe a percolation threshold as some extrinsic parameter is varied. Instead, percolation effects control the "rate-limiting steps", which determine the conductance of the system. Even a system, such as an inhomogeneous thin film, which does pass through a percolation threshold, will cross over from metallic conduction predicted by semi-classical theory to non-metallic Anderson localization (Anderson [59], and these lectures). This cross-over occurs whenever the conductivity is depressed by dilution to a value which violates the Ioffe–Regel criterion [60], and is particularly evident in 2D, where there is a near-universal minimum metallic conductivity [61]. The cross-over may or may not occur before the system has passed from concentrations where the conductivity is accurately predicted by effective medium theory to the region where $\sigma \propto (p - p_c)^t$ applies. Even when the critical region is not reached, the changes of the conductivity with composition in the effective medium region can reveal information about the microscopic structure of the material.

The characteristic $\exp(-[T_0/T]^{1/4})$ conductivity variation seen at low temperatures in impurity bands and some elemental amorphous semiconductors is a consequence of percolation (Ambegaokar et al. [62] and Thouless, these lectures). The following variational argument can be used to derive it, and generalizes to other models as well [7]. We model the actual material, with conductivity $\sigma(r)$ varying continuously, by a system in which σ takes on only two values, σ_0 and 0. Everywhere that $\sigma(r) < \sigma_c$ in the real material, the model has $\sigma = 0$. In the rest of the model, $\sigma = \sigma_c$. The fraction of volume which is conducting in the model, $p(\sigma_0)$, increases monotonically as σ_c decreases. If we choose σ_c too high there will not be enough conducting material in the model system for percolation $p(\sigma_0) < p_c$. Increasing $p(\sigma_0)$ past p_c gives a model conductance.

$$\sigma(\sigma_0) \propto \sigma_c(p(\sigma_0) - p_c)^t \,. \tag{4.3}$$

Equation (4.3) goes through a maximum when $p(\sigma_0)$ is slightly above the percolation threshold, since the decrease in σ_0 counters the increase due to the power law. At the maximum, (4.3) gives a fairly good approximation to the conductivity of the actual system, according to numerical tests on models of hopping conductors.

Equation (4.3) is essentially an estimate of where the current will flow in a highly inhomogeneous conductor, and predicts that most conduction processes are confined to slightly more than the volume needed for percolation. This aspect of the percolation picture is also in good agreement with numerical checks, and the argument leads to predictions of transport properties such as the thermopower in impurity band conductors, which depend on the typical energies by which carriers are excited from the Fermi level [7].

Rather successful effective medium theories have been constructed for conduction in random mixtures [63,64] and on random resistor networks [65]. Not surprisingly, when the simplest statistical assumptions are taken, the two theories are the same. In the continuum theory one treats the system as made of a distribution of small regions of known properties imbedded in an effective medium whose properties are to be determined. Inside the inclusions the electric field $\mathscr{E}_{\text{in}}$ is uniform (for inclusions with reasonable shapes) and related to the field $\mathscr{E}_{\text{m}}$ at a distance in the medium by

$$\mathscr{E}_{\text{in}} = \mathscr{E}_{\text{m}} - \frac{4\pi}{\text{d}} \boldsymbol{P} \,. \tag{4.4}$$

$\boldsymbol{P}$ is the polarization developed at the inclusion in order to guide the current around it. $\boldsymbol{P}$ is really the result of a charge density, ρ_n piling up at the surface of the inclusion, where

$$\rho_n = \hat{n}\cdot\boldsymbol{P}\,, \tag{4.5}$$

and $\hat{n}$ is a unit vector normal to the surface of the inclusion. Just outside the inclusion, the field $\mathscr{E}_{\text{out}}$ is given by

$$\mathscr{E}_{\text{out}} = \mathscr{E}_{\text{m}} + 4\pi\boldsymbol{P}\cdot\overleftrightarrow{\mathbf{L}}\,, \tag{4.6}$$

where $\overleftrightarrow{\mathbf{L}}$ is the depolarization tensor for the inclusion. Taking the simplest assumption, isotropy, we have

$$\overleftrightarrow{\mathbf{L}} = (\hat{n}\hat{n} - d\overleftrightarrow{\mathbf{1}})\,, \tag{4.7}$$

for $d = 3$ (spheres) or $d = 2$ (circles).

To relate $\boldsymbol{P}$ to $\mathscr{E}_{\text{m}}$ we need only require continuity of the current normal to the surface of the inclusion

$$\hat{n}\cdot\overleftrightarrow{\sigma}_{\text{in}}\cdot\mathscr{E}_{\text{in}} = \hat{n}\cdot\overleftrightarrow{\sigma}_{\text{m}}\cdot\mathscr{E}_{\text{out}}\,, \tag{4.8}$$

where σ_{in} is the conductivity tensor for the medium. In the absence of a magnetic field, we shall assume isotropic $\overleftrightarrow{\sigma}_{\text{in}}$, and thus obtain

$$\frac{4\pi}{3}\boldsymbol{P} = \frac{(\sigma_{\text{in}} - \sigma_{\text{m}})\mathscr{E}_{\text{m}}}{(d-1)\sigma_{\text{m}} + \sigma_{\text{in}}}\,. \tag{4.9}$$

If we have chosen σ_{m} properly, that is, if we have defined an effective medium whose properties are the same as a macroscopic sample, the average polarization obtained by averaging (4.8) over all possible values of σ_{in} appropriately weighted, must vanish. Thus σ_{m} is determined by:

$$\left\langle \frac{\sigma_{\text{in}} - \sigma_{\text{m}}}{(d-1)\sigma_{\text{m}} + \sigma_{\text{in}}} \right\rangle_{\text{in}} = 0\,. \tag{4.10}$$

Arguments in the same vein carried through for d-dimensional hypercubic resistor networks, where σ_{in} is the (random) value of the conductance of an individual bond, also lead to (4.10). In the case of networks, one can check the magnitude of higher-order corrections to the macroscopic conductance σ_{m}. One finds [6] that the first corrections do not occur until the fourth order in perturbation theory. Thus the solution to (4.10) will be quite accurate for all but the most extreme distributions of σ_{in}.

Solving (4.9) for the case $\sigma_{\text{in}} = \sigma_0$ (with probability p) or 0 (with probability $1 - p$) is well known [6, 64, 65] to predict a percolation threshold. The solution is

$$\sigma(p) = \sigma_0\left[1 - \frac{d}{d-1}(1-p)\right], \quad p > 1/d$$
$$= 0, \qquad p < 1/d. \tag{4.11}$$

Alternately one can solve (4.10) for a model with superconducting bonds of fraction p, and normal bonds of weight $(1 - p)$. The resulting resistance is:

$$\rho(p) = \rho_0[1 - \mathrm{d}p], \quad p < 1/d$$
$$= 0, \qquad p > 1/d. \tag{4.12}$$

and also varies linearly with composition.

Numerical calculations on resistor network models with bond percolation (Kirkpatrick [6] and work to be reported here) show that this straight line behavior, with the slope as calculated, continues until quite close to the critical concentration. Site percolation studies, however, show a conductivity which curves continuously [8]. For a wide range of probability, $\sigma_{\text{site}}(p) \propto p^2$. This can be accounted for by arguing that the sites at both ends of a given bond must be present for the bond to be present; thus the fraction of bonds present is p^2. To explain the coefficient of p^2, Watson and Leath [66] noted that the smallest defect removes two bonds at a time along each axis. The larger slope they calculated for this configuration is in excellent agreement with the results of simulations.

Shankland [67,68] has applied this observation to interpret the dependence of conductivity on porosity ρ, in rocks under geothermal (or laboratory) pressure. An old observation, known as "Archies Law" is that $\sigma \propto \rho^s$ where $1 \leqslant s \leqslant 2$, with values close to 1 or 2 the most common. He interprets s as the number of independent events necessary to open up a pore in the rock for ionic conduction, and was able to show that rocks in which pressure squeezes shut the individual pores have $s \approx 1$, while those with stable pores and joints which can close under pressure tend to have $s \approx 2$.

Since it contains a reasonable approximation to the actual percolation threshold, the effective medium theory is surprisingly accurate in treating systems with a wide range of local conductances. Given some $P(\sigma_{\text{loc}})$ one must solve for σ_{m} by setting $\langle \boldsymbol{P} \rangle = 0$:

$$0 = \int \mathrm{d}\sigma_{\text{loc}} P(\sigma_{\text{loc}}) \left\{ \frac{\sigma_{\text{loc}} - \sigma_{\text{m}}}{(d-1)\sigma_{\text{m}} + \sigma_{\text{loc}}} \right\}. \tag{4.13}$$

Approximating the integrand by $P(\sigma_{\mathrm{loc}})$ when $\sigma_{\mathrm{loc}} > \sigma_{\mathrm{m}}$ and by $-P(\sigma_{\mathrm{loc}})/(d-1)$ when $\sigma_{\mathrm{loc}} < \sigma_{\mathrm{m}}$ we obtain as a rough estimate for σ_{m} the implicit equation

$$\frac{1}{d-1}\int_0^{\sigma_{\mathrm{m}}} \mathrm{d}\sigma_{\mathrm{loc}} P(\sigma_{\mathrm{loc}}) = \int_{\sigma_{\mathrm{m}}}^{\infty} \mathrm{d}\sigma_{\mathrm{loc}} P(\sigma_{\mathrm{loc}})\,, \tag{4.14}$$

and defining

$$p(\sigma_{\mathrm{m}}) = \int_{\sigma_{\mathrm{m}}}^{\infty} \mathrm{d}\sigma_{\mathrm{loc}} P(\sigma_{\mathrm{loc}})\,, \tag{4.15}$$

we obtain from (4.13)

$$p(\sigma_{\mathrm{m}}) = 1/d\,. \tag{4.16}$$

Thus, just as in the variational argument I gave above, the value of conductance which is just sufficient for conduction is the value which characterizes the material. One can also obtain within the effective medium theory predictions for where current flow will take place. Using (4.4) and (4.8) we find

$$\boldsymbol{J}_{\mathrm{in}} = \sigma_{\mathrm{in}}\mathscr{E}_{\mathrm{in}} = \left[\frac{d\sigma_{\mathrm{m}}}{(d-1)\sigma_{\mathrm{m}} + \sigma_{\mathrm{in}}}\right]\sigma_{\mathrm{in}}\mathscr{E}_{\mathrm{m}}\,. \tag{4.17}$$

For $\sigma_{\mathrm{in}} \ll \sigma_{\mathrm{m}}$, $J_{\mathrm{in}} \to 0$, while for $\sigma_{\mathrm{in}} \gg \sigma_{\mathrm{m}}$ the current density saturates at $d\sigma_{\mathrm{m}}\mathscr{E}_{\mathrm{m}}$.

Finally, the tensor expressions for the effective medium theory ((4.4)–(4.7)) permit treating galvanomagnetic phenomena, where $\boldsymbol{\sigma}$ includes antisymmetric, field dependent terms [69] of the form:

$$\boldsymbol{j} = \sigma\mathscr{E} = \sigma\mathscr{E} + (u\sigma/c)\boldsymbol{H} \times \mathscr{E}\,. \tag{4.18}$$

Solving for $(4\pi/3)\boldsymbol{P}$ in the presence of a field one can expand the result in powers of H. To lowest order $(4\pi/3)P$ is again given by (4.8). One must again choose σ_{m} to make this term vanish. The term linear in $\boldsymbol{H}$, $(4\pi/3)\boldsymbol{P}_{\mathrm{anti}}$ is

$$\frac{4\pi}{3}\boldsymbol{P}_{\mathrm{anti}} = \frac{(d\sigma_{\mathrm{m}})\langle\mu_{\mathrm{in}}\sigma_{\mathrm{in}}\mu_{\mathrm{m}}\sigma_{\mathrm{m}}\rangle}{\langle(d-1)\sigma_{\mathrm{m}} + \sigma_{\mathrm{in}}\rangle^2}\frac{1}{c}\boldsymbol{H} \times \mathscr{E}_{\mathrm{m}}\,, \tag{4.19}$$

and requiring (4.17) to vanish on average determines μ_{m}. In two dimensions, using (4.19), one obtains $R_{\mathrm{m}} = \mu_{\mathrm{m}}/\sigma_{\mathrm{m}} = R(1)$, in agreement with the known result (Thouless, these lectures). In three dimensions, the effective medium theory predicts:

$$R_{\mathrm{m}} = R(1)[1 - \tfrac{3}{4}(1 - p)]^{-1}, \tag{4.20}$$

which increases steadily as p decreases, but remains finite as $p \to p_c$.

The effective medium prediction for R has the flavor of the transport calculations done in semiconductors by averaging contributions from all carrier energies, weighted by their current densities. Solving (4.19) explicitly gives:

$$\mu_{\mathrm{m}} = \left\langle \mu_{\mathrm{in}} \frac{d\sigma_{\mathrm{m}}\sigma_{\mathrm{in}}}{[(d-1)\sigma_{\mathrm{m}} + \sigma_{\mathrm{in}}]^2} \right\rangle \Big/ \left\langle \frac{d\sigma_{\mathrm{m}}^2}{[(d-1)\sigma_{\mathrm{m}} + \sigma_{\mathrm{in}}]^2} \right\rangle . \tag{4.21}$$

By comparison with (4.17) we see that μ_{m} is just an average of the possible μ_i, weighted by a factor which is essentially the current density associated with each. These formulae work well for describing transport properties in metal–ammonia solutions [70], but their application to some denser systems is suspect, in my opinion, because of the uncertain meaning of the several fitting parameters employed. Also there is evidence, which I will discuss below, that R should diverge weakly at p_c. The effective medium approximation fails to capture this.

4.3. Real-space RG

Resistor network models lend themselves quite nicely to real-space rescalings. Stinchcombe and Watson [46] were the first to treat the 2D square lattice, using a simple ad hoc rescaling sketched in fig. 13a. The new conductance, indicated by the dashed resistor joining sites 1 and 2, is the sum, in parallel, of the conductances of the two series paths 1–4–2 and 1–3–2. The scale factor is $\sqrt{2}$. They solved the necessary integral equation

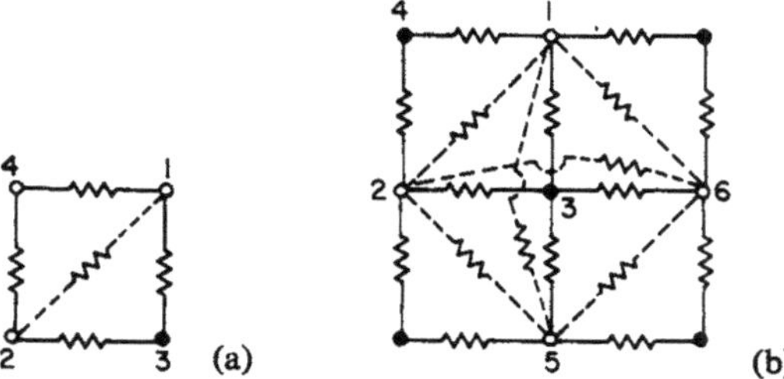

Fig. 13. Decimation procedures for resistor networks: (a) finite cell approximation, and (b) Gaussian elimination.

to obtain a fixed distribution $\tilde{g}^*(\sigma, p_c)$, and the eigenvalue λ of the rescaling. These quantities behave just as in the magnetic case ((3.4)–(3.5)). Identifying

$$\lambda = b^{t/\nu}, \tag{4.22}$$

they obtain $t = 1.33$.

There are a number of ambiguities in this apparently simple calculation. For example, the Migdal–Kadanoff recursion relation, when applied to the resistor network model using $b = 2$, gives a transformation identical to fig. 13a: two bonds in series, then two such units in parallel become the new bond. Since $b = 2$ instead of $\sqrt{2}$, the predictions for ν differ by a factor 2, but the prediction for t is the same, since b factors out. Since conductivity is dimensionless in 2D, there is no obvious way to determine which scale factor is appropriate.

One consistent approach to rescaling network problems is to use gaussian elimination, applied to the system of equations for the voltages at nodes of the network, as the analog of decimation by partial tracing. Given the equations

$$\mathbf{W}V = S, \tag{4.23}$$

where the S are surface contributions, one can break the matrix into blocks identifying two groups of voltages V_0 and V_1:

$$\begin{bmatrix} \mathbf{W}_{00} & \mathbf{W}_{01} \\ \mathbf{W}_{10} & \mathbf{W}_{11} \end{bmatrix} \begin{bmatrix} V_0 \\ V_1 \end{bmatrix} = \begin{bmatrix} S_0 \\ S_1 \end{bmatrix}. \tag{4.24}$$

Solving for the V_0 in terms of the V_1 gives

$$V_0 = -\mathbf{W}_{00}^{-1}[\mathbf{W}_{01}V_1 - S_0], \tag{4.25}$$

and substituting (4.25) back into (4.24) gives

$$[\mathbf{W}_{11} - \mathbf{W}_{10}[\mathbf{W}_{00}]^{-1}\mathbf{W}_{01}]V_1 = S_1 - \mathbf{W}_{00}^{-1}S_0. \tag{4.26}$$

If $\mathbf{W}_{00}$ is a diagonal matrix (no conductances joining any of the $\{V_0\}$) then W_{00}^{-1} is also diagonal, and the new conductances in (4.24) will connect any pair of nodes in $\{V\}$ which are neighbors of the same node in $\{V_0\}$. This is the situation envisioned in fig. 13b before the center and corner sites are eliminated. For the case where all interactions are initially uniform each new interaction will have conductivity $\frac{1}{4}\sigma$. There are two contributions to the new bond 1–2 (from eliminating sites 3 and 4), so its strength is $\frac{1}{2}\sigma$. Elimination of half the sites leads to a new problem, with interactions of two ranges and strengths; but for the uniform case one readily checks that the conductivity per cell is preserved (in fig. 13b).

To proceed beyond this stage it is necessary to approximate W_{00}^{-1} by some sort of expansion. Techniques similar to those used in statistical mechanics are easily invented to do this. In principle, the machinery of gaussian elimination permits study of conduction networks with longer-ranged interaction, such as those introduced by Webman et al. [71] to model the magneto-conductivity tensor, but as yet it has not been fully utilized.

None of the real-space RG calculations done on resistor networks to date has achieved obvious convergence to reliable results for critical exponents or even good agreement with numerical results from computer simulations. The Migdal–Kadanoff recursion in its differential form [33] gives $t = 1.13$ in 2D, which agrees well with numerical work (see below), but the result for ν, as we have already remarked, is sensitive to the scale factors b. In 3D, the M–K recursion relations give $t = 2.3$, $s = 0.45$. These do not agree very well with the finite-cell calculations reported by Straley [55] which in 3D give $t = 2.0$, $s = 0.8$, while in 2D $t = s = 1.3$. In contrast the most accurate computer calculations in 3D give $s =$ 0.7–0.8 [55] and $t = 1.62$ (see below).

Small cell real-space rescaling does seem to provide rather accurate results when applied to network models outside their critical regions, and may provide a powerful alternative to effective medium theory in dealing quantitatively with systems whose interactions are statistically correlated over moderate distances. I tried the Migdal recursion, with $b = 2$, on a model with a very wide spread of interactions ($\log_{10} \sigma_{ij}$ uniformly distributed from -3 to $+3$), and found agreement with the results of computer experiments [65] to within the roughly 5% accuracy of my calculations [33]. Bernasconi [30] has applied the "Wheatstone Bridge" transformation (fig. 13b) to bond percolation models in two and three dimensions and obtains agreement to within a percent or so throughout the non-critical regime ($|1p - p_c| \geqslant 0.1$) in both cases.

4.4. Computer simulations

Recent computer studies of the conduction threshold give results for the critical exponents which appear to be converging as larger samples are studied and better statistics are obtained. Furthermore there is reasonable agreement between the several different groups who have studied these questions. However, the numbers obtained are consistently lower than results of real-space RG, and the one series calculation which has been performed [44]. Since we will need accurate values of the conductivity exponent, t, to make quantitative tests of simple pictures in the next section, I will try to give enough detail in this section to let you form an opinion of the extent to which the computer simulations may be relied upon.

Early simulations of conduction just above the percolation threshold exhibited some obvious limitations. Studies were made of small or anisotropic samples without any systematic study of the effects of sample size or shape to allow accurate extrapolation. Some studies made with relatively large samples nonetheless used a population of only one or a few such samples, and could not assess the size of statistical errors. This is a particular drawback to analog calculations such as those of Last and Thouless [72] (holes punched in conducting paper), Adler et al. [73] (a 3D network of randomly soldered-together resistors), and Watson and Leath [66] (clipped hardware cloth, or "rat wire"). Numerical calculations using the classical Gauss–Seidel relaxation procedure (among others, Kirkpatrick [6,65] are poorly convergent close to p_c, and various tricks had to be invented to accelerate convergence. Even with under-relaxation and various sorts of smoothing techniques, it remains difficult to know whether such an iterative calculation has converged. Trying to fit a power law to include data outside the critical region for site percolation models may have accounted for the larger values of t reported in some early calculations. In particular, the discussion in Adler et al. [73] suggests that their value of t $(=2.0)$ is affected by this bias.

These limitations all seem to be avoidable using numerical simulation on modern computers with good algorithms. I can think of only two situations where analog simulations might still be preferable to computer simulation. The first is in study of randomly close-packed structures, which cannot be easily or accurately generated on a computer. The studies of pressed powders and small plastic pellets by the Marseilles group [74] is a promising step in this direction. The second is the question of the crossover to Anderson localization. The most accurate studies of the classical threshold to date have been obtained by exact solution of Kirchhoff's equations for the voltages at nodes of large (400×400 site, or $50 \times 50 \times 50$ site) networks. An alternate approach, studying diffusion processes on the random lattice and using an Einstein relation to obtain a conductivity [75], is also exact in principle but has yet to yield very accurate values of critical exponents.

Data from my calculations on bond percolation models in 2D and 3D are plotted in figs. 14a and 14b. In determining the conductance, periodic boundary conditions were imposed in all directions but one. In the

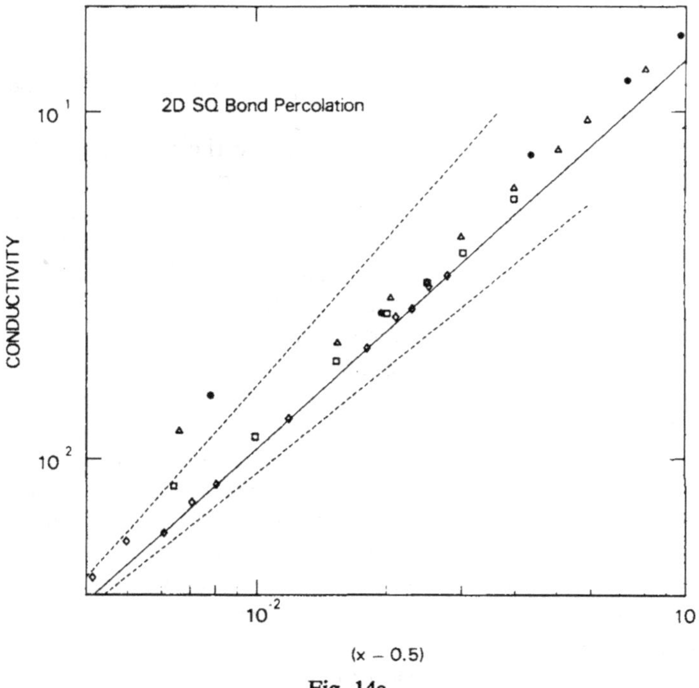

Fig. 14a.

remaining direction the two faces of the samples were held at differing constant potentials. I chose to study the bond percolation model because the distinction between critical and non-critical behavior is clear when the concentration dependence far from threshold is nearly linear. However, there is little difference between the critical behavior evident in figs. 14 and that observed by Straley [56] for site percolation models of up to 200^2 sites (2D) and 30^2 sites (3D).

When the exponent t is large, the finite size scaling form analogous to (2.28) is not very useful in analyzing size-dependence, since it predicts only very slow size-dependent variations close to threshold, and these will be susceptible to statistical error. Since positions of the infinite sample percolation thresholds are known exactly (2D) or to high accuracy (3D) for both models, in figs. 14 I have simply plotted $\log_{10}(\sigma(p))$ against $\log_{10}(p - p_c)$ for samples of various sizes, the sizes indicated by the different types of data points used.

The size effects can be estimated by eye. For each size of sample there appears to be a region of concentrations which gives a straight line in the

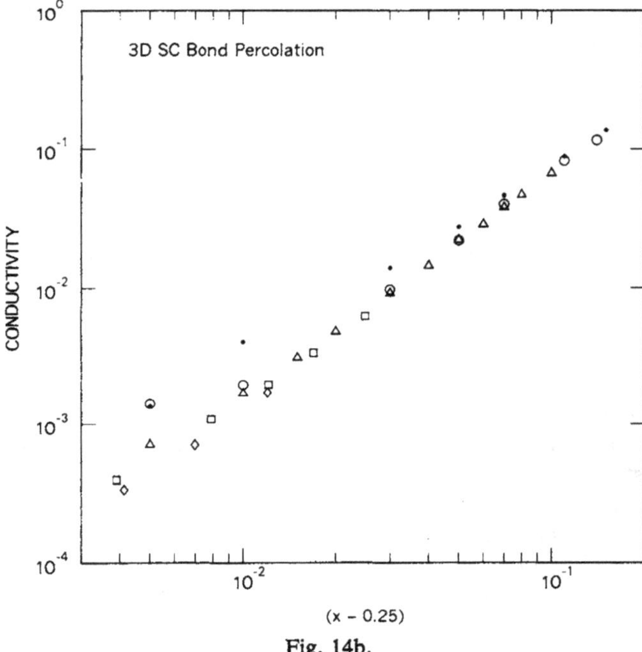

Fig. 14b.

Fig. 14. Log–log plots of the macroscopic conductance of finite bond percolation samples against distance to the percolation threshold. (a) 2D samples ranging in size from 40 × 40 (dots) to 400 × 400 (diamonds). The solid line has the slope $t = 1.1$; the dashed lines indicate two counter proposals, $t = 1$ and $t = \nu = 1.365$. (b) Preliminary data for 3D samples ranging in size from 8 × 8 × 8 to 50 × 50 × 50.

log–log plot. At higher concentrations the data curve, showing the non-critical behavior. At lower concentrations they curve up because of the finiteness of the sample. The straightline portions get longer and persist to smaller concentrations as the samples grow larger, but they appear to exhibit the same slope for all sample sizes.

From fig. 14a, I extract the exponent $t = 1.10 \pm 0.05$. For comparison, the dotted lines show two predicted values for t in 2D: $t = 1$ and $t = \nu = 1.365$, neither of which fits. I will discuss these predictions in the next section. De Gennes [9] has proposed that $\sigma(p) \propto (p - p_c)/|\ln(p - p_c)|$ in 2D. This expression can account for the data when $(p - p_c) \lesssim 0.04$, but is not properly dimensionless. The expression $x/|\ln x|$, where $x = (p - p_c)/p_c$, is in dimensionless form, but rises too steeply to fit the data in fig. 14a.

Also, the corrections to scaling which occur when an exponent has an integral value [140] take the form $x \ln x$, rather than $x/\ln x$. The data in fig. 14b may be improved soon by taking more samples etc., so my present estimate, $t = 1.62 \pm 0.05$ should be regarded as provisional.

A certain amount of algorithmic sophistication (i.e. complicated computer programs) is required to solve the network equations efficiently and obtain the conductivities of large samples. First, I take advantage of the observation [76] that the most efficient way to solve a system of linear equations of the form (4.23) is not a direct calculation of $\boldsymbol{W}^{-1}$ but a two step calculation. First one factors $\boldsymbol{W}$ into a product form:

$$\boldsymbol{W} = \boldsymbol{L}\boldsymbol{D}\boldsymbol{L}^T . \tag{4.27}$$

$\boldsymbol{L}$ is a lower-triangular matrix, with ones on the diagonal and zeroes in the upper right half, and $\boldsymbol{D}$ is diagonal. Solving (4.23) in two stages is easy. One first calculates the vector $U \equiv \boldsymbol{D}\boldsymbol{L}^T V$, one coefficient at a time, starting with the first, then from $\boldsymbol{D}^{-1} U$ obtains V, one coefficient at a time, starting with the last.

Calculating $\boldsymbol{L}$ takes far fewer operations than calculating $\boldsymbol{W}^{-1}$ would have, but the most important advantage of the method is that $\boldsymbol{L}$ can be kept reasonably sparse, while in principle all elements of $\boldsymbol{W}^{-1}$ are non-zero. For 2D problems in which all the sites participate in conduction, one can employ an elegant algorithm of George [77] to keep the number of $\boldsymbol{L}$'s non-zero elements down to order $N \ln_2 N$ and the number of multiplications required to perform the transformation proportional to $N^{3/2}$. This can be shown to be optimal for both properties [78]. The George procedure is rather similar in flavor to decimation (except that it is exact). The transformation by which L is constructed is formally just gaussian elimination. George's "nested dissection" prescribes a sequence for eliminating first sites which are not neighbors of each other, then blocks of sites which just avoid touching, then larger blocks, etc. At each stage this minimizes the number of new interactions which are generated by the elimination. The procedure can be generalized to 3D (Kirkpatrick, unpublished).

For sufficiently dilute models, the symmetry of the lattice is of less overriding importance. An efficient elimination sequence can be obtained by selecting at each step any site from among those with the fewest interactions at that stage of the calculation. Programs exist (called "MOOO" in the IBM SL-MATH scientific package and there is a similar set of programs "MA18A" available from Harwell) to find this sequence and carry out the factorization from $\boldsymbol{W}$ into $\boldsymbol{L}$ and $\boldsymbol{D}$.

Considerable reduction of effort is possible for dilute systems by expressing the problem in terms of the fewest possible node voltages. Naturally, only voltages at sites in the infinite cluster need be considered, but one can also eliminate sites in 1D chains which dangle off the main body of the cluster, since these will all beat the same potential at their point of attachment. By the same argument, one can discard all sites which occur

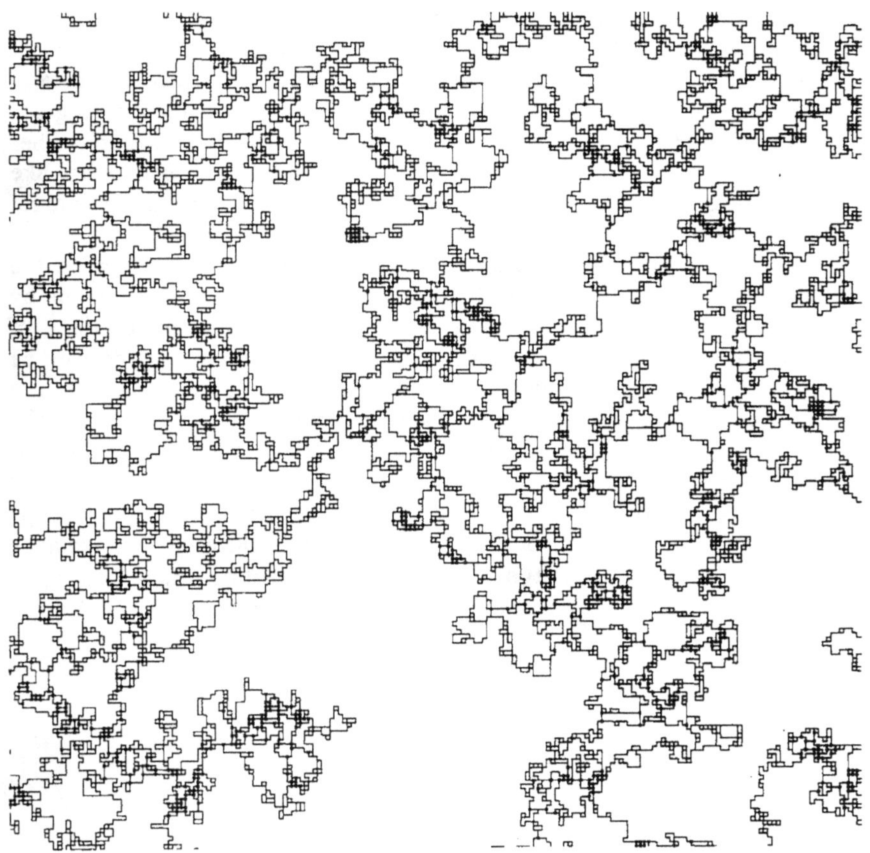

Fig. 15. Backbone (largest multiply-connected component) of the infinite cluster in a 2D bond percolation model of 200 × 200 sites with periodic boundary conditions. In this case (p = 0.514) 0.195 of the bonds and 0.319 of the sites were part of the backbone, yet the reductions described in the text left only 1165 sites (0.029 of the total) at which it was necessary to solve for the node voltage.

in more complicated configurations with a single attachment point, or articulation point, the removal of which will disconnect the small cluster from the larger one. To be precise, one need consider only the sites in the largest multiply-connected component of the infinite lattice, which I shall call the backbone.

There exists an O(N) algorithm for detecting articulation points and separating multiply-connected components in graphs, due to Tarjan [79]. It is an extension of the depth-first search procedure I described earlier for exploring clusters. The search sequence defines a numbering of the sites in the cluster. For each site one can develop information about how far back along the search tree that site can see by a path of bonds not already employed in the search. Articulation points can be detected during the backtracking part of the search as points whose horizon extends no further than themselves. Using a version of this algorithm, I find backbones like that shown in fig. 15, which are considerably sparser than the infinite cluster itself, and especially so close to p_c.

Even fig. 15 contains more sites than are necessary to determine the conductivity of the sample, so further local simplifications were made by eliminating node voltages and replacing simple local configurations of bonds by effective conductances. Thus, a 1D chain of bonds can be replaced by a single conductance. We need know only the voltages at the two ends. This process leads to cases where more than one bond links the same pair of nodes. These duplicate bonds are simply combined in parallel and further 1D reductions are usually then possible. Finally, configurations such as that shown in fig. 16a can be simplified by the star-triangle transformation (which encounters no difficulties for the linear conductance problem). The result, fig. 16b, has one fewer interaction and a more open structure, which usually permits further simplification.

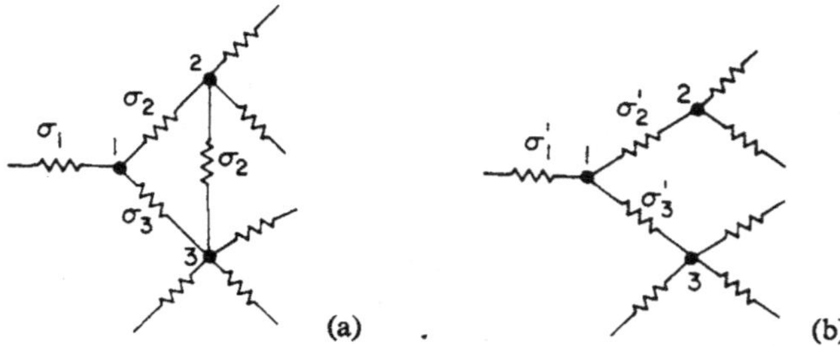

Fig. 16. Use of the star–triangle transformation to reduce the local complexity of a resistor network.

Table 1

p	Number of sites in infinite cluster	Number of sites in backbone	Number of sites in reduced network	Number of multiplications required	Storage (words) required
0.57	37,000	24,000	6,000	260,000	33,000
0.52	32,000	14,500	1,650	25,000	6,000
0.50	25,000	5,400	360	4,500	1,300

Table 1 summarizes my experience in applying these preconditioning transformations, followed by the MOOO subroutine package, to some bond percolation networks with 200 × 200 sites each.

The savings obtained were quite dramatic close to p_c. The same procedures were even more effective in 3D, perhaps because 3D backbones are sparser than 2D. For bond percolation samples with 50 × 50 × 50 sites, I was able to generate and solve each case in a total of $\approx$12 s at $p \approx 0.25$ and $\approx$20 s for $p = 0.27$. This made it possible to average over 40 or more samples per data point in figs. 14.

5. Fractal effects in magnets and conductors just above p_c

In the previous sections we developed quantitative information about percolation thresholds in general, as characterized by $\xi(p)$ and its exponent ν_p, and as they affect magnets and conductors, characterized by their respective exponents t. Now we can ask if these features can be accounted for, or related to one another, by simple geometrical models like those sketched in figs. 1a–c.

The channel picture of 1a asserts that exponents β and t are the same. If we think of the infinite cluster as essentially one-dimensional, and assume that the only effect of the disorder is to remove the isolated clusters, then the conductivity is reduced, at concentration p, by the factor

$$\sigma(p) = \sigma(1)P(p)\,, \tag{5.1}$$

which expresses a diminished cross-section for transport. Equation (5.1) disagrees hopelessly with experiments, which show $P(p)$ vanishing at threshold with infinite slope ($\beta < 1$) while $\sigma(p)$ vanishes with zero slope ($t > 1$). We might replace $P(p)$ by the fraction of sites in the backbone of the infinite cluster, $B(p)$, where the backbone was defined in the previous section as the largest multiply-connected component, since the dead ends or dangling material counted in P but not in B do not contribute to transport. However, $B(p)$ also vanishes with infinite slope at p_c [80] so this does not resolve the discrepancy.

What is missing from the channel picture is the microscopic length, $\xi(p)$, characterizing the fluctuations in connectivity. Even if conduction takes place in regions which look locally like 1D channels, these channels will intersect. Since the material must look homogeneous on scales $\gg\xi$, the intersections between channels cannot be much further apart than $\xi(p)$. The network picture (fig. 1b) builds in these intersections, called "nodes", which are assumed to be spaced more-or-less regularly at intervals of $\xi(p)$. Joining the nodes are links, assumed to be approximately 1D, of length L. Clearly L cannot be less than ξ, but an intriguing possibility is that $L(p)$ may diverge more rapidly than $\xi(p)$ as $p \to p_c^+$, the links becoming increasingly tortuous. Thus we define:

$$L(p) \propto (p - p_c)^{-\zeta}, \tag{5.2}$$

where

$$\zeta \geqslant \nu_p. \tag{5.3}$$

Calculations of various threshold properties within the nodes and links model is straightforward. To treat a dilute Ising magnet, we first replace each line of L interactions in series by an effective interaction, K_{eff}:

$$\tanh K_{\text{eff}} = (\tanh K)^L \sim (1 - 2\,\mathrm{e}^{-2K})^L \sim (1 - 2L\,\mathrm{e}^{-2K} + \cdots). \tag{5.4}$$

After this replacement, the magnetic system is roughly homogeneous, so we can set $K_{\text{eff}} = K_c$ for the pure system to identify $K_c(p)$:

$$\mathrm{e}^{-2K_c(p)} = \mathrm{e}^{-2J/kT_c(p)} = \text{const} \times L^{-1}(p). \tag{5.5}$$

Thus for Ising magnetic thresholds, the nodes and links picture predicts $t = \zeta \geqslant \nu_p$.

The conductance in this picture is estimated by joining each adjacent pair of nodes by a conductance $\sigma_0 L^{-1}$. The macroscopic conductivity of the resulting network is:

$$\sigma(p) \propto \sigma_0 L^{-1} \xi^{(2-d)} \propto (p - p_c)^{(d-2)\nu_p + \zeta}. \tag{5.6}$$

Using (5.3) we obtain the inequality

$$t \geqslant (d-1)\nu_p, \tag{5.7}$$

for the conduction exponent.

Finally, the density of material contained in the links joining the nodes is readily estimated, and gives an estimate for the backbone fraction, $B(p)$. Each cell of side ξ contains a finite number of links of length L. The density of material in the links is thus:

$$B(p) \sim L\xi^{-d} \propto (p - p_c)^{d\nu - \zeta} . \tag{5.8}$$

The predictions of this model are much closer to experiment than was (5.1), but there are still small discrepancies in 3D and serious errors in 2D. The network picture may give a correct description of percolation thresholds in 4–6D [80]. The magnetic threshold exponent we have seen is equal to 1 regardless of d [41]. Since in 2D $\nu_p = 1.365 \pm 0.015$ exceeds 1, this is inconsistent with the network prediction. The conductivity inequality (5.7) is also violated. In 2D, $t = 1.10 \pm 0.05$ is less than ν_p while in 3D only the error bars keep $t = 1.62 \pm 0.05$ from being strictly less than $2\nu_p = 1.69 \pm 0.03$. Equally accurate data for 4D are not available but one can estimate [80] that $3\nu_p \sim 2.0 \pm 0.15$ while $t \sim 2.2$ [44], so here the inequality is satisfied.

Inspecting the backbones of the infinite clusters found in large-sample simulations gives one a clear suggestion of what features are missing from the nodes and links picture. Figure 15 is a good example. You can also look at a series of side-by-side comparisons of infinite clusters and their backbones at various concentrations published in reference [80]. There is no evidence from the picture that any of the links in the network becomes violently tortuous as $p \to p_c^+$. Therefore one must assume that for the longest conducting links $L \approx \xi$. More important, although many regions of the backbone do look locally one-dimensional, the intersections usually come much closer together than $\xi(p)$, so that almost all of the 1D links are much shorter than $\xi(p)$. ξ characterizes the largest defects or longest links in the backbone, but not the typical ones.

What is present on scales shorter than ξ is a sort of sparse, graduated texture of conducting material. As the third simple picture of a conduction threshold (awkwardly sketched in fig. 1c), I will now try to characterize this texture in a way which permits calculation of various threshold properties.

First, the backbone is a vanishingly small fraction of the infinite cluster close to p_c. Numerical studies [80] show that $B(p)$ has a critical exponent $\beta_B > \beta_p$. The results were $\beta_B = 0.5$–0.6 (2D), 0.9 ± 0.1 (3D), and 1.1 ± 0.1 (4D). The contribution of the longest links to $B(p)$ will vanish at threshold with exponent $(d - 1)\nu$, which is roughly twice β_B in each case. Thus the longest links are in turn a vanishingly small part of the backbone close to p_c.

Since the backbone must be homogeneous on scales $\gg\xi$ we shall model it as made up of cells of linear dimension ξ, each cell different but subject to the same statistical characterization. Study of pictures of backbones suggests two simplifying features which we can build into the model. First, the texture appears to be self-similar. The coarse features of fig. 15 are roughly similar to the features of small portions of the same figure which have been enlarged. This self-similarity suggests that a single parameter may be used to characterize the texture on all length scales. It would be simplest if the parameter characterizing the texture did not depend upon $(p - p_c)$. Observation of a series of pictures of backbones in which (L/ξ) is kept constant by increasing the sample sizes as p tends towards p_c suggests that this is true. The coarse (scale $\approx \xi$) features of the backbone remain similar, while increasing both L and ξ together simply adds finer and finer details to the texture.

Geometrical objects, other than lines and planes, whose features are statistically self-similar from some maximum scale of interest down to some finest scale of possible resolution have been termed "scaling fractals" by Mandelbrot [81]. The term fractal is intended to indicate that the measure of such an object is a power of its linear dimension intermediate between its topological dimension (it must be made up of points, lines, etc.) and the dimension of the space in which it is embedded. The fractal dimension is in general not an integer. In a classic example, the coastline of Britain, measured on two maps which differ in resolution by a factor of two, proves to be more than twice as long on the higher resolution map because increased resolution permits new fine details to come into view. The process can be carried on for several decades of resolution. It is found that apparent length of coast $\propto$ (resolution)$^{-1.2}$. Mandelbrot therefore argues that the coast of Britain is really a curve with fractal dimension 1.2.

The backbone viewed on scales $\leqslant\xi$ is a fractal object. The largest dimension of interest, ξ, and the resolution, or lattice constant, a, are the two length scales in the problem. The measure of the backbone is the volume of material in a cell of side ξ, equal to $\xi^d B(p) \propto \xi^{d-\beta_B/\nu_p}$ when ξ is expressed in units of a. We identify the fractal dimension of the backbone, d_B,

$$d_B = d - \beta_B/\nu_p\,, \tag{5.9}$$

hence β_B/ν_p is the codimension, an index of the extent to which the backbone fails to fill space and of its increasing sparsity close to p_c. From the numerical results for β_B reported above, we have $d_B \approx 1.6$ (2D), ≈ 2.0 (3D).

This argument is quite general, and gives a geometrical interpretation to the density of any order parameter X, which vanishes at a conventional second order phase transition. Viewed on scales less than the appropriate correlation length, the regions where $\langle X(r) \rangle$ is instantaneously nonvanishing will form a fractal of codimension β_X/ν. Such textures are a natural consequence of the scaling, or RG picture of critical phenomena, because this picture distinguishes the length scales $\lesssim \xi$ over which correlation functions decay algebraically, as determined by the immediate neighborhood of some unstable fixed point of a rescaling transformation, from lengths $\gg \xi$ where correlations decay exponentially, and behavior is controlled by some stable homogeneous fixed point.

[A digression: You will find in the literature [82, 83] the statement that $d_p = \gamma_p/\nu_p$ for percolation. By the scaling law $\gamma = d\nu - 2\beta$, this would imply $d_p = d - 2\beta_p/\nu_p$, which is in disagreement with (5.9). This assignment is based on the idea that the appropriate measure for percolation is the mean cluster size, which diverges as $\zeta^{\gamma/\nu}$ at threshold. However, this quantity includes all finite clusters as well as some largest cluster which will become the infinite cluster above p_c. Stanley notes that a different fractal dimension can be extracted by expressing the mean size of the largest cluster at p_c as a power of ξ, since Stauffer (private communication) has shown by droplet model arguments that this diverges with a different exponent. Employing Stauffer's Ansatz, Stanley obtains a "$d_p^{\dagger}$" which can be shown to be identical to ours. This modification recovers the result that the codimension $= \beta_X/\nu$.

Which dimensionality gives the correct geometrical insight will depend on the property one is studying. I feel that the definition and line of arguments employed here are the more general, because they are not restricted to p_c. They also have the advantage of not requiring the use of the droplet model Ansatz.]

One possible construction of a random fractal multiply-connected backbone with noninteger dimensionality is shown in fig. 17. (This construction has an obvious family resemblance to the curdled fractals exhibited in plates 156–159 of Mandelbrot [81], and discussed on pp. 169–173 of that work.) The black lines should be thought of as conducting paths with some constant resistance per unit length. The figure represents a cell of dimensions $\xi(p)$. Viewed on the largest scale some part of the backbone is defective, and contains no further connections. To see what happens in the fraction, f, of the material which has further connections we shall have to increase the resolution. In fig. 17, $f = \frac{3}{4}$ for convenience in drawing. Viewed on a scale two times finer, we again assume that $(1 - f)$ of the regions of dimensions $\frac{1}{2}\xi$ arranged at random are defects, while the rest will contain finer scale connections. The procedure continues, using the same parameter f at each stage to yield self-similarity, until the scale considered is equal to the lattice constant.

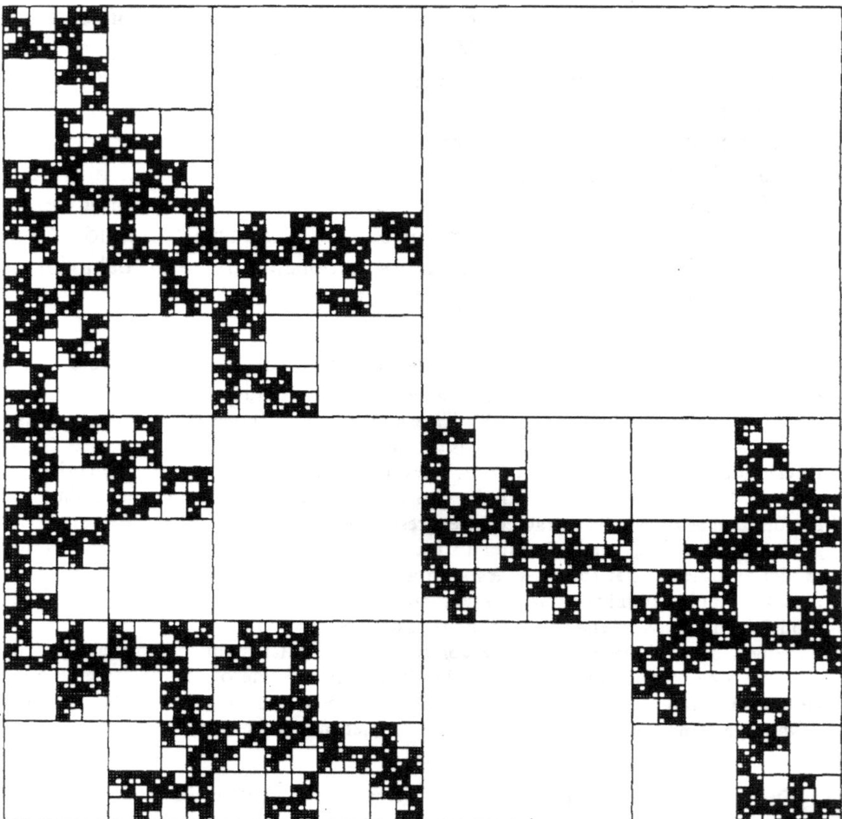

Fig. 17. An alternative to the homogeneous "nodes and links" picture of the backbone. The construction is described in the text. Current-carrying material is indicated by the black lines, the white space representing insulating, isolated, or "dead end" material.

To fix the parameter f, and to relate it to the fractal dimensionality of the backbone, we must calculate $B(p)$ for the model of fig. 17. The amount of material in a cell of side ξ can be evaluated by summing contributions from each scale. The longest links contribute an amount proportional to ξ. Of the 2^d cells of side $\xi/2$ a fraction f will contribute to B, while f^2 of the 2^{2d} cells of side $\xi/4$ will contribute, and so forth. There will be $\ln_2 \xi(p)$ scales to be summed over. The final result must be normalized to the volume ξ^d.

We obtain

$$B(p) \sim [\text{const} \times \xi^{1-d}][1 + 2^{d-1}f + (2^{d-1}f)^2 + \cdots + (2^{d-1}f)^{\ln_2 \xi}]\,. \tag{5.10}$$

If $(2^{d-1}f) > 1$ the geometric series is divergent, and dominated by its last term.

$$B(p) \propto \xi^{1-d}(2^{d-1}f)^{\ln_2 f} = \xi^{\ln_2 f} \propto (p - p_c)^{|\ln_2 f|/\nu_p}\,, \tag{5.11}$$

and since $|\ln_2 f|$ proves to be the codimension,

$$|\ln_2 f| \equiv \beta_B/\nu_p\,, \tag{5.12}$$

we observe that $(2^{d-1}f) = 2^{d_B-1}$. Since the backbone is connected, $d_B > 1$ (otherwise it would have to be a set of disconnected pieces). Thus our assumption that $(2^{d-1}f) > 1$ is consistent. Knowing β_B and ν_p from computer experiments we have $f \approx \frac{3}{4}$ (2D) and $\approx \frac{1}{2}$ (3D).

I do not know how to calculate the conductivity of fig. 17 or its 3D analogs exactly but will estimate it by a real-space RG calculation. (For a discussion of some inequalities which the conductivity of the fractal backbone will satisfy, which agree with experiment, see Kirkpatrick [80].) As in the discussion of RG's for dilute magnets and conductors, we divide the backbone, viewed on the microscopic or lattice constant scale into regions with non-zero conductance and regions of zero conductance (material not in the backbone). Consider only the non-zero conductances, σ_0. These will occur in groups of three conducting and one non-conducting cells (fig. 18). By an elementary calculation, such a group of 4 cells has a conductance $0.66\,\sigma_0$. Repeating the process with 4 large cells, three of them with

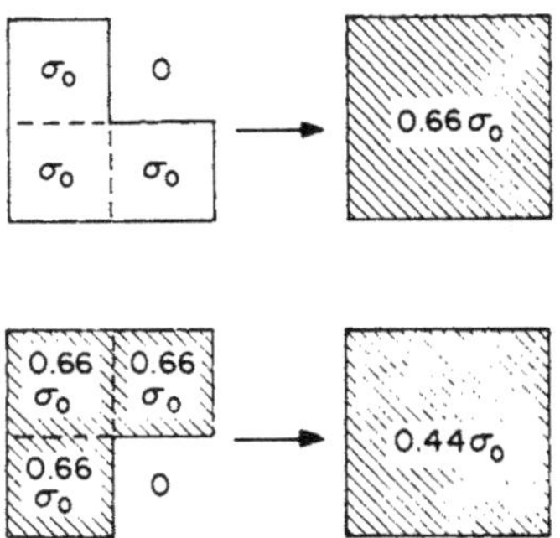

Fig. 18. Iterative construction for calculating the conductance of the fractal backbone in fig. 17.

effective conductance 0.66 σ_0 gives $\sigma = 0.44\,\sigma_0$ (fig. 18). Continuing in this fashion we obtain

$$\sigma(p)/\sigma(1) \approx 0.66^{\ln_2 \xi} = \xi^{\ln_2 0.66} \propto (p - p_c)^{\sim 0.9}, \tag{5.13}$$

which is reasonably close to $t = 1.1$.

A similar construction can be made in 3D. A cube of 8 subcells, half of which have conductivity σ_0 and the other half zero, will have a conductivity which is approximately $\frac{9}{32}\sigma_0$, once one averages over the positions of the conducting subcells. Thus in 3D we predict

$$\sigma(p)/\sigma(1) \equiv \xi^{\ln_2(9/32)} \propto (p - p_c)^{\sim 1.6}, \tag{5.14}$$

in good agreement with the computer results. I conclude, therefore, that the concept of the backbone as made up of fractal textures is not only a natural hypothesis, but also quantitatively justified.

Before concluding this section, I will discuss one more transport property, the Hall coefficient, R_H. The Leningrad group has predicted, from a nodes-and-links picture, that R_H diverges weakly as $p \to p_c^+$ in 3D [84,85]. Expressing

$$R_H = E_z/j_x H_z, \tag{5.15}$$

they argue that $j_x \propto \xi(p)^{(1-d)}$, while ξ_z will be reduced by the ratio of the microscopic cross section, a, of a conducting link to ξ. Thus they predict

$$R_H \propto \xi^{d-2}. \tag{5.16}$$

This gives R_H = const in 2D, as it must, and $R_H \propto \xi(p)$ in 3D.

Computer simulations by Swenumson et al. [86] and by Webman et al. [71] were not sufficiently accurate to test this prediction, as they were severely limited by sample size and geometry. Analog studies by Levinshtein et al. [84] were more successful. Rather than attempt to determine a critical exponent for R_H, they studied the quantity $R^2\sigma$, and found that it did tend to a constant, rather than to zero or infinity as $p \to p_c^+$. (An interesting feature of this analog investigation was that many samples were obtained by varying the stacking sequence of some 15 sheets of conducting paper, each with holes punched in it at random.)

The texture picture preserves the result $R^2\sigma \to$ constant, at least if we take a fairly naive interpretation of the consequences of texture. If we view the texture as adding to each link in the network picture a cross-sectional area A which diverges as $p \to p_c^+$, then the same arguments give $j_x(p) \propto$

$A(p)\xi(p)^{1-d}$, and $E_z(p) \propto A^{1/(d-1)}/\xi(p)$. In 2D, R_H = const again, while in 3D,

$$R_H = A^{-1/2}(p)\xi(p)\,. \tag{5.17}$$

The predicted divergence is less rapid, but the quantity $R^2\sigma \to$ const nonetheless:

$$R_H^2\sigma \propto (A^{-1/2}\xi)^2 A\xi^{-2} \to \text{const}\,. \tag{5.18}$$

More accurate study of $R_H(p)$, either by analog methods or by numerical real space rescaling, would seem to be desirable.

6. Systems with frustration or interference

In this section I will discuss localization and spin glasses. I have asserted at several points that Anderson localization will occur before any mixture of metallic and insulating material can reach its classical percolation threshold. Several approaches to the study of this crossover will be considered next. None of these has been pursued very far. The lectures by Anderson and by Thouless contain a broader discussion. I shall have more to say about spin glasses, since many of the speculations which are current about the microscopic causes of the spin glass state can only be resolved by recourse to explicit models. Also, given the incomplete and inconsistent state of present theories, it should be useful to review what is presently known about model systems.

6.1. Percolation–localization crossovers

First consider a system with a random potential, $V(r)$, in which we can consider the position of the Fermi level, E_F, to be an experimentally adjustable parameter (e.g., a Si inversion layer). Start with a very low value of E_F, such that only a very small fraction, p, of the system is classically allowed. Increasing E_F, we reach a critical value, E_c, at which $p(E_c) = p_c$, the threshold volume fraction. In a semi-classical approximation, this defines the mobility edge. When E_F is below E_c, the size of the classically-allowed regions should vary as $\xi_p \propto (E_c - E_F)^{-\nu_p}$. Quantum-mechanical effects modify this picture by shifting E_c to some new value, E_c^*, and may cause a crossover from classical to quantum-mechanical behavior as E_c^* is approached. If one assumes [87] that there is also a characteristic length $\xi_{QM} \propto (E_c^* - E_F)^{-\nu_{QM}}$, and uses the simplest "self-avoiding walk" estimate for $\nu_{QM} = 3/(d + 2)$ (Thouless' lectures), then $\nu_p < \nu_{QM}$ for $d = 2, 3, 4$. Thus there must be some crossover energy at which the two lengths are equal, and above which $\xi_{QM} \gg \xi_p$. To estimate $E_c^* - E_c$ is more difficult, since there are two competing effects. The possibility of tunnelling through potential barriers decreases E_c^*, while the kinetic energy cost of confining an electron in narrow quarters reduces the electron density near saddle points of the potential, increasing E_c^*. Numerical studies of model systems have usually shown the mobility edges to occur at energies where the density of states is rather large [88], so apparently the second effect is the dominant one.

One appealing direction in which to pursue these questions would be to construct a limiting process in which the quantum mechanical effects can be added as a weak perturbation to the semiclassical picture, for example, by letting $\hbar \to 0$ [89]. This limit, unfortunately, is a very delicate one. Even though both tunnelling and scattering of wave functions decrease with the ratio $\hbar^2/2m$, the minimum metallic conductivity, $\sigma_{min} = e^2/2\pi\hbar$, diverges as $\hbar \to 0$, and does not have an obvious semi-classical limit.

A model hamiltonian which permits more detailed analysis of the percolation to localization crossover is:

$$\mathscr{H} = \sum_i \varepsilon_i c_i^\dagger c_i + \sum_{\langle ij\rangle} t(c_i^\dagger c_j + c_j^\dagger c_i)\,, \tag{6.1}$$

where the random diagonal elements $\varepsilon_i = 0$ on a fraction, p, of "allowed sites", and $= \infty$ on the remaining, "forbidden sites". For the pure system, with all sites allowed, (6.1) gives rise to a tight-binding band of width no greater than $2zt$, centered on zero energy. For $p < 1$, this band gets narrower, and contains only p states per site. If we restrict the hopping matrix elements, t, to connect only nearest neighbors, the model has a percolation threshold at p_c for site percolation on the same lattice. This model was introduced by de Gennes et al. [90,91] at roughly the same time as Anderson's original paper [59], but has received little attention. We shall see that it possesses Anderson localization in addition to the evident percolation threshold.

The model (6.1) has some pathological features. As many as 0.1 to 0.15 states per site are found to be localized with energies precisely equal to 0, $\pm t$, and similar special values [92]. These give rise to delta function components of the density of states in addition to the continuum. The states are essentially the eigenstates of isolated clusters of A atoms. However, cluster states of high symmetry, in particular those with $E = 0$, have many nodes in their wavefunctions. When there are nodes at the edges of the cluster the state remains an eigenstate even when the bounding sites are A atoms rather than B's. Thus some of the same highly localized eigenstates are found in odd corners of the infinite cluster as are present in finite isolated clusters. At moderate concentrations, these anomalous localized states may account for roughly half the states in the spiky portion of the spectrum.

One does not expect to find localized and extended states occupying the same volume at the same energy, since they should mix and produce resonances. Instead, in this model there is numerical evidence for narrow dips in the continuum density of states around the special frequencies. A plausible interpretation is that the density of states tends to zero there just as it approaches zero in the usual band tail at the edge of a band. Confirming this interpretation, one finds that the continuum states which lie closest to the special frequencies are localized. As p decreases from unity, therefore, Anderson transitions will occur in various parts of the band through the merging of pairs of internal mobility edges [92].

To study Anderson localization in the continuum states of this model we must examine states relatively far from any special energies. In figs. 19, I have used Thouless' method [133] (described more fully in his lectures) of studying changes in the energy of the states under changes in boundary

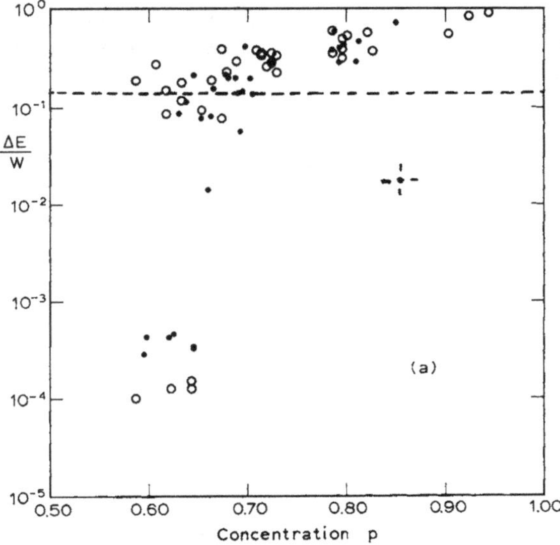

Fig. 19a.

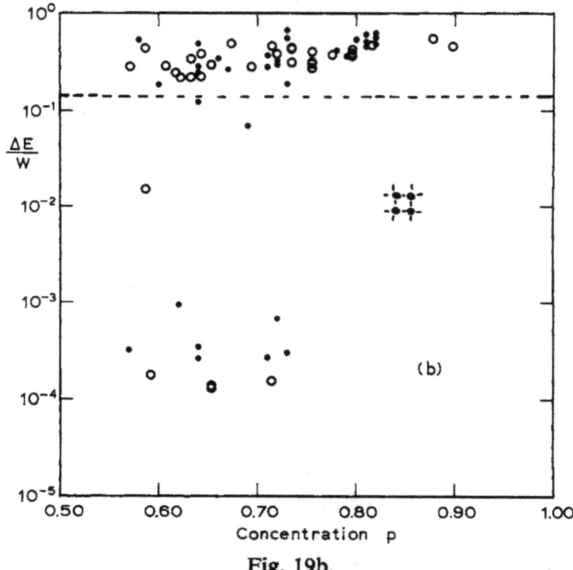

Fig. 19b.

Fig. 19. Ratio of the average energy shift upon changing from periodic to anti-periodic boundary conditions to the average spacing between levels (averaging over 20 states with energies between t and $2t$) for the dilute tight binding hamiltonian (6.1). Sample sizes are 14 × 14 (circles) and 20 × 20 (dots). In (a), individual sites were present with probability p. In (b), squares of four adjacent sites were simultaneously either present or absent.

conditions, using states close to $E = 1.5t$ in the 2D version of (6.1). The ratio of the energy shift produced to the spacing between levels (the vertical axis in figs. 19) is proportional to the conductivity which would be obtained if the Fermi energy were at E. The dashed line in each figure is the ratio corresponding to the minimum metallic conductivity for a 2D system.

In fig. 19a, the conductivity is seen to decrease smoothly as the fraction of allowed sites is decreased from 1 to roughly 0.75. Below this, the predicted conductivity becomes comparable to $\sigma_{\min}$, and there is evidence for the onset of localization. The data points become scattered, and the conductivity appears to decrease steadily with increasing sample size (the solid points lie consistently below the circles). In fig. 19b, I modified the statistics of the model by letting groups of four adjacent atoms be either allowed or forbidden. This reduces the kinetic energy cost of forcing an electron through a narrow channel, and should push the quantum mechanical crossover closer to the percolation threshold. Since the samples have fewer statistical degrees of freedom, the scatter is greater in fig. 19b than in fig. 19a, but the rate of decrease of the conductivity does appear to be less and the mobility edge, as identified by the appearance of size dependence in σ, still comes when $\sigma \approx \sigma_{\min}$.

The calculations in figs. 19 were performed using a much less efficient program than the Lanczos algorithm of Edwards et al. [21] so relatively poor quantitative accuracy was obtained. Since this case is complementary to the gaussian limit studied by Thouless and Elzain [88], in which the scatterers are weak and uncorrelated over distances comparable to the wavelengths of the unperturbed electronic states, more accurate calculations would be desirable. It would be of particular interest to extend these calculations to 3D, to determine whether $\sigma_{\min}$ remains proportional to the Fermi wave-vector, or becomes inversely proportional to the coherence length for the percolation fluctuations.

6.2. Spin glasses – general discussion

Recent experimental work on spin glasses has tended to blur, rather than clarify our picture of the essential features of the spin glass state. While measurements at very low (dc) fields on amorphous intermetallic films [93,94] have nicely supported the original Cannella and Mydosh [95] observation of a sharp transition from paramagnet to spin glass, the latest (unpublished) data from the Grenoble group, using the classic materials and ac techniques, show a broader transition which shifts with frequency. One of the earliest predictions of theory in spin glasses is a phase diagram with paramagnetic, ferromagnetic, and spin glass phases, all separated by lines of continuous transitions. Experimentalists have obligingly found alloy systems which should exhibit all three phases as composition and temperature are varied, but they find it extremely difficult to distinguish the low temperature magnetic properties of the highly disordered ferromagnet on one side of the proposed phase boundary from those of the nearly ferromagnetic spin glass on the other (see Heiman and Lee [96], McGuire et al. [97], Sarkissian and Coles [98]). As a result, I shall have to discuss quite a number of different models of possible spin glasses, with the hope of determining what microscopic details make a difference in macroscopic properties of the materials, and perhaps of providing sharper questions for experimental test.

It is sufficient to consider only models with random exchange interactions [102],

$$\mathscr{H} = \sum_{\langle ij \rangle} J_{ij} \boldsymbol{S}_i \cdot \boldsymbol{S}_j \,, \tag{6.2}$$

where the spins may have either Ising or Heisenberg symmetry. Although there are certainly systems in which random local anisotropy fields should play a role, we showed in sect. 3 that such systems will look Ising-like on a sufficiently large length scale. One could conceive of models with spins confined to orientations in a plane [99], but these seem rather unphysical, since the disorder which gives rise to the random exchange is also likely to cause easy axes in the plane of spin orientations, and thus a crossover to Ising behavior. In fact, almost all the study of concrete models to date has been done on Ising systems, but see Walker and Walstedt [100] for a rather realistic simulation of a classic RKKY spin glass alloy.

Suppose that we let the sum in (6.2) run over all i and all j, so that each spin is coupled to the $N - 1$ other spins. This procedure will make sense only if at the same time we weaken each individual interaction so that the sum of all exchange interactions from a given spin remains extensive on average. Since this average might vanish, fluctuations about the mean interaction strength must also be appropriately scaled. To do this we define

$$\tilde{J}_0 \equiv N\langle J_{ij}\rangle\,, \qquad \tilde{J} \equiv N\langle J_{ij}^2 - \langle J_{ij}\rangle^2\rangle^{1/2}\,. \tag{6.3a,b}$$

This effectively infinite-ranged model was introduced (Sherrington and Kirkpatrick [101], hereafter abbreviated SK) in the hope that it would prove soluble, and that the solution could shed light on the validity of the mean field theory of Edwards and Anderson [102], just as the infinite-ranged pure ferromagnetic model ($\tilde{J} = 0$) is soluble, and its solution proves identical to the Weiss mean field theory.

The model (6.3) is not trivial, since (6.3b) requires the variance of the individual interactions to increase without limit, with respect to the mean, as $N \to \infty$, and no complete solution is known. However, it is clear beyond any reasonable doubt (Thouless et al. [103], hereafter TAP, and Kirkpatrick and Sherrington [104]) that the model has a phase transition when

$$nkT_c = \max\{\tilde{J}_0, \tilde{J}\}\,, \tag{6.4}$$

where n is the number of spin components. The low temperature phase which is formed when $\tilde{J} > \tilde{J}_0$ has many novel properties and is not well understood at present. Anomalous slow relaxation phenomena are observed in the infinite-ranged model at all temperatures below T_c. Instead of one ground state, the model has many states of almost equally low energy, and an unusual class of what appear to be localized low-energy excitations. Both properties are discussed below in more detail. The value of the infinite-ranged models lies in the fact that these novel effects are also seen in more realistic models and appear to be experimental characteristics of actual spin glasses.

For more realistic models of spin glasses, the evidence for the existence of phase transition from conventional phases into a spin glass phase is much less clear. The question of when and why a spin glass phase cannot occur is the central theme of much recent work on the problem. This literature is confusing to the uninitiated because several general mechanisms for the destruction of spin glass order have been proposed, and they are not always clearly distinguished.

The upper critical dimension for spin glasses seems to be 6. Calculation schemes based upon replicas (Anderson, these lectures) suggest this strongly, since the resulting operators can form cubic invariants, just as occur in the Potts model treatment of percolation theory, for which the ucd. is 6. Analysis of a high temperature expansion for a generalized susceptibility [105],

$$\chi_{sg} = \sum_{ij} \overline{\langle S_i S_j \rangle^2}\,, \tag{6.5}$$

confirms this identification. χ_{sg} diverges as $|T - T_c|^{-\gamma}$ with $\gamma \approx 1$ at and above 6D. Since Fisch and Harris obtained the terms of their power series for (6.5) as explicit functions of d (on hypercubic lattices), they were able to study $\gamma(d)$ as d decreased continuously from 6D to 4D. They found that γ increases with decreasing d, and $\to\infty$ as $d \to 4^+$, but that the transition temperature remained finite at 4D. Below 4D, they were not able to analyze the series. Their calculation clearly indicates a change in the nature of the spin glass state, but does not give any prediction of the presence or nature of a spin glass phase below 4D.

Essentially the same behavior is seen between 6D and 4D in the recent extension to the replica theory carried out by Bray and Moore [106]. However, there are doubts about the internal consistency of the replica theory, especially below 6D [107,108], so I find the series work more convincing at the moment.

The identification of the lower critical dimension for spin glass order, whether 4D or lower (if spin glasses with different properties can exist in 3D or 2D), is quite controversial at present. There are two paradigms for the effects associated with the lower critical dimensionality in conventional phase transitions, neither of which is quite appropriate for thinking about possible spin glass behavior below 4D. First, long ranged order can be eliminated by divergent contributions from long wavelength spin fluctuations. This sort of infrared divergence sets the lcd of most pure systems with a continuous order parameter. Alternately, short wavelength fluctuations resulting from localized defects may occur with sufficiently little energy cost that entropy considerations favor their presence at all non-zero temperatures. The second mechanism may destroy long ranged order, as is the case in the 1D Ising model, or may introduce more subtle phenomena, as occurs in the 2D *xy* model. In the *xy* model, one class of defects (bound vortex pairs) is present at all temperatures, so that there can be no long ranged order of the conventional sort, but a second class of defects (free vortices) appears at a finite temperature to give rise to the observed phase transition [34,109,110].

There are difficulties in applying either paradigm to spin glasses, for these have many ground states, and the magnetic order is hidden. It is not clear whether long wavelength spin fluctuations or possible defects will merely take a spin glass from one of its ground states to another, or will destroy some of the ordering that is common to all the ground states. Recent attempts to describe the hidden order in a spin glass include the introduction of the concept of frustration [111] and the construction of gauge-invariant descriptions of simple models with weakly broken local symmetries [12,112]. These approaches are elegant and promising, and will be discussed below, but concrete results from them are still disappointingly few.

There is evidence for the existence of spin glass ordered phases in 2D and 3D from both real-space RG and Monte Carlo simulations. I will discuss the computer simulations in more detail below. The simplest RG argument tests the rigidity of the arrangement of spins in a macroscopic sample [113,114]: Consider a block of d-dimensional magnetic material, L lattice constants on a side, at a very low temperature. For a uniform system, the energy cost of rotating the spins on one face of the sample through some small angle, while holding those on the opposite face fixed will be proportional to the cross-sectional area, L^{d-1}, and inversely proportional to the length, L, since the spin deviations can relax over this length. Since the rigidity energy $\Delta E \propto L^{d-2}$, we see that the existence of an ordered state of a system with continuous spins at any finite temperature is problematical in 2D, and identify the lcd as 2. For Ising systems, one calculates ΔE for reversal of the spins on one face of the sample. The resulting interface energy is still proportional to the sample cross-section, but is independent of the length, since the interface is only one lattice constant thick. The resulting $\Delta E \propto L^{d-1}$, so the lcd for Ising spins is identified as 1.

When the exchange is random in sign, it is reasonable (although unproven) to assume that ΔE will be proportional to the square root of the sample cross-section. Thus for continuous spins, $\Delta E \propto L^{(d-1)/2-1}$, and the lcd = 3. If the interface energy is independent of sample length in the Ising case, this argument still gives lcd = 1. However, it is more likely that increasing the sample length while keeping the cross-section fixed will permit the interface to reduce ΔE by spreading out in some convoluted fashion. If the resulting relaxed ΔE depends upon a power of the length, any lcd between 1 and 3 is possible. This argument is essentially the Migdal construction for a large scale factor, L. Its weak point is that summing the contribution of all bonds which act in parallel in a given cross-section of the sample can introduce spurious cancellations between interactions of opposite sign which can in fact be simultaneously satisfied.

More detailed real-space RG calculations have been performed for Ising models, some of them using decimation approximations which respect frustration and thus do not introduce spurious cancellations. All of these extend the description (3.4) by considering distributions $\tilde{g}(J)$ of non-zero interactions containing both ferromagnetic and antiferromagnetic bonds. A spin glass low temperature fixed point is then identified as one at which $\langle J \rangle = 0$ while $\langle J^2 \rangle \to \infty$ upon repeated rescaling. Young and Stinchcombe [115] applied the simplest possible 2D decimation (that sketched in fig. 13a) to the Ising model, replacing the actual distribution of transformed interactions by delta functions at $\pm J_{\text{eff}}$ and 0 at each iteration. Cancellations mapped some of the interactions into the delta function at $J_{\text{eff}} = 0$ at each stage, so that only a paramagnetic phase was found. However, a similar procedure, applied in 3D [116], yields a spin glass fixed point with the expected properties. Jayaprakash et al. [117] used the Migdal recursion relation to study models with varying fractions of antiferromagnetic bonds, restricting the distribution of interactions to delta functions at 0, $\pm J_{\text{eff}}$. They obtained phase diagrams with a spin glass phase in addition to the usual ferro-, antiferro-, and paramagnetic phases. However, since their choice of scale factor, $b = 3$, artificially eliminates cancellations, the calculation probably overestimates the stability of the spin glass fixed point.

A more important limitation of all the above calculations is the fact that considering only one non-zero exchange interaction strength neglects weaker correlations which may be essential to the spin glass state. A simple example shows how such correlations arise. Consider the configuration in fig. 20. We are interested in possible correlations between spins 1 and 2, and will therefore trace over spin 3 and combine the resulting effective interaction with the direct interaction between 1 and 2 to make a single

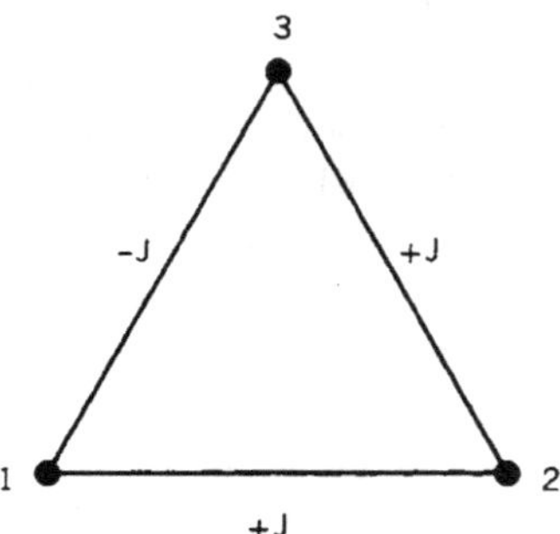

Fig. 20. Frustrated configuration of Ising spins in which there are, nonetheless, equilibrium correlations between pairs of spins.

effective bond between 1 and 2. The (unnormalized) density matrix for spins 1 and 2 is

$$\begin{aligned}\rho(1, 2) &= \mathrm{Tr}_3(1 - \tanh K\sigma_1\sigma_3)(1 + \tanh K\sigma_3\sigma_2)(1 + \tanh K\sigma_1\sigma_2) \\ &= (1 - \tanh^2 K\sigma_1\sigma_2)(1 + \tanh K\sigma_1\sigma_2)\,. \end{aligned} \tag{6.6}$$

At low temperatures, $\tanh K \approx \tanh^2 K \approx 1$, and the interactions along the two branches of fig. 20 are in opposition, but it would be a mistake to treat them as exactly cancelled. After expanding (6.6),

$$\rho(1, 2) = 1 - \tanh^3 K + \tanh K(1 - \tanh K)\sigma_1\sigma_2\,, \tag{6.7}$$

we can factor $(1 - \tanh K)$ out of both terms. In the limit $T \to 0$ this term corrects the free energy of the cluster for the fact that one bond of the three must always be unsatisfied. What is left, after normalizing so that the constant term is unity, is

$$\rho(1, 2) = 1 + \frac{\tanh K}{1 + \tanh K + \tanh^2 K}\sigma_1\sigma_2\,, \tag{6.8}$$

which has the low temperature limit,

$$\rho(1, 2) \sim 1 + \tfrac{1}{3}\sigma_1\sigma_2\,. \tag{6.9}$$

Eq. (6.9) shows there is a residual interaction $J_{\mathrm{eff}}(1, 2)$ between the spins, but that $J_{\mathrm{eff}}(1, 2) \propto T$ as $T \to 0$. An approximation which neglects such weaker interactions will also miss correlations such as that given by (6.9):

$$\langle\sigma_1\sigma_2\rangle \sim \tfrac{1}{3}\,. \tag{6.10}$$

A fairly systematic real-space RG calculation which confronts this problem is given in Kirkpatrick [33]. There the Migdal infinitesimal rescaling is applied to the actual invariant distribution of interactions, $g^*(J_{\text{eff}})$ obtained at low temperatures. The effects of the infinitesimal transformations on $\langle J_{\text{eff}}^2 \rangle$ were

$$\mathscr{L}_{\parallel}[\langle J_{\text{eff}}^2 \rangle] = +\langle J_{\text{eff}}^2 \rangle , \tag{6.11a}$$

$$\mathscr{L}_{\text{series}}[\langle J_{\text{eff}}^2 \rangle] \approx -1.5 \langle J_{\text{eff}}^2 \rangle , \tag{6.11b}$$

so, using (2.21),

$$\mathscr{L}[\langle J_{\text{eff}}^2 \rangle] \approx (d - 2.5) \langle J_{\text{eff}}^2 \rangle . \tag{6.12}$$

If $d = 2$, rescaling always weakens $\langle J_{\text{eff}}^2 \rangle$, but if $d = 3$, there is a spin glass phase near $T = 0$. In this approximation, the lcd for Ising spin glasses is 2.5.

One can level two criticisms at the RG calculations. First, all employ decimation, so that the spins remaining after each rescaling are the bare spins. A block spin transformation in which new spins, introduced at each step, reflect the most rigidly aligned local configurations, might give a better description of the spin glass ground states. The second, and more serious objection is that such calculations can predict infrared divergences in which local spin fluctuations renormalize long-ranged stiffnesses, but not the effects of extended defects. For example, Migdal decimation, applied to the pure 2D *xy* model, describes the renormalization of correlations by spin waves exactly, but treats effects of vortices only approximately [34].

Vannimenus and Toulouse [118] have proposed that domain walls act as low energy defects to bring about the transition with increasing disorder at low temperatures from ferromagnetic to spin glass. A domain wall between two regions of spins may cost a low energy when one region is reversed because the net energy change is the difference between the strengths of the bonds which were satisfied before the reversal and those which are satisfied after. They further proposed [119] that changes in the nature of the low energy interfaces, such as roughening, could provide a mechanism for a phase transition to the paramagnetic state with increasing temperature. For pure Ising ferromagnets, interface roughening occurs at zero temperature in 2D, but at a finite temperature below T_c in 3D. Toulouse et al. speculated by analogy that Ising spin glasses would have a finite T_{sg} in 3D, but that $T_{sg} = 0$ in 2D.

Bray and Moore [120] have argued that the presence of low energy boundaries implies the absence of a phase transition, or at least of the conventional sort of long ranged order. The computer simulations which they present in support of this view are discussed in the next section. They also give a general argument for the absence of order, stating that [121] for every configuration in which two spins separated by such an interface are parallel, there will be another configuration of equal energy in which they are antiparallel. Thus they claim that in equilibrium such correlations will vanish. The loophole in this argument should be evident from fig. 20. One can draw a zero energy interface between spins 1 and 2 in several ways. Each such interface leaves one bond unsatisfied in one chain and all bonds satisfied in the other. Since there are two bonds which can be broken in the upper chain, and only one in the lower chain, the ferromagnetic alignment favored by the lower bond is twice as likely. Hence the observed non-vanishing $\langle\sigma_1\sigma_2\rangle$. The differing degeneracy of the two states of an interface can be thought of as an interface entropy [122]. Periodic models can be constructed which have zero energy contours with this sort of associated entropy difference. They are found to have phase transitions at finite temperatures [123], in contradiction to the Bray–Moore argument.

It is appealing to speculate that the spin glass phase in 2D and 3D has a line of critical points instead of a conventional unstable fixed point [124], since such a phase diagram is consistent with the experimental observation that observable properties, such as the induced magnetization, decay to their equilibrium values with non-exponential time dependences below T_c. For a system with a more conventional low temperature fixed point structure, one would expect exponential decays. That speculation is equivalent to proposing that on all sufficiently large scales, only incomplete correlations of the type shown in fig. 20 survive. This kind of scaling will imply power law decay of all spin correlations. If this is the correct low-temperature behavior of a spin glass in 2D or 3D, it seems quite unlikely that any of the familiar real space RG schemes will describe it accurately.

6.3. Spin glasses – computer simulations

The natural system to consider first is the infinite-ranged Ising spin glass defined in (6.3), since the transition temperature (6.4) is well established, and there are the theories of SK and TAP with which to compare the results. I will discuss the time decay of spin correlations under non-conserving, independent spin–flip processes, since this is a reasonably physical model and is readily implemented on the computer. Take the microscopic time constant for spin–flip attempts (due to interactions with some kind of thermal fluctuations) to be unity. Under fairly weak assumptions, such as detailed balance, one can derive master equations for the time evolution of any initial configuration of spins, and from them obtain simple kinetic equations for the decay of any correlation function (see Suzuki and Kubo [125] for a clear discussion). When $T > T_c$, such equations take the form

$$d/dt\langle S_i S_i(t)\rangle = -\langle S_i S_i(t)\rangle + \langle S_i \beta h_i(t)\rangle , \tag{6.13}$$

where the first term describes the incoherent decay of the autocorrelation, and the second contains the restoring effect of a molecular field, $h_i(t)$, due to the other spins ($\beta \equiv 1/kT$).

Some care is required in constructing the appropriate reorienting field. Before calculating the interaction of spin S_i with its neighbors, we must first subtract off that part of the expectation value $\langle S_j(t)\rangle$ which was induced by $\langle S_i(t)\rangle$ and will therefore reverse if S_i reverses [126]. The result is

$$h_i(t) = \sum_j J_{ij}[S_j(t) - \chi_{ij} S_i(t)] . \tag{6.14}$$

When (6.14) is substituted into (6.13) the two terms each make contributions of order unity. If we use the high-temperature form for the susceptibility, $\chi_{ij} = \beta J_{ij}$, and specialize to the case $\tilde{J}_0 = 0$, the second term is $-\beta\tilde{J}^2 S_i(t)$ plus fluctuations which will be $O(1/N)$ and can be disregarded. The kinetic equation now becomes

$$(1 + \beta^2\tilde{J}^2 + d/dt)\langle S_i S_i(t)\rangle = -\beta\sum_j J_{ij}\langle S_i S_j(t)\rangle , \tag{6.15}$$

and can be solved by expanding formally in terms of eigenfunctions of the random matrix whose ijth element is J_{ij} [127]. The spectrum of this matrix is known to extend from $-2\tilde{J}$ to $+2\tilde{J}$, so the most slowly decaying modes will show an effective time constant

$$\tau_{\text{eff}} = (1 - \beta\tilde{J})^2 \propto (T - T_{\text{sg}})^{-2} . \tag{6.16}$$

The critical point is reached when this mode no longer decays exponentially, at $\beta\tilde{J} = 1$. Notice that without the second term in (6.14) we would have overestimated T_{sg} by a factor of 2. Solving for $\langle S_i\rangle$ in equilibrium, using the molecular field (6.14) leads to the random mean field equations derived by TAP, who have also given a diagrammatic derivation of this result. The dependence on $(T - T_{\text{sg}})^{-2}$ is also unusual, since the normal sorts of mean field theories give a simple pole for the critical slowing-down (however, see [143]). It is in agreement, though, with computer simulations

[104]. The linearized mean field kinetic equation (6.15) still gives interesting predictions just at T_{sg}. Integrating over the spectrum of eigenmodes of the J_{ij} matrix [104, 127], we obtain

$$\langle S_i S_i(t) \rangle \propto (1 + \alpha t)^{-1/2} . \tag{6.17}$$

Such power law decays can result from the non-linear terms in a kinetic equation, but here they are generated solely by the continuous distribution of relaxation modes extending down to zero frequency. To check the validity of the prediction (6.17) I have plotted some Monte Carlo data obtained with an infinite-ranged model (6.3), in fig. 21a. The form given in (6.17) gives a good account of the data, although any exponent near $\frac{1}{2}$ would work. The small sample size which could be studied limits the accuracy

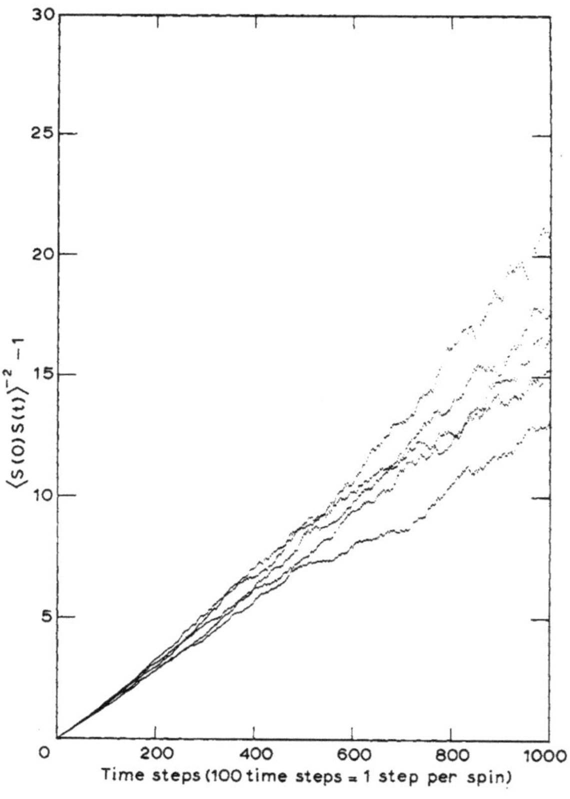

Fig. 21a.

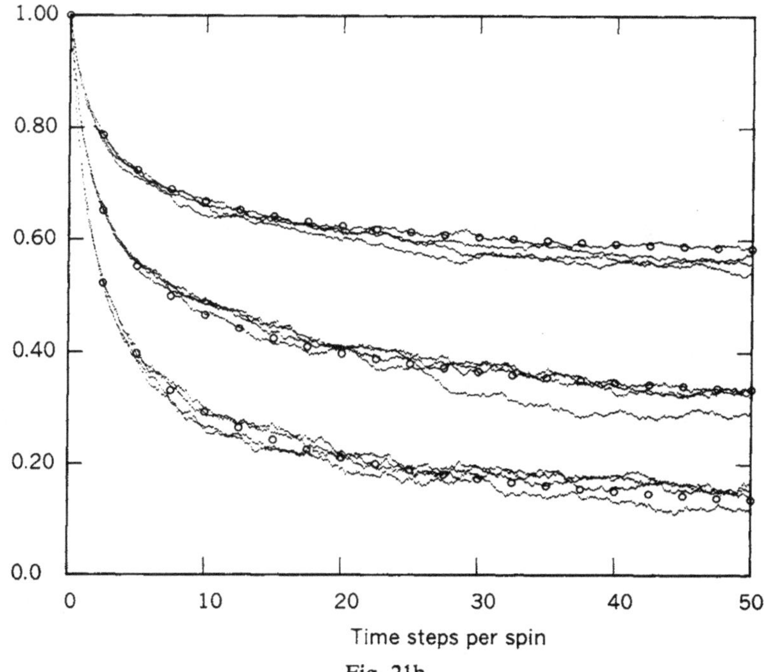

Fig. 21b.

Fig. 21. Graphical analysis of autocorrelation data for infinite ranged Ising spin glass model at $T = T_{sg} = \tilde{J}$. (a) To test for the $t^{-1/2}$ dependence predicted in (6.17), I have plotted $\langle S(0)S(t)\rangle^{-2} - 1$ versus t. (b) Autocorrelation data from four 800-spin samples at temperatures of T_{sg}, 0.8 T_{sg}, and 0.6 T_{sg} are fitted to the phenomenological expression, $\langle S(0)S(t)\rangle = q(T) + (1 - q(T))/(1 + \alpha t)^{1/2}$.

with which one can test (6.17). The limitation on sample size is due to the large number of interactions which must be stored in the computer to describe such a model.

The dynamics in the low-temperature phase of an infinite-ranged model spin glass is quite unlike that of conventional magnetic systems. Figure 21b shows the decay of autocorrelations at T_{sg} and two lower temperatures. All three are power law decays, and have been fitted to an expression which generalizes (6.17), keeping the exponent, $\frac{1}{2}$. In a conventional system, one would expect that relaxation towards equilibrium would again be exponential in time below the critical temperature.

There is no theoretical understanding at present of these non-exponential decays although they resemble the frequency-dependent correlations calculated by Ma and Rudnick [136] for a different model of a random magnet. TAP have given plausibility arguments that the free energy must have a saddle point at all temperatures below T_{sg} as well as at T_{sg}, and thus that the non-linear coupling between relaxation modes will be present below T_{sg} as well as at T_{sg}. However, even in the linearized theory a continuous distribution of relaxation rates extending down to zero is sufficient to produce decays like that in figs. 21, so the power law decay is not sufficient evidence to prove that the free energy remains at a saddle point.

The static properties predicted for the low temperature phase by the original SK mean field theory are clearly incorrect, but comparison with computer simulations is instructive to see the magnitude of the discrepancy and to test alternative proposals. The SK expression for the free energy yields an entropy which goes to a negative constant as $T \to 0$, which is an impossible result for an Ising system. (Free energies derived by replica mean-field theories for more realistic (3D) models also have this kind of pathological behavior at low temperatures, although this does not seem to be discussed in the literature.)

To treat low temperatures, TAP make an additional assumption, postulating that the principal low-lying excitations of the system are associated with single spin reversals and thus may be characterized by a distribution of molecular fields, $p(h_i)$. The SK solution gives a specific heat which is proportional to T for low T. This would be compatible with a $p(h)$ which tends to a constant limit, $p(0) \neq 0$, as $h \to 0$. TAP show that an infinite ranged model cannot have $p(0)$ nonzero, arguing that if it did, a macroscopic fraction of the spins could flip to lower their energies. Stability requires, in fact, that $p(h)$ increase from zero no faster than $\propto h$. TAP assume that $p(h)$ is linear in h for small h,

$$p(h) \sim \alpha h \,, \tag{6.18}$$

and fix α to be the smallest value for which a solution of their mean field equations exists. This choice of the constant minimizes the ground state energy, E_0, since in

$$|E_0| = \tfrac{1}{2} \int \mathrm{d}h p(h) \,, \tag{6.19}$$

decreasing the probability of finding small values of h will increase the number of spins in the larger molecular fields. With this prescription, TAP predict a specific heat $\propto T^2$, susceptibility $\propto T$, and entropy which goes to zero as T^2. All three results, as well as the presumed form for $p(h)$, are in reasonably good qualitative agreement at sufficiently low temperatures with recent computer simulations [104]. For example, the entropy obtained numerically by integrating the internal energy,

$$S(T) = S(\infty) - \int_T^{\infty} (\mathrm{d}U/\mathrm{d}T)\,\mathrm{d}T/T\,, \tag{6.20}$$

is compared in fig. 22 with the TAP and SK theories.

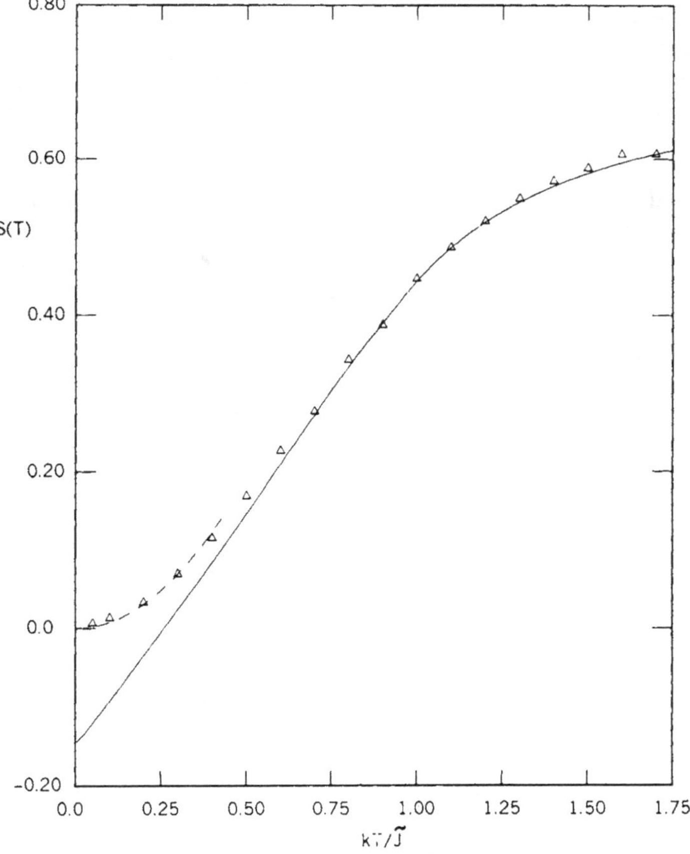

Fig. 22. Entropy of the infinite-ranged Ising spin glass as a function of temperature, from Monte Carlo simulations (triangles, 500 spin samples, four cases) and as predicted by the SK theory (solid line) and TAP (dashed line) low temperature approximation. The data are consistent with an entropy tending to zero as $T \to 0$.

The observation that deviations from mean field theory, in the form of local spin relaxation, will modify the excitation spectrum at low energies holds for models with finite-ranged interactions as well. Klein and Brout [128] and Klein [129] have studied a model of randomly positioned Ising spins with interactions of random sign and magnitude $1/r^3$, intended to reproduce the properties of randomly situated dipolar impurity ions in an ionic crystal. Their mean field analysis gives a Lorentzian form for $p(h)$, with the maximum value at $p(0)$. Stability requirements, however, require $p(0) = 0$ for this model as well, and computer simulations are consistent with the proposal [130] that $p(h)$ is proportional to h to a power less than 1. If the interactions are nearest-neighbor or finite-ranged, it seems likely that $p(0) \neq 0$ can be stable. However, mean field treatments of the Klein–Brout type produce a $p(h)$ with its maximum at $h = 0$, while the result of simulations for such models (Kirkpatrick, unpublished; Binder [131]) is depressed at $h = 0$, and has its maximum at some finite h.

Finally we consider those features which might be different in the 2D or 3D spin glass models than in the infinite ranged model. The Monte Carlo method is particularly well-suited to study of domain walls, since direct study of microscopic details of the system is possible, and gedanken experiments, such as changing the sign of a subset of interactions, can be carried out. However, it is very difficult to obtain accurate estimates of the wall energy itself. The wall energy is measured by finding the ground state of a particular finite sample of spin glass, then changing that sample in such a way that at least one wall which cuts the sample is present. This can be done by changing the sign of all the interactions in one plane, or by imposing different boundary conditions. The ground state of the modified system is then found. By comparing the two ground states, one can identify the wall as a continuous surface (or line, in 2D) of bonds which were satisfied in one ground state and unsatisfied in the other. The wall energy is the difference between the two ground state energies.

The fundamental difficulty is that finding the ground state of a spin glass, even after employing various heuristic tricks to accelerate the search, is hard, and there seems to be no way short of exhaustive enumeration to prove that a given state is truly the ground state. Thus finding two ground states and accurately determining the difference in their energies is harder still. You should therefore interpret the statement that domain walls are observed to have zero energy in a spin glass simulation to mean that their energy is less than the spread in energy between those nearly degenerate ground states which one can find in a reasonable length of computing time.

For a given distribution, $p(J)$, of interactions, the observed wall energy will fluctuate from sample to sample. Little is known at the moment about the sample size dependence of this fluctuation. Also, if the average wall energy is calculated by changing the signs of the bonds on a cutting surface, it must vanish for distributions with the property $p(J) = p(-J)$, because the modified configuration generated in this way is a perfectly valid member of the original ensemble of configurations. Vannimenus and Toulouse reported careful hand calculations on relatively small 2D samples in which J had constant magnitude and random sign. They found that E_{wall} decreased to zero when the concentration of antiferromagnetic bonds exceeded roughly 10%. My own computer calculations (unpublished) on larger samples (80×80 sites) support their data for large wall energies, but suggest that E_{wall} vanishes with zero slope at a concentration a few per cent higher. Reed et al. [121] studied wall energies in 2D, 3D, and 4D models, and concluded that at 4D, $E_{wall} \rightarrow 0$ only for a symmetric distribution of interactions, while in 2D and 3D low energy walls were observed for a range of unsymmetric distributions. The presence of a finite density of such walls in equilibrium would appear to distinguish the nature of correlations in any 2D or 3D low temperature spin glass phase from those possible in models of higher dimensionality. However, even at and above 4D there will always be low energy walls in models with a symmetric distribution of exchange interactions. Their consequences are not clear at present.

As the energy of a domain wall becomes small its shape changes. The walls become more contorted and increase in area to take advantage of the antiferromagnetic interactions. When $E_{wall} = 0$, a domain wall of linear dimension L may no longer have a limiting dimension of order L^{d-1}, but can grow to fill the sample. This is the mechanism Vannimenus and Toulouse [118] suggest brings about the transition from ferromagnet into spin glass. It is illustrated quite graphically in figs. 23a–c.

Computer experiments have yet to provide a conclusive answer to the ill-posed question of whether the low temperature state observed in 2D and 3D models of spin glasses represents a distinct equilibrium phase or a non-equilibrium effect. Bray and Moore [120] favor the latter view. They observed that when the susceptibility of a spin glass is calculated by averaging fluctuations in the total magnetization, it continues to increase for surprisingly long sample times at low temperatures. (They waited 90,000 Monte Carlo time steps per spin (MCS) in one example.) They propose that if one could wait long enough χ would increase to a limiting value, $1/T$, characteristic of the paramagnetic phase.

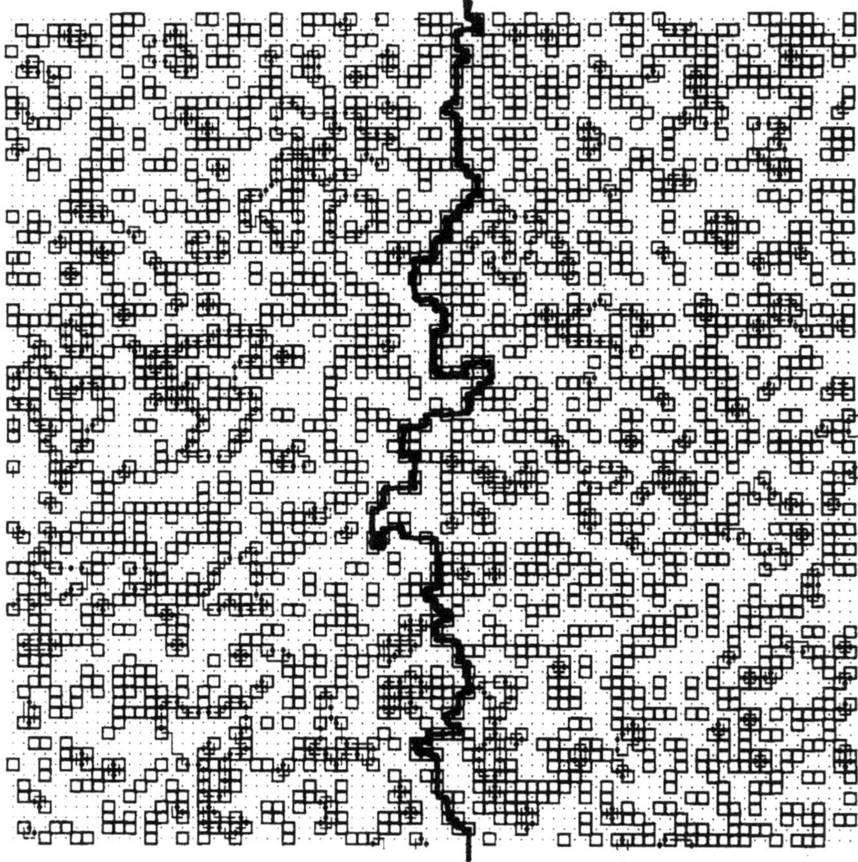

Fig. 23a.

However, the slow convergence of χ is not a feature restricted to models in less than 4D. Long relaxation times (the corollary of non-linear decays) present in the infinite ranged model lead to similar behavior. In calculating $\chi(T)$ for $T < T_{sg}$ in the infinite ranged model I have found that χ is still increasing over averaging times sufficiently long that the energy and observed energy fluctuations have already reached characteristic equilibrium values. Integrating to still longer times (several tens of thousands of MCS), I found that $\chi(T)$ did saturate at values in reasonable agreement with theory. In 2D and 3D it seems advisable to study convergence in the $\pm J$

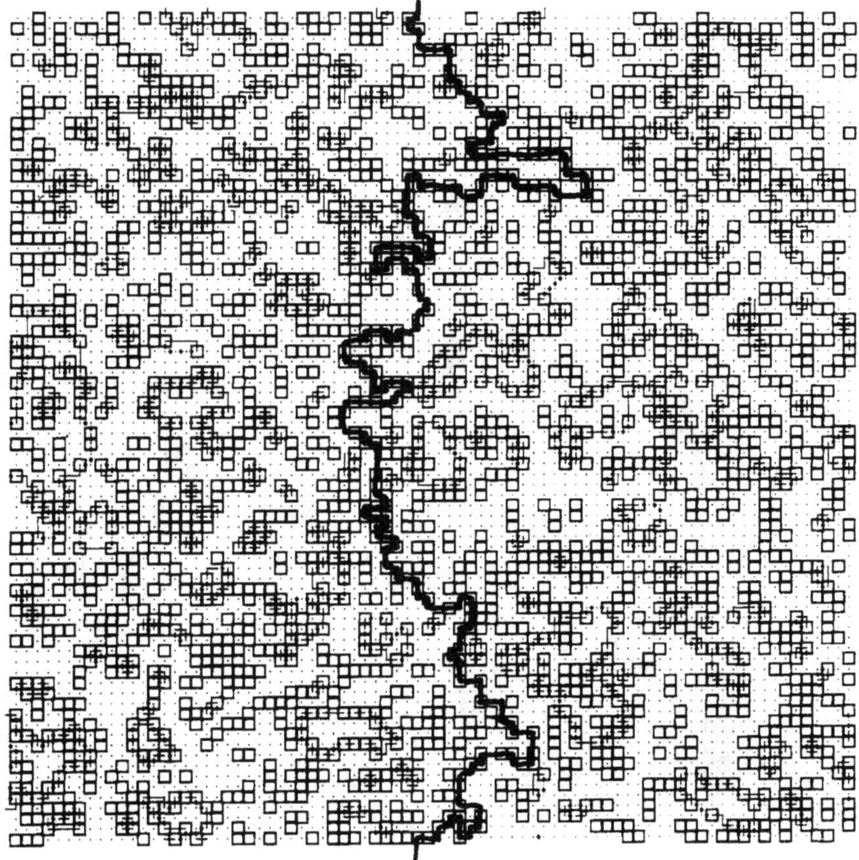

Fig. 23b.

Ising models, rather than in the model with gaussian distribution of interactions, studied by Bray and Moore. The $\pm J$ models have a finite density (5–10%) of spins free to flip at low temperatures [33]. Thus some part of phase space can be explored relatively rapidly in these models while the corresponding processes are progressively frozen out at low temperatures in any model with a continuous distribution of interactions. In the $\pm J$ models (either 2D or 3D) I found that $\chi(T)$ generally increases for as long as tens of thousands of MCS before saturating, sometimes overshooting and relaxing back to a lower value, but saturation values of χ clearly less

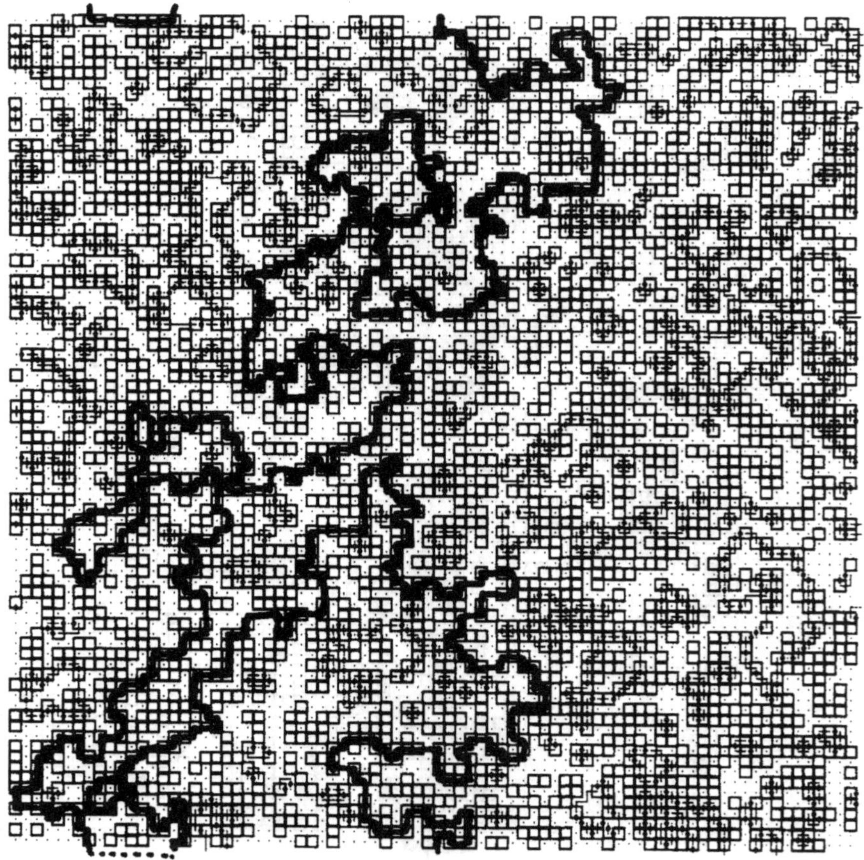

Fig. 23c.

Fig. 23. Characteristic shapes and growth of domain walls in a 2D Ising magnet with concentration $p = 0.14$ (a), 0.16 (b), and 0.5 (c) of antiferromagnetic bonds. Each 80×80 sample is shown in the dual space representation [111] in which dots mark the centers of unfrustrated squares, and the frustrated squares are drawn in. Defect strings (lines bounding regions in which the spins can be revised with little energy cost) are identified by comparing two ground states, before and after the signs of all the horizontal bonds in the center column have been changed. The defect string which cuts the sample is marked with a broad line. After the column of bonds is changed, this string is a straight line, but it relaxes by taking advantage of bonds previously unsatisfied. Case (a) is within the ferromagnetic phase, case (b) lies on the border, and case (c) should be in the middle of any spin glass phase. In case (c), the defect line has no energy per unit length, and thus will continue to grow without limit as the simulation is extended.

than $1/T$ were obtained. Again the energy and specific heat reach limiting values in relatively short times, well before the magnetic expectation values.

The source of the slow relaxation and erratic averaging of $\chi(T)$ is the motion of domain walls like those in fig. 23c. These walls can move in equilibrium because there is no energy cost associated, but their motion generally requires unpinning to take place at specific regions, and this unpinning is an activated process. Accompanying domain wall motion between pinning events are large fluctuations in the total magnetization.

My observations suggest that no amount of waiting will allow a spin glass to respond like a paramagnet. The rather subtler question of whether the low temperature spin glass is a respectable distinct phase or something rigid but more like a structural glass will probably not be resolved until we have a deeper understanding of both systems.

Acknowledgements

I should like to express my gratitude to the Groupe de Physique des Solides, Ecole Normale Supérieure, for their hospitality and a chance to prepare and present a preliminary version of these lectures. I am indebted to Dietrich Stauffer for many detailed comments on the original version of this manuscript.

References

[1] B. Abeles, P. Sheng, M. D. Coutts, and Y. Arie, Advan. Phys. 24 (1975) 407.
[2] N. T. Liang, Y. Shan, and S. Wang, Phys. Rev. Letters 37 (1976) 526.
[3] R. C. Dynes, J. P. Garno, and J. M. Powell, Phys. Rev. Letters 40 (1978) 479.
[4] V. K. S. Shante and S. Kirkpatrick, Advan. Phys. 20 (1971) 325.
[5] R. Zallen and H. Scher, Phys. Rev. B4 (1971) 4771.
[6] S. Kirkpatrick, Rev. Mod. Phys. 45 (1973) 570.
[7] S. Kirkpatrick, Lecture Notes of Kyoto Symp. on Electrons in Disordered Materials, available as Tech. Rept. B15 of the Institute of Solid State Physics, Tokyo (1973).
[8] A. B. Harris and S. Kirkpatrick, Phys. Rev. B16 (1977) 542.
[9] P. G. de Gennes, J. Phys. Lettres (Paris) 37 (1976) L1.
[10] D. Stauffer, Z. Physik B22 (1975) 161.
[11] A. Skal and B. I. Shklovskii, Soviet Phys.–Semiconductors 8 (1975) 1029.
[12] S. Kirkpatrick, Phys. Rev. B16 (1977) 4630.
[13] K. G. Wilson and J. Kogut, Phys. Rept. 12C (1974) 76.
[14] S. K. Ma, Modern Theory of Phase Transitions (Benjamin, New York, 1976).
[15] K. G. Wilson, Rev. Mod. Phys. 47 (1975) 765.
[16] L. P. Kadanoff, Ann. Phys. (NY) 100 (1976) 559.
[17] J. W. Essam, in: Phase Transitions and Critical Phenomena, Vol. 2, eds. C. Domb and M. S. Green (Academic Press, London, 1972) ch. 6.
[18] D. Knuth, The Art of Computer Programming, Vols. I–III (McGraw-Hill, New York, 1968–1973).
[19] A. Aho, J. Hopcroft, and R. Ullman, The Design and Analysis of Computer Algorithms (Addison-Wesley, Reading, MA, 1974).
[20] D. J. Rose and R. A. Willoughby, eds., Sparse Matrices and Their Applications (Plenum, New York, 1972).
[21] J. T. Edwards, D. C. Licciardello, and D. J. Thouless, J. Inst. Math. Appl., to be published.
[22] M. L. Garey and D. S. Johnson, Computers and Intractibility, a Guide to the Theory of NP-Completeness (W. H. Freeman, San Francisco, 1979).
[23] H. L. Frisch and J. M. Hammersley, J. Soc. Ind. Appl. Math. 11 (1963) 894.
[24] G. Toulouse, Nuovo Cimento B23 (1974) 234.
[25] P. W. Kasteleyn and C. M. Fortuin, J. Phys. Soc. Japan 26 (1969) 11.
[26] I. Syozi, in: Phase Transitions and Critical Phenomena, Vol. 1, eds. C. Domb and M. S. Green (Academic Press, London, 1972) p. 269.
[27] M. Giri and M. Stephen, J. Phys. C11 (1978) L541.
[28] H. J. Hilhorst, Phys. Letters 56A (1976) 156.
[29] P. J. Reynolds, W. Klein, and H. E. Stanley, J. Phys. C10 (1977) L167.
[30] J. Bernasconi, Phys. Rev. B18 (1978) 2185.
[31] P. J. Reynolds, H. E. Stanley, and W. Klein, J. Phys. A12 (1978) L199.
[32] A. A. Migdal, Soviet Phys.–JETP 42 (1976) 413, 743.
[33] S. Kirkpatrick, Phys. Rev. B15 (1978) 1533.
[34] J. José, L. P. Kadanoff, S. Kirkpatrick, and D. R. Nelson, Phys. Rev. B16 (1977) 1217.
[35] P. J. Dean and N. F. Bird, Mathematics Division report MA 61, available from National Physical Laboratory, Teddington, Middlesex, England.

[36] P. Dean and N. F. Bird, Proc. Cambridge Phil. Soc. 63 (1967) 477.
[37] J. Roussenq, J. Clerc, G. Giraud, E. Guyon, and H. Ottavi, J. Physique 37 (1976) L99.
[38] A. Sur, J. L. Lebowitz, J. Marro, M. Kalos, and S. Kirkpatrick, J. Statistical Phys. 15 (1976) 345.
[39] J. Hoshen and R. Kopelman, Phys. Rev. B14 (1976) 3438.
[40] A. B. Harris, J. Phys. C7 (1974) 1671.
[41] E. Domany, J. Phys. C11 (1978) L337.
[42] L. J. de Jongh and A. R. Miedema, Advan. Phys. 23 (1974) 1.
[43] D. J. Breed, K. Gilijamse, J. W. E. Sterkenburg, and A. R. Miedema, Physica 68 (1973) 303.
[44] A. B. Harris and R. Fisch, Phys. Rev. Letters 38 (1977) 796.
[45] T. C. Lubensky, Phys. Rev. B11 (1975) 3573.
[46] R. B. Stinchcombe and B. P. Watson, J. Phys. C9 (1976) 3221.
[47] C. Jayaprakash, E. K. Riedel, and M. Wortis, Phys. Rev. B18 (1978) 2244.
[48] D. Zobin, Phys. Rev. B18 (1978) 2387.
[49] D. E. Khmel'nitskii, Soviet Phys.-JETP 41 (1976) 981.
[50] Y. Imry and S. K. Ma, Phys. Rev. Letters 35 (1976) 1399.
[51] R. S. Alben, J. J. Becker, and M. C. Chi, J. Appl. Phys. 49 (1978) 1653.
[52] J. P. Straley, J. Phys. C9 (1976) 783.
[53] D. J. Bergman, Phys. Rev. B14 (1976) 4304.
[54] D. J. Bergman, AIP Conf. Proc. 40 (1978) 46.
[55] J. P. Straley, AIP Conf. Proc. 40 (1978) 118.
[56] J. P. Straley, Phys. Rev. B15 (1977) 5733.
[57] M. J. Stephen, Phys. Rev. B17 (1978) 4444.
[58] C. Dasgupta, T. C. Lubensky, and A. B. Harris, Phys. Rev. B17 (1978) 1375.
[59] P. W. Anderson, Phys. Rev. 109 (1958) 1492.
[60] N. F. Mott and E. A. Davis, Electronic Processes in Non-Crystalline Materials (Oxford Univ. Press, London, 1971).
[61] D. C. Licciardello, preprint (1977).
[62] V. Ambegaokar, B. I. Halperin, and J. S. Langer, Phys. Rev. B4 (1971) 2612.
[63] R. W. Landauer, J. Appl. Phys. 23 (1952) 779.
[64] R. W. Landauer, AIP Conf. Proc. 40 (1978) 2.
[65] S. Kirkpatrick, Phys. Rev. Letters 27 (1971) 1722.
[66] B. P. Watson and P. Leath, Phys. Rev. B9 (1974) 4893.
[67] T. J. Shankland, J. Geophys. Rev. 79 (1974) 4863.
[68] T. J. Shankland, Phys. Earth and Planetary Interiors 10 (1975) 209.
[69] M. H. Cohen and J. Jortner, Phys. Rev. Letters 30 (1973) 696.
[70] M. H. Cohen, J. Jortner, and I. Webman, AIP Conf. Proc. 40 (1978) 63.
[71] I. Webman, J. Jortner, and M. H. Cohen, Phys. Rev. B15 (1977) 1936.
[72] B. J. Last and D. J. Thouless, Phys. Rev. Letters 27 (1971) 1719.
[73] D. Adler, L. P. Flora, and S. D. Senturia, Solid State Commun. 12 (1973) 9.
[74] H. Ottavi, J.-P. Clerc, G. Giraud, J. Roussenq, E. Guyon, and C. D. Mitescu, AIP Conf. Proc. 40 (1978) 372.
[75] C. D. Mitescu, H. Ottavi, and J. Roussenq, AIP Conf. Proc. 40 (1978) 377.
[76] G. Forsythe and C. B. Moler, Computer Solution of Linear Algebraic Systems (Prentice-Hall, Englewood Cliffs, NJ, 1967).
[77] J. A. George, SIAM J. Numerical Analysis 10 (1973) 345.

[78] A. J. Hoffman, N. S. Martin, and D. J. Rose, SIAM J. Numerical Analysis 10 (1973) 364.
[79] J. Hopcroft and R. E. Tarjan, Commun. ACM (Algorithm 447) 16 (1973) 372.
[80] S. Kirkpatrick, AIP Conf. Proc. 40 (1978) 99.
[81] B. B. Mandelbrot, Fractals – Form, Chance and Dimension (Freeman, San Francisco, 1977).
[82] H. E. Stanley, J. Phys. A10 (1977) L211.
[83] B. B. Mandelbrot, in: STATPHYS 13, Intern. IUPAP Conf., Haifa, 1977, eds. D. Cabib, C. G. Kuper, and J. Riess, Ann. Israel Phys. Soc. 2 (1978) 226.
[84] M. E. Levinshtein, M. S. Shur, and A. L. Efros, Soviet Phys.–JETP 42 (1976) 1120.
[85] B. I. Shklovskii, Soviet Phys.–JETP 45 (1977) 152.
[86] R. D. Swenumson and J. C. Thompson, Phys. Rev. B14 (1976) 5142.
[87] G. Toulouse, Compt. Rend. (Paris) B280 (1975) 629.
[88] D. J. Thouless and M. E. Elzain, J. Phys. C11 (1978) 3425.
[89] J. Friedel, preprint (1976).
[90] P. G. de Gennes, P. Lafore, and J. P. Millot, J. Phys. Chem. Solids 11 (1959) 105.
[91] P. G. de Gennes, P. Lafore, and J. P. Millot, J. Phys. Radium 20 (1959) 624.
[92] S. Kirkpatrick and T. P. Eggarter, Phys. Rev. B6 (1972) 3598.
[93] T. Mizoguchi, T. R. McGuire, R. J. Gambino, and S. Kirkpatrick, Phys. Rev. Letters 38 (1977) 89.
[94] J. M. D. Coey and S. von Molnar, Lettres au J. de Phys. 39 (1978) L327.
[95] V. Cannella and J. A. Mydosh, Phys. Rev. B6 (1972) 4220.
[96] N. Heiman and K. Lee, AIP Conf. Proc. 34 (1976) 319.
[97] T. R. McGuire, T. Mizoguchi, R. J. Gambino, and S. Kirkpatrick, J. Appl. Phys. 49 (1977) 1689.
[98] B. V. B. Sarkissian and B. R. Coles, Commun. Phys. 1 (1977) 17.
[99] J. Villain, J. Phys. C10 (1977) 4793.
[100] L. R. Walker and R. E. Walstedt, Phys. Rev. Letters 38 (1977) 544.
[101] D. Sherrington and S. Kirkpatrick, Phys. Rev. Letters 35 (1975) 1792.
[102] S. F. Edwards and P. W. Anderson, J. Phys. F5 (1975) 965.
[103] D. J. Thouless, P. W. Anderson, and R. G. Palmer, Phil. Mag. 35 (1977) 593.
[104] S. Kirkpatrick and D. Sherrington, Phys. Rev. B17 (1978) 4384.
[105] R. Fisch and A. B. Harris, Phys. Rev. Letters 38 (1977) 785, and Phys. Rev. B18 (1978) 416.
[106] A. J. Bray and M. A. Moore, J. Phys. C, to be published.
[107] J. R. L. de Almeida and D. J. Thouless, J. Phys. A11 (1978) 983.
[108] E. Pytte and J. Rudnick, Phys. Rev. B19 (1979) 3603.
[109] J. M. Kosterlitz and D. J. Thouless, J. Phys. C6 (1973) 1181.
[110] J. M. Kosterlitz, J. Phys. C7 (1974) 1046.
[111] G. Toulouse, Commun. Phys. 2 (1977) 115.
[112] E. Fradkin, B. A. Huberman, and S. H. Shenker, Phys. Rev. B18 (1978) 4789.
[113] P. W. Anderson and C. M. Pond, Phys. Rev. Letters 40 (1978) 903.
[114] A. P. Young, preprint (1978).
[115] A. P. Young and R. B. Stinchcombe, J. Phys. C9 (1976) 4419.
[116] B. W. Southern and A. P. Young, J. Phys. C10 (1977) 2179.
[117] C. Jayaprakash, E. K. Riedel and M. Wortis, Phys. Rev. B18 (1977) 1495.

[118] J. Vannimenus and G. Toulouse, J. Phys. C10 (1977) L537.
[119] G. Toulouse, J. Vannimenus, and J.-M. Maillard, Lettres au J. de Phys. 38 (1977) L459.
[120] A. J. Bray and M. A. Moore, J. Phys. F7 (1977) L333.
[121] P. Reed, M. A. Moore, and A. J. Bray, J. Phys C11 (1978) L139.
[122] B. Derrida, J.-M. Maillard, J. Vannimenus, and S. Kirkpatrick, Lettres au J. de Physique 39 (1978) L465.
[123] G. André, R. Bidaux, J. P. Carton, R. Conté and L. de Seze, preprint. See also, B. Derrida, J.-M. Maillard, J. Vannimenus and S. Kirkpatrick, Lettres au J. de Phys. 39 (1978) L465.
[124] P. W. Anderson, in: Amorphous Magnetism II, eds. R. A. Levy and R. Hasegawa (Plenum, New York, 1977) p. 1.
[125] M. Suzuki and R. Kubo, J. Phys. Soc. Japan 24 (1968) 51.
[126] R. Brout and H. Thomas, Physics 3 (1967) 317.
[127] W. Kinzel and K. H. Fischer, Solid State Commun. 23 (1977) 687.
[128] M. W. Klein and R. Brout, Phys. Rev. 132 (1963) 124.
[129] M. W. Klein, Phys. Rev. B14 (1976) 5008.
[130] S. Kirkpatrick and C. M. Varma, Solid State Commun. 25 (1978) 821.
[131] K. Binder, Z. Physik B26 (1977) 339.
[132] V. E. Dubrov, M. E. Levinshtein and M. S. Shur, Sov. Phys.–JETP 43 (1976) 1050 (Zh. Exsp. and Teor. Fiz. 70 (1976) 2014).
[133] J. T. Edwards and D. J. Thouless, J. Phys. C5 (1972) 807.
[134] A. L. Efros and B. I. Shklovskii, Phys. Status Solidi (b) 76 (1976) 475.
[135] J. C. Garland and D. B. Tanner, eds., Electrical Transport and Optical Properties of Inhomogeneous Media, AIP Conf. Proc. 40 (1978) (American Inst. Physics, New York).
[136] S. K. Ma and J. Rudnick, Phys. Rev. Letters 40 (1978) 589.
[137] W. Klein, H. E. Stanley, P. J. Reynolds, and A. Coniglio, Phys. Rev. Letters 41 (1978) 1145.
[138] M. E. Levinshtein et al., Sov. Phys.–JEPT 42 (1976) 197.
[139] R. J. Birgeneau, R. A. Cowley, G. Shirane and H. J. Guggenheim, Phys. Rev. Lett. 37 (1977) 940; R. A. Cowley, G. Shirane, R. J. Birgeneau and E. C. Svensson, Phys. Rev. Lett. 39 (1978) 894.
[140] Wegner et al., Phys. Rev. B5 (1972) 4529.
[141] C. Domb and E. Stoll, J. Phys. A10 (1977) 1141; C. Domb, T. Schnieder and E. Stoll, J. Phys. A8 (1975) L90; E. Stoll and C. Domb, preprint.
[142] M. A. Klenin, Phys. Rev. B19 (1979) 3586.
[143] A mean relaxation time of experimental interest, τ_m, can be defined by integrating $\langle S_i S_i(t)\rangle$

$$\tau_m = \int_0^\infty dt \langle S_i S_i(t)\rangle \ .$$

This contains contributions from the more slowly decaying modes as well, and has the behavior $\tau_m \propto (T - T_c)^{-1}$ near the critical temperature. See S. Kirkpatrick, in "Proceedings of NATO Adv. Study Inst., Geilo, Norway" (1979), T. Ristl, ed. (Plenum Press, New York) to be published.

COURSE 6

THERMAL AND GEOMETRICAL CRITICAL PHENOMENA IN RANDOM SYSTEMS*

Tom C. LUBENSKY

Department of Physics and Laboratory for Research in the Structure of Matter, University of Pennsylvania, Philadelphia, Pennsylvania 19104, USA

* Supported in part by NSF grant No. DMR 76-21073, NSF MRL program grant No. DMR 76-00678, and ONR grant No. N00014-76-C-0106.

Contents

R. Balian et al., eds.
Les Houches, Session XXXI, 1978 – La matière mal condensée/Ill-condensed matter

Introduction

The study of critical phenomena in homogeneous systems has in a sense reached maturity. Over the past several years, our understanding of such properties as non-analytic temperature dependence of thermodynamic functions and scaling in the vicinity of continuous phase transitions has increased enormously. There is satisfactory agreement between theory and experiment in almost all cases – near simple second-order phase transitions as well as near more esoteric multi-critical points which are too numerous to enumerate. Randomness adds a whole new dimension to types of critical behavior that can occur in the neighborhood of continuous transitions. The study of critical phenomena in random systems is currently very active and is the subject of these lectures.

The Wilson renormalization group [1,2] (RG) was instrumental in increasing our understanding of critical properties of homogeneous systems. It put previously developed heuristic scaling theories on a sound mathematical footing and provided the tools to actually calculate non-analytic scaling functions. It also provided the framework for the systematic study of multi-critical points. The effect of the Wilson renormalization group on the study of critical phenomena in random systems was in some sense more profound. Prior to the RG, very little was known about how randomness would affect a continuous phase transition. It was vaguely felt that randomness should somehow smear the transition. This view was buttressed by an exact calculation showing a smeared transition in a two-dimensional Ising model with a pecular type of disorder [3]. There was, however, no convincing reason for accepting this view or any other for random systems of the type one commonly encounters in the lab. The RG provided a workable language with which to discuss disorder at critical points. The first concrete result to emerge was that a phase transition in a weakly random system is sharp [4,5] (rather than smeared) if the transition in the pure system is sharp. The critical exponents in the disordered system may, however, differ from those of the pure system. This first application of the RG to the study of critical properties of random systems in a sense opened the door to the vast activity that we see in this field today.

The study of critical properties in the vicinity of continuous thermally driven phase transitions in quenched random systems leads naturally to the study of geometrically driven phase transitions. In particular, in randomly diluted magnetic systems, there is no magnetic phase transition if there is no infinite connected cluster of magnetic ions. Thus, in these systems, one must consider the geometrical problem of the formation of an infinite cluster as a function of the concentration of magnetic ions. This is the phenomenon of percolation [6]. The study of percolation immediately leads to a consideration of other geometrically driven transitions such as gelation and vulcanization [7] or the properties of random resistor networks. A large part of these lectures will be devoted to the critical properties of these geometrically driven transitions.

These lectures are not meant to be in any way complete, nor are they to be considered as a review of the field. They draw most heavily on the work done at the University of Pennsylvania over the last few years and represent my personal interests at the moment. They strongly reflect the ideas and approach of A. B. Harris with whom I worked very closely and who is directly responsible for the formative ideas on much of the material presented here that was done at Penn. Our graduate students, Jing-Huei Chen, Chandan Dasgupta, Ronald Fisch, and Joel Isaacson also made substantial contributions to this body of work. Because these lectures were prepared under rather severe time constraints, they are disjoint with sections often appearing to end in mid-sentence. I apologize for their lack of clarity. I hope, nevertheless, that they convey a useful overview of the modern approach to the problem of critical behavior of random systems.

I have divided these lectures into two broad chapters. The first chapter is largely descriptive and heuristic. It describes the problems of interest and extracts what information can be obtained with a minimum of calculation. The second chapter emphasizes model details and calculations using field theory and the Wilson–Fisher ε-expansion. The development of the first chapter is approximately historical, beginning with weakly random systems, then moving to percolation and spin-glasses [8] and finally to branched polymers, gelation, and percolation. The second chapter presents a grab-bag of mathematical tricks that have proven useful in the treatment of disordered systems. These include the infamous "$n = 0$ trick" [5,8,9] and a demonstration of how the statistics of percolating clusters can be obtained from the one-state Potts model [10]. It also deals with mean field theories and critical exponents near the upper critical dimension of the problems considered.

1. Descriptive overview

A. Definitions and Weak Randomness

1.1. Types of disorder

It is customary to distinguish two broad categories of disorder in solid systems, *substitutional* and *structural* (or *topological*) disorder. In a substitutionally disordered system impurities occupy lattice sites of a regular lattice. Translational periodicity of the lattice is not destroyed. Solid alloys such as $Rb_2Mg_{1-x}Mn_xF_4$ or $Mn_xZn_{1-x}F_2$ are examples [11] of this type of disorder. Structurally disordered systems, on the other hand, have no regular lattice structure. The most familiar examples of such systems are the amorphous semi-conductors. For the study of critical phenomena, the most important examples are the so-called metglasses [12] such as $Fe_{32}Ni_{36}Cr_{14}P_{12}B_6$ or $Fe_{29}Ni_{49}P_{14}B_6Si_2$.

In substitutionally disordered systems, one further distinguishes *annealed* and *quenched* disorder. In an annealed system, the impurity degrees of freedom $\{I\}$ are in thermal equilibrium with other degrees of freedom (such as spin) $\{S\}$. The variables $\{S\}$ and $\{I\}$ are treated on the same footing, and the partition function is

$$Z = \mathrm{Tr}_{\{S\},\{I\}} \exp(-\beta\mathscr{H}[\{S\},\{I\}]) , \tag{1.1.1}$$

where $\beta^{-1} = T$ where T is the temperature and where we have chosen units so that the Boltzmann constant is unity.

An effective hamiltonian in terms of the variables $\{S\}$ that participate actively in the phase transition can be introduced via

$$\exp(-\beta\mathscr{H}_{\mathrm{eff}}[\{S\}]) = \mathrm{Tr}_{\{I\}} \exp(-\beta\mathscr{H}[\{S\},\{I\}]) , \tag{1.1.2}$$

so that

$$Z = \mathrm{Tr}_{\{S\}} \exp(-\beta\mathscr{H}_{\mathrm{eff}}[\{S\}]) . \tag{1.1.3}$$

Thus the variables $\{I\}$ are treated in the same way as thermal variables, such as phonons, that are not of immediate interest to the transition of interest. In fact, the presence of annealed impurities leads to a Fisher renormalization [13] of critical exponents if the specific heat exponent, α, is positive, just as phonons lead to such a renormalization in appropriately constrained isotropic systems [14]. We will not be concerned further with annealed disorder in these lectures.

In quenched random systems, the impurities are frozen into a non-equilibrium but random configuration. This might be accomplished by heating a sample to some temperature T_i, allowing the impurities to reach thermal equilibrium and then suddenly cooling to a much lower temperature T_f. The impurities are then frozen into a configuration appropriate to temperature T_i. They are inhibited by high potential barriers from diffusing to equilibrium positions characteristic of the temperature T_f. In this case the variables $\{I\}$ and $\{S\}$ must be treated on a different footing. In fact, for any given configuration of $\{I\}$, one can calculate a free energy via

$$F[\{I\}] = -\beta^{-1} \ln Z[\{I\}], \qquad (1.1.4)$$

where

$$Z[\{I\}] = \mathrm{Tr}_{\{S\}} \exp[-\beta\mathscr{H}(\{S\}, \{I\})]. \qquad (1.1.5)$$

F in eq. (1.1.4) depends on all of the co-ordinates $\{I\}$ and is computationally unmanageable. One can, however, as pointed out by Brout [15], exploit the randomness in $\{I\}$ to simplify (1.1.4). To do this, we divide the sample up into a large number of sub-volumes, each with linear dimensions much larger than a correlation length, ξ. Each sub-volume can be viewed as an independent member of an ensemble of systems characterized by the random variables $\{I\}$. The observable free energy is then the average of F over all sub-volumes or, alternatively, the ensemble average over $\{I\}$:

$$F = [F[\{I\}]]_{\mathrm{av}}. \qquad (1.1.6)$$

Similar considerations apply to derivatives of F with respect to temperature and external fields. One word of caution regarding the applicability of eq. (1.1.6) is in order. It only applies so long as the system can be divided into a large number of sub-volumes each larger than a correlation volume. Since the correlation length, ξ, diverges in the vicinity of a second-order phase transition, this condition becomes more difficult to satisfy as the transition is approached. In practice, this effect is unimportant at experimentally controllable temperatures. In order to have at least 10^6 cells in a 1 mm^3 sample, ξ must be less than 10^5 Å. ξ is $at^{-\nu}$ where a is a length typically of order a few Angstroms, t is the reduced temperature and $\nu \sim \frac{2}{3}$. Thus in order for ξ to be $\geqslant 10^5$ Å, t must be $\leqslant 10^{-7}$ which is beyond the realm of experimental control in a typical laboratory. Eq. (1.1.6) also assumes that $\{I\}$ are truly random variables. If there are any macroscopic homogeneities such as an overall concentration gradient, it does not apply.

The problem of thermodynamics in quenched random systems has now been reduced to the problem of averaging F over the random distribution of impurities. This is a mathematically difficult average to perform. It is often carried out using the "$n = 0$ trick" to be discussed in the next section.

Structural disorder is by its nature quenched since presumably there exists a crystalline state that is lower in energy. The effects of structural disorder on electronic properties is a subject of intense research at the moment. In these lectures, we will be concerned only with its effects on magnetic phase transitions.

In studying phase transitions in homogeneous systems it is often convenient to consider models that describe only the essential features of the transition in question and to ignore the many irrelevant variables that are present in a real system. In random systems, similar models are studied. In addition, random potentials rather than randomly substituted impurities are often more convenient to study. Thus if there is a set of random potentials (which may be nearest neighbor exchange anisotropy potentials, etc.), $\{v\}$, with probability distribution $\mathscr{P}(\{v\})$ the free energy as a function of external fields $H(\boldsymbol{x})$ at sites $\boldsymbol{x}$ is

$$F(\beta, H(\boldsymbol{x})) = -\beta^{-1}\int \mathrm{d}\{v\}\mathscr{P}(\{v\})\ln(\mathrm{Tr}\exp(-\beta\mathscr{H}[\{v\}, H(\boldsymbol{x})])) . \tag{1.1.7}$$

1.2. Weakly random systems

Before discussing the effects of disorder on critical phenomena, let us review briefly some relevant facts about continuous phase transitions in homogeneous systems. It is well established [2] that the free energy density, $\mathscr{F}(F = \int \mathrm{d}^d\chi\mathscr{F}(x))$, in d dimensions obeys a homogeneity relation of the form

$$\mathscr{F}(t, \{g_a\}) = b^{-d}\mathscr{F}(b^{1/\nu}t, \{b^{\lambda_a}g_a\}) = t^{2-\alpha}\mathscr{F}(1, \{g_a t^{-\phi_a}\}) , \tag{1.2.1}$$

where t is the reduced temperature $(T - T_c)/T$, g_a are external potentials, ν is the correlation length exponent, $\alpha = 2 - d\nu$ is the specific heat exponent and $\phi_a = \lambda_a\nu$ are crossover exponents for the fields g_a. The local order parameter at site $\boldsymbol{x}$, $S(\boldsymbol{x})$, develops a non-vanishing thermal average which satisfies

$$M(t, \{g_a\}) \equiv \langle S\rangle = t^{\beta}\Phi_M(\{g_a t^{-\phi_a}\}) , \tag{1.2.2}$$

where $\langle\ \rangle$ signifies an equilibrium thermal average. The non-local order parameter susceptibility $\chi(\boldsymbol{x}, \boldsymbol{x}')$ is related to correlations in S via

$$\chi(\boldsymbol{x}, \boldsymbol{x}') = \beta[\langle S(\boldsymbol{x})S(\boldsymbol{x}')\rangle - \langle S(\boldsymbol{x})\rangle\langle S(\boldsymbol{x}')\rangle] \equiv \beta G^{c}(\boldsymbol{x}, \boldsymbol{x}') . \qquad (1.2.3)$$

At times, we will also use the notation

$$G(xx') = \langle S(\boldsymbol{x})S(\boldsymbol{x}')\rangle . \qquad (1.2.4)$$

Tensor subscripts may appear on both χ and G. The Fourier transform of $\chi(\boldsymbol{x}, \boldsymbol{x}')$

$$\chi(t, \{g_{\text{a}}\}, \boldsymbol{q}) = \sum_{\boldsymbol{x}'} \exp[-i\boldsymbol{q}\cdot(\boldsymbol{x} - \boldsymbol{x}')]\chi(\boldsymbol{x}, \boldsymbol{x}') , \qquad (1.2.5)$$

and the correlation-length, ξ, satisfy similar homogeneity relations

$$\chi(t, \{g_{\text{a}}\}, \boldsymbol{q}) = t^{-\gamma}\Phi_{\chi}(\{g_{\text{a}}t^{-\phi_{\text{a}}}\}, qt^{-\nu}) = q^{2-\eta}\tilde{\Phi}_{\chi}(\{g_{\text{a}}t^{-\phi_{\text{a}}}\}, qt^{-\nu}) , \qquad (1.2.6)$$

$$\xi(t, \{g_{\text{a}}\}) = t^{-\nu}\Phi_{\xi}(\{g_{\text{a}}t^{-\phi_{\text{a}}}\}) . \qquad (1.2.7)$$

If a field g_{a} is increased from zero, it will influence functions such as F, χ, and ξ only when it is of order $t^{1/\phi_{\text{a}}}$. This gives rise to a characteristic temperature

$$t_{\text{a}} \sim g_{\text{a}}^{1/\phi_{\text{a}}} , \qquad (1.2.8)$$

at which the influence of g_{a} is felt. Thus, for example, if g_{a} represents a uniaxial anisotropy field taking the system from a Heisenberg to Ising symmetry, there will be a shift in transition temperature proportional to $g_{\text{a}}^{\phi_{\text{a}}}$ for small g_{a}.

In disordered systems, the fields $g_{\text{a}}(\boldsymbol{x})$ can be regarded as spatially varying variables with variance Δ_{a}. In *weakly random systems*, Δ_{a} is small enough that no new critical point or phase boundary appears separating a phase with the same ground state symmetry as the pure system from some new phase. If a phase boundary appears, the system is *strongly random*. In some cases, as we shall see, even an infinitesimal Δ_{a} will lead to a strongly random system under certain conditions. A heuristic argument due to Harris [16] can be used to determine whether or not a small addition of a random field to an otherwise homogeneous system will change the nature (i.e., change the universality class) of the phase transition. Divide the system up into volumes with linear dimension proportional to the coherence length, ξ. Each volume will have a "transition temperature" and the ensemble of volumes will have a variance, Δt_a, in transition temperatures which will be given from eq. (1.2.8) by

$$\Delta t_a \sim (\delta g_a)^{1/\phi_a}, \tag{1.2.9}$$

where δg_a is the value of the field g_a averaged over the coherence volume. In order for the transition to maintain the character of the pure system, it is necessary for Δt_a to be smaller than t, the reduced temperature measured with respect to the self-consistent transition temperature in the presence of the impurities:

$$\Delta t_a < t. \tag{1.2.10}$$

If $g_a(\boldsymbol{x})$ is an independent random variable at each site $\boldsymbol{x}$, we can estimate δg_a from the central limit theorem

$$\delta g_a \sim \Delta_a \xi^{-d/2}, \tag{1.2.11}$$

where $\xi \sim t^{-\nu}$. Therefore, the condition that the character of the transition be maintained is

$$t^{d\nu/2\phi_a - 1} < 1, \tag{1.2.12}$$

or

$$d\nu - 2\phi_a > 0. \tag{1.2.13}$$

Now let us consider some specific examples. We begin with the Heisenberg hamiltonian for a classical m-component spin of length m, $\boldsymbol{S}(\boldsymbol{x})$, with nearest neighbor exchange

$$\mathscr{H}_H = -\sum_{\langle \boldsymbol{x}, \boldsymbol{x}' \rangle} J(\boldsymbol{x}, \boldsymbol{x}')\boldsymbol{S}(\boldsymbol{x})\cdot\boldsymbol{S}(\boldsymbol{x}'), \tag{1.2.14}$$

where $J(\boldsymbol{x}, \boldsymbol{x}')$ is a random variable with average $[J]_{av}$ and variance $[(\delta J)^2]_{av}^{1/2}$. In this case, $g_a(\boldsymbol{x}, \boldsymbol{x}') = J(\boldsymbol{x}, \boldsymbol{x}') - [J]_{av}$ and ϕ_a is unity, and eq. (1.2.13) becomes $d\nu - 2 = -\alpha > 0$. Thus, phase transitions in systems with $\alpha < 0$ (such as the three-dimensional Heisenberg Model) are unaffected by the addition of small amounts of impurities. If, on the other hand $\alpha > 0$, one can expect disorder to lead to new behavior. Renormalization group calculations to be presented in sect. 2 verify this result and show that weakly random systems with $\alpha > 0$ still have a sharp phase transition with new critical exponents. Note that this result depends only on the sign of α and *not* on the details of the probability distribution for J.

The random anisotropy axis model of Harris, Plishke, and Zuckerman treats the major effects of structural disorder. It consists of $\mathscr{H}_H$ plus an additional term

$$\mathscr{H}_A = -\sum_x D(\boldsymbol{n}(\boldsymbol{x})\cdot\boldsymbol{S}(\boldsymbol{x}))^2 , \tag{1.2.15}$$

where $\boldsymbol{n}(\boldsymbol{x})$ is a random unit vector specifying the local preferred anisotropy axis. In this case, $g_a(\boldsymbol{x}) = Dn_i(\boldsymbol{x})n_j(\boldsymbol{x})$ and ϕ_a is the anisotropy crossover exponent [18] which is $1 + \{m/2(m + 8)\}\varepsilon$ in $4 - \varepsilon$ dimensions [19] and of order [20] 1.25 in 3 dimensions. In both cases $d\nu - 2\phi_a$ is negative and one can expect such disorder to lead to critical behavior that differs from that of the pure system. Exactly what happens is still open to question. There appears to be no long range ferromagnetic order of the usual type for this model below four dimensions [21]. A type of spin-glass ordering may be possible. Alternatively, there may be a type of ferromagnetic order in which the spins are locked into the plus or minus direction of the local anisotropy axis. In this case even an infinitesimal value for d will lead to strong randomness. Above four dimensions, $\phi_a = 1$ and $d\nu - 2\phi = \frac{1}{2}(d - 4) > 0$ and the homogeneous behavior remains unaltered for sufficiently small D.

Eqs. (1.2.9) and (1.2.10) can also be used to assess the importance of correlations in the random potential g_a. For example, consider an ensemble of systems with $J(\boldsymbol{x}, \boldsymbol{x}')$ uniform in each system. If the free energy is averaged over such an ensemble, the transition will clearly be smeared. Δt_a in this case is clearly just $[(\delta J)^2]^{1/2}$ and is always greater than t, indicating a departure from pure system behavior as required. Similarly if there is striped randomness such as was considered by McCoy and Wu [3] in which J is constrained to be uniform along a line and uncorrelated in the other $d - 1$ directions, we have

$$\delta g_a \sim \Delta_a \xi^{-(\frac{1}{2}(d-1))} , \tag{1.2.16}$$

leading to

$$(d - 1)\nu - 2 = -\alpha - \nu , \tag{1.2.17}$$

which is always negative for n-component Heisenberg systems.

B. Strongly Random Systems

In *strongly random systems*, the disorder destroys the order of the homogeneous system even at zero temperature. The nature of the critical point or line terminating the homogeneous-like phase depends on the nature of the disorder. We will, therefore, consider three commonly studied types of strongly disordered systems: randomly diluted systems, Mattis spin-glasses [24,25] and Edwards and Anderson (EA) spin-glasses [8].

1.3. Random dilution

Here we consider the Heisenberg hamiltonian of eq. (1.2.12). Bonds of strength J are present with probability p and absent with probability $1 - p$ (bond dilution). Alternatively sites are occupied with magnetic ions with probability p and non-magnetic ions with probability $1 - p$ (site dilution). Unless otherwise specified, we will consider only bond dilution in these lectures. The phase diagram for dilution is shown in fig. 1. A phase transition cannot occur unless there is an infinite connected cluster of spins. If p is less than a critical value, p_c, there is no infinite cluster and $T_c(p)$ is zero. At p_c, an infinite cluster appears. This is the phenomenon of *percolation*. For $p > p_c$, $T_c(p)$ is greater than zero. The point A in fig. 1 is a multicritical point [26–28] at which the thermally driven magnetic phase transition and the geometrically driven percolation transition meet. Before considering scaling in the vicinity of this point in more detail, let us first review some of the properties of percolation [6]. For $p > p_c$, a fraction $\mathscr{P}(p)$ of the sites is in an infinitely large cluster connected by bonds J. $\mathscr{P}(p)$ goes continuously to zero at p_c so that we have

$$\mathscr{P}(p) \sim \begin{cases} 0 & \text{if } p < p_c \\ (p - p_c)^{\beta_p} & \text{if } p > p_c \,. \end{cases} \tag{1.3.1}$$

Thus $\mathscr{P}(p)$ is analogous to the order parameter in thermodynamic phase transitions. The analog of the non-local susceptibility is the pair connectedness function

$$G_p^c(\boldsymbol{x}, \boldsymbol{x}') = [C(\boldsymbol{x}, \boldsymbol{x}')]_{av} - \mathscr{P}^2(p)\,, \tag{1.3.2}$$

where $C(\boldsymbol{x}, \boldsymbol{x}') = 1$ if $\boldsymbol{x}$ and $\boldsymbol{x}'$ are in the same cluster and zero otherwise. $[\ ]_{av}$ represents an average over all configurations. $G_p^c(\boldsymbol{x}, \boldsymbol{x}')$ becomes long range at p_c, and its Fourier transform satisfies a homogeneity relation similar to that of $\chi(\boldsymbol{q})$:

$$G_p^c(\boldsymbol{q}, p) = |p - p_c|^{-\gamma_p}\Phi(q\xi_p) = q^{2-\eta_p}\tilde{\Phi}(q\xi_p)\,, \tag{1.3.3}$$

where the correlation length, ξ, diverges at p_c

$$\xi \sim |p - p_c|^{-\nu_p}\,. \tag{1.3.4}$$

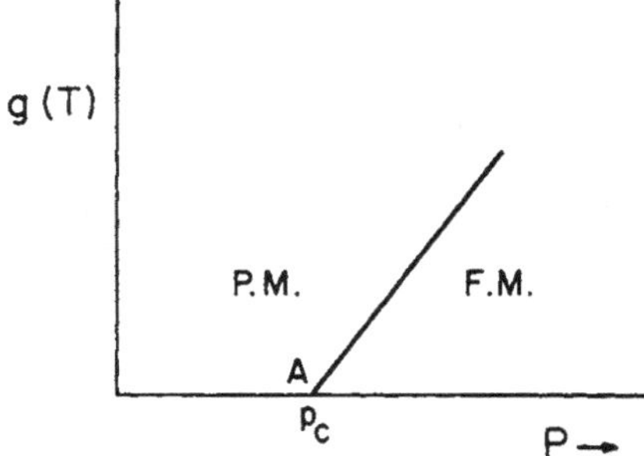

Fig. 1. Phase diagram for a randomly diluted ferromagnet. Active bonds (sites) are present with probability p and absent with probability $1 - p$. p_c is the percolation probability. $g(T)$ is a thermal variable that is zero at $T = 0$. For Ising spins $g(T) \sim e^{-2J/T}$ and for m-component classical spins ($m > 1$) $g(T) \sim T$. The point A ($g(T) = 0$, $p = p_c$) is a multi-critical point. The line $g(T_c(p))$ separating the paramagnetic (PM) from the ferromagnetic (FM) phase is straight near A in accord with the prediction that $\zeta = 1$.

$G_p^c(\boldsymbol{q} = 0, p)$ is, apart from a non-critical part involving the infinite cluster, the mean square size of finite clusters,

$$S(p) = \frac{1}{N} \sum_{\boldsymbol{x},\boldsymbol{x}'} G_p^c(\boldsymbol{x}, \boldsymbol{x}') = \sum_{n_s}{}' n_s^2 \mathscr{K}(n_s) , \tag{1.3.5}$$

where $\mathscr{K}(n_s)$ is the number per site of clusters containing n_s sites, N is the number of sites in the lattice, and the sums are over all finite clusters.

The non-local spin susceptibility is the configurational average of eq. (1.2.3)

$$T\chi(\boldsymbol{x}, \boldsymbol{x}') = [\langle \boldsymbol{S}(\boldsymbol{x}) \cdot \boldsymbol{S}(\boldsymbol{x}') \rangle - \langle \boldsymbol{S}(\boldsymbol{x}) \rangle \cdot \langle \boldsymbol{S}(\boldsymbol{x}') \rangle]_{\text{av}} . \tag{1.3.6}$$

At zero temperature, all of the spins in a given cluster are parallel so that $\langle \boldsymbol{S}(\boldsymbol{x}) \cdot \boldsymbol{S}(\boldsymbol{x}') \rangle = C(\boldsymbol{x}, \boldsymbol{x}')$, and $[\langle \boldsymbol{S}(\boldsymbol{x}) \rangle \cdot \langle \boldsymbol{S}(\boldsymbol{x}') \rangle]_{\text{av}} = \mathscr{P}^2(p)$ implying that

$$\lim_{T \to 0} T\chi(\boldsymbol{x}, \boldsymbol{x}') = G_p^c(\boldsymbol{x}, \boldsymbol{x}') . \tag{1.3.7}$$

This relates spin correlations as p is carried along the line $T = 0$ in fig. 1 to percolation.

We now inquire into the behavior of χT when temperature is increased in the vicinity of point A. A reasonable ansatz which has been verified explicitly within the context of the ε-expansion [29] is that χT is a scaling function of $p - p_c$ and some field g depending on T (and possibly $p - p_c$) and equal to zero when $T = 0$. Since $\chi T = S(p)$ at $T = 0$, we should have

$$\chi T = |p - p_c|^{-\gamma_p} \Phi\left(\frac{g}{|p - p_c|^{\phi}}\right) , \tag{1.3.8}$$

where ϕ is the crossover exponent for the field g. Similar scaling relations can be produced for the free energy, specific heat and magnetization.

A simple geometrical interpretation of the field g and the exponent x can be obtained from a heuristic picture of the infinite cluster above p_c due to de Gennes [30] and Skal and Shklovskii [31]. They remove all "dangling ends" from the infinite cluster and view the resulting backbone as a superlattice of nodes defined as points where there are more than two disjoint paths to infinity. This construction [32] is shown schematically in fig. 2. Two nodes are adjacent if there is a path connecting them which does not pass through another node. A length L_{ij} is then defined as being the smallest number of bonds in a path connecting adjacent nodes i and j. A length L may then be defined as a typical or average value of L_{ij}. Evidently, the real space distance between the nodes of the superlattice is of order the pair connectedness correlation length, ξ_p, whereas L is greater than or equal to ξ_p. For high enough spatial dimensionality, d ($d < 6$ as we shall see later), the critical behavior of the percolation statistics is mean field-like,

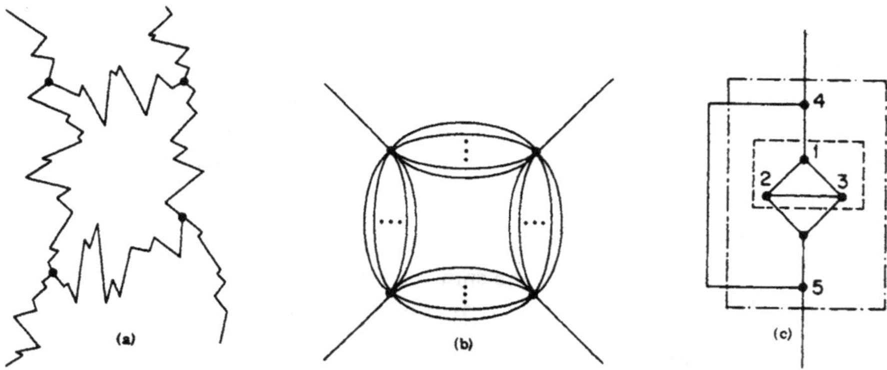

Fig. 2. (a) Schematic representation of a randomly diluted network near the percolation threshold. A path length between nodes (. . .) is of order L whereas the real space separation between nodes is of order ξ. This picture holds for $d > d_*$ when parallel paths between nodes can be neglected. (b) The same as (a) except for $d < d_*$ when parallel paths between nodes cannot be neglected. (c) Here we illustrate the difficulty in defining nodes in two dimensions by the criterion that it be a point with more than two independent paths to "infinity". Considering only the inner rectangle (dashed line) vertices 1, 2, and 3 appear to be nodes. Examination of the larger rectangle (dot-dashed lines) shows that 1, 2, and 3 are not nodes but that 4 and 5 appear to be. Finally, from consideration of the entire diagram, we see that even these are not nodes. For a two-dimensional graph, it is believed that this argument can be extended indefinitely.

i.e., there are no self-avoidance constraints, and the path between nodes is a random walk so that $L \sim \xi^2$ as $p \to p_c^+$.

As d is lowered one would expect L to be less than or equal to the number of steps in a self-avoiding walk between the nodes, i.e., $L \leqslant \xi_p^{1/\nu_{saw}}$ where ν_{saw} is the correlation length exponent for the self-avoiding walk. Defining an exponent ν via,

$$L \sim |p - p_c|^{-\zeta} . \tag{1.3.9}$$

we must have

$$\nu_p \leqslant \zeta \leqslant \nu_p/\nu_{saw} . \tag{1.3.10}$$

This picture of the super-lattice is almost certainly true for d large enough as predictions based on this construction have been explicitly verified in $6 - \varepsilon$ dimensions [29]. Below some critical dimension d_*, however, parallel paths may become important so that nodes are connected not by single strands but by rather complicated parallel networks as depicted in fig. 2b. In this regime one could distinguish various values for L. For instance, in addition to the purely geometrical length defined above, another length L_m can be defined from the effective exchange, J_{eff}, between nodes. For an Ising model, we have

$$\tanh \beta J_{eff} = [\tanh \beta J]^{L_m} . \tag{1.3.11}$$

Similar expressions for L_m can be obtained for m-component classical spins. For $d > d_*$, $L \sim L_m$. For $d > d_*$, L_m is less than L. For $d = 2$, it is likely that nodes cannot even be defined (cf. fig. 2c).

We now return to the scaling field g [27]. Correlations between spins with separation less than L are essentially one dimensional in character. Thus if the correlation length, $\xi_1(T)$, of a one-dimensional system is less than the number of steps L between nodes (we drop the possible distinction between L and L_m for simplicity), the system behaves like a collection of one-dimensional segments. If $\xi_1(T)$ is greater than L, the true d-dimensional character of the network is felt and ferromagnetic ordering can occur. It is thus appropriate to compare $\xi_1(T)$ to L

$$\chi T = |p - p_c|^{-\gamma_p} \Phi\left(\frac{\xi_1(T)}{L}\right) = |p - p_c|^{-\gamma_p} \tilde{\Phi}\left(\frac{\xi_1^{-1}(T)}{|p - p_c|^{\zeta}}\right) , \tag{1.3.12}$$

from which we obtain $\phi = \zeta$ and $g = \xi_1^{-1}(T)$. For m-component classical systems, $\xi_1^{-1}(T)$ is a function of Bessel functions of imaginary argument [33] which reduces at low temperatures to

$$\xi_1^{-1}(T) = \begin{cases} e^{-2J/T}, & m = 1 \\ \dfrac{m-1}{2m}\left(\dfrac{T}{J}\right), & m > 1 . \end{cases} \tag{1.3.13}$$

General symmetry arguments [34] show that ζ is unity to all orders in perturbation theory in agreement with previous ad hoc guesses. Explicit calculations show it to be unity to second-order in $(6 - d)$ and first-order in $(d - 1)$. Using $\zeta = 1$ and eqs. (1.3.12) and (1.3.13), we obtain

$$T_c(p) \sim \begin{cases} \dfrac{2J}{\ln|p - p_c|}, & m = 1 \\ \dfrac{2m}{m-1} J|p - p_c|, & m > 1 . \end{cases} \tag{1.3.14}$$

1.4. Spin-glasses

Magnetic ions in dilute magnetic alloys such as Mn in Cu interact via an RKKY potential that oscillates in sign. This means that the interaction between two spins alternates between ferromagnetic and antiferromagnetic with distance, and a given spin will receive contradictory signals for the direction it should align from different spins in the sample. The long range oscillating nature of the RKKY interaction is somewhat unpleasant to deal with theoretically. Arguing that the alternating sign of the exchange is the essential feature of a spin-glass, Edwards and Anderson [8] introduced a model with random short range exchange, J, with probability distribution $P(J)$ giving non-zero weight to both positive and negative J. We will investigate the properties of this model in this section.

Consider first the case $[J]_{\text{av}} = \int \mathrm{d}J JP(J) = 0$. The assumption is that there exists a freezing temperature, T_f, below which spins are locked into some specific orientation. Thus the thermal average, $\langle \boldsymbol{S}(\boldsymbol{x})\rangle_{\{J\}}$ of the spin at site $\boldsymbol{x}$ is non-zero for $T < T_f$ for any configuration, $\{J\}$, of bonds. On the other hand, $[\langle \boldsymbol{S}(\boldsymbol{x})\rangle_{\{J\}}]_{\text{av}}$ is zero since $\langle \boldsymbol{S}(\boldsymbol{x})\rangle_{\{J\}}$ is as likely to point up as down in any configuration. Thus, $[\langle \boldsymbol{S}(\boldsymbol{x})\rangle_{\{J\}}]_{\text{av}}$ is not an acceptable order parameter for the spin-glass transition. The square of $\langle \boldsymbol{S}(\boldsymbol{x})\rangle_{\{J\}}$ is, however, positive in every configuration for $T < T_f$. Following Edwards and Anderson, we, therefore, introduce the spin-glass order parameter

$$Q_{ij} = [\langle S_i(\boldsymbol{x})\rangle_{\{J\}} \langle S_j(\boldsymbol{x})\rangle_{\{J\}}]_{\text{av}} = (Q/m)\delta_{ij} , \tag{1.4.1}$$

where

$$Q = [\langle S(x)\rangle_{\{J\}}\cdot\langle S(x)\rangle_{\{J\}}]_{av}\,. \tag{1.4.2}$$

Q is expected to grow as $[(T_f - T)/T]^{\beta_{SG}}$ below T_f. The susceptibility associated with the order parameter Q is

$$\chi_{SG}(x, x') = [\langle S(x)\cdot S(x')\rangle^2_{\{J\}}]_{av} - (1/m)Q^2\,, \tag{1.4.3}$$

where we have not included irrelevant factors of temperature. χ_{SG} $(q = 0)$ should diverge as $|T - T_f|^{-\gamma_{SG}}$ near T_f. High temperature series [35] have verified this for $d > 4$. The high T results for γ_{SG} are summarized in table 1. Note that the series yields nonsense for $d \leqslant 4$ indicating a possible breakdown of the Edwards and Anderson state below four dimensions.

If $[J]_{av} = 0$, $[\langle S(x)\cdot S(x')\rangle]_{av} = \delta_{x,x'}$. This can be seen most easily by considering a high temperature expansion of this expression. Each bond has zero average and appears once and only once in such an expansion implying that all diagrams linking x to x' for $x \neq x'$ will vanish upon averaging. Similarly, $[\langle S(x)\rangle\cdot\langle S(x')\rangle]_{av} = Q\delta_{x,x'}$. This implies that the magnetic susceptibility will obey a Curie law for $T > T_f$ and have a cusp at T_f since

$$\begin{aligned}\chi &= \frac{1}{T}\sum_{x'}[\langle S(x)\cdot S(x')\rangle - \langle S(x)\rangle\cdot\langle S(x')\rangle]_{av} \\ &= \begin{cases}1/T\,, & T > T_f \\ (1 - Q)/T\,, & T < T_f\,.\end{cases}\end{aligned} \tag{1.4.4}$$

A cusp is in fact observed in the susceptibility of experimental spin glasses [36]. It must be observed, however, that a cusp is also predicted for the Mattis model [24,25] to be discussed in the next section.

If $P(J)$ is such that $[J]_{av}$ is greater than zero, there is competition between spin glass and ferromagnetic ordering, and there will be a critical point where the paramagnetic, spin-glass and ferromagnetic phases meet [22,37]. To be specific, consider a two sublattice system (such as simple cubic or bcc) with bonds of strength $+J$ present with probability p and of strength $-J$ present with probability $1 - p$. The phase diagram for this system is shown schematically in fig. 3. At $p = \frac{1}{2}$, $P(J) = P(-J)$, and the previous discussion applies. As p is increased, ferromagnetic clustering manifested in finite range correlations in $[\langle S(x)\cdot S(x')\rangle]_{av}$ and $[\langle S(x)\rangle\cdot\langle S(x')\rangle]_{av}$ occurs in the spin-glass phase. At the ferromagnetic

Table 1
High T series results for γ_Q as a function of dimensionality d [35].

d	γ_Q [a])	γ_Q [b])
6.00	1.32	1.00
5.75	1.40	1.25
5.50	1.52	1.53
5.25	1.69	1.84
5.00	1.95	2.23
4.75	2.4	2.7
4.50	3.4	3.7
4.25	7.0	7.1
4.00	∞	∞

[a]) No corrections to scaling assumed.
[b]) Corrections to scaling as predicted by renormalization group calculations assumed.

phase boundary, these functions develop infinite range correlations and the spin-glass order parameter, Q, goes to zero. The clustering effect for $[J]_{av} \neq 0$ can be used to explain a shift in the peak of the diffuse neutron scattering cross-section [38] as a function of wave number $\boldsymbol{q}$.

The phase diagram shown in fig. 3 and associated critical exponents can

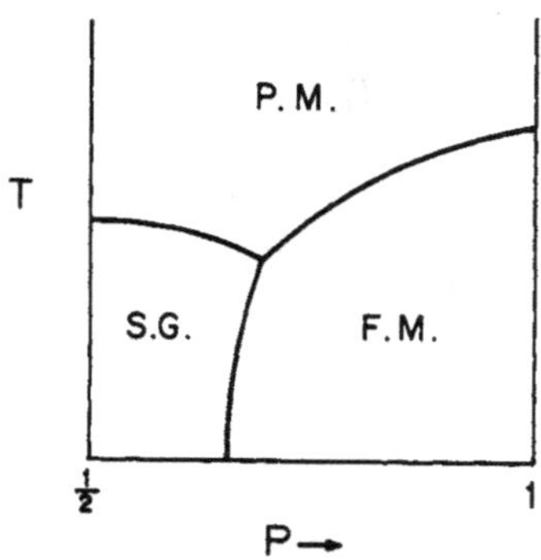

Fig. 3. Phase diagram for a spin-glass in which bonds with strength $+J$ and $-J$ are present respectively with probability p and $1 - p$. For $p = \frac{1}{2}$ there is a paramagnetic phase (PM) at high temperature and a spin-glass (SG) phase at low temperature. As p is increased at $T = 0$, there is a transition from the spin-glass phase to a ferromagnetic phase (FM). Values of $p < \frac{1}{2}$ have not been included in this diagram because what happens depends on the type of lattice.

be calculated in $6 - \varepsilon$ dimensions using the replica trick to be discussed. The paramagnetic phase is almost certainly described correctly by this technique. Questions still remain, however, about the spin-glass phase [37,39–41]. We will return to this question in sect. 2.6. It is expected that fig. 3 is qualitatively correct for dimension above four. What happens below four dimensions is still a subject of controversy.

The description of the Edwards and Anderson spin-glass presented so far is the one that lends itself most readily to the field theoretic formulation to be given in the next section. It does not, however, efficiently emphasize some important and unique features of this state such as the high ground state degeneracy. An alternate description of the spin-glass exploiting the local gauge invariance [42–44] of the EA model is useful in discussing ground state properties. The free energy, F, of the EA model is invariant under the gauge transformation $S(\boldsymbol{x}) \to \varepsilon(\boldsymbol{x})S(\boldsymbol{x})$, $J(\boldsymbol{x}, \boldsymbol{x}') \to \varepsilon(\boldsymbol{x})J(\boldsymbol{x}, \boldsymbol{x}') \times \varepsilon(\boldsymbol{x}')$ where $\varepsilon(\boldsymbol{x}) = \pm 1$. F can, therefore, only be a function of gauge invariant quantities. In a model in which $J(\boldsymbol{x}, \boldsymbol{x}')$ can only take on values $\pm J$, an appropriate gauge invariant quantity is

$$\Phi = \prod_C \frac{J(\boldsymbol{x}_i, \boldsymbol{x}_{i+1})}{|J|}, \tag{1.4.5}$$

where C is the circumference of a fundamental plaquet. A plaquet with $\Phi = -1$ is frustrated. Each frustrated plaquet can be viewed as the origin of a string passing through dissatisfied bonds and terminating in another frustrated plaquet or at the boundary of the sample. A ground state of the system is a state with the minimum possible length of string. The degeneracy

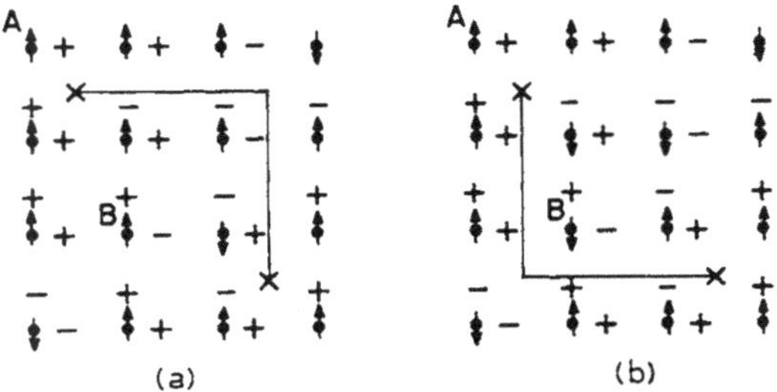

Fig. 4. A section of a 2 × 2 square lattice occupied with $+J$ bonds (+) and $-J$ bond (−). The two plaquets marked (×) are frustrated. Figs. 4a and 4b represent two different configurations of spins (marked ↑ or ↓) with the same energy. The frustrated plaquets are connected by strings passing through the dissatisfied bonds. The energy of a given spin configuration is proportional to the length of string in that configuration. Note that the spin A is up in both figs. 4a and 4b whereas spin B is up in fig. 4a and down in fig. 4b.

of the ground state arises from the fact that there are many configurations with the same total length of string. An example of this effect is shown in fig. 4. Fig. 4 is also useful in emphasizing another essential feature of the EA spin-glass. Concentrate on spins A and B in this figure, and specify that spin A will be up. For one configuration of string, spin B is up whereas for the other, it is down. One cannot determine the sign of $S(\boldsymbol{x}')$ from the sign of $S(\boldsymbol{x})$ even at zero temperature without knowing the signs of spins on intermediate sites between $\boldsymbol{x}$ and $\boldsymbol{x}'$. In other words local information provided in the vicinity of $\boldsymbol{x}$ is not sufficient to determine $S(\boldsymbol{x})$. When there is a continuous distribution of J's, it is not so obvious that there will be a degenerate ground state, though there should be many low lying excited states.

1.5. The Mattis model [24,25,45]

This model begins with the Heisenberg hamiltonian as does the EA model. In the EA model, $J(\boldsymbol{x}, \boldsymbol{x}')$ is an independent random variable at each bond $\boldsymbol{x} - \boldsymbol{x}'$. In the Mattis model, the randomness in $J(\boldsymbol{x}, \boldsymbol{x}')$ is controlled by site variables. Each site, $\boldsymbol{x}$, can be occupied at random with atoms A or B with probabilities p_A and $p_B = 1 - p_A$. There is an exchange J_{AA} between neighboring A spins, J_{BB} between B spins and J_{AB} between neighboring A and B spins. $J(\boldsymbol{x}, \boldsymbol{x}')$ is then given by [25]

$$J(\boldsymbol{x}, \boldsymbol{x}') = J_1 + J_2\varepsilon(\boldsymbol{x})\varepsilon(\boldsymbol{x}') + J_3[\varepsilon(\boldsymbol{x}) + \varepsilon(\boldsymbol{x}')] , \tag{1.5.1}$$

where $\varepsilon(\boldsymbol{x}) = +1$ if $\boldsymbol{x}$ is occupied by an A atom and -1 if it is occupied by a B atom and

$$\begin{aligned} J_1 &= \tfrac{1}{4}(J_{AA} + J_{BB} + 2J_{AB}) , \\ J_2 &= \tfrac{1}{4}(J_{AA} + J_{BB} - 2J_{AB}) , \\ J_3 &= \tfrac{1}{4}(J_{AA} - J_{BB}) . \end{aligned} \tag{1.5.2}$$

The phase diagram for this model for general J_{AA}, J_{AB}, J_{BB} and p_A is quite complex. If $J_{AA} = J_{BB} = -J_{AB}$, the transformation $\boldsymbol{\tau}(\boldsymbol{x}) = \varepsilon(\boldsymbol{x})\boldsymbol{S}(\boldsymbol{x})$ converts the Hamiltonian into a homogeneous hamiltonian in terms of $\boldsymbol{\tau}(\boldsymbol{x})$:

$$\mathcal{H} = - \sum_{\langle \boldsymbol{x}, \boldsymbol{x}' \rangle} J_2 \boldsymbol{\tau}(\boldsymbol{x}) \cdot \boldsymbol{\tau}(\boldsymbol{x}') . \tag{1.5.3}$$

The partition function, free energy and specific heat of this model are identical to those of the m-component Heisenberg model

$$Z[\{\varepsilon(x)\}, J] = \mathrm{Tr}_s \exp(-\beta\mathcal{H}[\{\varepsilon(\boldsymbol{x})\}, S]) \equiv \mathrm{Tr}_\tau \exp[-\beta\mathcal{H}(\tau)] . \tag{1.5.4}$$

The magnetization and correlation functions, on the other hand, depend on p_A:

$$[\langle \boldsymbol{S}(\boldsymbol{x})\rangle]_{av} = [\varepsilon(\boldsymbol{x})]_{av}\langle \boldsymbol{\tau}(\boldsymbol{x})\rangle = 0 \quad \text{for} \quad p_A = \tfrac{1}{2}, \tag{1.5.5a}$$

$$\begin{aligned}[\langle \boldsymbol{S}(\boldsymbol{x})\rangle \cdot \langle \boldsymbol{S}(\boldsymbol{x}')\rangle]_{av} &= [\varepsilon(\boldsymbol{x})\varepsilon(\boldsymbol{x}')]_{av}\langle \boldsymbol{\tau}(\boldsymbol{x})\rangle \cdot \langle \boldsymbol{\tau}(\boldsymbol{x}')\rangle \\ &= \delta_{\boldsymbol{x},\boldsymbol{x}'}|\langle \boldsymbol{\tau}(\boldsymbol{x})\rangle|^2 \quad \text{for} \quad p_A = \tfrac{1}{2},\end{aligned} \tag{1.5.5b}$$

$$\begin{aligned}T\chi &= \sum_{\boldsymbol{x}'} [\varepsilon(\boldsymbol{x})\varepsilon(\boldsymbol{x}')]_{av}[\langle \boldsymbol{\tau}(\boldsymbol{x})\cdot \boldsymbol{\tau}(\boldsymbol{x}')\rangle - \langle \boldsymbol{\tau}(\boldsymbol{x})\rangle \cdot \langle \boldsymbol{\tau}(\boldsymbol{x})\rangle] \\ &= \langle |\boldsymbol{\tau}(\boldsymbol{x})|^2\rangle - |\langle \boldsymbol{\tau}(\boldsymbol{x})\rangle^2| \,.\end{aligned} \tag{1.5.5c}$$

One can easily see from eq. (1.5.5c) that the magnetic susceptibility χ has a cusp at T_c, just as the EA spin-glass. The models are, however, fundamentally different. The exponents for the Mattis model with $J_{AA} = J_{BB} = -J_{AB}$ are those of the homogeneous m-component Heisenberg model. This model is classical above four dimensions whereas the EA model becomes classical only above six dimensions as demonstrated by high T series [35] and the ε-expansion [46] of the next section.

The fundamental difference between the two models is that there is no frustration in the Mattis model when $J_1 = J_3 = 0$:

$$\Phi = \prod_C \frac{J_2(\boldsymbol{x}_i, \boldsymbol{x}_{i+1})}{|J_2|} = \prod_C \varepsilon(\boldsymbol{x}_i)\varepsilon(\boldsymbol{x}_{i+1}) = 1$$

for every plaquet. In fact since we know $\langle \tau(\boldsymbol{x})\rangle = 1$ in the ground state, we know that every A spin will be up and every B spin will be down. We can determine the sign of the spin at site $\boldsymbol{x}$ merely by knowing the type of spin that occupies that site, i.e., only local information is needed to determine the spin in the Mattis model whereas global information is required for the EA model. Note, however, the Mattis model with J_1 and $J_3 \neq 0$ can have frustration.

If the constraint that $J_{AA} = J_{BB} = -J_{AB}$ is relaxed, the phase diagram of the Mattis model becomes quite complex. In general, the order parameter will be a linear combination of $\boldsymbol{S}(\boldsymbol{x})$ and $\tau(\boldsymbol{x})$ and critical exponents will be those appropriate to a weakly random m-component system. Special points, such as $J_{AB} = 0$ which represents two non-interacting sublattices, also occur.

C. Geometrical Critical Problems

In considering the problem of phase transitions in randomly diluted magnets, we were led naturally to a consideration of the problem of percolation. Percolation is the process of formation of an infinite connected cluster in a randomly diluted network. Since there are no dynamical variables such as spin that play a role in the transition, percolation is often referred to as a geometrically driven transition. There are other examples of critical problems that are geometrical in origin which are naturally described in the language of phase transitions and that can be treated by the analytical techniques to be discussed in the next chapter. We will describe some of these problems in following sections.

1.6. Random resistor networks

A regular network of resistors is formed by placing a resistor of conductance σ at each bond on a lattice. In a random resistor network, each bond, $\boldsymbol{x} - \boldsymbol{x}'$, has a random conductance, $\sigma(\boldsymbol{x}, \boldsymbol{x}')$ governed by some probability distribution. In these lectures, we will consider in detail only the diluted lattice in which $\sigma(\boldsymbol{x}, \boldsymbol{x}')$ has value σ with probability p and zero with probability $1 - p$. More generally [47] one might consider systems in which $\sigma(\boldsymbol{x}, \boldsymbol{x}')$ is $\sigma_>$ with probability p and $\sigma_<$ with probability $1 - p$. The limit $\sigma_> \to \infty$ corresponds to randomly connected superconducting links. The quantity of greatest interest is the macroscopic conductivity Σ which relates the macroscopic current density $\boldsymbol{J}$ to gradients in the voltage V:

$$\boldsymbol{J} = \Sigma \nabla V . \tag{1.6.1}$$

The divergence of $\boldsymbol{J}$ satisfies

$$\nabla \cdot \boldsymbol{J}(\boldsymbol{r}) = \sum_{\boldsymbol{x}} I(\boldsymbol{x}) \delta(\boldsymbol{r} - \boldsymbol{x}) , \tag{1.6.2}$$

where $I(\boldsymbol{x})$ is the outward current at the site $\boldsymbol{x}$. Thus, we have

$$\Sigma \nabla^2 V(\boldsymbol{r}) = \nabla \cdot \boldsymbol{J} , \tag{1.6.3}$$

and

$$V(\boldsymbol{q}) = -\frac{1}{\Sigma q^2} \mathscr{T}(\boldsymbol{q}) , \tag{1.6.4}$$

where

$$V(\boldsymbol{q}) = \int \mathrm{d}^d r \exp(-\mathrm{i} \boldsymbol{q} \cdot \boldsymbol{r}) V(\boldsymbol{r}) , \tag{1.6.5}$$

and

$$\mathscr{T}(\boldsymbol{q}) = \int \mathrm{d}^d r \exp(-\mathrm{i} \boldsymbol{q} \cdot \boldsymbol{r}) \nabla \cdot \boldsymbol{J}(\boldsymbol{r}) = \sum_{\boldsymbol{x}} \exp(-\mathrm{i} \boldsymbol{q} \cdot \boldsymbol{x}) I(\boldsymbol{x}) . \tag{1.6.6}$$

The currents in a resistor network satisfy Kirchhoff's laws. In particular the sum of the currents leaving a node at $\boldsymbol{x}$ must be zero. The current in the resistor at bond $\boldsymbol{x} - \boldsymbol{x}'$ is $\sigma(\boldsymbol{x}, \boldsymbol{x}')[V(\boldsymbol{x}) - V(\boldsymbol{x}')]$ so that

$$\sum_{\boldsymbol{x}'} \sigma(\boldsymbol{x}, \boldsymbol{x}')[V(\boldsymbol{x}) - V(\boldsymbol{x}')] = -I(\boldsymbol{x}), \tag{1.6.7}$$

where $I(\boldsymbol{x})$ is the current flowing from $\boldsymbol{x}$ through an external wire attached at that point. Eq. (1.6.7) can be rewritten as

$$\sum_{\boldsymbol{x}'} B(\boldsymbol{x}, \boldsymbol{x}')V(\boldsymbol{x}') = -I(\boldsymbol{x}), \tag{1.6.8}$$

where

$$B(\boldsymbol{x}, \boldsymbol{x}') = \delta_{\boldsymbol{x},\boldsymbol{x}'}\left(\sum_{\boldsymbol{x}''} \sigma(\boldsymbol{x}, \boldsymbol{x}'')\right) - \sigma(\boldsymbol{x}, \boldsymbol{x}'), \tag{1.6.9}$$

so that

$$V(\boldsymbol{x}) = -\sum_{\boldsymbol{x}'} B^{-1}(\boldsymbol{x}, \boldsymbol{x}')I(\boldsymbol{x}'). \tag{1.6.10}$$

In homogeneous systems, $B(\boldsymbol{x}, \boldsymbol{x}')$ depends only on $\boldsymbol{x} - \boldsymbol{x}'$ so that we can introduce its Fourier transform

$$B_{\boldsymbol{q}} = \sum_{\boldsymbol{x}} \exp(-\mathrm{i}\boldsymbol{q}\cdot\boldsymbol{x})B(\boldsymbol{x}, 0). \tag{1.6.11}$$

Similarly, we define

$$A_{\boldsymbol{q}} = \sum_{\boldsymbol{x}} \exp(-\mathrm{i}\boldsymbol{q}\cdot\boldsymbol{x})A(\boldsymbol{x}), \tag{1.6.12a}$$

$$A(\boldsymbol{x}) = \frac{1}{N}\sum_{\boldsymbol{q}} \exp(\mathrm{i}\boldsymbol{q}\cdot\boldsymbol{x})A_{\boldsymbol{q}} = \Omega_0 \int_0 \frac{\mathrm{d}^d q}{(2\pi)^d} \exp(\mathrm{i}\boldsymbol{q}\cdot\boldsymbol{x})A_{\boldsymbol{q}}, \tag{1.6.12b}$$

for $A(\boldsymbol{x}) = V(\boldsymbol{x})$ or $I(\boldsymbol{x})$ where $\Omega_0 = a^d$ is the volume of a unit cell. Thus, $A(\boldsymbol{q}) = \Omega_0 A_{\boldsymbol{q}}$. For a homogeneous network consisting only of nearest neighbor conductances σ we have

$$B_{\boldsymbol{q}} = \sigma \sum_{\boldsymbol{\delta}} (1 - \exp[-\mathrm{i}\boldsymbol{q}\cdot\boldsymbol{\delta}]) \tag{1.6.13a}$$

$$= \sigma(qa)^2 + \mathrm{O}((qa)^4), \tag{1.6.13b}$$

where $\boldsymbol{\delta}$ are the nearest neighbor vectors and eq. (1.6.13a) is appropriate to a d-dimensional hypercubic lattice. Fourier transforming eq. (1.6.10), we have

$$V(\boldsymbol{q}) = -\Omega_0(I_{\boldsymbol{q}}/B_{\boldsymbol{q}}) = -(B_{\boldsymbol{q}}/\Omega_0)^{-1}\mathscr{I}(\boldsymbol{q}), \tag{1.6.14}$$

where we used eq. (1.6.5). In the long wavelength limit, this must agree with eq. (1.6.4) so that

$$\Sigma = \lim_{q\to 0}(B_q/\Omega_0 q^2) = a^{-d+2}R_0^{-1}\,, \tag{1.6.15}$$

where $R_0 = \sigma^{-1}$ is the fundamental resistance of a bond.

In random systems, homogeneity is restored only after averaging. Thus, to find the macroscopic conductivity of a random resistor network, we need to average eq. (1.6.10) and then Fourier transform, i.e., we need to deal with the Fourier transform of $[B^{-1}(\boldsymbol{x}, \boldsymbol{x}')]_{\text{av}}$. We, therefore, define

$$[B^{-1}(\boldsymbol{q})]_{\text{av}} = \sum_{\boldsymbol{x}} \exp(-\mathrm{i}\boldsymbol{q}\cdot\boldsymbol{x})[B^{-1}(\boldsymbol{x}, \boldsymbol{x}')]_{\text{av}}\,. \tag{1.6.16}$$

(Since we will never deal with a situation where we first Fourier transform and then average, there should be no confusion with the above notation.) The macroscopic conductivity of a random resistor network is, therefore, determined by

$$\Sigma = \lim_{q\to 0}\frac{1}{q^2}\frac{1}{\Omega_0}[B^{-1}(\boldsymbol{q})]_{\text{av}}^{-1}\,. \tag{1.6.17}$$

At times, it is convenient to introduce external frequency dependent fields [48] by introducing a capacitor at each site connecting to a potential $U(\boldsymbol{x}, t) = U(\boldsymbol{x})\exp(-\mathrm{i}\omega t)$. Then Kirchhoff's laws read

$$V(\boldsymbol{x}) = \sum_{\boldsymbol{x}'} B^{-1}(\boldsymbol{x}, \boldsymbol{x}'; \omega)(\mathrm{i}\omega U(\boldsymbol{x}))\,, \tag{1.6.18}$$

where we have set the capacitance at each site equal to unity and where

$$B(\boldsymbol{x}, \boldsymbol{x}'; \omega) = B(\boldsymbol{x}, \boldsymbol{x}') - \mathrm{i}\omega\delta_{\boldsymbol{x},\boldsymbol{x}'}\,. \tag{1.6.19}$$

The dc conductivity can then be expressed as a Kubo formula

$$\Sigma = \lim_{\omega\to 0}\lim_{q\to 0}\Omega_0\frac{\omega^2}{q^2}\,\mathrm{Re}[B^{-1}(\boldsymbol{q}, \omega)]_{\text{av}}\,, \tag{1.6.20}$$

where the order of limits is important.

In a randomly diluted network, the macroscopic conductivity will be zero for $p < p_c$ and greater than zero for $p > p_c$

$$\Sigma \sim \begin{cases} 0 & p < p_c \\ (p-p_c)^{\mu} & p > p_c\,. \end{cases} \tag{1.6.21}$$

The Skal–Shkovskii–de Gennes picture [30,31] of the diluted lattice above p_c and eq. (1.6.15) can be used to relate μ to γ_p and ζ [30]. The diluted lattice is viewed as a regular network of nodes separated by ξ_p. Between each pair of nodes, the fundamental resistance is $L_r R_0$. Therefore, using eq. (1.6.15), we have

$$\Sigma \sim \xi^{-d+2} L_r^{-1}, \tag{1.6.22}$$

and

$$\mu = (d-2)\nu_p + \zeta_r . \tag{1.6.23}$$

Near six dimensions where parallel paths are unimportant the magnetic and resistive lengths L_m and L_r are equal. Near two dimensions there is some possibility that they differ. It may in fact be impossible to define nodes in two dimensions [32]. Nevertheless, general symmetry arguments indicate that $\zeta_r = \zeta_m = 1$ to all orders in perturbation theory [34].

The macroscopic conductivity is only defined for $p > p_c$. It is usually easier to calculate critical exponents in disordered phases. It is, therefore, of some interest to find some quantity defined for $p < p_c$ which can lead to the exponent μ by scaling. One such quantity is the resistive susceptibility [49] defined as

$$\chi_r(\boldsymbol{x}, \boldsymbol{x}') = [R(\boldsymbol{x}, \boldsymbol{x}') C(\boldsymbol{x}, \boldsymbol{x}')]_{av} , \tag{1.6.24}$$

where $R(\boldsymbol{x}, \boldsymbol{x}')$ is the resistance between points $\boldsymbol{x}$ and $\boldsymbol{x}'$. In a homogeneous system where $C(x, x') = 1$, $\chi_r(x, x')$ is easily calculated

$$\chi_r(\boldsymbol{x}, \boldsymbol{x}') = \int_0 \frac{d^d q}{(2\pi)^d} \frac{1 - \exp[i\boldsymbol{q}\cdot(\boldsymbol{x} - \boldsymbol{x}')]}{\Sigma q^2}$$

$$\sim \frac{1}{\Sigma a^{d-2}} \quad \text{as} \quad |\boldsymbol{x} - \boldsymbol{x}'| \to \infty , \tag{1.6.25}$$

where the integral is over the first Brillouin zone. In a randomly diluted system, the factor $C(\boldsymbol{x}, \boldsymbol{x}')$ restricts $\boldsymbol{x}$ and $\boldsymbol{x}'$ to be in the same cluster and

$$\chi_r(\boldsymbol{x}, \boldsymbol{x}') \to \mathscr{P}^2(p)/\Sigma \xi^{d-2} , \tag{1.6.26}$$

as $|x - x'| \to \infty$. Below p_c, the resistive susceptibility should diverge:

$$\chi_r = \sum_{\boldsymbol{x}'} \chi_r(\boldsymbol{x}, \boldsymbol{x}') \sim |p - p_c|^{-\gamma_r} . \tag{1.6.27}$$

Assuming $\chi_r(\boldsymbol{x}, \boldsymbol{x}')$ satisfies the usual homogeneity relations, we have

$$\chi_r(\boldsymbol{x}, \boldsymbol{x}') = |\boldsymbol{x} - \boldsymbol{x}'|^{-d+\gamma_r/\nu_p} \Phi_r\left(\frac{|\boldsymbol{x} - \boldsymbol{x}'|}{\xi_p}\right) . \tag{1.6.28}$$

This scaling form and eq. (1.6.26) then imply

$$\gamma_{\mathrm{r}} = \gamma_{\mathrm{p}} + \mu - (d-2)\gamma_{\mathrm{p}} = \gamma_{\mathrm{p}} + \zeta_{\mathrm{r}}\,. \tag{1.6.29}$$

A low density series for χ_{r} can be calculated for hypercubic lattices in any dimensions. This series shows that $\gamma_{\mathrm{r}} = 2$ for $d \geqslant 6$ and that it increases for $d < 6$. This implies that $\mu = 3$ in six dimensions. Calculations of Σ on a Cayley tree also imply $\mu = 3$ in high dimension [30,50]. A summary of the series results for γ_{r} is given in table 2.

If conductances $\sigma_>$ are present with probability p and $\sigma_<$ with probability $1-p$, then Straley [47] has pointed out that the macroscopic conductivity should obey a homogeneity relation of the form

$$\Sigma = |p - p_{\mathrm{c}}|^{\mu}\sigma_>\Phi_{\Sigma}\left(\frac{\sigma_</\sigma_>}{|p-p_{\mathrm{c}}|^{\phi_<}}\right). \tag{1.6.30}$$

If $\sigma_> \to \infty$ (superconducting links), then

$$\Sigma \sim |p - p_{\mathrm{c}}|^{-s} \quad \text{as} \quad p \to p_{\mathrm{c}}^{-}\,, \tag{1.6.31}$$

where $s = \phi_< - \mu$. The elegant treatment of this problem by Stephen [48] indicates that

$$\phi_< = \zeta_{\mathrm{r}} + \Delta\,, \tag{1.6.32}$$

where Δ is the gap exponent $d\nu_{\mathrm{p}} - \beta_{\mathrm{p}}$. In six dimensions $\Delta = 2$, $\phi_< = 3$ and $s = 0$.

Before closing this section, we should emphasize that analogs of the random resistor network appear in a variety of physical situations. Any quantity which obeys an equation of the form of eq. (1.6.7) that reduces to Laplace's equation with a source (eq. 1.6.3) in the long wavelength limit gives rise to a stiffness analogous to the conductivity Σ [51]. Thus the spin wave stiffness A in a randomly diluted ferromagnet will grow as $(p-p_{\mathrm{c}})^{\mu}$

Table 2 [49]

Results for $\gamma^{(r)}$ as a function of dimensionality, d, obtained from a series in p, the probability that a bond is active. These numbers do not take into account corrections to scaling.

d	7	6	5	4	3	2
$\gamma^{(r)}$	2.04	2.09	2.19	2.45	2.78	3.8

for $p > p_c$ [52]. The spin wave frequencies are then proportional to Dq^2 where $D = |p - p_c|^{\mu-\beta}$. Similarly the shear modulus just above the gel point of a gel [53] should be proportional to $|p - p_c|^{\mu}$.

1.7. Polymers, gelation, and vulcanization

There are a number of problems related to the statistics of large polymers that can be fruitfully described in the language of critical phenomena. The most familiar of these is that of the configurational statistics of an isolated linear polymer consisting of a large number, N, of individual monomers. If the monomers are freely joined and can pass through each other, the probability, $P(\boldsymbol{x}, N)$, that one endpoint is at $\boldsymbol{x}$ given that the other is at the origin is just the probability that x is reached after N steps in a random walk. It obeys the diffusion equation

$$\partial P/\partial N = \nabla^2 P \,, \tag{1.7.1}$$

with solution

$$P(\boldsymbol{x}, N) = (1/4\pi N)^{1/2} \exp(-x^2/4N) \,. \tag{1.7.2}$$

Thus, the average end to end separation is proportional to $\sqrt{N}$:

$$\xi = \langle x^2 \rangle^{1/2} \sim N^{1/2} \,. \tag{1.7.3}$$

If there is a repulsive interaction between monomers, one would expect ξ to be larger for a given N. In fact the polymer undergoes a self-avoiding walk of N steps and

$$\xi \sim N^{\nu_{\mathrm{saw}}} \,, \tag{1.7.4}$$

where ν_{saw} is the self-avoiding walk exponent. We will see in sect. 2.9 that this exponent is the correlation length exponent for the $m = 0$ Heisenberg model. The radius of gyration

$$R_G^2 = \frac{1}{N} \left\langle \sum_{i<j} |\boldsymbol{x}_i - \boldsymbol{x}_j|^2 \right\rangle , \tag{1.7.5}$$

where the sum is over all joints in the polymer also diverges as $N^{\nu}{}_{\mathrm{saw}}$.

If the concentration, c_p, of linear polymers in a solution is large enough, neighboring polymers will entangle even though the total volume occupied by polymeric monomers may still be small. The crossover monomer concentration, c, separating the dilute and semi-dilute can easily be calculated [55,56] by equating the average volume occupied by a polymer in the dilute regime to the inverse polymer concentration $c_p^{-1} = Nc^{-1}$

$$R_G^d \sim N^{d\nu_{saw}} \sim Nc^{*-1} . \tag{1.7.6}$$

Thus

$$c^* \sim N^{1-d\nu_{saw}} . \tag{1.7.7}$$

Since $d\nu_{saw} > 1$ for all dimensions of interest, c^* is much less than unity for large N. This crossover behavior gives rise to many interesting phenomena which will be treated in more detail in the lectures by P. G. de Gennes.

A monomer or bifunctional unit consists of two reactive endgroups that are strongly bound. For example, there may be a single type of reactive group, A, which can react with other groups to form a bound pair. A monomer would then have two A units denoted by A–A. A linear polymer is produced when these monomers react to produce chains $[A\text{–}A]_N$. When polyfunctional units $R\text{–}A_f$ are introduced, non-linear or branched polymers are produced. One can then inquire as to the dependence of the average endpoint separation or the radius of gyration of a non-linear polymer as a function of the number of monomers. Of particular interest is the case where the number of poly-functional units N_f is proportional to the number of monomers. Presumably, R_G is again proportional to N^ν where ν is some new correlation length exponent. In sect. 2.9, we will see that $\nu = \frac{1}{4}$ for gaussian branched polymers [57,58] with no loops. We will also calculate ν in an ε-expansion [59].

When branching is allowed, a new phenomena, not present in linear polymers, occurs. This is the phenomenon of gelation [7] which is intimately related to percolation. Gelation is the formation of an infinite molecule, the analog of the formation of an infinite cluster in percolation. A mean field theory for this process was formulated years ago by Flory [7]. We repeat his calculation here. For simplicity we consider only f and bi-functional units. Let α be the probability that a chain of monomers originating at a polyfunctional unit will terminate at another poly-functional unit. If $\alpha > 1/(f-1)$, each polyfunctional unit will lead on the average to at least one other unit and the molecule will be infinite. The condition for gelation is then

$$\alpha = 1/(f-1) . \tag{1.7.8}$$

If loop formation is neglected, α can easily be calculated. Let p_A be the fraction of reacted end groups and ρ be the fraction of endgroups on

f-functional units. α is the probability that an f-functional unit is connected directly to another f-functional unit plus the probability that it is connected to a monomer that is in turn connected to another f-functional unit, and so on. Therefore, we have

$$\alpha = p_A \rho \sum_{k=0}^{\infty} [p_A(1 - \rho)]^k = \frac{p_A \rho}{1 - p_A(1 - \rho)} . \tag{1.7.9}$$

If one assumes equal reactivity of all endgroups, regardless of their placement on polymers, p_A and ρ can easily be related to measurable quantities of the system. To do this, we first introduce some additional notation. Let

$$\begin{aligned} c &= N/\Omega = \text{concentration of monomers,} \\ c_p &= N_p/\Omega = \text{concentration of polymers,} \\ c_f &= N_f/\Omega = \text{concentration of } f\text{-functional units,} \\ c_v &= N_v/\Omega = \text{concentration of unreacted endgroups, and} \\ c_L &= N_L/\Omega = \text{concentration of loops.} \end{aligned}$$

If N_f f-functional units and N monomers are tossed into a melt, the initial concentration of unreacted endpoints is

$$c_v^0 = fc_f + 2c . \tag{1.7.10}$$

Each reaction reduces the number of endpoints by 2 so that

$$c_v = c_v^0 - 2R , \tag{1.7.11}$$

where R is the number of reactions divided by the volume. c_v can also be expressed in terms of c_p with the aid of the fundamental relation

$$\begin{aligned} N_v &= 2N_p + (f - 2)N_f - 2N_L , \\ c_v &= 2c_p + (f - 2)c_f - 2c_L . \end{aligned} \tag{1.7.12}$$

The fraction of reacted end groups satisfies

$$p_A = 2R/c_v^0 . \tag{1.7.13}$$

If the number of loops is assumed to be zero, this reduces to

$$p_A = 1 - \frac{2c_p + (f - 2)c_f}{2c + fc_f} . \tag{1.7.14}$$

The fraction of endgroups on f-functional units is trivially

$$\rho = \frac{fc_f}{2c + fc_f} . \tag{1.7.15}$$

c, c_p, c_f are measurable quantities. Eqs. (1.7.8), (1.7.9), (1.7.14), and (1.7.15) then determine at what values of these quantities gelation occurs. Experiments agree with these predictions to within ten or fifteen percent which is remarkably good considering that closed loops have been neglected. If $c \gg c_f$ or $c \gg c_p - c_f$, the condition for gelation reduces to

$$c_p = \tfrac{1}{2}(f-1)(f-2)c_f\,. \tag{1.7.16}$$

Vulcanization differs slightly from gelation. In this process, one starts with a "spaghetti" of long linear polymers. Cross linking of the polymer chains is then brought about by the introduction of 4-functional units that react with the polymers. Common methods of crosslinking include the introduction of sulfur or irradiating with light of the correct wavelength to cause photo-chemical reaction between polymers. At a critical concentration, p_c, of crosslinks an infinite molecule will form. Thus, vulcanization also is a percolation-like process.

It is obvious that gelation is nearly the same thing as percolation. For example the lattice model of gelation introduced by Stauffer [60] in which each site of a d-dimensional hypercube is occupied by a $2d$-functional unit and in which reactions of endgroups of neighboring sites occur with probability p is identical to bond percolation. When bi-functional units are introduced and molecules are allowed to move freely about, it is at least

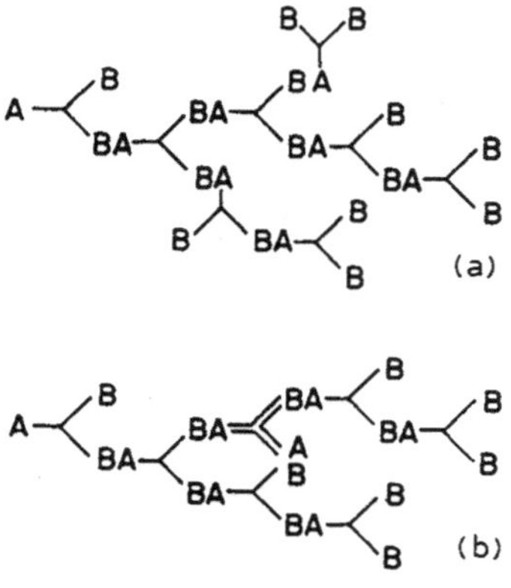

Fig. 5. (a) Typical polymer formed from tri-functional units ARB_2 in which only AB bonds are permitted. Note that these molecules can have at most one loop formed by reacting the initial A group with any of the terminal B's. (b) This figure shows how the addition of a single A_2RB unit (marked with double lines) can lead to the formation of an additional loop in a polymer composed of ARB_2 units with only AB bonds permitted.

plausible that gelation, vulcanization, and percolation remain in the same universality class. We shall see in sect. 2.9 that this is true in most cases, though it is possible in theory to change the universality class of gelation by controlling the number of loops [59].

Since the loop number will play an important part in the field theory of gelation to be introduced in sect. 2.9, it is worth pointing out here that it is possible to construct non-linear molecules with a limited number of loops. For example, suppose there are two types of endgroups A and B and suppose that A–B interactions can occur but A–A and B–B interactions cannot. Then a mixture of f-functional units A–R–B_{f-1} and monomers B–A will produce polymers with at most one loop per polymer as can be seen from fig. 5. In the loopless approximation, the quantity α is easily seen to be

$$\alpha = p/(f-1) \tag{1.7.17}$$

so that gelation cannot be reached even if $p \to 1$. Each poly-functional unit A_2–R–B_{f-2} added to a polymer consisting only of A–B's and A–R–B_{f-1}'s can increase the number of loops in that polymer by at most one.

2. Models and computational details

2.1. *Construction of field theories* [61,62]

In all systems that we consider, there will be a set of variables $S_l(\boldsymbol{x})$ defined on lattice site $\boldsymbol{x}$. $S_l(\boldsymbol{x})$ will vary from problem to problem, but it will usually be strongly constrained. For example, $S_l(\boldsymbol{x})$ might be the lth component of an m-component classical spin satisfying $\sum_l |S_l(\boldsymbol{x})|^2 = m$, or $S_l(\boldsymbol{x})$ might be a vector constrained to point along the axes of an m-dimensional hypercube. The hamiltonian is a function of $S_l(\boldsymbol{x})$ and can be expressed as a sum of three parts.

$$\beta\mathscr{H} = \beta\mathscr{H}_0 + \beta\mathscr{H}_a + \beta\mathscr{H}_{ext} \,. \tag{2.1.1}$$

$\beta\mathscr{H}_{ext}$ is the energy of interaction with an external field $H_l(\boldsymbol{x})$:

$$\beta\mathscr{H}_{ext} = -\sum_{\boldsymbol{x},l} H_l(\boldsymbol{x})S_l(\boldsymbol{x}) \,, \tag{2.1.2}$$

and $\beta\mathscr{H}_0$ is quadratic in $S_l(\boldsymbol{x})$,

$$\beta\mathscr{H}_0 = -\tfrac{1}{2} \sum_{\boldsymbol{x},\boldsymbol{x}',l} K_l(\boldsymbol{x},\boldsymbol{x}')S_l(\boldsymbol{x})S_l(\boldsymbol{x}') \,, \tag{2.1.3}$$

where the sum is over all sites x and x'. $TK_l(x, x')$ is often a nearest neighbor exchange $J(x, x')$. $\beta\mathscr{H}_a$ represents anharmonic contributions to $\beta\mathscr{H}$. Statistical properties of the interacting variables $S_l(x)$ are determined by the partition function

$$Z[\{H_l(x)\}] \equiv \exp(W[H]) = \mathrm{Tr}_s \exp(-\beta\mathscr{H}) , \tag{2.1.4}$$

where $W[H] = -\beta F[H]$ where F is the free energy.

Z can be expressed in terms of a functional integral over continuous fields with the aid of the Hubbard–Stratanovitch identity [63]

$$\exp\left\{\tfrac{1}{2} \sum_{x,x',l} K_l(x, x')S_l(x)S_l(x')\right\}$$
$$= \prod_{x,l} \int_{-\infty}^{\infty} \mathrm{d}\varphi_l(x) \exp\left\{-\tfrac{1}{2} \sum_{x,x',l} K_l^{-1}(x, x')\,\varphi_l(x)\varphi_l(x') + \sum_{x,l} \varphi_l(x)S_l(x)\right\} . \tag{2.1.5}$$

Thus, we have

$$Z = \prod_{x,l} \int_{-\infty}^{\infty} \mathrm{d}\varphi_l(x) \exp\left\{-\tfrac{1}{2} \sum_{x,x',l} K_l^{-1}(x, x')\varphi_l(x)\varphi_l(x')\right\}$$
$$\times \left(\mathrm{Tr}_s \exp(-\beta\mathscr{H}_a) \exp\left\{\sum_{x,l} (H_l(x) + \varphi_l(x))S_l(x)\right\}\right) , \tag{2.1.6}$$

or

$$Z = \exp\left\{-\tfrac{1}{2} \sum_{x,x',l} H_l(x)K_l^{-1}(x, x')H_l(x')\right\}\bar{Z}[H] . \tag{2.1.7}$$

In this expression,

$$\bar{Z}[H] = \prod_{x,l} \int_{-\infty}^{\infty} \mathrm{d}\psi_l(x)\left\{\exp\left\{-\tfrac{1}{2} \sum_{x,x',l} \psi_l(x)K_l(x, x')\psi_l(x')\right\}\right.$$
$$\left.\times \exp\left\{\sum_{x,l} H_l(x)\psi_l(x)\right\} \exp\left\{\sum_{x} \mathscr{S}[y_l(x)]\right\}\right\}$$
$$\equiv \exp\{\bar{W}[H]\} , \tag{2.1.8}$$

where

$$\sum_{x'} K_l(x, x')\psi_l(x') = H_l(x) + \varphi_l(x) \equiv y_l(x) , \tag{2.1.9}$$

and

$$\exp\{\mathscr{S}[y_l(x)]\} = \mathrm{Tr}_s \exp(-\beta\mathscr{H}_a) \exp\{y_l(x)S_l(x)\} . \tag{2.1.10}$$

It will often be more convenient in what follows to deal with $\bar{Z}$ rather than Z. From eqs. (2.1.7) and (2.1.4), we have

$$W[H] = \bar{W}[H] - \tfrac{1}{2} \sum_{x,x'l} H_l(x) K_l^{-1}(x, x') H_l(x') , \tag{2.1.11}$$

so that

$$\langle S_l(x) \rangle = \frac{\partial W[H]}{\partial H_l(x)} = \langle \psi_l(x) \rangle - \sum_{x'} K_l^{-1}(x, x') H_l(x') , \tag{2.1.12}$$

where

$$\langle \psi_l(x) \rangle = \frac{\partial \bar{W}[H]}{\partial H_l(x)} . \tag{2.1.13}$$

Similarly

$$\begin{aligned} G^c_{ll'}(x, x') &= \frac{\partial^2 W[H]}{\partial H_l(x) \partial H_{ll'}(x')} \\ &= \frac{\partial^2 \bar{W}[H]}{\partial H_l(x) \partial H_{ll'}(x')} - K_l^{-1}(x, x') \delta_{ll'} . \end{aligned} \tag{2.1.14}$$

Hence $\langle S_l(x) \rangle = \langle \psi_l(x) \rangle$ if $H_l(x) = 0$, and non-local susceptibilities for $\psi_l(x)$ and $S_l(x)$ differ only by a constant $K_l^{-1}(x, x')$. If we are interested in properties near a critical point where $\chi_{ll'}(x, x')$ becomes long range, the susceptibilities for $S_l(x)$ and $\psi_l(x)$ are essentially the same. In what follows, we will, therefore, usually deal with $\psi_l(x)$ and its associated generating function $\bar{W}[H]$. We will also often use the Legendre transformed functions

$$\Gamma[\langle S_l(x) \rangle] = -W + \sum_{x,l} H_l(x) \langle S_l(x) \rangle , \tag{2.1.15a}$$

and

$$\bar{\Gamma}[\langle \psi_l(x) \rangle] = -\bar{W} + \sum_{x,l} H_l(x) \langle \psi_l(x) \rangle . \tag{2.1.15b}$$

$\mathcal{S}$ can be expanded in a power series in $y_l(x)$. Usually, $S_l(x)$ will satisfy

$$\begin{aligned} \langle S_l(x) \rangle_0 &= e^{-S_0} \operatorname{Tr} S_l(x) = 0 , \\ \langle S_l(x) S_{l'}(x') \rangle &= e^{-S_0} \operatorname{Tr} S_l(x) S_{l'}(x') = \delta_{l,l'} , \end{aligned} \tag{2.1.16}$$

where $e^{S_0} = \operatorname{Tr} 1$ so that

$$\mathcal{S}(y) = S_0 + \tfrac{1}{2} y^2 + \mathcal{S}_1(y) . \tag{2.1.17}$$

To obtain a continuum field theory, we introduce Fourier transformation

$$\psi_l(\boldsymbol{q}) = \frac{1}{\sqrt{N}}\sum_{\boldsymbol{x}} \exp(-\mathrm{i}\boldsymbol{q}\cdot\boldsymbol{x})\psi_l(\boldsymbol{x}), \tag{2.1.18}$$

and

$$K_l(\boldsymbol{q}) = \sum_{\boldsymbol{x}-\boldsymbol{x}'} \exp\{-\mathrm{i}\boldsymbol{q}\cdot(\boldsymbol{x}-\boldsymbol{x}')\}K_l(\boldsymbol{x},\boldsymbol{x}'). \tag{2.1.19}$$

If $K_l(\boldsymbol{x},\boldsymbol{x}')$ is equal to K for nearest neighbor bonds on a hypercubic d dimensional lattice with lattice constant, a,

$$K_l(\boldsymbol{q}) = K\sum_{\boldsymbol{\delta}} \exp(\mathrm{i}\boldsymbol{q}\cdot\boldsymbol{\delta}) = K(z - q^2a^2 + \cdots), \tag{2.1.20}$$

where $z = 2d$ is the number of nearest neighbors labeled by $\boldsymbol{\delta}$ and a is the lattice constant. The quadratic part of $\beta\mathscr{H}$ (which includes a contribution from $\mathscr{S}$) is then

$$\begin{aligned}\tfrac{1}{2}\sum_{\boldsymbol{q},l} K_l(\boldsymbol{q})(1 - K_l(\boldsymbol{q}))\psi_l(\boldsymbol{q})\psi_l(-\boldsymbol{q})\\ = \tfrac{1}{2}zK\sum_{\boldsymbol{q},l}[1 - Kz + (2K - 1/z)q^2a^2]\psi_l(\boldsymbol{q})\psi_l(-\boldsymbol{q}).\end{aligned} \tag{2.1.21}$$

This determines the mean field transition temperature $zK_c = zJ/T_c = 1$. If we restrict ourselves to the vicinity of the critical point, we can, therefore, replace the coefficient of q^2 by $K_c = 1/z$. Higher order terms in $\mathscr{S}$ can be expressed in terms of $K(\boldsymbol{q})\psi_l(\boldsymbol{q})$ which in the long wave-length limit can be replaced by $zK\psi_l(\boldsymbol{q})$. As a final step to the continuum limit, we replace sums over $\boldsymbol{x}$ by integrals, and rescale $\psi_l(\boldsymbol{x})$: $\psi_l(\boldsymbol{x}) \to K^{-1/2}a^{(d/2-1)}\psi_l(\boldsymbol{x})$. We then find

$$\bar{Z} = \int \mathscr{D}\psi_l(\boldsymbol{x}) \exp\{-\bar{\mathscr{H}}[\psi_l(\boldsymbol{x})]\}, \tag{2.1.22}$$

where

$$\bar{\mathscr{H}} = \int \mathrm{d}^dx\mathscr{F}[\psi_l(\boldsymbol{x})] - \int H_l(\boldsymbol{x})\psi_l(\boldsymbol{x})\,\mathrm{d}^dx,$$

and

$$\mathscr{F}[\psi_l(\boldsymbol{x})] = \tfrac{1}{2}\sum_l r_l\psi_l^2(\boldsymbol{x}) + \tfrac{1}{2}\sum_l(\nabla\psi_l(\boldsymbol{x}))^2 + a^{-d}\mathscr{S}_1(zK^{1/2}a^{(d/2-1)}\psi_l(\boldsymbol{x})), \tag{2.1.23}$$

where

$$r_l = za^{-2}(1 - zK_l) \equiv za^{-2}\left(\frac{T - T_c}{T}\right) \quad \text{if} \quad K_l = K \quad \text{for every } l.$$

If $S_l(x) \equiv S_i(x)$ is an m-component spin, we have

$$\mathscr{F}[\psi_i(x)] = \tfrac{1}{2}r \sum_i \psi_i^2(x) + \tfrac{1}{2} \sum_i (\nabla\psi_i(x))^2 + u\Big(\sum_i \psi_i^2(x)\Big)^2 + \cdots, \tag{2.1.24}$$

where

$$u = \frac{1}{12}\frac{m+3}{m+2} K^{-2} a^{d-4}.$$

Thus the field theory for an m-component spin is what is commonly called a ϕ^4 field theory.

2.2. *Mean field theory and its demise*

Mean field theory can be derived in various ways. In the present context, it is most easily derived by evaluating the integrals in Z by steepest descent

$$\overline{W}[H(x)] = -\int d^d x \mathscr{F}[\psi(x)] + \int H(x)\psi(x)\, d^d x, \tag{2.2.1}$$

where

$$H(x) = \frac{\partial \mathscr{F}}{\partial \psi(x)} = \frac{\delta}{\delta\psi(x)} \int d^d x \mathscr{F}[\psi(x)]. \tag{2.2.2}$$

We have dropped the subscript l and we have replaced $\langle\psi(x)\rangle$ by $\psi(x)$ for notational convenience. It follows from eq. (2.1.15) that

$$\overline{\Gamma}_{MF}(\psi(x)) = \int d^d x \mathscr{F}[\psi(x)]. \tag{2.2.3}$$

Expanding F in powers of ψ, we have

$$\mathscr{F} = \tfrac{1}{2}r\psi^2 + \tfrac{1}{2}(\nabla\psi)^2 + w\psi^3 + u\psi^4 + \cdots. \tag{2.2.4}$$

In equilibrium, eq. (2.2.2) is satisfied so that when $H(x) = 0$, ψ is uniform in space and obeys

$$r\psi + 3w\psi^2 + 4u\psi^4 = 0. \tag{2.2.5}$$

We will encounter two cases in what follows. In the case of the common ψ^4 field theory, $w = 0$ and

$$\psi \sim \begin{cases} 0 & \text{if } T > T_c \\ \sqrt{-r/4u} & \text{if } T < T_c, \end{cases} \tag{2.2.6}$$

implying that $\beta = \frac{1}{2}$. In some case such as the percolation problem, w is non-zero and for small r,

$$\psi \sim \begin{cases} 0 & \text{if} \quad T > T_c \\ -r/3w & \text{if} \quad T < T_c \,, \end{cases} \tag{2.2.7}$$

so that $\beta = 1$.

Correlation functions are also easily obtained from eq. (2.2.3).

$$G^{c^{-1}}(\boldsymbol{x}, \boldsymbol{x}') = \frac{\delta^2 \overline{\Gamma}}{\delta\psi(\boldsymbol{x})\delta\psi(\boldsymbol{x}')} = \Gamma^{(2)}(\boldsymbol{x}, \boldsymbol{x}') \,, \tag{2.2.8}$$

so that in mean field theory,

$$G^c(\boldsymbol{q}) = (r + q^2)^{-1} \,, \quad T > T_c \,, \tag{2.2.9}$$

implying that the correlation length ξ is $r^{-1/2}$ and that

$$\gamma = 1 \,, \qquad \nu = \tfrac{1}{2} \,, \qquad \eta = 0 \,, \tag{2.2.10}$$

in mean field theory. One can also easily verify that there is a jump in the specific heat implying $\alpha = 0$.

The steepest descent evaluation of $\overline{W}$ neglects inhomogeneous configurations of $\psi(\boldsymbol{x})$ when $H(\boldsymbol{x}) = 0$. This is a good approximation so long as the energy associated with non-uniform configurations is high (much greater than unity in the present units). Spatial variations in $\psi(\boldsymbol{x})$ lead to higher values of $\mathscr{F}(x)$ only through the terms $\frac{1}{2}(\nabla\psi)^2$. One can estimate the energy associated with this term [64] by assuming that ψ is correlated within a coherence volume and that the scale of spatial variations is set by ξ so that

$$\int \mathrm{d}^d x (\nabla\psi)^2 \sim \xi^{d-2}\psi^2 \sim \xi^{d-2-2\beta/\nu} \,. \tag{2.2.11}$$

The last step follows from eqs. (2.2.6) and (2.2.7). Thus paths involving non-uniform variations of $\psi(\boldsymbol{x})$ make a negligible contribution to $\overline{W}$ so long as

$$d > d_c = 2 + 2\beta/\nu = 2 + 4\beta \,. \tag{2.2.12}$$

For ψ^4 field theories, $\beta = \frac{1}{2}$ and $d_c = 4$. For ψ^3 theories $\beta = 1$ and $d_c = 6$. d_c is called the upper critical dimension. Mean field theory does not provide an adequate description for $d < d_c$. It can, however, be used as a starting point for perturbation calculations of critical properties in powers of $\varepsilon = d - d_c$, using the Wilson–Fisher ε-expansion [1,2] to be discussed in the next section.

At times, it is useful to know the results for mean field theory at all temperatures. These can also be obtained by a steepest descent evaluation of $\overline{W}[\psi]$ assuming a uniform ψ. Using eq. (2.1.8)–(2.1.10) and $y_l = zK\psi_l$, we obtain

$$\frac{1}{N}\frac{\partial \overline{W}}{\partial \psi_l} = -zK\psi_l + H_l + zK\mathscr{S}'(zK\psi_l) = 0\,, \tag{2.2.13}$$

which implies

$$\langle S_l\rangle = \mathscr{S}'(H_l + zK\langle S_l\rangle)\,, \tag{2.2.14}$$

and

$$\frac{1}{N}\Gamma(\langle S_l\rangle) = -\tfrac{1}{2}zK\langle S_l\rangle^2 + zK\langle S_l\rangle\psi_l(\langle S_l\rangle) - \mathscr{S}(zK\psi_l(\langle S_l\rangle))\,, \tag{2.2.15}$$

where $zK\psi_l = H_l + zK\langle S_l\rangle$ is determined by eq. (2.2.11). Note that this differs from

$$\frac{1}{N}\overline{\Gamma}(\psi_l) = \tfrac{1}{2}zK\psi_l^2 - \mathscr{S}(zK\psi_l)\,, \tag{2.2.16}$$

even though both yield the same equation for $\langle S_l\rangle$ when minimized.

2.3. *The renormalization group* [1,2]

The hamiltonian $\mathscr{H}$ can be characterized by its potentials $\{v\} = r, w, u$, etc. The renormalization group is used to map $\mathscr{H}$ from a calculationally intractable region in the space of potentials into a region where normal perturbation theoretic methods can be used. The renormalization group transformation R can be defined as follows

$$R_b = R_b^{\mathrm{s}} R_b^{\mathrm{i}}\,, \tag{2.3.1}$$

where b is a number > 1, R_b^{i} is an integration over fields $\psi(\boldsymbol{q})$ with $b^{-1} < q < 1$, and R_b^{s} is a change of scale $\boldsymbol{q} \to b\boldsymbol{q}$, $\psi(\boldsymbol{q}) \to b^{(d+2-\eta)/2}\psi(b\boldsymbol{q})$. The rescaling of $\psi(\boldsymbol{q})$ is determined by the homogeneity relation at the critical point $G(q) = b^{2-\eta}G(bq)$ where $\langle\psi(q)\psi(q')\rangle = G(q)\delta^d(q + q')$. To be more explicit about R_b, let

$$\psi(q) = \begin{cases} \psi^<(q) & \text{if} \quad 0 < |q| < b^{-1} \\ \psi^>(q) & \text{if} \quad b^{-1} < |q| < 1 \,. \end{cases} \tag{2.3.2}$$

Then

$$\exp\{-R_b^0\bar{\mathscr{H}}(\psi^<(q))\} = \int \mathscr{D}\psi^>(q) \exp\{-\bar{\mathscr{H}}(\psi^<(q) + \psi^>(q))\}\,, \tag{2.3.3}$$

and

$$\bar{\mathscr{H}}'(\psi(q')) = R_b\bar{\mathscr{H}}(\psi^<(q)) = R_b^1\bar{\mathscr{H}}(b^{(d+2-\eta)/2}\psi^<(b^{-1}q'))\,, \tag{2.3.4}$$

where $q' = bq$. η is chosen so that the gradient term in $\bar{\mathscr{H}}'$ is equal to

$$\tfrac{1}{2}\int_0^1 \frac{d^dq'}{(2\pi)^d}\, q'^2|\psi(q')|^2\,.$$

These relations give rise to a new set of potentials $\{v'\}$. If we let $b = e^l$ and remove an infinitesimal shell at each iteration, the recursion relations can be expressed as first order non-linear differential equations [65]

$$dv_i/dl = \Theta_i(\{v_i\})\,. \tag{2.3.5}$$

At a critical point, $\{v_i^*\}$, v_i is stationary.

$$\Theta_i(\{v_i^*\}) = 0\,.$$

Critical exponents are obtained by linearizing eq. (2.3.5) about its fixed points:

$$\frac{d}{dl}(v_i - v_i^*) = \sum_j \left(\frac{\partial\Theta_i}{\partial v_j}\right)_{\{v_i\}=\{v_i^*\}} (v_j - v_j^*)\,. \tag{2.3.6}$$

The eigenvalues, $\lambda\alpha$, of the matrix $\partial\Theta_i/\partial v_j = \lambda_{ij}$ are the critical exponents. The exponent associated with changing all r_i's by the same amount is ν^{-1}. Usually, there is only one other exponent associated with quadratic potentials, and it is the exponent $\lambda_a = \nu^{-1}\phi_a$ for turning on exchange anisotropy. In most cases of interest to us, the other exponents will be negative and give information only about corrections to scaling.

2.4. *The replica trick* [5,8,9]

The replica trick has been used widely in the study of quenched random systems. It is based on the identity

$$[\ln Z]_{av} = \lim_{n\to 0}(1/n)\ln[Z^n]_{av}\,. \tag{2.4.1}$$

$[Z^n]_{av}$ can be evaluated for integral n by replicating the hamiltonian,

$$\begin{aligned}[Z^n]_{av} &= \mathrm{Tr}_{s^\alpha}\left[\exp\left(-\beta\sum_{\alpha=1}^{n}\mathscr{H}^\alpha\right)\right]_{av} \equiv \mathrm{Tr}[\exp(-\beta\mathscr{H}^{\mathrm{r}})]_{av} \\ &\equiv \mathrm{Tr}_{s^\alpha}\exp(-\beta\mathscr{H}_{\mathrm{eff}})\,.\end{aligned} \tag{2.4.2}$$

For example the replicated Heisenberg model is

$$\beta\mathscr{H}^{\mathrm{r}} = -\tfrac{1}{2}\sum_{x,x'\alpha,i}K(x,x')S_i^\alpha(x)S_i^\alpha(x') - \sum_{x,\alpha,i}H_i(x)S_i^\alpha(x)\,, \tag{2.4.3}$$

which is bilinear in S_i^α. If $K(x, x')$ is a nearest neighbor random variable, $\beta\mathscr{H}_{\mathrm{eff}}$ can easily be evaluated in terms of the cumulants $C_p(x)$ of $K(x, x')$

$$\beta\mathscr{H}_{\mathrm{eff}} = -\sum_{\langle x,x'\rangle,p}(1/p!)C_p(K)\left(\sum_{\alpha,i}S_i^\alpha(x)S_i^\alpha(x')\right)^p - \sum_{x,\alpha,i}H_i(x)S_i^\alpha(x)\,, \tag{2.4.4}$$

where $\langle x, x'\rangle$ represents all bonds each of which appears only once in the sum, and $C_0 = [K]_{av}$, $C_2 = [K^2] - [K]_{av}^2, \ldots$.

The average spin, and correlation functions can also be evaluated in terms of replica variables. For example

$$\begin{aligned}[\langle S_i(x)\rangle]_{av} &= [Z^{-1}\,\mathrm{Tr}\exp(-\beta\mathscr{H})S_i(x)]_{av} \\ &= \lim_{n\to 0}[Z^{n-1}\,\mathrm{Tr}\exp(-\beta\mathscr{H})S_i(x)]_{av} \\ &= \lim_{n\to 0}\mathrm{Tr}\exp(-\beta\mathscr{H}_{\mathrm{eff}})S_i^\alpha(x) \equiv \langle S_i^\alpha(x)\rangle\,,\end{aligned} \tag{2.4.5}$$

where α is any replica index $1, \ldots, n$; and

$$\begin{aligned}&[\langle S_i(x)\rangle\langle S_j(x')\rangle]_{av} \\ &= [Z^{-2}(\mathrm{Tr}\exp(-\beta\mathscr{H})S_i(x))(\mathrm{Tr}\exp(-\beta\mathscr{H})S_j(x'))]_{av} \\ &= \lim_{n\to 0}[Z^{n-2}(\mathrm{Tr}\exp(-\beta\mathscr{H})S_i(x))(\mathrm{Tr}\exp(-\beta\mathscr{H})S_j(x'))]_{av} \\ &= \lim_{n\to 0}\mathrm{Tr}\exp(-\beta\mathscr{H}_{\mathrm{eff}})S_i^\alpha(x)S_j^\beta(x') \\ &= \langle S_i^\alpha(x)S_j^\beta(x')\rangle\,,\quad \alpha\neq\beta\,.\end{aligned} \tag{2.4.6}$$

Thus, the spin-glass order parameter is $\langle S_i^\alpha(x)S_j^\beta(x)\rangle$, $\alpha\neq\beta$, in the replica language.

Formally, the replica trick is very appealing. It always gives the same results as other techniques in the paramagnetic phase, usually with a great deal less work. The fundamental ambiguity arising from the fact that the replicated hamiltonian is only defined for integral $n \geqslant 1$ whereas the relation eq. (2.4.1) applies only for continuous n appears to lead to problems in ordered phases, especially in spin-glass phases. These problems have not yet been fully resolved, and we will not give them too much further attention.

2.5. *Weakly random systems* [4,5]

Calculations of the critical properties of the m-vector model with weak bond randomness or small impurity concentration have been done in two ways, and it is worth mentioning both. Impurities destroy the translational invariance of the hamiltonian so that the reduced hamiltonian eqs. (2.1.22) and (2.1.24) become

$$\bar{\mathscr{H}} = \tfrac{1}{2}\int_{q_1,q_2} v_2(q_1, q_2)\psi(q_1)\cdot\psi(q_2) + \int_{q_1,q_2,q_3,q_4} v_4(q_1, q_2, q_3, q_4)(\psi(q_1)\cdot\psi(q_2))(\psi(q_3)\cdot\psi(q_4)), \quad (2.5.1)$$

where

$$\psi(q_1)\cdot\psi(q_2) = \sum_{i=1}^{m} \psi_i(q_1)\psi_i(q_2)$$

and where v_2 and v_4 are random potentials with probability distribution $P(\{v_l\})$. The renormalization group yields recursion relations $\{v_l'\} = R\{v_l\}$ for any realization of the potentials. From these, one can renormalize the probability distribution [4] via

$$P'(\{v_l'\}) = \int \mathrm{d}\{v_l\}\delta(\{v_l'\} - R\{v_l\})P(\{v_l\}). \quad (2.5.2)$$

Near the upper critical dimension (4 in this case), it is more convenient to deal with the cumulants of P rather than P itself,

$$\begin{aligned} [v_2(q, q')]_{\mathrm{av}} &= (r + q^2)\delta^d(q + q'), \\ [\delta v_2(q_1, q_2)\delta v_2(q_3, q_4)]_{\mathrm{av}} &= \Delta\delta^d(q_1 + q_2 + q_3 + q_4), \\ [v_4(q_1, q_2, q_3, q_4)]_{\mathrm{av}} &= u\delta^d(q_1 + q_2 + q_3 + q_4). \end{aligned} \quad (2.5.3)$$

Near four dimensions, all cumulants other than r, Δ, and u are irrelevant. Recursion relations for these variables can be developed along the lines outlined in sect. (2.3) and discussed in detail in ref. [4]. To first order in $\varepsilon = 4 - d$, they are

$$\frac{\mathrm{d}r}{\mathrm{d}l} = 2r + K_4 \frac{4(m+2)u - \Delta}{1+r}, \tag{2.5.4a}$$

$$\frac{\mathrm{d}u}{\mathrm{d}l} = \varepsilon u - K_4[4(m+8)u^2 - 6u\Delta], \tag{2.5.4b}$$

$$\frac{\mathrm{d}\Delta}{\mathrm{d}l} = \varepsilon\Delta - K_4[8(m+2)u\Delta - 4\Delta^2], \tag{2.5.4c}$$

where $k_{d'}^{-1} = \Omega_d/(2n)^d$ where Ω_d is the solid angle subtended by the unit sphere in d dimensions.

An alternate way [5] of deriving these equations is to use the replica procedure to derive a translationally invariant effective hamiltonian

$$\begin{aligned}\mathscr{H}_{\text{eff}} &= \tfrac{1}{2}\int \mathrm{d}^d x\Big[r\sum_\alpha \psi^\alpha\cdot\psi^\alpha + \sum_\alpha \nabla\psi^\alpha\cdot\nabla\psi^\alpha\Big] \\ &\quad - \tfrac{1}{8}\Delta\int \mathrm{d}^d x\Big(\sum_\alpha \psi^\alpha\cdot\psi^\alpha\Big)^2 + u\int \mathrm{d}^d x\sum_\alpha (\psi^\alpha\cdot\psi^\alpha)^2 .\end{aligned} \tag{2.5.5}$$

Recursion relations for r, Δ, and u can be derived as for any homogeneous system. In the limit $n \to 0$, they reduce to eqs. (2.5.4).

Let us now analyze the fixed point structure implied by eqs. (2.5.4). There are four fixed points with fixed point values of u and Δ and stability exponents λ as follows:

(i) Gaussian

$$\begin{aligned} u^{\mathrm{G}} &= \Delta^{\mathrm{G}} = 0, \\ \lambda_u^{\mathrm{G}} &= \lambda_\Delta^{\mathrm{G}} = \varepsilon . \end{aligned} \tag{2.5.6}$$

(ii) m-component Heisenberg

$$\begin{aligned} u^{\mathrm{H}} &= \frac{\varepsilon}{4K_4(m+8)}, &\quad \Delta^{\mathrm{H}} &= 0, \\ \lambda_u^{\mathrm{H}} &= -\varepsilon, &\quad \lambda_\Delta^{\mathrm{H}} &= \frac{4-m}{m+8}\varepsilon . \end{aligned} \tag{2.5.7}$$

(iii) Unphysical (or $m = 0$ Heisenberg)

$$\begin{aligned} u^{\mathrm{U}} &= 0, &\quad \Delta^{\mathrm{U}} &= -\frac{\varepsilon}{4K_4}, \\ \lambda_u^{\mathrm{U}} &= -\tfrac{1}{2}\varepsilon, &\quad \lambda_\Delta^{\mathrm{U}} &= -\varepsilon . \end{aligned} \tag{2.5.8}$$

Table 3 [4]

Critical Exponents for the non-Gaussian Fixed Points for the Random Exchange m-vector Model.[a]

	Heisenberg	Random		Unphysical
γ	$\frac{1}{2}+\frac{m+2}{4(m+8)}+\frac{(m+2)(m^2+23m+60)}{8(m+2)}\varepsilon^2$	$\frac{1}{2}+\frac{3m}{32(m-1)}\varepsilon+\frac{m(127m^2-572m-32)}{4096(m-1)^3}\varepsilon^2$	$(m \neq 1)$	$\frac{1}{2}+\frac{1}{16}\varepsilon-\frac{27}{512}\varepsilon^2$
		$\frac{1}{2}+\frac{1}{4}\sqrt{\frac{6\varepsilon}{53}}+\frac{1}{4}\left(\frac{535-756\zeta(3)}{5618}\right)\varepsilon$	$(m = 1)$	
η	$2\frac{m+2}{2(m+8)^2}\varepsilon^2$	$\frac{5m-8}{256(m-1)^2}\varepsilon^2$	$(m \neq 1)$	$\frac{1}{64}\varepsilon^2$
		$-\frac{1}{106}\varepsilon+O(\varepsilon^{3/2})$ $(m = 1)$		
α	$\frac{4-m}{2(m+8)}\varepsilon-\frac{(m+2)(m^2+30m+56)}{4(m+8)^3}\varepsilon^2$	$\frac{m-4}{8(m-1)}\varepsilon-\frac{m(31m^2-380m-128)}{1024(m-1)^3}\varepsilon^2$	$(m \neq 1)$	$\frac{1}{4}\varepsilon+\frac{35}{128}\varepsilon^2$
		$-\sqrt{\frac{6}{53}}\sqrt{\varepsilon}+\frac{1137+378\zeta(3)}{(53)^2}\varepsilon$		

[a]) The $O(\varepsilon)$ results for $m = 1$ were taken from C. Jayaprakash and H. J. Katz, Phys. Rev. B16 (1977) 3987.

(iv) Random

$$u^{\mathrm{R}} = \frac{\varepsilon}{16K_4(m-1)}, \qquad \Delta^{\mathrm{R}} = \frac{(4-m)\varepsilon}{8K_4(m-1)},$$
$$\lambda_1^{\mathrm{R}} = -\varepsilon, \qquad \lambda_2^{\mathrm{R}} = \frac{(m-4)\varepsilon}{4(m-1)}. \tag{2.5.9}$$

Fixed point (iii) is labeled unphysical because Δ, which is a positive definite quantity, is negative at this fixed point. It is exactly the $m = 0$ Heisenberg fixed point which, as we shall see, describes the statistics of self avoiding walks. It is inaccessible to random m-component systems with $m \geqslant 1$. Bray and Moore [66], however, have argued that it is accessible to the random $m = 0$ system, in which case it describes self-avoiding walks on a random lattice. Critical exponents ν, η, and α are listed to second order in ε in table 3. Note that the m-component Heisenberg fixed point is stable for $m > 4$ to first order in ε. More generally it is stable for $\alpha_{\mathrm{H}} < 0$ as predicted by the heuristic arguments in sect. 1. If $\alpha_{\mathrm{H}} > 0$, the random fixed point becomes stable. Thus the heuristic arguments predict correctly that the pure system behavior breaks down when $\alpha_{\mathrm{H}} > 0$. They fail, however, to predict that the transition will nevertheless be continuous with new critical exponents. The case $m = 1$ deserves special attention. Both u^{R} and Δ^{R} appear to diverge at $m = 1$. As pointed out by Khmelnilski [67, 5], this leads to an expansion of critical exponents in powers of $\sqrt{\varepsilon}$ rather than in ε at $m = 1$. The interesting critical exponents are listed in table 3.

2.6. *Spin-glasses* [8,22,46]

In this section, we will analyze the EA spin-glass within the context of the $n = 0$ field theory. Though there remain some fundamental difficulties in applying this technique, its concrete predictions can be reproduced in almost all cases by other techniques [68]. There are, however, some disquieting problems regarding the EA spin-glass itself that remain unresolved. A naive mean field calculation [37] for the infinite range spin-glass yields an unacceptable negative entropy at zero temperature. More sophisticated treatments [39] do remove this difficulty. In addition, the ordered phase of the EA spin-glass appears [40,41] to have unphysical negative mass modes which break the replica symmetry. These would seem to indicate that the EA order parameter provides an inadequate description of the ordered phase. Ways of resolving this problem have been proposed [41], though no general understanding has emerged. In this section we will not consider these problems further.

For simplicity, we will deal with a model with a gaussian distribution of bond strengths with

$$\mathscr{P}(J) = (2\pi\Delta)^{-1/2} \exp\{-(J - J_0)^2/2\Delta\} . \tag{2.6.1}$$

This distribution may lead to pathologies since there is finite probability of having a cluster of arbitrary size with exchange integrals less than (or greater than) any specified value. These pathologies are most important at low temperature and can probably be ignored for large enough d near the phase boundaries we are considering. The effective hamiltonian eq. (2.4.4) with $H_i = 0$ is then

$$\begin{aligned}\beta\mathscr{H}_{\text{eff}} = &- \sum_{\langle xx'\rangle} \beta J_0 \sum_{\alpha i} S_i^\alpha(x) S_i^\alpha(x') \\ &- \tfrac{1}{2} \sum_{\langle xx'\rangle} \beta^2\Delta \sum_{\substack{\alpha\neq\beta\\ i,j}} S_i^\alpha(x) S_j^\beta(x) S_i^\alpha(x') S_j^\beta(x') \\ &- \tfrac{1}{2} \sum_{\langle xx'\rangle} \beta^2\Delta \sum_{\alpha,i,j} S_i^\alpha(x) S_i^\alpha(x') S_j^\alpha(x) S_j^\alpha(x') .\end{aligned} \tag{2.6.2}$$

Introducing Hubbard–Stratanovich fields $\psi_i^\alpha(x)$ and $Q_{ij}^{\alpha\beta}(x)$ ($\alpha \neq \beta$) and treating the fourth term as a contribution to $\beta\mathscr{H}_{\text{a}}$, we obtain a reduced effective hamiltonian

$$\overline{\mathscr{H}}_{\text{eff}} = \overline{\mathscr{H}}_\psi + \overline{\mathscr{H}}_Q + \overline{\mathscr{H}}_{\psi Q} , \tag{2.6.3}$$

where $\overline{\mathscr{H}}_\psi$ is given by eq. (2.5.5) and

$$\begin{aligned}\overline{\mathscr{H}}_Q = &\tfrac{1}{4} \int \mathrm{d}^d x \{ r_Q \operatorname{Tr} Q^2(x) + \operatorname{Tr}(\nabla Q(x) \cdot \nabla Q(x))\} \\ &- w \int \mathrm{d}^d x \operatorname{Tr} Q^3(x) + \text{higher order terms} ,\end{aligned} \tag{2.6.4}$$

$$\overline{\mathscr{H}}_{\psi Q} = w_1 \int \mathrm{d}^d x \sum_{\substack{\alpha,\beta\\ i,j}} Q_{ij}^{\alpha\beta}(x) \psi_i^\alpha(x) \psi_j^\beta(x) , \tag{2.6.5}$$

where

$$\operatorname{Tr} Q^2 = \sum_{\substack{\alpha,\beta\\ i,j}} Q_{ij}^{\alpha\beta} Q_{ji}^{\beta\alpha} \quad \text{and} \quad \operatorname{Tr} Q^3 = \sum_{\substack{\alpha,\beta,\gamma\\ i,j,k}} Q_{ij}^{\alpha\beta} Q_{jk}^{\beta\gamma} Q_{ki}^{\gamma\alpha} .$$

$\overline{\mathscr{H}}_Q$ is positive definite for large Q even though the fourth-order term is negative, at least for $n = 0$. r_Q changes sign at the mean field spin-glass freezing temperature. For Ising spin-glasses, $T_f = \sqrt{z\Delta}$ and

$$r_Q = z\left(1 - \frac{\Delta z}{T^2}\right) = z\left(\frac{T + T_f}{T}\right)\left(\frac{T - T_f}{T}\right). \tag{2.6.6}$$

The potential $w_1 = \frac{1}{2}z^2\beta^2 J_0\sqrt{\Delta}$ is zero when $J_0 = 0$.

We now consider the pure Ising spin-glass, $m = 1$, $J_0 = 0$. Using eq. (2.2.12) and setting

$$\langle Q^{\alpha\beta}\rangle = \langle Q\rangle(1 - \delta^{\alpha\beta}) = \langle S^\alpha S^\beta\rangle\,, \tag{2.6.7}$$

we obtain the mean field theory results

$$\frac{1}{N}\Gamma_{SG}(\langle Q\rangle) = n(n-1)\left\{\frac{1}{4}r_Q\langle Q\rangle^2 - \frac{1}{3!}(n-2)\langle Q\rangle^3 + \tfrac{1}{32}(3n^2 - 19n + \tfrac{80}{3})\langle Q\rangle^4\right\} \tag{2.6.8}$$

$$\frac{1}{\Omega}\overline{\Gamma}_{SG}(Q) = \overline{\mathscr{H}}_Q = n(n-1)\{\tfrac{1}{4}r_Q Q^2 - (n-2)wQ^3 + \cdots\}\,.$$

Notice that if $\langle Q\rangle$ is appropriately rescaled, $\Gamma_{SG}(\langle Q\rangle)$ and $\overline{\Gamma}_{SG}(Q)$ are identical to order Q^3. This will be of some use in what follows.

Equilibrium is obtained when $\Gamma(\langle Q\rangle)$ is an extremum. For $n \geqslant 1$, the extremum is obviously a minimum. For $n < 1$, however, it becomes a maximum because of the overall factor of $n(n - 1)$. This is an embarrassing problem that is not fully resolved. What one usually does is to choose the correct extremum (minimum) for $n > 1$ and then stay in this extremum as $n \to 0$. In mean field theory, this is equivalent to staying in a minimum of $[1/n(n-1)]\Gamma_{MF}(\langle Q\rangle)$. This makes some sense since one might argue that the proper quantity to study is the free energy per degree of freedom, and there are $\frac{1}{2}n(n-1)$ degrees of freedom in $Q^{\alpha\beta}$. $[1/n(n-1)]\Gamma_{MF}(\langle Q\rangle)$ is sketched in fig. 6 for $n > 2$ and for $n < 2$. For $n > 2$, a minimum develops at $\Gamma(\langle Q\rangle) = 0$ and $Q > 0$ before $\Gamma_Q = 0$ leading to first-order transition as is usually predicted by a Landau theory with a cubic potential. For $n < 2$, this first order minimum appears at negative $\langle Q\rangle$. $\langle Q\rangle$ is by definition a positive definite quantity so that this extremum is physically prohibited. A secondary minimum develops at positive $\langle Q\rangle$ for $T < T_f$. This is the minimum of physical interest. To lowest order in r_Q, $\langle Q\rangle$ is independent of fourth-order potentials and is identical (apart from scale

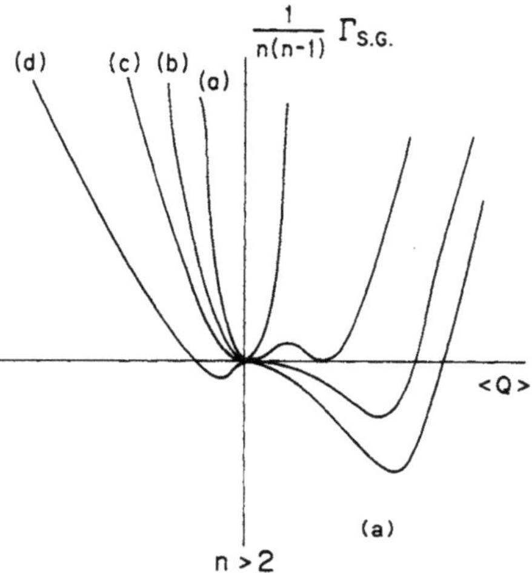

Fig. 6a. Plot of $[1/n(n-1)]\Gamma_{SG}(\langle Q\rangle)$ truncated at order $\langle Q\rangle^4$ in the mean field approximation for various values of r_Q for $n > 2$. Curve (a) is for $r_Q \gg 0$, curve (b) for $r_Q = r_Q^0$ at which a first-order transition occurs, curve (c) for $r_Q < 0$. Note that there are two minima for $r_Q < 0$.

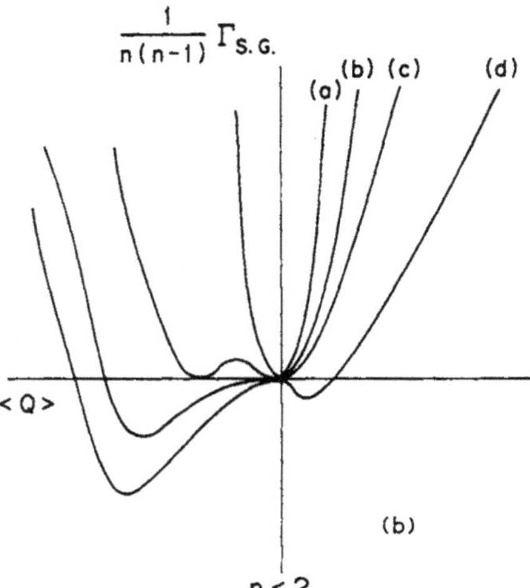

Fig. 6b. Same as fig. 6a but with $n < 2$. Note that this figure can be obtained from fig. 6a by reflection about the vertical axis. If the order parameter $\langle Q\rangle$ is restricted to be positive, the phase transition does not occur until $r_Q = 0$, and it is second order.

factors) to Q obtained by minimizing $\bar{\Gamma}(Q)$. To reduce notation in what follows, we will use $\bar{\Gamma}(Q)$ and Q. We obtain

$$Q = -\frac{r_Q}{6w} \quad \text{or} \quad \beta_{SG} = 1\,. \tag{2.6.9}$$

If J_0 is not zero, there is potential competition between ferromagnetic and spin-glass ordering. If we set $\psi^\alpha = \psi$ for all replicas, we obtain the following mean field free energy:

$$\frac{1}{n\Omega}\bar{\Gamma}(Q,\psi) = \frac{1}{n\Omega}\bar{\Gamma}_{SG}(Q) + \tfrac{1}{2}r\psi^2 + (u - \tfrac{1}{8}n\Delta)\psi^4 - (n-1)wQ\psi^2\,. \tag{2.6.10}$$

In the $n = 0$ limit, the extremum conditions for $\bar{\Gamma}$ yield

$$\begin{gathered}(r + 4u\psi^2 + 2w_1 Q)\psi = 0 \\ -(\tfrac{1}{2}r_Q Q + 6wQ^2 - w_1\psi^2) = 0\,.\end{gathered} \tag{2.6.11}$$

These equations produce the phase diagram shown in fig. 3, and are analyzed fully in ref. [22]. We note here that the multi-critical point where the paramagnetic, spin-glass and ferromagnetic phases meet has $d_c = 6$ with exponents that can be calculated in $6 - \varepsilon$ dimensions. At this critical point, both $Q_{ij}^{\alpha\beta}$ and ψ_i^α are critical and both r and r_Q become zero. There are two critical exponents, λ_{r_1} and λ_{r_2}, associated with r and r_Q. One is $\nu^{-1} = \lambda_{r_1}$ and the other defines a crossover exponent $\varphi = \nu\lambda_{r_2}$. w and w_1 have stability exponents λ_1 and λ_2.

Recursion relations for r, r_Q, w and w_1 can be derived in $6 - \varepsilon$ dimensions. To first-order in ε, they are

$$\frac{dr}{dl} = (2-\eta)r - 4K_6 m(n-1)\frac{w_1^2}{(1+r)(1+r_Q)}\,, \tag{2.6.12a}$$

$$\frac{dr_Q}{dl} = (2-\eta_{SG})r_Q - 36K_6 m(n-2)\frac{w^2}{(1+r_Q)^2} - 4K_6\frac{w_1^2}{(1+r)^2}\,,$$

$$\frac{dw}{dl} = \tfrac{1}{2}(\varepsilon - 3\eta_{SG})w + 36K_6[(n-3)m+1]w^3 + \tfrac{4}{3}K_6 w_1^3\,, \tag{2.6.12b}$$

$$\frac{dw_1}{dl} = \tfrac{1}{2}(\varepsilon - \eta - \tfrac{1}{2}\eta_{SG})w_1 + 4K_6 w_1^3 + 12K_6(n-2)mww_1^2\,, \tag{2.6.12c}$$

$$\eta_{SG} = K_6[12(n-2)mw^2 + \tfrac{4}{3}w_1^2]\,, \qquad \eta = \tfrac{4}{3}(n-1)mK_6 w_1^2\,. \tag{2.6.12d}$$

Table 4 [22]
Exponents for $m = 2$ and $m = 3$ non-Gaussian fixed point in $6 - \varepsilon$ dimensions for the paramagnetic–ferromagnetic–spin-glass multi-critical point.

	$\lambda_{r1,2} - 2$	λ_1	λ_2	η_1	η_2
$m = 2$	$-(1.5101 \pm 0.3247i)\varepsilon$	$-\varepsilon$	-1.079ε	-0.2451ε	-0.2149ε
$m = 3$	$-(0.9707 \pm 0.2539i)\varepsilon$	$-\varepsilon$	-0.8686ε	-0.2253ε	-0.1960ε

For $n = 1$, these equations yield critical exponents

$$\nu = \tfrac{1}{2} + \tfrac{2}{3}\varepsilon\,, \qquad \phi = 1 + \tfrac{1}{2}\varepsilon\,,$$
$$\eta = \eta_{SG} = -\tfrac{1}{3}\varepsilon\,, \tag{2.6.13}$$
$$\lambda_1 = -\varepsilon\,, \qquad \lambda_2 = -\tfrac{5}{3}\varepsilon\,.$$

For $n = 2$ and 3, the thermal exponents λ_{r_1} and λ_{r_2} are complex. They are listed in table 4. This is a rather strange result. A careful analysis of the trajectories implied by eqs. (2.6.12) for $n = 2$ and 3 shows that the phase boundary separating any two phases is an infinite spiral with ever decreasing radius ending at the critical point [68]. Thus if the critical point is approached along constant Δ or J_0, an infinite number of phase boundaries will be crossed.

2.7. *The Potts model and percolation* [10,69–74]

The Potts model [69] is a generalization of the two state Ising model. It is important for the study of critical properties of quenched random systems because the zero and one state limits describe the statistics of trees [74] and percolating clusters [10]. At each site, $\boldsymbol{x}$, on a lattice of N sites and $N_B = \tfrac{1}{2}zN$ bonds, there is a variable $\sigma(\boldsymbol{x})$ that can be in any of s different states. If neighboring sites are in the same state their associated bond has one energy. If they are in different states, it has another. The hamiltonian for the Potts model in the presence of an external field is

$$\beta\mathscr{H} = -\sum_{\langle \boldsymbol{x},\boldsymbol{x}'\rangle} K(s\delta^{\sigma(\boldsymbol{x})\sigma(\boldsymbol{x}')} - 1) - \sum_{\boldsymbol{x}} H(s\delta^{\sigma(\boldsymbol{x})1} - 1)\,. \tag{2.7.1}$$

The partition function is thus

$$Z = s^{-N}\,\mathrm{Tr}_\sigma \exp(-\beta\mathscr{H}) \equiv \exp\{(s-1)[KN_B + HN]\}Z'\,, \tag{2.7.2}$$

where the factor s^{-N} has been introduced to facilitate the limit $s \to 0$, and

$$\begin{aligned} Z' &= s^{-N} \sum_{\{\sigma\}} \exp\left\{ sK \sum_{\langle xx' \rangle} (\delta^{\sigma(x)\sigma(x')} - 1) + sH \sum_{x} (\delta^{\sigma(x)1} - 1) \right\} \\ &= s^{-N}(1-p)^{N_B} \sum_{\{\sigma\}} \prod_{\langle xx' \rangle} \left[1 + \frac{p}{1-p} \delta^{\sigma(x)\sigma(x')} \right] \\ &\quad \times \exp\left\{ sH \sum_{x} (\delta^{\sigma(x)1} - 1) \right\}, \end{aligned} \tag{2.7.3}$$

where

$$p = 1 - e^{-sK}. \tag{2.7.4}$$

In the one state limit, p will be the probability that a bond is occupied and $q = 1 - p$ the probability that it is absent. Each term in the expansion of eq. (2.7.3) can be represented by a graph $\mathscr{G}$ on the lattice in which a factor $(p/(1-p))\delta^{\sigma(x)\sigma(x')}$ is represented by a bond and the factor unity by the absence of a bond. Potts variables in connected clusters must be in the same state because of the factors $\delta^{\sigma(x)\sigma(x')}$ so that a cluster with n_b bonds and n_s sites is weighted by a factor

$$\left(\frac{p}{1-p} \right)^{n_b} [1 + (s-1)\, e^{-sHn_s}], \tag{2.7.5}$$

after the trace over σ-variables is taken. Let $N\mathscr{K}(\mathscr{G}, n_b, n_s)$ be the number of clusters of n_s sites and n_b bonds in the graph $\mathscr{G}$. We then have

$$\begin{aligned} Z' &= s^{-N}(1-p)^{N_B} \sum_{\mathscr{G}} \left\{ \left(\frac{p}{1-p} \right)^{N \sum_{n_b, n_s} \mathscr{K}(\mathscr{G}; n_b, n_s) n_b} \right\} \\ &\quad \times \prod_{n_b, n_s} [1 + (s-1)\, e^{-sHn_s}]^{N\mathscr{K}(\mathscr{G}; n_b, n_s)}. \end{aligned} \tag{2.7.6}$$

Using

$$\sum_{n_b, n_s} n_b N\mathscr{K}(\mathscr{G}; n_b, n_s) = N_b(\mathscr{G}), \tag{2.7.7}$$

the total number of occupied bonds in $\mathscr{G}$, we obtain

$$\begin{aligned} Z' &= s^{-N} \sum_{\mathscr{G}} p^{N_b(\mathscr{G})}(1-p)^{N_B - N_b(\mathscr{G})} \\ &\quad \times \prod_{n_b, n_s} [1 + (s-1)\, e^{-sHn_s}]^{N\mathscr{K}(\mathscr{G}; n_b, n_s)} \\ &= s^{-N} \sum_{\mathscr{G}} P(\mathscr{G}) \prod_{n_b, n_s} [1 + (s-1)\, e^{-sHn_s}]^{N\mathscr{K}(\mathscr{G}; n_b, n_s)}, \end{aligned} \tag{2.7.8}$$

where $P(\mathscr{G})$ is the probability of occurrence of the configuration $\mathscr{G}$.

Eq. (2.7.8) is our fundamental result for the Potts model. We will now consider the one and zero state limits which yield respectively the generating functions for bond percolation and for the statistics of trees on a lattice.

It is easy to see from eq. (2.7.8) and the binomial theorem that in the limit $s \to 1$, $Z = 1$, and $\ln Z = 0$. We, therefore, define the dimensionless free energy per site, f, as follows

$$f_1 = \lim_{N\to\infty} \lim_{s\to 1} \frac{1}{N(s-1)} \ln Z = \lim_{N\to\infty} \frac{1}{N} \left[\frac{\partial \ln Z}{\partial s}\right]_{s=1} . \tag{2.7.9}$$

Using eqs. (2.7.8) and (2.7.2), we obtain

$$\begin{aligned} f_1 &= -1 + \tfrac{1}{2}zK + H + \sum_{\mathscr{G}} P(\mathscr{G}) \sum_{n_b, n_s} \mathscr{K}(\mathscr{G}; n_b, n_s)\, e^{-Hn_s} \\ &= -1 + \tfrac{1}{2}zK + H + \sum_{n_s} \mathscr{K}(n_s)\, e^{-Hn_s} , \end{aligned} \tag{2.7.10}$$

where

$$\mathscr{K}(n_s) = \sum_{\mathscr{G}} P(\mathscr{G}) \sum_{n_b} \mathscr{K}(\mathscr{G}; n_b, n_s) , \tag{2.7.11}$$

is the average number of clusters per site containing n_s clusters. The contribution of any cluster that becomes infinite in the limit $N \to \infty$ vanishes in eq. (2.7.10) so that all sums are restricted to finite clusters.

Eq. (2.7.10) is the basic result for the one state Potts model. When $H = 0$, it reduces to

$$f_1 = -1 - \tfrac{1}{2}z \ln(1-p) + \langle N_c \rangle , \tag{2.7.12}$$

where $\langle N_c \rangle = \sum_s \mathscr{K}(n_s)$ is the average number of clusters per site. The first derivative f_1 with respect to H yields the probability, $\mathscr{P}(p)$, that a site is in the infinite cluster:

$$\begin{aligned} \frac{\partial f_1}{\partial H} &= 1 - \sum_{n_s} n_s \mathscr{K}(n_s)\, e^{-Hn_s} \\ &\to 1 - \sum_{n_s} n_s \mathscr{K}(n_s) = \mathscr{P}(p) \quad \text{as} \quad H \to 0 . \end{aligned} \tag{2.7.13}$$

since $n_s \mathscr{K}(n_s)$ is the probability that a site is in a cluster containing n_s sites and the sum is over finite clusters.

The second derivative yields the mean square cluster size $(S(p))$:

$$\left.\frac{\partial^2 f_1}{\partial H^2}\right|_{H=0} = \sum_{n_s} n_s^2 \mathscr{K}(n_s) = S(p) . \tag{2.7.14}$$

Note that eq. (2.7.13) allows an in principle determination of $\mathscr{K}(n_s)$ by Laplace transformation

$$n_s\mathscr{K}(n_s) = \frac{1}{2\pi i}\int_{-i\infty}^{i\infty} dH\, e^{Hn_s}\left(1 - \frac{\partial f_1}{\partial H}\right). \tag{2.7.15}$$

In the limit that $s \to 0$, $p = sK + O(s^2)$. The fundamental lattice relation

$$N_b + N_c - N_L = N, \tag{2.7.16}$$

where N_L is the number of closed loops then implies that the $s = 0$ limit eq. (2.7.5) reduces to

$$\begin{aligned}\lim_{s\to 0} Z' &= \lim_{s\to 0}\sum_{\mathscr{G}} s^{N_L(\mathscr{G})}\prod_{n_b,n_s}(1 + Hn_s)^{N\mathscr{K}(\mathscr{G};n_b,n_s)} \\ &= \sum_{\mathscr{G}_t} K^{N_b(\mathscr{G}_t)}\prod_{n_s}(1 + Hn_s)^{N\mathscr{K}(\mathscr{G}_t,n_s)} \equiv Z_0'(H),\end{aligned} \tag{2.7.17}$$

where $\mathscr{G}_t$ refers to tree-graphs (i.e., graphs no closed loops) and $N\mathscr{K}(\mathscr{G}_t, n_s)$ is the number of tree graphs per site in configuration $\mathscr{G}_t$ containing n_s clusters. The free energy per site is

$$f_0 = \lim_{N\to\infty}\lim_{s\to 0}\frac{1}{N(s-1)}\ln Z = \tfrac{1}{2}KZ + H - \lim_{N\to\infty}\frac{1}{N}\ln Z_0'. \tag{2.7.18}$$

The first and second derivatives of f_0 with respect to H are respectively

$$\frac{\partial f_0}{\partial H} = 1 - \frac{1}{Z_0'}\sum_{\mathscr{G}_t}K^{N_b(\mathscr{G}_t)}\sum_{n_s} n_s\mathscr{K}(\mathscr{G}_t, n_s) = 1 - \langle n_s\rangle_t \tag{2.7.19}$$

$$\frac{\partial^2 f_0}{\partial H^2} = \frac{1}{Z_0'}\sum_{\mathscr{G}_t}K^{N_b(\mathscr{G}_t)}\sum_{n_s^2} n_s^2\mathscr{K}(\mathscr{G}_t, n_s) = \langle n_s^2\rangle_t, \tag{2.7.20}$$

where $\langle n_s\rangle_t$ and $\langle n_s^2\rangle_t$ are the mean and mean square cluster size of tree-like clusters averaged with respect to the probability distribution $Z_0'^{-1}K^{N_b(\mathscr{G}_t)}$.

To calculate the pair correlation function, $\langle s\delta^{\sigma(x)\sigma(x')} - 1\rangle$, we note first that eq. (2.7.8) reduces to

$$Z' = \sum_{\mathscr{G}} P(\mathscr{G})s^{N_c(\mathscr{G})-N}, \tag{2.7.21}$$

in the limit $H \to 0$. Given two sites $\boldsymbol{x}$, and $\boldsymbol{x}'$, any set of graphs, $\mathscr{G}$, can be divided into two disjoint sets: the set $\mathscr{G}_1$ in which $\boldsymbol{x}$ and $\boldsymbol{x}'$ are in the same cluster and the set $\mathscr{G}_2 = \mathscr{G} - \mathscr{G}_1$, in which $\boldsymbol{x}$ and $\boldsymbol{x}'$ are in different clusters.

In $\mathscr{G}_1$, multiplication by $\delta^{\sigma(x)\sigma(x')}$ changes nothing, whereas in $\mathscr{G}_2$, multiplication by $\delta^{\sigma(x)\sigma(x')}$ decreases the number of clusters by one. We, therefore, have

$$\begin{aligned}\langle\langle(s\delta^{\sigma(x)\sigma(x')} - 1)\rangle\rangle &= Z'^{-1}\Big[\sum_{\mathscr{G}_1}(s-1)P(\mathscr{G}_1)s^{N_c(\mathscr{G}_1)-N} \\ &\qquad + \sum_{\mathscr{G}_2}(ss^{-1}-1)P(\mathscr{G}_2)s^{N_c(\mathscr{G}_2)-N}\Big] \\ &= (s-1)Z'^{-1}\sum_{\mathscr{G}_1}P(\mathscr{G}_1)s^{N_c(\mathscr{G}_1)-N}C(x,x') \\ &\equiv (s-1)D_s(x,x'),\end{aligned} \tag{2.7.22}$$

where $C(x, x')$ is defined following eq. (1.3.2). This expression is valid for all s. The two limits of interest are

$$\begin{aligned}\lim_{s\to 1} D_s(x,x') &= \sum_{\mathscr{G}} P(\mathscr{G})C(x,x') = [C(x,x')]_{\text{av}}, \\ \lim_{s\to 0} D_s(x,x') &= Z_0'^{-1}\sum_{\mathscr{G}_t} K^{N_b(\mathscr{G}_t)}C(x,x') = [C(x,x')]'_{\text{av}},\end{aligned} \tag{2.7.23}$$

where $[\ ]'_{\text{av}}$ signifies an average with respect to the tree weighting function $Z_0'^{-1}K^{N_b(\mathscr{G}_t)}$. Thus the Potts model pair correlation function in the one state limit is simply related to the pair connectedness function $G^c_p(x, x')$ via eq. (1.3.2). The large K limit of eq. (2.7.23b) bears further scrutiny. When $K\to\infty$, only tree graphs with the maximum number of bonds survive. These are *spanning trees* or tree graphs that visit every site.

If a graph contains a spanning tree, it contains only one cluster and from eq. (2.7.16) has $N_B = N - 1$ bonds. Let $\mathscr{G}_s$ be the set of spanning tree graphs and let $\mathscr{G}_2$ be the set of graphs containing exactly two clusters obtained by removing a single bond in all possible ways from each member of $\mathscr{G}_s$. We then have

$$\begin{aligned}[C(x,x')]'_{\text{av}} &= \frac{N_{\text{sp}} + K^{-1}\sum_{\mathscr{G}_2} C(x,x') + \cdots}{N_{\text{sp}} + K^{-1}\sum_{\mathscr{G}_2} 1 + \cdots} \\ &= 1 - K^{-1}N_{\text{sp}}^{-1}\sum_{\mathscr{G}_2}(1 - C(x,x')),\end{aligned} \tag{2.7.24}$$

where N_{sp} is the number of spanning trees on the lattice. The factor $[1 - C(x, x')]$ is unity if x and x' are in different clusters and zero otherwise. Thus $\sum_{\mathscr{G}_2}(1 - C(x, x'))$ is just, $N_{\text{sp}}(x, x')$, the number of spanning trees in which there is a bond (not necessarily a nearest neighbor bond) connecting sites x and x'. If each bond carries a conductance σ, then the resistance $R(x, x')$ between points x and x' is given by

$$R(x, x') = \frac{1}{\sigma} \frac{N_{sp}(x, x')}{N_{sp}} . \tag{2.7.25}$$

This theorem was proven some years ago by Kirchhoff [75]. The general proof is quite lengthy, but the reader can easily check that it is true in simple configurations. We, therefore, have

$$[C(x, x'; H = \sigma J)]'_{av} = 1 - \frac{1}{J} R(x, x') + O\left(\frac{1}{J^2}\right) . \tag{2.7.26}$$

This result can be used to calculate the resistive susceptibility exponent γ_r (eq. 1.6.29) in $6 - \varepsilon$ dimensions.

In order to convert the discrete Potts model into a field theory, it is convenient to introduce a complete set of vectors of length s as follows [76]. Let e_0^σ be the s dimensional vector $(1, 1, \ldots, 1)$. It is of length s since $\sum_{\sigma=1}^{s} (e_0^\sigma)^2 = s$. Now, let e_l^σ, $l = 1, \ldots, s-1$ be the $s - 1$ orthogonal vectors that are orthogonal to e_0^σ so that

$$\sum_{l=0}^{s-1} e_l^\sigma e_l^{\sigma'} = s\delta^{\sigma\sigma'} , \qquad \sum_{\sigma=1}^{s} e_l^\sigma e_{l'}^\sigma = s\delta_{ll'} . \tag{2.7.27}$$

These relations in turn imply

$$\begin{aligned} \sum_{l=1}^{s-1} e_l^\sigma e_l^{\sigma'} &= s\delta^{\sigma\sigma'} - 1 , \\ \sum_{\sigma=1}^{s} e_l^\sigma &= 0 \quad \text{for} \quad l = 1, \ldots, s-1 . \end{aligned} \tag{2.7.28}$$

Note that

$$\langle e_l^\sigma e_{l'}^\sigma \rangle = \frac{1}{s}\left(\sum_\sigma e_l^\sigma e_{l'}^\sigma\right) = \delta_{ll'} .$$

The Potts hamiltonian is therefore

$$\beta\mathscr{H} = -K \sum_{\langle x, x' \rangle, l} e_l^{\sigma(x)} e_l^{\sigma(x')} - \sum_x H_l(x) e_l^{\sigma(x)} , \tag{2.7.29}$$

where $H_l(x) = H(x)e_l^1$. In terms of these variables, the pair connectedness function is

$$G_p^c(x, x') = \lim_{s \to 1} \frac{1}{s-1} \sum_l \frac{\partial^2 \ln Z}{\partial H_l(x) \partial H_l(x')} . \tag{2.7.30}$$

Mean field theory for the Potts model can easily be derived from eq. (2.7.29) along the lines outlined in sect. 2.1. Let

$$\langle e_i^\sigma \rangle = e_i^\prime \Psi \equiv e_i^\prime \left\langle \frac{s\delta^{\sigma 1} - 1}{s - 1} \right\rangle . \tag{2.7.31}$$

The mean field equation for Ψ is then

$$\Psi = \frac{\exp\{s(H + zK\Psi)\} - 1}{\exp\{s(H + zK\Psi)\} + s - 1}, \tag{2.7.32}$$

and the free energy per site is

$$\frac{1}{N(s-1)} \Gamma(\Psi) = \tfrac{1}{2}\left(\frac{T - T_c}{T_c}\right)\Psi^2 - \tfrac{1}{6}(s-2)\Psi^3 + \tfrac{1}{12}(s^3 - 3s + 3)\Psi^4 + \cdots, \tag{2.7.33}$$

where $T_c = zJ = zK_cT_c$. Note that the cubic term changes sign at $s = 2$ just as occurred at $n = 2$ in the spin-glass. At $s = 1$ ($s = 0$) Ψ is the probability that a site is in the infinite cluster (infinite tree) so that it is positive definite. We therefore find that there is a continuous transition with $\beta = 1$ leading to $d_c = 6$ as in the spin-glass.

At $s = 1$, eq. (2.7.32) reduces to

$$\Psi = 1 - \exp\{-(H + zK\Psi)\}. \tag{2.7.34}$$

The generating function $A(H) = 1 - \Psi = \sum n_s \mathscr{K}(n_s) \exp(-Hn_s)$ satisfies [71]

$$A = \exp(-zK - H)\exp(zKA), \tag{2.7.35}$$

which is solved:

$$A(H) = \sum_{A_s = 1}^{\infty} \frac{(n_s zK)^{n_s - 1}}{n_s!} \exp\{-n_s(H + zK)\}, \tag{2.7.36}$$

so that

$$n_s \mathscr{K}(n_s) = \frac{(zKn_s)^{n_s - 1}}{n_s!} \exp(-zKn_s). \tag{2.7.37}$$

Using $p = 1 - \exp(-K)$ and $zK_c = 1$, we can re-express eq. (2.7.37) in terms of

$$r_0 = 1 - zK = \frac{p_c - p}{(1 - p_c)\ln(1 - p_c)}. \tag{2.7.38}$$

For large n_s and small r_0, we obtain

$$n_s \mathcal{K}(n_s) = \left(\frac{1}{2\pi n_s^3}\right)^{1/2} \exp(-\tfrac{1}{2} n_s r_0^2) . \tag{2.7.39}$$

To obtain critical exponents in $6 - \varepsilon$ dimensions, we again develop a field theory from eq. (2.7.29). We obtain

$$\bar{\mathcal{H}} = \int \mathrm{d}^d x \left\{ \tfrac{1}{2} r \sum_l \psi_l^2 + \tfrac{1}{2} \sum_l (\nabla \psi_l)^2 - w_3 \sum_{l_1 l_2 l_3} \lambda^{(3)}_{l_1 l_2 l_3} \psi_{l_1} \psi_{l_2} \psi_{l_3} \right\} , \tag{2.7.40}$$

where $\lambda^{(3)}_{l_1 l_2 l_3} = (1/s) \sum_\sigma e^\sigma_{l_1} e^\sigma_{l_2} e^\sigma_{l_3}$. As in the spin-glass case, higher order terms in ψ are not needed to determine critical properties in the vicinity of six dimensions. The recursion relations to first order in $\varepsilon = 6 - d$ for r and w_3 are [72]

$$\begin{aligned} \frac{\mathrm{d}r}{\mathrm{d}l} &= (2 - \eta_\mathrm{p}) r - 18 K_6 (s - 2) \frac{w_3^2}{(1 + r)^2} , \\ \frac{\mathrm{d}w_3}{\mathrm{d}l} &= \tfrac{1}{2}(\varepsilon - 3\eta_\mathrm{p}) w_3 + 36 K_6 (s - 3) w_3^3 , \end{aligned} \tag{2.7.41}$$

where

$$\eta_\mathrm{p} = 6 K_6 w_3^2 (s - 2) . \tag{2.7.42}$$

These equations yield

$$K_6 (w_3)^2 = \frac{\varepsilon}{18(10 - 3s)} ,$$

and

$$\eta_\mathrm{p} = \frac{s - 2}{3(10 - 3s)} \varepsilon , \qquad \frac{1}{\nu_\mathrm{p}} = 2 + \frac{5(s - 2)}{3(10 - 3s)} \varepsilon . \tag{2.7.43}$$

To second-order in ε, the results for the one-state Potts model are [73,77]

$$\begin{aligned} \eta_\mathrm{p} &= -\tfrac{1}{21}\varepsilon - \frac{206}{(21)^3} \varepsilon^2 , \\ \frac{1}{\nu_\mathrm{p}} &= 2 - \tfrac{5}{21}\varepsilon - \frac{653}{2(21)^3} \varepsilon^2 . \end{aligned} \tag{2.7.44}$$

2.8. Diluted magnets near percolation and random resistor networks [29,32,78]

We argued in sect. 1.3 that a randomly diluted magnet will exhibit no long range order even at $T = 0$ unless the concentration of active bonds is greater than p_c. The point $p = p_c$, $T = 0$ is a type of multi-critical point which can be studied by an elegant application of the "$n = 0$" trick due to Stephen and Grest [29] and Bidaux et al. [78]. Consider the replicated Ising hamiltonian

$$\beta\mathcal{H}^r = -K \sum_{\langle x,x'\rangle,\alpha} (\sigma^\alpha(x)\sigma^\alpha(x') - 1) - H \sum_{\alpha,x} \sigma^\alpha(x)\,. \tag{2.8.1}$$

The average of Z^n for randomly diluted bonds is then

$$\begin{aligned}[][Z^n]_{\text{av}} &= \mathrm{Tr}_{\sigma^\alpha} \prod_{\langle x,x'\rangle} \left(1 + v \exp\left[K \sum_{\alpha=1}^{n} (\sigma^\alpha(x)\sigma^\alpha(x') - 1)\right]\right) \\ &\quad \times \exp\left\{H \sum_\alpha \sigma^\alpha(x)\right\} \\ &= \mathrm{Tr}_{\sigma^\alpha} \exp(-\beta\mathcal{H}_{\text{eff}})\,,\end{aligned} \tag{2.8.2}$$

where $v = P/(1 - p)$ and where we have dropped an overall factor of $(1 - p)^{N_B}$. $\beta\mathcal{H}_{\text{eff}}$ can conveniently be expressed in terms of projection operators onto the states with all replicas having $\sigma^\alpha(x) = \sigma^\alpha(x')$, one replica with $\sigma^\alpha(x) = -\sigma^\alpha(x')$ and so on:

$$\beta\mathcal{H}_{\text{eff}} = \sum_{\substack{k=0 \\ \langle x,x'\rangle}}^{\infty} A_k P_k(x, x') + H \sum_{\alpha,x} \sigma^\alpha(x)\,, \tag{2.8.3}$$

where

$$P_0(x, x') = (\tfrac{1}{2})^n \prod_\alpha (1 + \sigma^\alpha(x)\sigma^\alpha(x'))\,, \tag{2.8.4a}$$

$$P_1(x, x') = (\tfrac{1}{2})^n \sum_\alpha (1 - \sigma^\alpha(x)\sigma^\alpha(x')) \prod_{\beta\neq\alpha} (1 + \sigma^\beta(x)\sigma^\beta(x'))\,, \tag{2.8.4b}$$

$$\vdots$$

and

$$A_k = \ln(1 + v \exp\{-2kK\})\,. \tag{2.8.5}$$

Note that $A_k = 0$ for $k \geqslant 1$ when $T = 0$ ($K = \infty$) so that $\mathcal{H}_{\text{eff}}$ becomes a 2^n state Potts model. In the limit $n \to 0$, this is a one-state Potts model that describes percolation as it should.

The projection operators (eqs. (2.8.4) can be expanded in terms of spin operators to yield

$$P_k(x, x') = \sum_{p=0}^{n} a_p^k \mu^{(p)}(x, x') , \tag{2.8.6}$$

where

$$\mu^{(1)}(x, x') = \sum_{\alpha} \sigma^\alpha(x)\sigma^\alpha(x') ,$$
$$\mu^{(2)}(x, x') = \sum_{\alpha<\beta} \sigma^\alpha(x)\sigma^\beta(x)\sigma^\alpha(x')\sigma^\beta(x') , \tag{2.8.7}$$
$$\vdots$$

and where

$$a_p^k = \binom{n}{k}\binom{n}{p}^{-1} \sum_{l=0}^{p} (-1)^l \binom{k}{l}\binom{n-k}{p-l} . \tag{2.8.8}$$

$\binom{n}{k}^{-1}\binom{n}{p}a_p^k$ is easily seen to be the coefficient of x^p in an expansion of $(1 - x)^k(1 + x)^{n-k}$. By considering the coefficient of x^p in $[(1 + e^\lambda) + (1 - e^\lambda)x]^n$, one can verify that

$$\sum_k a_p^k \, e^{\lambda k} = [1 + e^\lambda]^{n-p}[1 - e^\lambda]^p . \tag{2.8.9}$$

Thus $\beta\mathscr{H}_{\text{eff}}$ can be expressed as an expansion in terms of $\mu^{(p)}(x, x')$:

$$\beta\mathscr{H}_{\text{eff}} = - \sum_{\langle x,x'\rangle} \tilde{K}_p \mu^{(p)}(x, x') - H \sum_{\alpha,x} \sigma^\alpha(x) , \tag{2.8.10}$$

where

$$\tilde{K}_p = \sum_k A_k a_p^k = \sum_{k=0}^{n} \sum_{l=1}^{\infty} \frac{(-1)^l}{l} [v \exp(-2kK)]^l a_p^k$$
$$= \sum_{l=1}^{\infty} \frac{(-1)^l}{l} v^l [1 + \exp(-2lK)]^{n-p}[1 - \exp(-2lk)]^p . \tag{2.8.11}$$

This expression as a well-defined analytic to continuation to $n = 0$ for all p. In the limit $\exp(-2K) \ll 1$, we have

$$\tilde{K}_p = \ln(1 + v) - 2p \exp(-2K) . \tag{2.8.12}$$

Thus, we see that at $T = 0$, all of the variables $\sigma^{(\alpha)}$ where $\sigma^{(1)} = \sigma^\alpha$, $\sigma^{(2)} = \sigma^\alpha\sigma^\beta \cdots$ order simultaneously at $p = p_c$. Temperature breaks the degeneracy of these variables, i.e., it behaves as an anisotropy field in the complicated space of replicated order parameters.

$\beta\mathscr{H}_{\text{eff}}$ can be converted into a field theory by introducing Hubbard–Stratanovitch transformations for each $\sigma^{(\alpha)}$. The result is

$$\bar{\mathscr{H}} = \tfrac{1}{2}\int \mathrm{d}^d x \sum_{(\alpha)} (r_{(\alpha)}(\psi^{(\alpha)})^2 + (\nabla\psi^{(\alpha)})^2) - H\int \mathrm{d}^d x \psi^{\alpha}(x)$$

$$- w\int \mathrm{d}^d x \sum \lambda_{(\alpha)(\beta)(\gamma)}\psi^{(\alpha)}\psi^{(\beta)}\psi^{(\gamma)}, \tag{2.8.13}$$

where $\lambda_{(\alpha)(\beta)(\gamma)}$ is unity if each replica $1, 2, \ldots, n$ appears twice, once in each of two different indices, or not at all and zero otherwise, and where $r_\alpha = r_1 = r_0 - \mathrm{e}^{-2k}$, $r_{\alpha\beta} = r_2 = r_0 - 2\mathrm{e}^{-2k}, \ldots$ where $r_0 \sim p_c - p$. Recursion relations for the potentials in $\bar{\mathscr{H}}$ can be derived in $6 - \varepsilon$ dimensions. $T = 0$ reproduces exactly the results for percolation when $n \to 0$. The crossover exponent associated with e^{-2K} is

$$\zeta = 1 + \frac{2^n - 1}{10 - 3\cdot 2^n}\varepsilon. \tag{2.8.14}$$

When $n \to 0$, $\zeta = 1$. This result remains true to all orders in renormalized perturbation theory. Note that e^{-2K} is precisely the correlation length for the low temperature one dimensional Ising model so that the ζ in eq. (2.8.14) is precisely the same quantity as was introduced in sect. 1.3.

The random resistor network can be studied in much the same way as the randomly diluted magnet [32]. We learned in sect. 2.7 that the zero state limit of the Potts correlation function yields the resistance between two points. This result applies for any cluster, finite or otherwise. In particular, it applies to any cluster in a percolating network:

$$\lim_{s\to 0} D_s(x, x') = C(x, x')\left[1 - \frac{1}{J}R(x, x')\right]. \tag{2.8.15}$$

Averaging this over the percolation probability distribution, we find

$$\lim_{s\to 0}[D_s(x, x')]_{\text{av}} = [C(x, x')]_{\text{av}} - \frac{1}{J}\chi_r(x, x'), \tag{2.8.16}$$

where $\chi_r(x, x')$ is the resistive susceptibility introduced in sect. (1.6). Scaling should apply in the vicinity of $p = p_c$, $J = \infty$ so that

$$\lim_{s\to 0}\sum_x [D_s(x, x')]_{\text{av}} = |p - p_c|^{-\gamma_p}\Phi_p(J|p - p_c|^{\zeta_r}), \tag{2.8.17}$$

so that $\gamma_r = \gamma + \zeta_r$. ζ_r can be calculated by replicating the s state Potts model to obtain an s^n (rather than a 2^n) state Potts model. The same techniques that Stephen and Grest used to obtain ζ for the randomly diluted magnet can be used to obtain ζ_r. Not surprisingly ζ_r is also unity to all orders in perturbation theory.

We close this section with the observation that the techniques discussed in this chapter have also been applied [79,80] to the problem in which bonds of strength J are present with probability p_1, bonds of strength $-J$ with probability p_2 and bonds of strength zero with probability $p_3 = 1 - p_1 - p_2$.

2.9. *Field theory for branched polymers, gelation, and vulcanization* [59]

The primary purpose of this section is to develop a field theory capable of treating the effects of branching on the statistics of polymers. Before proceeding to branched polymers, however, we will review the connection between the $n = 0$ n-vector model and the statistics of linear polymers first derived by de Gennes [54] and extended by des Cloizeaux [56].

We begin by returning to the diffusion equation (eq. 1.7.1) for the gaussian random walk. $P(N, \boldsymbol{x})$ can be Fourier transformed in $\boldsymbol{x}$ and Laplace transformed in N to produce

$$G^0(r, \boldsymbol{k}) = \int_0^\infty \mathrm{d}N \int \mathrm{d}^d x \exp(+Nr) \exp(-\mathrm{i}\boldsymbol{k}\cdot\boldsymbol{x}) P(N, \boldsymbol{x}) = \frac{1}{r + k^2}\cdot \tag{2.9.1}$$

$G^0(r, \boldsymbol{k})$ is simply the propagator $\langle \psi(\boldsymbol{k})\psi(-\boldsymbol{k})\rangle$ for a gaussian field theory with $\beta\mathscr{H} = \frac{1}{2}\sum_k (r + k^2)\psi(\boldsymbol{k})\psi(-\boldsymbol{k})$. A self-avoiding walk can be pictured as a sequence of gaussian walks interrupted by repulsive encounters. A perturbation theory for the self-avoiding walk generator $G(r, \boldsymbol{k})$ can be developed by using a path integral formulation of the diffusion equation in the presence of a repulsive interaction u [81]. In real space, diagrams for $G(r, \boldsymbol{x}, \boldsymbol{x}')$ are depicted in fig. 7a. Here, each solid line represents a G^0 propagator and each dotted line a repulsive potential. Note that closed loop Hartree-like graphs (fig. 7b) never appear in this expansion.

De Gennes was the first to realize that the above graphical expansion is exactly reproduced by the propagator $\langle \psi_1(\boldsymbol{x})\psi_1(\boldsymbol{x}')\rangle$ in the $n = 0$ limit of the n-component field theory with hamiltonian

$$\beta\overline{\mathscr{H}} = \int \mathrm{d}^d x \left\{ \tfrac{1}{2} r \sum_{i=1}^{n} \psi_i^2 + \tfrac{1}{2} \sum_{i=1}^{n} (\nabla\psi_i)^2 + u\left(\sum_i \psi_i^2\right)^2 - H\psi_1(\boldsymbol{x}) \right\}, \tag{2.9.2}$$

with $H = 0$, because the $n = 0$ limit eliminates all closed-loop graphs.

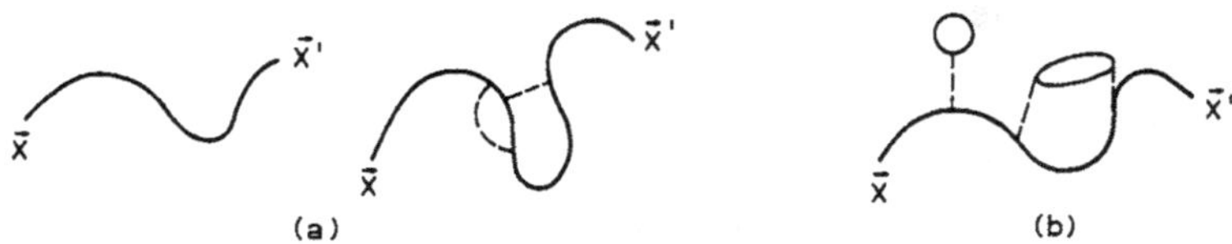

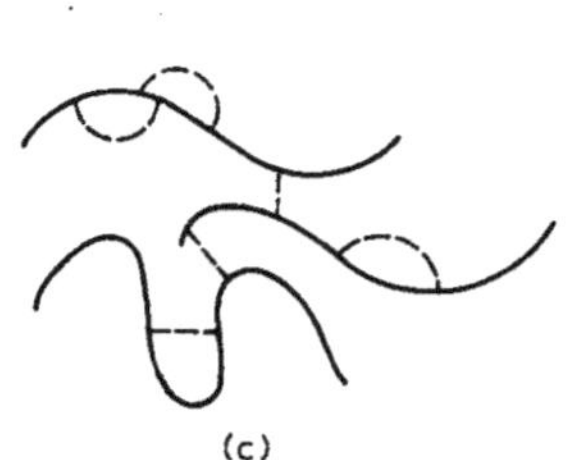

Fig. 7. (a) Typical diagrams for $G(r, \boldsymbol{x}, \boldsymbol{x}')$ for a self-avoiding walk. The solid lines represent the Gaussian random walk propagator $G^0(r, \boldsymbol{x}, \boldsymbol{x}')$ and the dotted lines represent repulsive potentials. (b) A typical diagram containing closed loops for the propagator for an n-component spin model which is eliminated in the $n = 0$ limit. (c) Diagrams for a semi-dilute solution of polymers. Each solid line represents a polymer. Dotted lines represent inter- and intra-polymer repulsive potentials.

This was an important advance because it alerted the physics community to the possibility of treating non-thermodynamic statistical problems not only with the same language but also with the same sorts of model hamiltonia that are used to describe phase transitions. Des Cloizeaux [56] pointed out that the external field H acts as a chemical potential for polymers. This can easily be seen by considering an expansion of the partition function for the hamiltonian of eq. (2.9.2). When $H = 0$, $Z \to 1$ in the limit $n \to 0$ because closed loops are eliminated. Each factor of H marks the endpoint of a polymer. Since there is no mechanism in this model for producing branched polymers, there will be one polymer for each factor of H^2. A term in the expansion of Z containing N_p polymers (i.e., N_p distinct G-lines) will have a prefactor H^{2N_p}. A typical term is shown in fig. 7c. Thus polydisperse solutions of polymers (i.e., solutions in which there is a distribution of molecular weights) can be described by the $n = 0$ n-vector model in a field in which the average number of monomers and polymers, $\langle N_b \rangle$ and $\langle N_p \rangle$ are determined by

$$\langle N_b \rangle = \frac{\partial \ln Z}{\partial r} = \frac{\partial \Gamma}{\partial r}, \tag{2.9.3a}$$

$$\langle 2N_p \rangle = H \frac{\partial \ln Z}{\partial H} = -M \frac{\partial \Gamma}{\partial M}, \tag{2.9.3b}$$

where $M = \langle \psi_1(\boldsymbol{x}) \rangle$. The correlation functions

$$G^c_{11}(\boldsymbol{x}, \boldsymbol{x}') = \langle\psi_1(\boldsymbol{x})\psi_1(\boldsymbol{x}')\rangle - \langle\psi_1(\boldsymbol{x})\rangle\langle\psi_1(\boldsymbol{x}')\rangle\,,$$
$$G^c_{22}(\boldsymbol{x}, \boldsymbol{x}') = \langle\psi_2(\boldsymbol{x})\psi_2(\boldsymbol{x}')\rangle\,, \tag{2.9.4}$$

then measure respectively correlations between endpoints on any polymer and endpoints on a single polymer. Both of these functions depend on r and H which are in turn related to $\langle N_b\rangle$ and $\langle N_p\rangle$ via eqs. (2.9.3).

To obtain a field theory that can describe solutions of branched polymers, two things must be added to the linear polymer theory. First, potentials representing polyfunctional units must be introduced. One can easily see that an f-functional unit can be represented by a potential of the form $w_f\psi_1^f$. Second, an additional degree of freedom must be introduced to distinguish between the number of external vertices and the number of polymers. In solutions of linear polymers, the number of external vertices is always equal to twice the number of polymers. In polydisperse solutions of branched polymers, this relation no longer holds. An external field H will mark endpoints. A chemical potential for polymer number can be obtained by using an nm component field ψ_{ij}, $i = 1, \ldots, n$, $j = 1, \ldots, m$ and an external field $H\sum_j \psi_{1j}$. An adequate field theory for branched polymers is then generated by the hamiltonian

$$\beta\overline{\mathscr{H}} = \int \mathrm{d}^d x\left\{\tfrac{1}{2}r\sum_{ij}\psi_{ij}^2 + \tfrac{1}{2}\sum_{ij}(\nabla\psi_{ij})^2 + u\left(\sum_{ij}\psi_{ij}^2\right)^2 + \left(v\sum_{ij}\psi_{ij}^2\right)^2 - \frac{1}{f!}\sum_f w_f\sum_j\psi_{1j}^f - H\sum_j\psi_{1j}\right\}, \tag{2.9.5}$$

where v represents the effect of three particle repulsive encounters. At the θ-point [83], u is zero in mean field theory. Again, closed loops associated with the potential, u, are eliminated by the limit $n \to 0$, and $\langle\psi_{ij}(\boldsymbol{x})\psi_{ij}(\boldsymbol{x})\rangle_{H=0,w_f=0}$ (for any ij) describes the self-avoiding walk of a single polymer. A typical term in the expansion of the partition function containing N_f f-functional units, N_v endpoints and N_p polymers will be weighted by a factor

$$X = \prod_f w_f^{N_f} H^{N_v} m^{N_p}\,. \tag{2.9.6}$$

Thus w_f, H, and m are fugacities for N_f, N_v, and N_p. A convenient alternate representation of this product can be derived from eq. (1.7.12) relating N_v, N_f, and N_p to the number of loops N_L:

$$X = \prod_f \Lambda_f^{N_f}\Lambda_p^{N_p}\Lambda_L^{N_L}\,, \tag{2.9.7}$$

where

$$\Lambda_f = w_f H^{f-2}, \qquad \Lambda_p = mH^2, \qquad \Lambda_L = H^{-2}. \tag{2.9.8}$$

A particularly striking result of this analysis is that

$$\Lambda_p \Lambda_L = m. \tag{2.9.9}$$

Since H couples to the $(1, 1, \ldots, 1)$ component of ψ_{1j}, it is useful to introduce the orthonormal basis vectors

$$a_i^l = (1/\sqrt{m}) e_i^l, \quad l = 0, 1, \ldots, m-1, \tag{2.9.10}$$

where e_i^l are the vectors introduced in eq. (2.7.30). The fields ψ_{ij} can be expressed in the new basis

$$\psi_{ij} = \sum_{l=0}^{m-1} a_j^l \psi_i^l, \qquad \psi_i^l = \sum_{j=1}^{m} a_j^l \psi_{ij}. \tag{2.9.11}$$

We now shift ψ_{ij}

$$\psi_{ij} = \delta_{i1} \sqrt{m} Q a_0^j + \varphi_{ij}, \tag{2.9.12a}$$

where $\langle \varphi_{ij} \rangle = 0$. The hamiltonian density $\mathscr{F}$ can be expressed as $\sum_k \bar{\mathscr{F}}_k$ where k specifies the power of φ_{ij}. We have

$$\bar{\mathscr{F}}_0 = \tfrac{1}{2} m r Q^2 - \sum_f \frac{1}{f!} m w_f Q^f + u m^2 Q^4 + v m^3 Q^6 - mHQ, \tag{2.9.12b}$$

$$\bar{\mathscr{F}}_1 = \left(rQ - \sum_f \frac{1}{(f-1)!} w_f Q^{f-1} + 4umQ^3 + 6vm^2 Q^5 - H \right) \sqrt{m} \varphi_1^0, \tag{2.9.12c}$$

$$\begin{aligned} \bar{\mathscr{F}}_2 = {} & \tfrac{1}{2}(r + 4umQ^2 + 6vm^2 Q^4) \sum_{i,j} \varphi_{ij}^2 + \tfrac{1}{2} \sum_{i,j} (\nabla \varphi_{ij})^2 \\ & - \tfrac{1}{2} \left(\sum_f \frac{1}{(f-2)!} w_f Q^{f-2} \right) \sum_j \varphi_{ij}^2 + (4umQ^2 + 6vm^2 Q^4)(\varphi_1^0)^2, \end{aligned} \tag{2.9.12d}$$

$$\begin{aligned} \bar{\mathscr{F}}_3 = {} & -\frac{1}{3!} \bar{w}_3^0 \sum_{l_i=0}^{m} \lambda_{l_1 l_2 l_3}^{(3)} \varphi_1^{l_1} \varphi_1^{l_2} \varphi_1^{l_3} + \sqrt{m} u_3^0 \varphi_1^0 \sum_{l=0}^{m=1} (\varphi_1^l)^2 \\ & + \frac{1}{3!} m^{3/2} v_3^0 (\varphi_1^0)^3, \end{aligned} \tag{2.9.12e}$$

$$\bar{\mathscr{F}}_4 = u\left(\sum_{ij} \varphi_{ij}^2\right)^2 - \frac{1}{4!}\,\bar{w}_4^0 \sum_{l_i=0}^{m} \lambda^{(4)}_{l_1 l_2 l_3 l_4} \varphi_1^{l_1}\varphi_2^{l_2}\varphi_3^{l_3}\varphi_4^{l_4}\,, \tag{2.9.12f}$$

where $\quad \bar{w}_3^0 = \sum_{f\geqslant 3} w_f Q^{f-3}\,, \qquad \bar{w}_4^0 = \sum_{f\geqslant 4} w_f Q^{f-4}\,, \qquad u_3^0 = 4uQ\,,$

and $\quad \lambda^{(\mathrm{p})}_{l_1,l_2,\ldots,l_p} = \sum_{j=1}^{m} a^j_{l_1}\ldots,a^j_{l_p}$

and where we have not written down terms for $k > 4$. Using the fact that $\lambda^p_{0,0,0,\ldots,l_p} = 0$ for $l_p \neq 0$. It is easy to see that

$$G^{\mathrm{c}}_{ll}(\boldsymbol{x},\boldsymbol{x}') = \langle \varphi_1^0(\boldsymbol{x})\varphi_1^0(\boldsymbol{x}')\rangle\,, \tag{2.9.13}$$

measures correlations between endpoints on any polymer and

$$G_\perp(\boldsymbol{x},\boldsymbol{x}') = \langle \varphi_1^l(\boldsymbol{x})\varphi_1^l(\boldsymbol{x}')\rangle \quad l \neq 0\,, \tag{2.9.14}$$

measure correlations between endpoints on the same polymer.

The variable Q is non-zero as long as H is non-zero. It can be expressed as

$$Q = H\chi_{\mathrm{T}} \equiv \Lambda_{\mathrm{L}}^{-1}\chi_{\mathrm{T}}\,, \tag{2.9.15}$$

where χ_{T} has a well defined large and small H limit. It, therefore, follows that

$$\begin{aligned} mQ^2 &= mH^2\chi_{\mathrm{T}}^2 = \Lambda_{\mathrm{p}}\chi_{\mathrm{T}}^2\,,\\ u_3^0 &= 4uQ = 4u\Lambda_{\mathrm{L}}^{-1/2}\chi_{\mathrm{T}}\,,\\ \sqrt{m}u_2^0 &= 4u\Lambda_{\mathrm{P}}^{1/2}\chi_{\mathrm{T}}\,,\\ \bar{w}_3^0 &= \sum_f w_f H^{f-3}\chi_{\mathrm{T}}^{f-3} = \Lambda_{\mathrm{L}}^{1/2}\left(\sum_f \Lambda_f\chi_{\mathrm{T}}^{f-3}\right). \end{aligned} \tag{2.9.16}$$

It is immediately apparent from these expressions and eq. (2.9.12f) that all m components of φ_{ij} are critical if and only if $\Lambda_f = 0$ *and* $\Lambda_{\mathrm{p}}u = 0$. If $\Lambda_f \neq 0$ and $\Lambda_{\mathrm{p}}u = 0$, only the m-components φ_{ij} are critical. Finally if both Λ_f and $\Lambda_{\mathrm{p}}u$ are non-zero, only the $m-1$ component φ_1^l, $l = 1,\ldots,m-1$ are critical. With this information, we consider some interesting special cases.

(a) $u = 0, v = 0, \Lambda_{\mathrm{L}} = 0, \Lambda_f = 0$ *for* $f > 3$. This is the case of the gaussian branched polymer with no closed loops considered by Zimm and Stockmayer [57] and by de Gennes [58]. In this case, $\bar{\mathscr{F}}_k \equiv 0$ for all $k \geqslant 3$, and the problem can be solved exactly. From eqs. (2.9.12c), (2.9.12d), and (2.9.15), we have

$$r\chi_T - \tfrac{1}{2}\Lambda_3\chi_T^2 - 1 = 0\,, \tag{2.9.17a}$$

$$G_{\parallel}^{c-1} = G_{\perp}^{c-1} \equiv G^{-1} = r - \Lambda_3\chi_T + q^2\,. \tag{2.9.17b}$$

Introducing the Laplace transform

$$G_{N_b}(\boldsymbol{q}) = \int_{-i\infty}^{i\infty} e^{N_b r} G(r, \boldsymbol{q})\, dr\,, \tag{2.9.18}$$

we can calculate the average separation between endpoints of a branched polymer containing N monomer units. We find

$$\xi^2 = -G_{N_b}^{-1}\frac{dG_{N_b}}{dq^2} \sim N_b^{1/2}\Lambda_3^{-1/4}\,. \tag{2.9.19}$$

This is the same result found by Zimm and Stockmayer [57] for the radius of gyration rather than for the average endpoint separation. Λ_3 is, of course, related to the average number of tri-functional units. The exact form this relation takes depends on the distribution of polymer sizes. If endpoints are considered indistinguishable, and all configurations with a given N_b and N_3 are equally probable,

$$\langle N_3\rangle = \frac{\Lambda_3}{Z_{N_b}}\frac{dZ_{N_b}}{d\Lambda_3} \sim \Lambda_3^{1/2}N_b\,, \tag{2.9.20}$$

where

$$Z_{N_b} = \int_{-i\infty}^{i\infty} e^{N_b r} Z(r)\, dr\,. \tag{2.9.21}$$

If, on the other hand, endpoints are considered distinguishable, the appropriate distribution to use is λ_T so that

$$\langle N_3\rangle = \frac{\Lambda_3}{\chi_T(N_b)}\frac{\partial\chi_T(N_b)}{\partial\Lambda_3} \sim \Lambda_3^{1/2}N_b\,, \tag{2.9.22}$$

where $\chi_T(N_b)$ is Laplace transform of $\chi_T(r)$. This is the distribution that was used by Zimm and Stockmayer [57] and de Gennes [58]. Two things can be learned from this. First a variety of probability distributions for branched polymers can be invented, some more closely related to what can actually be created in the laboratory than others. Second, the large N_b properties of correlation functions are fairly insensitive to the details of the distribution. Thus we can proceed with the reasonable confidence that the distribution that treats endpoints as indistinguishable, that we can treat the most easily, will be in the same universality class as other, perhaps more physical distributions.

A peculiarity [82] of the loopless gaussian branched polymers are perhaps worth mentioning. First the density of monomers in such a polymer is much higher at the outer radius than at the interior for $d < 4$. This can be seen as follows. Let $N_b(r) \sim r^4$ be the number of monomers contained in a sphere of radius r. Then the density of monomers $\rho(r)$ at radius r is related to N_b via

$$N_b(r) \sim \int_0^r r^{d-1}\rho(r)\,dr\,, \tag{2.9.23}$$

so that

$$\rho(r) \sim r^{4-d} \sim N_b^{(4-d)/4}\,. \tag{2.9.24}$$

This implies that for large N_b, any repulsive interactions become very important at the periphery, and that the relation $\xi \sim N_b^{1/4}$ must break down even at the θ point [83]. In fact, one can see that a ψ^6 term in $\beta\mathscr{H}$ will lead to a mass splitting of the φ_{ij} components.

(b) $\Lambda_p u \neq 0$ *and* $\Lambda_f \neq 0$ *for some* f. In this case, only the fields φ_1^l for $l = 1, \ldots, m-1$ are critical. Thus to study the critical properties of this case, a new effective hamiltonian for φ_1^l can be obtained by tracing over φ_1^0 and φ_{ij}, $i > 1$. The resulting hamiltonian is identical in form to the m-state Potts hamiltonian, eq. (2.7.40). *Thus gelation with* $\Lambda_p\Lambda_L = m$ *is in the same universality class as the m state Potts model, and critical exponents for gelation will vary continuously with* $\Lambda_p\Lambda_L$. If $\Lambda_L = 0$, there are no closed loops, and the gelation process is described by the zero state Potts model [74]. We will argue later that in most cases of experimental interest, $\Lambda_p\Lambda_L = 1$ implying that gelation is in the same universality class as percolation in agreement with previous arguments.

c) $\Lambda_p = 0$, v *and/or* $u = 0$. This corresponds to the dilute limit of branched polymers (with or without loops). In this limit, all fields φ_1^l, $l = 0, 1, \ldots, m-1$ become equivalent because they all measure correlations within a single polymer. In fact, as $m \to 0$, we have

$$G_{\parallel}^{c\,-1}(q=0) - G_{\perp}^{-1}(q=0) = m\tau + O(m^2) \tag{2.9.25}$$

where $\tau = 4uQ^2$ in mean field theory. Thus, the critical theory will involve m-fluctuating components. This theory is quite complex and is discussed in detail in reference [84]. Here, we will only outline some important results.

First, the upper critical dimension for a good solvent is eight, and an ε-expansion in $8 - \varepsilon$ dimensions is possible. In this theory, $w_3^2\tau$ is of order ε at the fixed point. In a θ solvent, $\tau = 0$, and d_c becomes 6. In $6 - \varepsilon'$ dimensions, there is a new fixed point with u_3 and v_3 of order $\sqrt{\varepsilon'}$. In both cases, values of critical exponents depend on whether Q or H is fixed. If Q is fixed, we have

1. Good solvents $\varepsilon = 8 - d$

$$\eta = -\tfrac{1}{9}\varepsilon\,, \qquad \nu = \tfrac{1}{2} + \tfrac{5}{36}\varepsilon\,, \qquad \gamma = 1 + \tfrac{1}{3}\varepsilon\,; \tag{2.9.26}$$

2. θ-solvents $\varepsilon' = 6 - d$

$$\eta = -0.073\varepsilon'\,, \qquad \nu = \tfrac{1}{2} + 0.049\varepsilon'\,, \qquad \gamma = 1 + 0.135\varepsilon'\,. \tag{2.9.27}$$

If H is constant, Q must be determined in terms of H via the equation of state. This constraint leads to a Fisher-like [85] renormalization of critical exponents. This renormalization depends on the exponent, μ_3, defined via

$$\frac{1}{\Omega}\frac{\partial^3\Gamma}{\partial Q^3} \sim (r - r_c(Q))^{\mu_3} \tag{2.9.28}$$

where

$$\mu_3 = \tfrac{1}{3}\varepsilon \text{ (good solvent)}\,, \qquad \mu_3 = 0.66\varepsilon' \text{ (θ-solvent)}\,. \tag{2.9.29}$$

The renormalized correlation length and susceptibility exponents satisfy

$$\gamma_r = \frac{\gamma}{2\gamma - \mu_3}\,, \qquad \nu_r = \frac{\nu}{2\gamma - \mu_3}\,. \tag{2.9.30}$$

Note that $\nu_r = \frac{1}{4}$ for $d > d_c$ in agreement with the result of Zimm and Stockmayer.

The natural variables for the model presented here are the fugacities for the number of polymers, Λ_p, number of f-functional units, Λ_f, etc., or alternatively the potentials w_f, H, and m. The most natural experimental variables are the densities c, c_p, c_v, c_L, and c_f. The densities can be expressed in terms of the potentials using

$$\begin{aligned} c_f &= \frac{1}{\Omega} w_f \frac{\partial W}{\partial w_f}\,, \qquad & c &= -\frac{1}{\Omega}\frac{\partial W}{\partial r}\,, \\ c_v &= -\frac{1}{\Omega} H \frac{\partial W}{\partial H}\,, \qquad & c_p &= \frac{1}{\Omega} m \frac{\partial W}{\partial m}\,, \end{aligned} \tag{2.9.31}$$

where $W = \ln Z$, and Ω is the volume. In polydisperse systems a Legendre transformation to $\Gamma(Q) = W + \int mHQ\, \mathrm{d}^d x$ is useful. In terms of Γ, we have

$$
\begin{aligned}
c_f &= -w_f \frac{1}{\Omega}\frac{\partial \Gamma}{\partial W_f}, & c &= \frac{1}{\Omega}\frac{\partial \Gamma}{\partial r}, \\
c_{\mathrm{v}} &= Q\frac{1}{\Omega}\frac{\partial \Gamma}{\partial Q}, & c_{\mathrm{p}} &= -\frac{m}{\Omega}\frac{\partial \Gamma}{\partial m} + mHQ\,.
\end{aligned}
\tag{2.9.32}
$$

In mean field theory, $(1/\Omega)\Gamma = \bar{\mathscr{F}}_0(Q)$. Eqs. (2.9.32) then imply that $c_{\mathrm{v}} = 2c_{\mathrm{p}} + (f-2)c_f$ so that the concentration of loops is zero as is expected in a mean field theory. Using eq. (2.9.12c) for $\mathscr{F}_2$ and eqs. (2.9.32), we obtain

$$
G_{\perp}^{-1} = \frac{1}{c}\left(c_{\mathrm{p}} - \tfrac{1}{2}(f-1)(f-2)c_f\right) + q^2\,. \tag{2.9.33}
$$

When $c_f = 0$, this agrees with previous results for polydisperse linear polymers. Gelation occurs when $c_{\mathrm{p}} = \frac{1}{2}(f-1)(f-2)c_f$. This agrees with the Floy result eq. (1.7.16). Eq. (2.9.37) also provides an explicit verification that the correlation length near the vulcanization threshold is given by

$$
\xi^2 = N_{\mathrm{b}}(\Delta\rho/\rho_{\mathrm{c}})\,, \tag{2.9.34}
$$

where $N_{\mathrm{b}} = 2c/c_{\mathrm{v}}$, the average number of monomers per primary polymer and $\rho_{\mathrm{c}} = \frac{1}{2}c_{\mathrm{v}}/c = N_{\mathrm{b}}^{-1}$. This relation was used by de Gennes [86] to show that the critical region for the vulcanization transition is small when N_{b} is large.

We now return to the question of what value m has in an actual gelation transition. In such a transition, one begins with a concentration c of bifunctional units, A–A, and c_f of f-functional units, R–A_f. The number of reactions R that occur is a monotonic function of time. Using eq. (1.7.11), one can replace $\Lambda_{\mathrm{p}}^{N_{\mathrm{p}}}\Lambda_f^{N_f}\Lambda_{\mathrm{L}}^{N_4}$ by $\Lambda_{\mathrm{p}}^{-R}(\Lambda_f\Lambda_{\mathrm{p}})^{N_f}(\Lambda_{\mathrm{p}}\Lambda_{\mathrm{L}})^{N_{\mathrm{L}}}$ in the partition sum. Since the number of loops in the reaction is not restricted, all values of N_{L} should be given equal weight. This is accomplished by setting $\Lambda_{\mathrm{p}}\Lambda_{\mathrm{L}} = m = 1$. Thus, it would appear that gelation should be in the same universality class as percolation. In the vulcanization process, N_{v} is fixed and 4-functional cross-linking units are added. Here, there are no restrictions in the partition sum other than $2N_{\mathrm{p}} \leqslant N_{\mathrm{v}} - 2N_4$. Thus m must be unity in the product $w_4^{N_4} H^{N_v} m^{N_p}$ in order to weight all permissible values of N_p equally. This puts vulcanization into the same universality class as percolation. These considerations assume that all permissible configurations occur with equal probability. It is not obvious that this is true in laboratory situations. In the gelation transition, in particular, loop formation may be limited purely by the kinetics of the reaction. This appears to be the case in linear polymers where the probability that two endpoints will undergo a close enough encounter to allow a loop forming reaction to occur is very small. Further study of this problem is clearly called for. Finally, we note that some control on the number of loops can be obtained by mixing ARB_2 and A_2RB tri-functional units with A–B monomers when only AB interactions are allowed.

References

[1] K. G. Wilson, Phys. Rev. B4 (1971) 3174; B4 (1971) 3184;
K. G. Wilson and J. Kogut, Phys. Repts. 12 (1974) 75;
K. G. Wilson and M. E. Fisher, Phys. Rev. Lett. 28 (1972) 240.

[2] There are a number of review articles and books on the renormalization group. A selected list includes:
Shang-keng Ma, Modern Theory of Critical Phenomena (Benjamin, Reading, Mass., 1976);
C. Domb and M. S. Green, eds., Phase Transitions and Critical Phenomena, Vol. VI (Academic Press, New York, 1976);
G. Toulouse and P. Pfeuty, Introduction to the Renormalization Group and to Critical Phenomena (Wiley, New York, 1977);
M. E. Fisher, Rev. Mod. Phys. 46 (1974) 587.

[3] B. M. McCoy and T. T. Wu, Phys. Rev. Lett. 23 (1968) 383.

[4] A. B. Harris and T. C. Lubensky, Phys. Rev. Lett. 33 (1974) 1540;
T. C. Lubensky and A. B. Harris, AIP Conf. Proc. 24 (1975) 311;
T. C. Lubensky, Phys. Rev. B11 (1975) 3573.

[5] G. Grinstein and A. Luther, Phys. Rev. B13 (1976) 1329.

[6] For a review of percolation theory, see:
V. K. S. Shante and S. Kirkpatrick, Advan. Phys. 20 (1971) 325;
J. W. Essam, in: Phase Transitions and Critical Phenomena, Vol. 2, eds. C. Domb and M. S. Green (Academic Press, New York, 1972) p. 197.

[7] P. J. Flory, Principles of Polymer Chemistry (Cornell University Press, Ithaca, 1969).

[8] S. F. Edwards and P. W. Anderson, J. Phys. F5 (1975) 965.

[9] V. J. Emery, Phys. Rev. B11 (1975) 239.

[10] P. W. Kasteleyn and C. M. Fortuin, J. Phys. Soc. Japan Suppl. 16 (1969) 11;
C. M. Fortuin and P. W. Kasteleyn, Physics (Utr.) 57 (1972) 536.

[11] R. J. Birgeneau, R. A. Cowley, G. Shirane and H. J. Guggenheim, Phys. Rev. Lett. 37 (1976) 940.
[12] See, e.g., R. Malmhall, K. V. Rao, G. Backstrom and S. M. Bhagat, Physica 86–88D (1977) 196;
E. Figueroa, L. Lundgren, O. Beckman and S. M. Bhagat, Solid State Commun. 20 (1976) 961.
[13] M. E. Fisher, Phys. Rev. 176 (1968) 257.
[14] J. Sak, Phys. Rev. B10 (1974) 3957;
M. A. DeMoura, T. C. Lubensky, Y. Imry and A. Aharony, Phys. Rev. B13 (1976) 2176;
D. J. Bergman and B. I. Halperin, Phys. Rev. B13 (1976) 2145.
[15] R. Brout, Phys. Rev. 115 (1959) 824.
[16] A. B. Harris, J. Phys. C7 (1974) 1671.
[17] A. Harris, M. Plishke and M. J. Zuckermann, Phys. Rev. Lett. 31 (1973) 160; J. Phys. (Paris) 35 (1973) C4-265.
[18] Amnon Aharony, Phys. Rev. B12 (1975) 1038.
[19] M. E. Fisher and P. Pfeuty, Phys. Rev. B6 (1972) 1889;
F. J. Wegner, Phys. Rev. B6 (1972) 1891.
[20] P. Pfeuty, D. Jasnow and M. E. Fisher, Phys. Rev. B10 (1974) 2088.
[21] R. A. Pelcovits, E. Pytte and J. Rudnick, Phys. Rev. Lett. 40 (1978) 476.
[22] Jing-Huei Chen and T. C. Lubensky, Phys. Rev. B16 (1977) 2106.
[23] S. Alexander and T. C. Lubensky, Phys. Rev. Lett. 42, 125 (1979).
[24] D. C. Mattis, Phys. Lett. A56 (1976) 421;
J. M. Luttinger, Phys. Rev. Lett. 37 (1976) 778.
[25] Amnon Aharony and Y. Imry, Solid State Commun. 20 (1976) 899.
[26] D. Stauffer, Z. Phys. B22 (1975) 161.
[27] T. C. Lubensky, Phys. Rev. B15 (1972) 311.
[28] H. E. Stanley, R. J. Birgeneau, P. J. Reynolds and J. Nicoll, J. Phys. C9 (1976) L553.
[29] M. J. Stephen and G. S. Grest, Phys. Rev. Lett. 38 (1977) 567.
[30] P. G. de Gennes, J. Phys. Lett. (Paris) 37 (1976) L1.
[31] A. S. Skal and B. I. Shklovskii, Fiz. Tekh Poluproudn. 8 (1974) 1582 [Sov. Phys. Semicond. 8 (1975) 1029].
[32] C. Dasgupta, A. B. Harris and T. C. Lubensky, Phys. Rev. B17 (1978) 1375.
[33] H. E. Stanley, Phys. Rev. 179 (1969) 570.
[34] D. J. Wallace and A. P. Young, Phys. Rev. B17 (1978) 2384.
[35] R. Fisch and A. B. Harris, Phys. Rev. Lett. 38 (1977) 785.
[36] V. Canella and J. A. Mydosh, Phys. Rev. B6 (1972) 4220;
J. A. Mydosh, AIP 24 (1975) 131.
[37] D. Sherrington and S. Kirkpatrick, Phys. Rev. Lett. 35 (1975) 1792.
[38] C. M. Soukoulis, G. S. Grest and K. Levin, Chicago preprint.
[39] D. J. Thouless, P. W. Anderson and R. G. Palmer, Phil. Mag. 35 (1977) 1792.
[40] J. R. L. de Almeida and D. J. Thouless, J. Phys. A11 (1978) 983.
[41] J. Rudnick and E. Pytte, IBM preprint;
A. Bray and M. A. Moore, Phys. Rev. Lett. 41 (1978) 1068; J. Phys. C12 (1979) 19.
[42] G. Toulouse, Comm. Phys. 2 (1977) 115;
J. Villain, J. Phys. C10 (1977) 1717.

[43] S. Kirkpatrick, Phys. Rev. B16 (1977) 4630.
[44] E. Fradkin, B. A. Huberman and S. H. Shenker, Phys. Rev. B18 (1978) 4189.
[45] A. B. Harris and T. C. Lubensky, Phys. Rev. B16 (1977) 2141.
[46] A. B. Harris, T. C. Lubensky and Jing-Huei Chen, Phys. Rev. Lett. 36 (1976) 415.
[47] J. P. Straley, J. Phys. C9 (1976) 783.
[48] M. J. Stephen, Phys. Rev. B17 (1978) 4444.
[49] A. B. Harris and R. Fisch, Phys. Rev. Lett. 38 (1977) 796;
R. Fisch and A. B. Harris, Phys. Rev. B18 (1978) 416.
[50] R. B. Stinchcombe, J. Phys. C7 (1974) 179.
[51] S. Kirkpatrick, Rev. Mod. Phys. 45 (1973) 574.
[52] A. B. Harris and S. Kirkpatrick, Phys. Rev. B16 (1977) 542.
[53] P. G. de Gennes, J. Physique Letters 37 (1976) L-1.
[54] P. G. de Gennes, Phys. Lett. 38A (1972) 339.
[55] M. Daoud et al., Macromolecules 8 (1975) 804.
[56] J. Des Cloizeaux, J. Physique 36 (1975) 281.
[57] B. H. Zimm and W. H. Stockmayer, J. Chem. Phys. 17 (1949) 1301.
[58] P. G. de Gennes, Biopolymers 6 (1968) 715.
[59] T. C. Lubensky and J. Isaacson, Phys. Rev. Lett. 41 (1978) 829; 42 (1979) 410 (E).
[60] D. J. Stauffer, Chem. Soc. Faraday Transactions II 72 (1972) 1354.
[61] E. Brezin, J. C. LeGuillou and J. Zinn-Justin, in: Phase Transitions and Critical Phenomena, Vol. 6, eds. C. Domb and M. Greene (Academic Press, New York, 1977).
[62] D. J. Amit, Field Theory, Renormalization Group and Critical Phenomena (McGraw Hill, New York, 1978).
[63] J. Hubbard, Phys. Rev. Lett. 3 (1954) 77;
R. L. Stratanovich, Doklady Akad. Nauk SSSR 115 (1957) 1097 [Soviet Phys. Doklady 2 (1958) 416].
[64] V. L. Ginzburg, Fizi Tverd. Tela. 2 (1960) 2031 [Soviet Phys. Solid State 2 (1961) 1824].
[65] F. J. Wegner and A. Houghton, Phys. Rev. A8 (1973) 401.
[66] A. J. Bray and M. A. Moore, Manchester preprint.
[67] D. E. Khmel'nitsky, Zh. Eksp. Teor. Fiz. 68 (1975) 1960 [Sov. Phys. JETP 41 (1976) 981].
[68] J. Rudnick, unpublished.
[69] R. B. Potts, Proc. Camb. Phil. Soc. 48 (1952) 106.
[70] C. Dasgupta, unpublished thesis. The treatment here follows this thesis very closely.
[71] M. J. Stephen, Phys. Rev. B15 (1977) 5674;
M. R. Giri, M. J. Stephen and G. S. Grest, Phys. Rev. B16 (1977) 4971.
[72] A. B. Harris, T. C. Lubensky, W. K. Holcomb and C. Dasgupta, Phys. Rev. Lett. 35 (1975) 327.
[73] R. G. Priest and T. C. Lubensky, Phys. Rev. B13 (1976) 4159.
[74] M. J. Stephen, Phys. Lett. A56 (1976) 149.
[75] G. Kirchhoff, Ann. Physik 72 (1847) 497.
[76] R. K. P. Wallace and D. J. Wallace, J. Phys. A8 (1975) 1495.
[77] D. J. Amit, J. Phys. A9 (1976) 1441.
[78] R. Bidoux, J. P. Carton and G. Sarma, J. Phys. A9 (1976) L87.

[79] M. Giri and M. J. Stephen, J. Phys. C11 (1978) L541.
[80] A. Aharony, J. Phys. C11 (1978) L457.
[81] M. Fixman, J. Chem. Phys. 23 (1955) 1657;
H. Yamakawa et al., J. Chem. Phys. 45 (1966) 1938.
[82] P. Pincus, private communication.
[83] Ref. [7] pp. 425, 523.
[84] T. C. Lubensky and J. Isaacson, to be published in Phys. Rev.
[85] M. E. Fisher, Phys. Rev. 176 (1968) 257.

COURSE 7

(*Summary*)

A SHORT GUIDE TO POLYMER PHYSICS

Pierre-Gilles DE GENNES

Collège de France, 75231 Paris Cedex 05, France

R. Balian et al., eds.
Les Houches, Session XXXI, 1978 – La matière mal condensée/Ill-condensed matter

(1) The first aim of these talks has been to define what is a polymer chain; to what extent it can be made linear (i.e., free of branch points); with what precision one can fix its molecular weight. The best reference for this part is still the beginning of the book by Flory [1]; standard methods for separating chains of different length are described by Carrol [2].

(2) Another essential group of concepts concerns the notion of *chain flexibility*: they are discussed extensively by Birshtein and Ptitsyn [3] and in a second book by Flory [4]. The present talks have been entirely restricted to *flexible* chains.

(3) The statistics of interacting chains has been analyzed here at three different levels:

(a) Through the notion of a self-consistent field, as introduced by Flory, Edwards, Lifshitz and others (for a simplified discussion see de Gennes [5]).

(b) Through a certain correspondence between polymer problems and the n-vector model for phase transitions, exploited in particular by J. des Cloizeaux. A simple presentation of this aspect is found in ref. [6] by Sarma.

(c) Through a renormalization group method, which I call "decimation along the chemical sequence": This amounts to grouping the monomers into submits of g monomers, and then to iterate the process, going to new submits with $g^2, g^3, \ldots, g^p$ monomers. The principles are given by de Gennes [7]. More recent applications to the collapse transition of one chain in a poor solvent are discussed in ref. [8], and in ref. [9] by Joanny.

(4) The static properties of polymer solutions have been discussed at some length: here the central result (due to Flory) is that the chains, in spite of their strong interactions, are essentially ideal. Correlations in a melt have been measured first by neutron methods (Jannink and coworkers at Saclay, Kirste and Fischer in Germany). They have been discussed here by a simple random phase technique, introduced in ref. [10].

(5) The crossover between single chain behaviour and melt behaviour occurs in the so-called semi-dilute solutions, for which the most detailed reference (as regards both experiments and theory) is ref. [6].

(6) One separate talk dealt with the interfacial properties of polymer solutions. When in contact with a repulsive wall, these solutions exhibit a depletion layer of thickness equal to one correlation length ξ; this notion, recently introduced by Joanny and Leibler [11], leads to a number of interesting predictions. In particular, a colloidal suspension of grains in a polymer solution may show a new type of attraction between grains of range ξ and of controllable magnitude (through the concentration). Various features of polymer coils trapped into small tubes have also been discussed, see ref. [12a, b]. The problem of polymer absorption is discussed by de Gennes [13]: the first part of the paper (single chain adsorption) is valid, but the second part (many chains) contains a glaring mistake and should deserve a much deeper theoretical effort.

(7) The properties of polymer gels have been analyzed here only from a restricted point of view – namely to the sol–gel transition and its relation to percolation. The basic references are: (a) Stauffer [14] for the cluster size distribution; (b) de Gennes [15] for the relation between the gel Young's modulus and percolation conductance; (c) Lubensky [16] for a field theoretical analysis. The special case of vulcanization (cross-linking of pre-existent long chains in a melt) is discussed by de Gennes [17].

(8) A very different, but possibly related problem is found not in polymer physics, but in the *hydrodynamics of suspensions*. Experimental data on viscous flow of macroscopic hard spheres in a liquid matrix, show that, beyond a certain concentration threshold, the flows are qualitatively modified (plug flows); see Cox and Mason [18]. The present author was convinced long ago that this was related to a percolation process, but did not understand at the time the nature of clusters (the spheres do not attract each other). Recently, this became clear: while in thermal equilibrium, one sphere in a hard sphere gas is almost *never* in contact with other spheres; in the present system (macroscopic spheres, no Brownian motion), a shear flow field often forces two spheres to remain in contact for a *finite time*. I then assume that beyond a certain critical concentration, a finite fraction of the spheres belongs to an infinite cluster of spheres in contact. This cluster is constantly renewed, and is anisotropic, but we may still assume something like a percolation transition (possibly with different exponents). The hydrodynamic consequences of this model are analyzed in a forthcoming paper [19].

(9) The dynamics of flexible chains in solutions or in melts is a highly complex field.

(a) The classical views on the single chain matrix are beautifully explained in a course by Stockmayer in this same school [20]. More recent thoughts

can be found in ref. [21]. See also Brochard [22] for a clear explanation of dynamical scaling.
(b) For melts, the essential effects are quite different, and are well described in a book by Ferry [23] (see also his more recent lecture notes Les Houches lecture notes [24]). The main theoretical attempts towards an explanation of these remarkable facts are summarized by Graessley [25]. One useful concept in this field is *reptation*, introduced in ref. [26] and generalized in recent work by Edwards and Doi [27].

(10) A qualitative summary of recent joint activities at Strasbourg (CRM), Saclay, and Collège de France (STRASACOL) can be found in ref. [28]. As mentioned in this note, one of the most pleasant aspects of this French polymer work was the close cooperation between very different groups!

References

[1] P. Flory, Principles of Polymer Chemistry, 8th printing (Cornell Univ. Press, Ithaca, NY, 1971).
[2] B. Carrol, ed., Practical Methods of Macromolecular Chemistry (Dekker, New York, 1972).
[3] T. Birshtein and O. Ptitsyn, Conformations of Macromolecules (Wiley, New York, 1966).
[4] P. Flory, Statistical Mechanics of Chain Molecules.
[5] P.-G. de Gennes, Rept. Progr. Phys. 32 (1969) 187.
[6] G. Sarma, appendix to paper by M. Daoud et al., Macromolecules 8 (1975) 804.
[7] P.-G. de Gennes, Nuovo Cimento Rivista (September 1977).
[8] P.-G. de Gennes, J. Physique, to be published.
[9] J. F. Joanny, Thèse 3ème Cycle, Paris (1978).
[10] J. Physique 31 (1970) 235.
[11] J. F. Joanny and L. Leibler.
[12a] M. Daoud and P.-G. de Gennes, J. Physique 38 (1977) 85.
[12b] F. Brochard and P.-G. de Gennes, J. de Physique Lettres (1979) to be published.
[13] P.-G. de Gennes, J. Physique 38 (1977).
[14] D. Stauffer, J. Chem. Soc. (Faraday Trans.) 72 (1976) 1354.
[15] P.-G. de Gennes, J. Physique Lettres 37 (1976) L61.
[16] T. Lubensky, this volume.
[17] P.-G. de Gennes, J. Physique Lettres 38 (1977) L355.
[18] Cox and Mason, Ann. Rev. Fluid Mechanics 3 (1971).
[19] P.-G. de Gennes, to be published in J. de Phys. 40 (Aug. 1979).
[20] W. Stockmayer, Les Houches Lecture Notes (Gordon and Breach, New York, 1976).
[21] P.-G. de Gennes, in: Proc. 30th Anniversary of the Strasbourg Center, 1977.
[22] F. Brochard, J. Chem. Phys. 67 (1977) 52.

[23] J. Ferry, Viscoelastic Properties of Polymers (Wiley, New York, 1970).
[24] J. Ferry, Les Houches Lecture Notes (Gordon and Breach, New York, 1976).
[25] W. Graessley, Advan. Polymer Sci. 16 (1974) 1.
[26] Edwards and Doi, J. Chem. Phys. 55 (1971) 572.
[27] Edwards and Doi, Faraday Trans., in press.
[28] J. Polymer Sci. (Polymer Letters) 15 (1977) 623.

SOME OF THE SEMINARS GIVEN DURING THE SESSION

ATOMIC STRUCTURE OF AMORPHOUS BODIES

André GUINIER

Laboratoire de Physique des Solides,
Université Paris-Sud, 91405 Orsay, France

The description of the structure of an amorphous body begins by its geometry, more specifically the Short Range Order of the atoms: it is, of course, only a part of the study, but it is indispensable, in order to understand the electronic properties.

Experimentally, the fundamental tool is diffraction by X-rays, neutrons and electrons. The theory is simple, at least for monoatomic substances. From the value of the diffracted intensity measured against the angle of diffraction, 2θ, or the function $I(q)$, with $q = (2\sin\theta)/\lambda$, one gets through a Fourier transformation, the *radial distribution function* which gives the partition of the lengths of the atomic pairs.

Two kinds of difficulties are encountered:

(1) Experimentally, it is impossible to determine the whole $I(q)$ function, because q is limited to a maximum value of $2/\lambda$. The cut-off of I produces serious perturbations in the Fourier transform, distortions and artefacts. On the other hand the measured values must be corrected and normalized; some corrections are difficult and rather uncertain. With the neutrons, the correction due to the existence of the inelastic scattering (Placzek correction) is still controversial.

(2) Theoretically, the maximum information which can be drawn from experimental data, supposed perfect and complete, is the pair correlation function: it is insufficient to determine the structure. For instance, this function is the same for a diatomic molecule with a separation a and a

R. Balian et al., eds.
Les Houches, Session XXXI, 1978 – La matière mal condensée/Ill-condensed matter

tetrahedral molecule with edge a. The triplet, quadruplet, ... correlation functions, which intervene in the theory of the disordered structures, cannot be reached by the diffraction phenomena.

Progress in the domain of structure of amorphous structures has been rather slow: the accuracy of the measurements has been considerably improved, but one finds that there are significant discrepancies between the distribution functions published by different authors.

Recently, a new approach to the problem has been used. Instead of trying to deduce directly the structure from the experimental data, one builds *atomic models* and one compares the calculated diffracted intensity with the observed one. When a good fit has been found, the model is *one* possible solution.

For amorphous metals, the atoms are represented by hard spheres, the radius of which is determined by the lattice parameter of the metallic crystal. The spheres are irregularly packed together to attain maximum density. The experiment was made at first with an ensemble of steel balls in a bag and their arrangement was observed. Now the positions of the centres of the spheres are calculated by a computer; different algorithms have been proposed to realize a random close packing with the highest possible density, but the results of the different authors are in general agreement.

Let us consider at first the two-dimensional case: identical spheres laid on a plane are pressed against each other; spontaneously, one obtains an hexagonal crystalline arrangement, each sphere being tangential to 6 others, in mutual contact. But the three-dimensional case is different: around one sphere it is possible to place 12 tangent spheres and, in addition, *a small empty space is left*. That means that one can move the spheres a little, still keeping the contact with the central sphere.... If the 12 spheres are disposed as regularly as possible, they form a regular *icosahedron*, which has the remarkable property of possessing 6 rotation fivefold axes. Such an axis is not crystallographic (i.e. incompatible with a group of translations). So it is not possible to build a crystal by adding spheres to the first 13. There are unavoidable holes of various volumes at different places. One can also say that it is impossible to fill the space with tetrahedra of 4 tangent spheres. The result is a *random close packing* which is a good model for the amorphous metal. It has been found empirically that the maximum density which can be reached is 0.63 (ratio of the volumes of the spheres to the occupied volume); for the close packed crystal, it is 0.73, and the experimental density of the amorphous metal is comprised between these two figures.

It is possible to arrange the 12 spheres of the first layer as a nucleus of a close packed crystal (fcc or hc): 6 spheres are placed in a plane and the six others form two triangles, above and below the plane. This nucleus may be enlarged indefinitely by adding spheres according to the crystalline pattern.

These experiments show that atoms may form random close packings and that, on the other hand, they can crystallize if an initial nucleus is present. When the number of atoms is very large, the regular crystalline arrangement is the most dense, and thus the most stable. But for a limited number of atoms (< 1000), there are non-crystalline aggregates which are stable with icosahedral arrangement. Such small aggregates have been effectively observed by electron diffraction and microscopy.

For covalent atoms (germanium, for instance), the base of the model is the tetrahedron of 4 neighbours. In crystals, these tetrahedra are linked in a perfectly regular order. But it is possible to cancel the long distance order, with only small deformations of the tetrahedron.

The structure of amorphous metallic alloys raises some difficulties: theoretically in the case of a binary alloy, different pair correlation functions intervene for AA, AB, BB pairs: the three functions cannot be determined by a single experiment. One must use combinations of X-ray and neutron diffraction, or neutron diffraction with alloys of the same composition but with different isotopes having different scattering lengths. The experiments are delicate and the results sometimes uncertain.

Now the studies of amorphous materials are very active, because new materials have been produced by various technologies: splat cooling, vapor quenching, sputtering, ...

References

Diffraction methods

B. E. Warren, X-ray Diffraction (Addison-Wesley, Reading, MA, 1969).

H. P. Klug and L. E. Alexander, X-Ray Diffraction Procedures, 2nd ed. (Wiley, New York, 1974).

Methods of correction of experimental results

H. A. Levy, P. A. Agron, and M. P. Danford, J. Appl. Phys. 30 (1959) 2012.

R. Kaplow, S. L. Strong, and B. L. Averbach, Phys. Rev. 138 (1965) A1336.

J. Krogh-Moe, Acta Cryst. 9 (1956) 951.

N. Norman, Acta Cryst. 10 (1957) 370.

B. E. Warren and R. L. Mozzi, Acta Cryst. 21 (1966) 459.

C. W. Dwiggins and D. A. Park, Acta Cryst. A17 (1971) 264.

Corrections of neutron diffraction
G. Placzek, Phys. Rev. 86 (1952) 377.
J. G. Powles, Mol. Phys. 36 (1978) nos. 3m, 4m, 5.

Models of amorphous metals
J. D. Bernal, Proc. Roy. Soc. (London) A280 (1964) 299.
J. L. Finney, J. Physique Colloq. 36 (1975) C2, 1.
J. F. Sadoc, J. Dixmier, and A. Guinier, J. Non-Crystalline Solids 12 (1973) 46.
G. S. Cargill, J. Appl. Phys. 41 (1970) 12.

Small aggregates
M. R. Hoare and P. Pal, J. Crystal Growth 17 (1972) 77.
J. Farges, B. Raoult, and G. Torchet, J. Chem. Phys. 59 (1973) 3545.

ABOUT SPIN GLASSES AND OTHER GLASSES

Abstract

Jean SOULETIE

Centre de Recherches sur les Très Basses Températures, CNRS,
BP 166X, 38042 Grenoble Cedex, France

We give hereafter an abstract of the subjects which we have discussed in the seminar. A detailed version of this discussion and a list of references can be found in: J. Souletie, J. Physique Colloque 39, C2 (1978) 3. The two figures and the references in the present text correspond to recent interesting developments which outdate this paper.

R. Balian et al., eds.
Les Houches, Session XXXI, 1978 – La matière mal condensée/Ill-condensed matter

The name spin glass was given around 1971 to the systems of magnetic impurities diluted in metal matrices and interacting through the Rudermann Kittel interaction (CuMn, AuFe, AgMn, . . .), after the observation of a sharp cusp was reported in the low field ac susceptibility.

From the studies prior to 1971 a few main points should be retained:

– An exact result (the scaling laws in H/c, T/c, where c is the concentration) which can be deduced by a purely geometrical consideration from the fact that the interaction decreases like r^{-3} (r is the distance).

– At low temperatures, a molecular field approach taking into account correlations of a limited spatial extent which aimed in particular at the justification of a linear term in the low temperature specific heat. Questions arise, however, when a generalization is attempted from the Ising to the Heisenberg case and so far this important problem remains unsolved.

– A high temperature (high field) approach which determines exactly the first terms in the developments in c/T (c/H) of the thermodynamic quantities. Estimations for the interactions can be made which result in general agreement with the theory.

Experimental studies since 1971 have been increasingly concerned with the memory effects apparent at temperatures below that of the cusp. A rather general type of phenomenological explanation seems to apply which is relevant to a whole class of non-ergodic phenomena (vitreous transitions, low temperature properties of glasses, magnetization in the Rayleigh domain, fine magnetic particles). It amounts to the assumption of a random distribution of potential barriers separating different levels of energy allowed to the excitations and determining a limitation (depending on the temperature) of the phase space accessible to an experiment realized in a finite time. Very crude assumptions, such as that of a constant density in energy of these barrier heights, imply definite conclusions which we check experimentally. If, on another hand, the same model applies to, say, glasses and spin glasses, a correspondence of some sort should appear at the level of the experimental properties. We present a set of data showing the parallelism that indeed exists between the magnetic properties of a spin glass and the plastic properties of a rubber around its glass transition. The ultimate in these considerations boils down to the following question:

Is the cusp itself, despite its unexplained sharpness, characteristic of some type of glass transition temperature, or would we rely enough on this sharpness to assess that it manifests some type of critical phenomenon? Interesting new developments are on their way with the occurrence of data at different frequencies. Unfortunately, with only two data, we have so far two answers:

– The result of Löhneysen et al. [1] on $(La_{1-x}Gd_x)Al_2$, shown in fig. 1, defines as a glass transition the frequency dependent maximum observed in this system.

– On AgMn, on the contrary, Dahlberg et al. [2] did not observe a significant variation of T_g over a wide range in frequencies (fig. 2) and their result, then, seems to give support to some theoretical speculations which have come around to the idea of a new type of phase transition. Clearly, experiments in the same line will have to establish whether the apparent contradiction between those two results can be resolved by extending the data to a still larger range of frequencies and whether or not some limiting

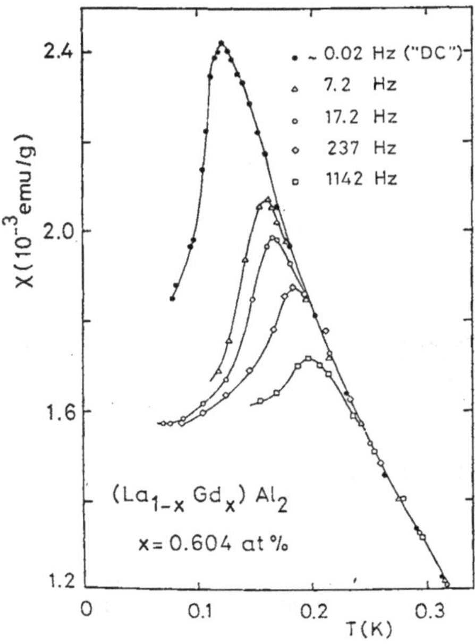

Fig. 1. The position of the peak in the ac susceptibility of a $(La_{1-x}Gd_x)Al_2$ alloy is seen to depend on the frequency of the measurement [1].

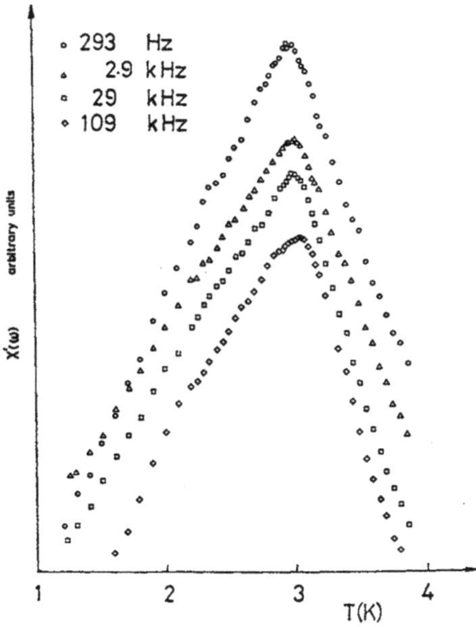

Fig. 2. No significant shift is observed in the position of the peak in the ac susceptibility of a dilute AgMn alloy when the frequency is varied over 3 decades [2].

value of T_g is reached when the characteristic time of the experiment is modified. Also, we still lack a satisfactory connection between traditional magnetization data and results of dynamic measurements using less conventional techniques.

In any case, for the time being, the experimental data and the theoretical speculations agree at least on one point: this phase transition, if its existence was granted, would be of a completely new type and would present many uncommon properties. The goal represented by the clarification of this point seems, therefore, to justify the utilization of unconventional methods: computer simulations have been used with particular success; such new concepts as that of frustration, or the definition of an order parameter measuring the correlation in time (rather than in space) have been introduced. Those notions may well turn out to be of interest in a domain much broader than that of the spin glasses for which they have been defined. They may, hopefully, be relevant to the understanding of the amorphous state in general, a subject for which a great deal of interest has surfaced in the last few years.

References

[1] H. v. Löhneysen, J. L. Tholence, and R. Tournier, J. Physique Colloque 39, C6 (1978) 922.

[2] E. D. Dahlberg, M. Hardiman, and J. Souletie, J. Physique Lettres 39 (1978) L389.
E. D. Dahlberg, M. Hardiman, R. Orbach, and J. Souletie, Phys. Rev. Lett., 42 (1979) 401.

STUDY OF THE DELOCALIZATION OF THE ELECTRONS IN n-CdS AROUND THE MOTT TRANSITION BY SPIN FLIP RAMAN SCATTERING AND FARADAY ROTATION

Robert ROMESTAIN

Laboratoire de Spectrométrie Physique, USMG, BP 53, 38041 Grenoble Cedex, France

and

S. GESCHWIND and G. DEVLIN

Bell Laboratories, Murray Hill, New Jersey 07974, USA

The Mott transition in semiconducting CdS occurs around $N = 2.4 \times 10^{17}\ \mathrm{cm}^{-3}$. In an external magnetic field H_0, the Spin Flip Raman Scattering (SFRS) line at $\Delta V = g\mu_B H$ is broadened by Doppler effect and motionally narrowed by collisions [1]. Thus the diffusional motion occurring above the transition can be measured by the linewidth of the SFRS [2]. It is found that in compensated crystals the linewidth can be calculated by assuming a free electron model with the effective mass m^* of the conduction band.

Measurement of the magnetic susceptibility by Faraday Rotation [3] indicates a composite behaviour which can be interpreted as a superposition of clusters of delocalized donors (Pauli like) and of isolated donors (Curie–Weiss like).

R. Balian et al., eds.
Les Houches, Session XXXI, 1978 – La matière mal condensée/Ill-condensed matter

However, the SFRS lineshape reveals no structure which indicates that electrons are completely delocalized, continuously tunnelling between clusters and localized states.

References

[1] J. F. Scott, T. C. Damen, and P. A. Fleury, Phys. Rev. B6 (1972) 3856.
[2] S. Geschwind, R. Romestain, and G. E. Devlin, J. Physique Colloq. 37, C4 (1976) 313;
R. Romestain, NATO Advanced Studies Institutes B29 (1977) 351.
[3] R. Romestain, S. Geschwind, and D. E. Devlin, Phys. Rev. Letters 35 (1975) 803.

A QUADRUPOLAR GLASS PHASE IN SOLID HYDROGEN

Michel DEVORET

Service de Physique du Solide et de Résonance Magnétique, Orme des Merisiers, BP 2, 91190 Gif-sur-Yvette, France

The problem of orientational order in solid hydrogen bears strong resemblance to the spin glass problem. Solid hydrogen, even in pure monocrystalline form, is a random alloy of ortho-hydrogen molecules ($J = 1$ state) and para-hydrogen molecules ($J = 0$, ground state). The latter act as a diluant of ortho-molecules which are the active particles for cooperative phenomena and whose quadrupolar moments play the same role as the magnetic moments in spin glasses. However, the interaction between ortho-molecules is dominantly electrostatic quadrupole–quadrupole which is short-ranged and anisotropic in contrast to the RKKY interaction in spin glasses. Thus, the problem of orientational order in solid hydrogen is practically that of spin 1 particles distributed at random on lattice sites and coupled by bond-independent nearest neighbour anisotropic interactions. In this system, the concentration of spins varies continuously as metastable ortho-molecules slowly convert into para-molecules and one can therefore make measurements at different concentrations on the same sample.

Sullivan et al. have found evidence in this system [1] for an ordered phase in which the quadrupolar moments are frozen at random. This quadrupolar glass phase completes the phase diagram which previously consisted of an orientationally disordered phase at high temperatures and a

R. Balian et al., eds.
Les Houches, Session XXXI, 1978 – La matière mal condensée/Ill-condensed matter

long range ordered phase at low temperatures and for ortho-concentrations greater than 55%. With the addition of a quadrupolar glass phase at low temperatures and for ortho-concentrations less than 55%, the phase diagram closely resembles the phase diagram of magnetic alloys. As in the spin glass problem, the existence of the quadrupolar glass phase at reduced ortho-concentration is not just a mere consequence of the underlying disorder in occupancy of lattice sites, since frustration effects [2] also play a crucial role. When no disorder is present (pure ortho-hydrogen), as a result of the particular form of the quadrupolar interaction and the close-packed lattice of solid hydrogen, the molecules adopt an orientational configuration which is totally frustrated, i.e., it minimizes the total energy without minimizing the energy of each bond. We believe that the appearance of the quadrupolar glass phase below the concentration threshold of 55% is due to the onset of availability of lowest energy configurations in which the frustration is reduced at the expense of global translational symmetry by taking advantage of local disorder.

References

[1] N. S. Sullivan, M. Devoret, B. P. Cowan, and C. Urbina, Phys. Rev. B17 (1978) 5016.

[2] G. Toulouse, Commun. Phys. 2 (1977) 115.

AMORPHOUS SILICON AND ITS APPLICATION TO SEMICONDUCTOR DEVICES

Ionel SOLOMON

Laboratoire de Physique de la Matière Condensée, Ecole Polytechnique, 91128 Palaiseau Cédex, France

Many species of amorphous silicon appear in the literature. The situation has been somewhat clarified in the last two years and three types of material can be clearly distinguished:

(1) *Pure amorphous silicon.* This is a well-defined material, with a remarkably constant density of localized defects as detected by magnetic resonance: For the whole range of deposition temperatures T_s or annealing temperatures T_A ($20°C < T_s, T_A < 400°C$) the concentration of defects (broken bonds?) remains, within a factor of 2, close to the value 5×10^{19} cm^{-3}.

These defects are responsible for a high density of states in the gap and the electrical conductivity is mostly carried by variable range hopping:

$$\sigma \approx \exp[-(T_0/T)^{1/4}] .$$

The material is very sensitive to contamination, and reproducible properties have been obtained only for the films deposited under ultra-high vacuum conditions.

The basic study of this material can be found in ref. [1].

(2) *Highly hydrogenated silicon.* At the other extreme, the material obtained by decomposition of silane SiH_4 at low deposition temperature ($T_s < 70°C$) is a material containing a large proportion of $(SiH_2)_n$ chains.

R. Balian et al., eds.
Les Houches, Session XXXI, 1978 – La matière mal condensée/Ill-condensed matter

The hydrogen-to-silicon ratio in this material is very large (30% or more). Its density is much lower than that of the other types of silicon and its mechanical as well as electrical properties can be considered as those of a polymer ("polysilene").

(3) *Silicon–hydrogen alloy* ("*I–IV compound*"). The most interesting material as regards applications (solar cells, efficiency close to 6%) is the hydrogenated material deposited at medium or high temperature ($T_s > 150°C$). It can be obtained by decomposition of SiH_4 or by sputtering in the presence of hydrogen.

A very recent review has been given on this material by Spear [2].

The properties of this material are summarized in the following list:

– The hydrogen-to-silicon ratio varies from 2% to about 25%.

– The optical gap is a well-defined quantity (linear extrapolation of $\sqrt{\alpha h\nu}$) and varies from 1.6 to 1.85 eV with the hydrogen content.

– The density-of-states in the gap is remarkably low: 10^{16} cm^{-3} eV^{-1}.

– The conductivity is activated:

$$\sigma = \sigma_0 \exp(-E_a/kT) ,$$

with σ_0 nearly constant: $\sigma_0 = 10^4\ (\Omega\ \text{cm})^{-1}$.

– The Fermi level for the undoped material is pinned at a constant distance ($\simeq 0.93$ eV) from the valence band.

– The variation of the activation energy is thus entirely determined by the variation of the optical gap.

All this suggests a picture of a "covalent crystal" (hydrogen–silicon alloy or "I–IV semiconductor"), where the disordered character is of little importance.

(This aspect of the hydrogenated amorphous silicon is to appear in ref. [3].)

References

[1] P. A. Thomas, M. H. Brodsky, D. Kaplan, and D. Lepine, Phys. Rev. B, 18 (1978) 3059.

[2] W. E. Spear, Advan. Phys. 26 (1977) 811.

[3] I. Solomon, J. Perrin, and B. Bourdon, in: Proc. 14th Intern. Conf. on Physics of Semiconductors, Edinburgh, 1978.

EXCITATIONS IN THE HEISENBERG–MATTIS MODEL: NON-TRIVIAL OSCILLATIONS ON A FLAT BACKGROUND

Abstract

David SHERRINGTON

Institut Laue–Langevin, BP 156, 38042 Grenoble Cedex, France

The Heisenberg–Mattis model is one which can exhibit frozen magnetism without average long-range order, features common to a spin glass. Unlike the spin glass its classical ground state and classical thermodynamics are trivially characterizable. Its excitation structure has, however, several features believed common to spin glasses, but open to explicit analytic calculation. These features are extended states at low energies, localized at high, within a harmonic oscillation approximation. The scattering function $S(q, \omega)$ coupling to a probe of small laboratory wavevector is shown in a three-dimensional system to exhibit well-defined peaks as a function of ω with means proportional to q, widths proportional to q^2. One-dimensional magnets, however, exhibit strong damping of response to probes with sharp laboratory frame wavevectors.

R. Balian et al., eds.
Les Houches, Session XXXI, 1978 – La matière mal condensée/Ill-condensed matter

1. Introduction

There is currently much interest in disordered magnetic systems which can exhibit frozen magnetism without average long-range order; mathematically these features may be characterized by [1]

$$\text{frozen magnetism:} \quad \overline{|\langle \boldsymbol{S}_i \rangle|} \neq 0 \,, \tag{1.1}$$

$$\text{no average l.r.o.:} \quad \lim_{\boldsymbol{R}_i - \boldsymbol{R}_j \to \infty} \overline{\langle \boldsymbol{S}_i \rangle \cdot \langle \boldsymbol{S}_j \rangle} = 0 \,, \tag{1.2}$$

where $\langle\ \rangle$ refers to a thermodynamic average or an average over the time of an experiment, and the bar refers to a spatial average. Systems with these properties are now realized as falling into two categories depending upon whether the system also exhibits a sufficient random degree of "frustration" [2]; frustration is the feature that the system cannot choose a ground state in which the ordering tendencies of all the exchange bonds (and perhaps local fields) are satisfied.

Spin glasses are systems exhibiting all three of the above features and their study is extremely active at present. The interest is largely due to the fact that even within classical thermodynamics a randomly frustrated system cannot be transformed into an unfrustrated one and presents new physics. By contrast, a system which satisfies (1.1) and (1.2) but is unfrustrated is generally considered essentially trivial or "flat" [2]. We shall show here, however, that when one goes beyond classical thermodynamics even an unfrustrated model satisfying (1.1) and (1.2) can be non-trivial and show features in common with frustrated systems.

The system we consider is the Heisenberg version of a model discussed originally by Mattis [3] and characterized by the hamiltonian [4]

$$H = -\sum J_{ij} \xi_i \xi_j \boldsymbol{S}_i \cdot \boldsymbol{S}_j \,, \tag{1.3}$$

where the ξ_i are independently randomly ± 1, with probability c that $\xi_i = -1$. J_{ij} is not disordered. The triviality of the classical thermodynamics of this model is immediately apparent from the observation that the transformation

$$\boldsymbol{\tau}_i = \xi_i \boldsymbol{S}_i \tag{1.4}$$

changes (1.2) into a pure system hamiltonian

$$H = -\sum J_{ij}\boldsymbol{\tau}_i\boldsymbol{\tau}_j . \tag{1.5}$$

The commutation relations of the τ are, however,

$$[\tau_i^x, \tau_i^y] = i\hbar\xi_i\tau_i^z . \tag{1.6}$$

It follows that only in the limit $\hbar \to 0$ can all the $\{\tau\}$ be considered spin operators. Hence, although the classical thermodynamics is trivial, the quantum thermodynamics is not. Similarly the excitations are non-trivial even in the so-called classical approximation where a Poisson bracket takes the place of the commutator.

The standard starting point for a study of the elementary excitations (magnons) of a magnetic system is a specification of the classical ground state. For a spin glass such specification for ground or metastable states is extremely difficult, and theoretically only very limited information is available (essentially just that the energy is locally minimal). Analytic study of magnons in spin glasses has thus been restricted, involving poorly controlled approximations, and has been limited to low energies [5,6]. However, computer simulations have been performed [7,8] on finite systems within a linearized equation of motion approximation, exhibiting extended state behaviour at low energies, localization at high.

The Heisenberg–Mattis model, on the other hand, has a readily-specified classical ground state

$$\langle S_i^z\rangle = \xi_i S , \tag{1.7}$$

and the excitations can be studied analytically in a controlled manner. Furthermore, it is possible to relate the excitations to observations in the laboratory (as opposed to spin) frame. Such modes are found to exhibit several of the features believed to be characteristic of magnons in spin glasses, such as quasi-propagating modes at low energies, localized at high. It is with such a study that this paper is concerned.

2. Scattering function

In a Heisenberg system with colinear ground state structure, such as is the case here, a scattering experiment measures the quantity [9].

$$S(\boldsymbol{q}, \omega) = \lim_{\delta\to 0} \operatorname{Im} \Delta(\boldsymbol{q}, \omega + i\delta) , \tag{2.1}$$

where

$$\Delta(\boldsymbol{q}, E) = N^{-1} \sum_{ij} \Delta_{ij}(E) \exp(i\boldsymbol{q}\cdot(\boldsymbol{R}_i - \boldsymbol{R}_j)) \tag{2.2}$$

$$= \sum_{j} \overline{\Delta_{ij}(E)} \exp(i\boldsymbol{q}\cdot(\boldsymbol{R}_i - \boldsymbol{R}_j)), \tag{2.3}$$

and

$$\Delta_{ij}(E) = \langle\langle S_i^+ ; S_j^- \rangle\rangle_E + \langle\langle S_i^- ; S_j^+ \rangle\rangle_E, \tag{2.4}$$

using Zubarev notation. Δ turns out to be very convenient for perturbation analysis. Linearizing the Green function equations of motion in the standard RPA fashion, substituting from (1.6) for the $\langle S_i^z \rangle$ and rearranging, there results the self-consistency equation for Δ

$$\Delta_{ij}(E) = \Delta_{ij}^0(E) + \sum_{pl} \Delta_{ip}^0(E) V_{pl} \Delta_{lj}(E), \tag{2.5}$$

where

$$V_{pl} = -\tfrac{1}{4} J_{pl}(\xi_p \xi_l - \phi), \tag{2.6}$$

$\Delta_{ij}^0(E)$ is the Fourier transform of

$$\Delta^0(\boldsymbol{q}, E) = \frac{4S^2(J(0) - J(\boldsymbol{q}))}{(E^2 - S^2(J(0) - J(\boldsymbol{q}))(J(0) - \phi J(\boldsymbol{q})))}, \tag{2.7}$$

and ϕ is arbitrary. We have not yet made any assumption about the distribution of the ξ. (2.5) is a linear eigenvalue equation and its unaveraged solutions yield as poles of Δ a set of well-defined magnon normal modes. The wavevector $\boldsymbol{q}$ is not, however, a good quantum number to describe these modes and their evaluation is not a simple task. Three primary interesting questions concern (i) their density of states, (ii) their coupling to a laboratory frame, (iii) the existence of localized solutions. The first two of these questions relate to the investigation of $\Delta(\boldsymbol{q}, E)$. Localization requires different analysis [10] and discussion is deferred until sect. 4.

Iterating, averaging, and resumming (2.5) yields

$$\Delta(\boldsymbol{q}, E) = \left((\Delta^0(\boldsymbol{q}, E))^{-1} - \sum(\boldsymbol{q}, E)\right)^{-1}, \tag{2.8}$$

where $\sum(\boldsymbol{q}, E)$ is the infinite sum of terms of the form $\overline{V\Delta^0 \cdots V}$ which are irreducible with respect to the ξ-averages. It is evident that the standard results for pure nearest neighbour ferromagnet or simple antiferromagnet follow immediately from the choices $\phi = 1$ or $\phi = -1$ respectively (V being zero with these choices). Perturbation theory for small perturbations from these pure cases is straightforward; for the basically ferromagnetic case, to linear order in c, $\sum$ is directly given from the local self-energy of a single $\xi = -1$ impurity in a ferromagnetic host and yields to this order an $S(\boldsymbol{q}, \omega)$ with peaks increased in energy by $(1 + 2c)$ with respect to the pure ferromagnet and with widths proportional to cq^{d+2}, where d is the lattice dimensionality.

Of particular interest, however, is the case $c = \frac{1}{2}$ for which (1.1) and (1.2) hold. In this case, the convenient choice is $\phi = 0$. A simplifying feature for all $(\boldsymbol{q}, \omega)$ is that not only is $\xi_i^2 = 1$ but also the average of any odd power of ξ is zero. For the small ω limit an additional simplification ensues from the fact that $\Delta_{ij}^0(E)$ tends to δ_{ij} as $E \to 0$ for this choice of ϕ. This leads to the dominant behaviour of $\sum(q, E)$ for $(\boldsymbol{q}, E) \to 0$ being given by retaining only local Δ^0 in the infinite series for the self-energy. The resulting series is exactly summable. In three-dimensions there results an $S(\boldsymbol{q}, \omega)$ which for small $(\boldsymbol{q}, \omega)$ has well-defined peaks located at

$$\omega(\boldsymbol{q}) = \pm S[(J(0) - J(\boldsymbol{q}))J(0)/I]^{1/2}, \tag{2.9}$$

where I is the lattice sum

$$I = N^{-1} \sum_{\boldsymbol{q}} (1 - J(\boldsymbol{q})/J(0))^{-1}, \tag{2.10}$$

which for cubic lattices takes the value sc $I = 1.5164$, fcc $I = 1.3447$, bcc $I = 1.3932$. For nearest-neighbour interactions on a lattice of coordination number z the leading $\omega(\boldsymbol{q})$ behaviour may be re-written as

$$\omega(\boldsymbol{q}) = \pm zJSq/(2I)^{1/2}, \tag{2.11}$$

q being measured in units of the inverse lattice spacing. The corresponding half-widths of the peaks are

$$\Gamma(\boldsymbol{q}) = \frac{3\pi}{2\sqrt{2}} \frac{SJz}{I^{3/2}} \left(\frac{q}{q_m}\right)^2, \tag{2.12}$$

where q_m is the Debye radius. It is interesting to note that the q dependence of $\omega(\boldsymbol{q})$ is as obtained in the crude spin glass theories [5] and that the q-dependences of both $\omega(\boldsymbol{q})$ and $\Gamma(\boldsymbol{q})$ are as predicted by hydrodynamic analysis of a spin glass [6]; it should, however, be noted that the q^2 dependence of Γ here results only from disorder, from the fact that q is not a good quantum number, while in the hydrodynamic theory its physical origin lies in magnon–magnon interaction which is ignored in our analysis.

Our results can be compared with computer studies of $S(\mathbf{q}, \omega)$ for the same model [8]. Unfortunately, only one low q point was considered in the computer studies but for that point ($\pi/6(1, 1, 1)$) the position and width are in good agreement with our theoretical treatment.

The density of states follows from integrating $S(\mathbf{q}, \omega)$. It has the same power form as for a pure antiferromagnet, $\rho(\omega) \propto \omega^2$, leading to specific heat $\propto T^3$.

For the linear approximation to be justifiable, zero-point spin deviations must be small compared with S. This is true for large S, as for an ordinary antiferromagnet.

3. One-dimensional systems

In less than three space dimensions true thermodynamically stable co-operative magnetic order is impossible; (1.1) cannot hold for $\langle\ \rangle$ defined in a strict thermodynamic sense. Nevertheless, linearization of the equation of motion, replacing $S^{\pm}S^{z}$ by $S^{\pm}\langle S^{z}\rangle$, with $\langle S^{z}\rangle$ given by its classical ground state value, yields a one-body problem which is of interest in its own right through its non-trivial disorder. Without further analysis its relevance to a true one-dimensional random magnet cannot be assessed but we note that for a pure $d = 1$ antiferromagnet such an analysis [11] (i) leads to a dispersion in excellent accord with experiments for an $S = \frac{5}{2}$ chain (TMMC) [12], (ii) has the same functional form (but different coefficient) as found in an exact evaluation of the spectrum of an $S = \frac{1}{2}$ chain [13], itself verified experimentally [14]. It thus seems plausible that analysis of the linearized equation of motion may have relevance to random systems. The further low dimensional discussion here will be confined to $d = 1$.

For $c = \frac{1}{2}$ the relevant series for $\sum$ is the same as before except that care must be taken in letting E tend to zero since lattice sums such as (2.10) diverge for $d < 3$. We find a dominant small (q, ω) behaviour of $S(q, \omega)$

$$S(q, \omega) = 4JS^2q^2\omega^{-1/2}\{vq^2[(\omega^{3/2} - vq^2)^2 + v^2q^4]^{-1}\} \tag{3.1}$$

where $v = (\sqrt{2}JS)^{3/2}$. Thus, in this case, we find that the energies of the peak and the width of $S(q, \omega)$ are comparable, indicating strong damping of excitations of fixed (laboratory) q.

Another interest in $d = 1$ is that systems with finite-range interactions that would be frustrated in higher dimensionality are not for $d = 1$. For example, a system with nearest neighbour bonds able to take randomly the values $\pm J$ is the classic pure frustration model of Toulouse [2] but for $d = 1$ its classical ground state can be written down trivially and analysed as above. Thus we consider also the model characterized by

$$H = -J \sum_i \eta_i \boldsymbol{S}_i \cdot \boldsymbol{S}_{i+1} , \tag{3.2}$$

with the η_i taking randomly and independently the values ± 1 with probability $(1 - c)$, c. This model and that of (1.2) with $d = 1$ are only identical for $c = \frac{1}{2}$. They clearly differ for small c in that (1.2) leads to a classical ground state which is on average ferromagnetic with moment/site $S(1 - 2c)$, while (3.2) has no average long range order for any finite c.

Both (3.2) and (1.2) are soluble to order c in $\sum$. As mentioned in sect. 2, at long wavelength (1.2) leads simply to a modification of the usual q^2 dispersion, stiffening it by a factor $(1 + 2c \cdots)$ and broadening the peak proportionately to cq^{d+2}. Eq. (3.2), however, yields for long enough wavelengths $(q \ll c^4)$ an $S(q, \omega)$ with the same form as (3.1) but with $\nu = c(JS)^{3/2}$. For larger q and small c the usual ferromagnetic peaks occur. Thus, as expected, if there is no overall order and the wavelength is much larger than the range of ferromagnetic order, laboratory excitations do not propagate in one-dimension, although they appear to be quasipropagating in $d = 3$ or greater.

Notice that in all cases studied above, both $\omega(\boldsymbol{q})$ and $\Gamma(\boldsymbol{q})$ tend to zero as $q \to 0$. This is in accordance with Goldstone's theorem.

4. Localization

It is known that certain disorder can lead to localization of some of the states of a one-body system [10]. Any such one-body problem can be expressed in an equation of motion form but another common formulation of a disorder problem is the Anderson model [15]

$$H = \sum_i \varepsilon_i a_i^\dagger a_i + \sum_{ij} t_{ij} a_i^\dagger a_j . \tag{4.1}$$

In this site-representational model the ε are local potentials and the t_{ij} are interactions allowing hopping from site to site. (4.1) could apply to electrons in a tight binding model if a, $a^\dagger$ are Fermi operators. It could also describe a magnetic system with only ferromagnetic interactions treated in an approximation equivalent to that used above in this paper, the operators a, $a^\dagger$ being now Holstein–Primakoff boson spin deviation operators.

For a model such as (4.1) any ε disorder can lead to localization of some eigenstates. Disorder of the magnitude of t can lead to localization but disorder of the sign of t may not. Will disorder of the sign of the exchange as in (1.2) and (3.2) lead to localized states within the linearized equation of motion approximation? We shall argue "Yes".

Let us first express the linearization procedure in spin-deviation hamiltonian form. This is achieved by the Holstein–Primakoff transformation

$$S_i^z = \lambda_i(S - a_i^\dagger a_i)\,, \tag{4.2a}$$

$$S_i^+ \simeq (2S)^{1/2}\{(1 + \lambda_i)a_i/2 + (1 - \lambda_i)a_i^\dagger/2\}\,, \tag{4.2b}$$

$$S_i^- \simeq (2S)^{1/2}\{(1 + \lambda_i)a_i^\dagger/2 + (1 - \lambda_i)a_i/2\}\,, \tag{4.2c}$$

where $\lambda_i = \pm 1$ for local $\langle S_i^z\rangle$ in the $\pm z$ direction. For the Mattis model $\lambda_i = \xi_i$, for the model of (3.2)

$$\lambda_i = \prod_{j<i} \eta_j\,.$$

This transformation yields

$$H = \sum_i \varepsilon_i a_i^\dagger a_i + \sum_{ij} t_{ij}\{\tfrac{1}{2}(1 + v_{ij})a_i^\dagger a_j + \tfrac{1}{4}(1 - v_{ij})(a_i^\dagger a_j^\dagger + a_i a_j)\}\,, \tag{4.3}$$

where $\varepsilon_i = J(0)S$, $t_{ij} = -J_{ij}S$, and v_{ij} is 1 if the bond between i and j is ferromagnetic, -1 if it is antiferromagnetic; i.e.

$$v_{ij} = \xi_i\xi_j\,, \quad \text{or} \quad [2(\eta_i\delta_{j,i+1} + \eta_{i-1}\delta_{j,i-1}) - 1]\,. \tag{4.4a,b}$$

We see therefore, that for these simple models ε and t are not random although v may be. If all $v = 1$ we get for H the spin-deviation hamiltonian for a pure ferromagnet, for all $v = -1$ we get a pure antiferromagnet.

Let us now try to argue that v disorder alone can lead to localization of some states. In the absence of detailed analysis the argument will be by analogy.

Consider the model of (4.1). In the absence of the t term any Green functions are local;

$$G_{ij}^{(0)}(E) = \delta_{ij}(E - \varepsilon_i)^{-1}\,. \tag{4.5}$$

Interaction modifies the Green function so that it is no longer local and

$$G_{ii}(E) = (E - \varepsilon_i - S_i(E))^{-1}\,, \tag{4.6}$$

where $S_i(E)$ is the (generally complex) local self-energy. $S_i(E)$ may be expressed as a Feenberg perturbation series in terms of non-repeating paths [15]. Let us restrict consideration for the moment to $d = 1$ and nearest-neighbour interactions. In this case, the expression for the self-energy may be written in terms of repeated fractions since non-repeating paths simply propagate away from the site in question either to the right or the left;

$$S_i(E) = \cfrac{t_{i,i+1}^2}{E - \varepsilon_{i+1} + \cfrac{t_{i+1,i+2}^2}{E - \varepsilon_{i+2} + \cfrac{t_{i+2,i+3}^2}{E - \varepsilon_{i+3} + \cdots}}}$$

$$+ \text{ corresponding series to left}. \qquad (4.7)$$

Denoting the effective self-energies at n, $n + 1$ steps down the ladder (4.7) by $S_i^n(E)$, $S_i^{n+1}(E)$ we see

$$S_i^n(E) = \frac{t_{i+n,i+n+1}^2}{(E - \varepsilon_{i+n+1} - S_i^{n+1}(E))}. \qquad (4.8)$$

A condition for localized states [16] is that the sequence for $S_i^n(E)$ converges. If it does not converge then the corresponding states, of energy E, are extended.

For a pure system, all ε equal and all $t_{i,i\pm1}$ equal, convergence or otherwise is signalled by the location in the complex plane of roots of the equation obtained by replacing $S^{n+1}(E)$ by $S^n(E)$ in (4.8). Complex roots signal that excitations of energy E propagate (giving the usual band). Real roots indicate non-propagating modes.

In a disordered system the above procedure of replacing S^{n+1} by S^n can yield more than one set of self-consistent roots or fixed points. If this is the case and the sequence (4.7) chooses randomly between them, because of the randomness of ε and/or t, then it can be shown that all states are localized for $d = 1$ [16], except if all such points have no real part; this latter corresponds to $E = \varepsilon$ for the case of no ε-disorder [17] and is pathological. It is evident that pure sign-disorder of t does not give localization.

In the present problem for $d = 1$ (4.8) is replaced by

$$S_i^n(E) = t^2(E - \varepsilon - S_i^{n+1}(E)) \qquad \text{if} \quad \nu_{i+n,i+n+1} = 1 \qquad (4.9a)$$

$$= t^2(-E - \varepsilon - S_i^{n+1}(-E)) \quad \text{if} \quad \nu_{i+n,i+n+1} = -1. \qquad (4.9b)$$

Thus we have to follow the self-consistent solutions of both $S(E)$ and $S(-E)$. For the pure ferromagnet the equations for $S(E)$ and $S(-E)$ decouple. For the pure antiferromagnet they are interrelated but yield simple fixed points. In each case one finds periodic (non-convergent) behaviour for S^n within the relevant band. The fixed point locations are, however, different in the pure ferromagnet and pure antiferromagnet (and are not both on the imaginary axis for any E). It thus follows that a system with random ferromagnetic and antiferromagnetic bonds chooses randomly between two fixed points analogously to a system of type (4.1) with random-magnitude t and/or ε disorder. Its states are thus all localized in one dimension.

In three-dimensions we expect, on the basis of the above one-dimensional discussion, that higher energy eigenstates will be localized, separated from low energy propagating modes by a mobility edge [10]. In the Anderson model (4.1) the mobility edge(s) separating extended and localized states depend(s) upon the magnitude and form of ε and t-disorder, the mean t, and on lattice structure. In the present problem there is only one energy J so that the mobility edge simply scales as J with a factor which can depend only upon the concentration c and the lattice structure. It is thus a clean localization problem in the way that the $\pm J$ bond classical model is a clean one to study the thermodynamics of frustration.

Spin wave interactions (going beyond the linearized equation of motion approximation) may modify the localization properties but are relatively unstudied even in disordered ferromagnets.

References

[1] D. Sherrington, AIP Conf. Proc. 29 (1975) 224.
[2] G. Toulouse, Commun. Phys. 2 (1977) 115.
[3] D. C. Mattis, Phys. Letters A56 (1976) 421.
[4] The same Hamiltonian describes an idealized AB alloy with $J_{AA} = J_{BB} = -J_{AB}$, $S_A = S_B$, $\xi = -1$ corresponding to A, $\xi = -1$ to B.
[5] S. F. Edwards and P. W. Anderson, J. Phys. F6 (1976) 1927.
[6] B. I. Halperin and W. M. Saslow, Phys. Rev. B16 (1977) 2154.
[7] L. R. Walker and R. E. Walstedt, Phys. Rev. Letters 38 (1977) 514.
[8] W. Y. Ching, K. M. Leung, and D. L. Huber, Phys. Rev. Letters 39 (1977) 729.
[9] W. Marshall and S. W. Lovesey, Theory of Thermal Neutron Scattering (Oxford Univ. Press, London, 1971).
[10] See D. J. Thouless, Phys. Rept. 13C (1974) 93; and this volume.
[11] P. W. Anderson, Phys. Rev. 86 (1972) 694.
[12] M. T. Hutchings, G. Shirane, R. B. Birgeneau, and S. L. Holt, Phys. Rev. B5 (1972) 1999.
[13] J. des Cloizeaux and J. J. Pearson, Phys. Rev. 128 (1962) 2131.
[14] Y. Endoh, G. Shirane, R. J. Birgeneau, P. M. Richards, and S. L. Holt, Phys. Rev. Letters 32 (1974) 170.
[15] P. W. Anderson, Phys. Rev. 109 (1958) 1492.
[16] E. N. Economou and M. H. Cohen, Phys. Rev. B4 (1971) 396.
[17] G. Theodorou and M. H. Cohen, Phys. Rev. B13 (1976) 4597.

THRESHOLD CONDUCTION IN INVERSION LAYERS

C. John ADKINS

Cavendish Laboratory, Madingley Road, Cambridge CB3 0HE, England

After an introduction to inversion layers and a summary of the theoretical ideas concerning the expected effects of disorder in such systems, the observed behaviour near the threshold of conduction is reviewed. Although the system shows a classic metal–insulator transition, the detailed behaviour is not consistent with an independent-particle mobility edge model. A new model is proposed in which correlation causes the carriers to behave like a viscous liquid near the conduction threshold. The new model provides quantitative explanations of many observed properties. Other experiments are suggested by which it may be further tested.

References

C. J. Adkins, J. Phys. C11 (1978) 851.
C. J. Adkins, Phil. Mag. B 38 (1978) 535.

R. Balian et al., eds.
Les Houches, Session XXXI, 1978 – La matière mal condensée/Ill-condensed matter

ELECTRONIC PROCESSES IN GLASSY SEMICONDUCTORS*

D. C. LICCIARDELLO

Department of Physics, Princeton University, Princeton, New Jersey 08540, USA

and

Bell Laboratories, 600 Mountain Avenue, Murray Hill, New Jersey 07974, USA

Although the theory of microscopically disordered solids is in its embryonic stages compared with that of crystalline systems, some very general principles have begun to emerge, tempting theorists to venture into the glassy state. The most obvious property of glasses requiring attention is the presence, almost universally, of a well defined absorption edge (indeed, it is commonly true that the gap is sharper than in the crystalline phase of the same material). The problem is a fundamental one because the transparency of glassy systems suggests a smaller role for the intrinsic randomness than is indicated from very general considerations.

The thrust of modern ideas on the nature of disordered systems have elucidated two main principles: (1) that for many quantities of interest, probability distributions play a more important role than their averaged or most probable values; that is, from information about the latter only, one often misses the essential feature describing the way microscopic processes affect the system behavior. (2) The eigenstates of the system divide into two classes, localized and extended, which are *qualitatively*

* Work supported in part by NSF DMR 78-03015 and ONR N00014-77-C-0711.

R. Balian et al., eds.
Les Houches, Session XXXI, 1978 – La matière mal condensée/Ill-condensed matter

different. These states are separated, within the spectrum, by sharp energies (on the scale of $1/N$) called mobility edges. This dichotomy is specific to disordered systems and has no crystalline analogue. The interesting point is that both these theoretically well founded conceptions seem particularly difficult to establish experimentally in glasses.

Thus, in amorphous semiconductors, for example, in addition to having a sharp optical gap, there exists no firm experimental evidence for the existence of a localization edge. This has persuaded many workers to treat these materials as "effective crystals" where the absence of long-range structural order leaves the system, for most processes, qualitatively unchanged. Weaire has shown [1], for instance, within a simple model of covalent bonding, how real gaps can occur without invoking long-range order. This model arbitrarily truncates the range of the matrix elements which, of course, begs the effect and so leaves the more general question open.

Anderson [2] has attempted to preserve the general principles by introducing a third: strong lattice coupling. Basically it is argued that lattice relaxation near localized carriers leads to a net attraction between electrons in the same state (site). The model is further discussed by Anderson in these Les Houches lectures. Stein et al. [3] have shown how this model can lead to specific defects in chalcogenide glasses and the topology of these centers as well as the nature of the low-energy excitations is discussed.

In this seminar, we examine some important consequences of the approach taken in refs. [2] and [3]. In particular, we reconsider the question of the presence of sharp phenomena, e.g., absorption edges, in a model which presumes a continuous distribution of states. We ask where are the gap states and how can they be observed experimentally? Anderson has taken the view [2] that since a mobility edge is the only sharp energy which exists in a disordered system, the observed band edge must be a mobility edge. It is argued that localized states are eliminated by phonon self-trapping to below the gap.

In the linear model employed by Anderson the local lattice coupling energy W_i is quadratic in the occupation:

$$W_i = -n_i^2 C_i/2 , \tag{1}$$

where $C_i \equiv 2g^2/\omega_i$. Here n_i is the occupation at the site i, g is the electron–phonon coupling and ω_i is the local phonon frequency. Each center or bond is characterized by an electronic energy ε_i and a phonon frequency ω_i. The local lattice modes are displaced on occupation (with one or two electrons according to eq. (1)) and so the electron energies are renormalized. For $n = 1$, the effective electronic energy is, from eq. (1)

$$E_i = \varepsilon_i - \tfrac{1}{2}C_i , \qquad (2)$$

and for $n = 2$ the level eigenvalue is

$$2\varepsilon_i + U_i - 2C_i = 2E_i - (C_i - U_i) , \qquad (3)$$

where U_i is the contribution from the Coulomb interaction between the electrons. Thus we may recharacterize the local parameters with the effective electronic energy E_i and the effective electronic *attraction* $\tilde{U}_i \equiv C_i - U_i$ which acts when the site i is doubly occupied. Note that $\tilde{U}_i$ includes the Coulomb repulsion U_i between the particles which, of course, reduces the effect of the lattice interaction. It is presumed that for diamagnetic glasses the distribution function $P(\tilde{U}_i)$ favors positive values of $\tilde{U}_i$ for most sites near the Fermi level.

It is clear from eq. (3) that for $\tilde{U}_i > 0$ all sites are doubly occupied up to $E_i = \tfrac{1}{2}\tilde{U}_i$; sites with larger energies are empty. Assuming a distribution of site energies which does not deplete appreciably near $E_i = \tfrac{1}{2}\tilde{U}_i$, the Fermi level for the pair states is pinned at this value.

Absorption edges arise from one-electron excitations, so to study this question we consider the process by which we break a pair. One may think of the process as the creation of two particle states (as opposed to a particle and a hole) each of energy E_i out of a doubly-occupied state of energy $2E_i - \tilde{U}_i$. (Here we presume the two states are each characterized by the same parameters E_i, $\tilde{U}_i$, for simplicity, but we relax this restriction below.) The total energy for the process is $\tilde{U}_i$ and we remark that these considerations only apply to low frequency processes ($\omega \ll \omega_i$) when the two singly-occupied states have had time to relax to their effective energies E_i. For optical processes (Franck–Condon) there is no time for relaxation and the single-particle energies take the unrenormalized values ε_i. In this case the gap energy $2\varepsilon_i - (2E_i - \tilde{U}_i) = \tilde{U}_i + C_i$ which is larger by an amount C_i. Thus we see that an optical gap will occur even in the case of weak diamagnetism, i.e. $C_i \gtrsim U_i$.

Up to now, we have not introduced any correlations between the site parameters E_i and $\tilde{U}_i$. In fig. 1, we define the space of the parameters; in this diagram each bond (site) is represented by a point and, in general, no region is excluded. For simplicity, however, we examine the strongly-coupled case where some minimum $\tilde{U}_{\min}$ exists below which there are no

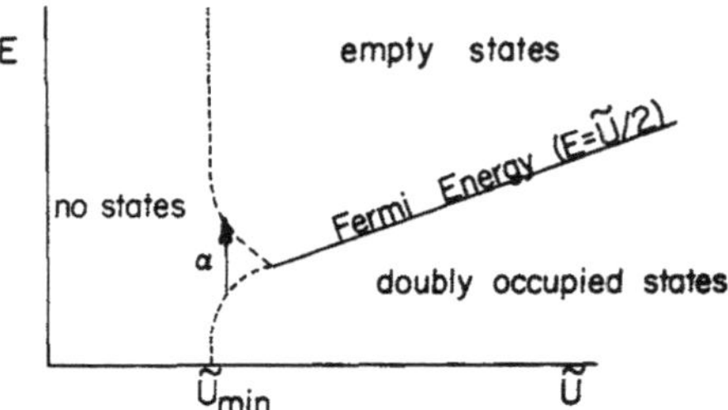

Fig. 1. Phase space for two-electron centers characterized by an effective energy E and an effective electronic attractive coupling $\tilde{U}$ which acts when the site is doubly occupied. There are no sites characterized by parameters to the left of the dashed boundary. The arrow indicates a possible minimum energy absorption process.

states.* The line $E = \frac{1}{2}\tilde{U}$ divides the existing states into ones which are doubly occupied and those which are empty. A one-electron process involves the excitation of an electron from a state i with $E_i < \frac{1}{2}\tilde{U}_i$ and $\tilde{U}_i > \tilde{U}_{\min}$ to a state j with $E_j > \frac{1}{2}\tilde{U}_j$ and $\tilde{U}_j > \tilde{U}_{\min}$. The energy for the process is $\Delta E = E_j + \tilde{U}_i - E_i$ and the absorption edge occurs at $\min(\Delta E)$. If the entire space is accessible, with the proviso that $U > U_{\min}$ for all i, the minimum will occur for two states at the Fermi level with absorption energy $\Delta E = \frac{1}{2}(\tilde{U}_i + \tilde{U}_j)$.

We assume for the moment that the space of fig. 1 is not uniformly occupied with sites but that states near the Fermi level have larger values of $\tilde{U}$ on average. This is indicated by the dotted boundary in the figure where the "no states' region protrudes to larger $\tilde{U}$ values near the Fermi energy. In this case, it is not clear that the first states to absorb will come from the Fermi level. In order to minimize the quantity $E_j + \tilde{U}_i - E_i$, larger values of $E_j - E_i$ may be necessary and a typical absorption process under such circumstances is indicated in the figure (labeled α). The possibility of strongly coupled pairs of electrons near E_F is an interesting one for two reasons. Firstly, the corresponding single-particle states could be pushed deep into the spectrum and may be in resonance with the mobile states of the valence band. Even more interesting is the possibility that probing deep into the spectrum for the minimum absorption pair E_i, E_j may reach the mobility edge of this spectrum. (It is presumed that some transfer matrix element T_{ij} operates to move an electron from site i to j.)

* This can be approximately so in glasses of low-averaged coordination number (e.g. amorphous chalcogenides) but a wider range of values of $\tilde{U}_i$ (extending into the negative region) is expected in the highly coordinated systems a-Si and a-Ge.

Whether or not the minimum absorption energy ΔE equals the mobility gap depends entirely on the exponent describing the coupling of the lattice to a localized state near E_c. This problem has been discussed by Licciardello and Thouless [4]. The situation in which absorption between mobile states occurs would confirm Anderson's conjecture that the observed absorption edge is indeed a mobility edge. The effect is enhanced by the T_{ij} term, as discussed by Anderson, since the extended states may lower their energy gap through kinetic broadening and thereby lower their absorption energy.

As yet no mechanism has been proposed to account for the enhanced coupling of pair states near E_F. Stein et al. have argued that the two-electron states near the Fermi level are chemically distinct from those deep in the valence band. An important effect not included in Anderson's strong coupling theory of glass is the overwhelming preference for atoms to satisfy their local valence requirements. This leads to "phonon sharing" and so a model which introduces only one frequency ω_i for each bond does not take account of the possibility that some lattice modes may extend over several bonds. This effect tends to reduce the coupling to the electronic system and prevails in regions where the local coordination is satisfied.

During glass formation (quenching) most atoms, in fact, manage to find the requisite number of neighbors to satisfy valence and so minimize their local free energy. A few, depending on the temperature from which the system is quenched, do not, however, and the defect so formed has been termed valence alternation [5,6]. These atoms have more or fewer chemically bonded neighbors than is indicated from normal valency. It is worth pointing out that it is not necessary to restrict these considerations to lone-pair or chalcogenide semiconductors, although for these systems coordination fluctuations are easier. For example, silicon can, in the amorphous phase, bind with only three neighbors.

Stein et al. have shown [2] how the excess energy of formation may be compensated by the configurational entropy to account for their presence in significant numbers. More importantly, it is argued that these centers are responsible for pinning the Fermi level of the two-electron spectrum. Perhaps without doing an injustice to the idea of continuous distributions we can argue that the centers near E_F tend to be more defect-like than those near the true bonding energy.

Although these centers are the weakest two-electron states, we argue here that they have the strongest coupling, i.e. $C_i \gg U_i$. As an example, we consider elemental glassy selenium with four-valence electrons in the three 4p states. The atoms prefer two-fold coordination (in either eight-member rings of N polymer chains) with the other two electrons in non-bonding

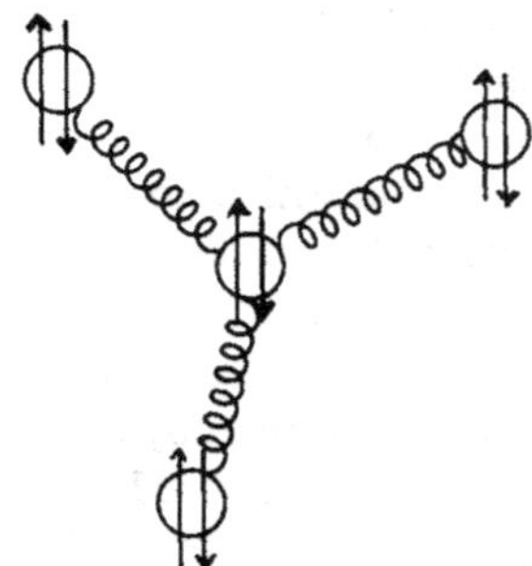

Fig. 2. The D^+ defect present in amorphous chalcogenide semiconductors. All sites (bonds) are occupied by electrons of both spin. The central ion in the defect is three-fold coordinated and has charge $+1$.

states. The coordination defect, then, is $z = 3$ or 1; the trigonally bonded case is illustrated in fig. 2. Here the orbitals of the defect atom are exhausted (p levels) and so the atom has room for only three electrons at the bonding energy with no non-bonding levels available. If we require double occupancy in every state (as shown in the figure), the center defect atom must divest itself of its fourth electron and so the defect has charge $+1$ (called D^+). By charge neutrality we require an equal number of defects in the $(-)$ charge state (D^-) where a selenium atom has only one bond and four electrons in non-bonding states. The defect illustrated then, is an empty two-electron state and we imagine, according to the arguments given above, that it exists at E_F. That is, the reaction $D^+ + 2e \Rightarrow D^-$ is a low-energy excitation as discussed by Stein et al.

We now consider the process of adding *one* electron. Kastner et al. speculate [5] that the D^0 center remains three-fold coordinated, although the presence of the σ^* electron should certainly weaken the bonds. The important point is that the addition of yet a second electron causes considerable softening of ω_t because one of the bonds now certainly breaks leaving one of the three neighbors with only one bond (it becomes the D^- center). An elementary argument gives a factor 2–3 lower frequency for the center and this softening could give rise to a large local distortion. No such softening can be expected to occur at other two-electron centers where the bonding is canonical and the phonons are large.

Thus we are left with strongly-coupled states in the gap which pins the Fermi level against doping but which eludes optical probes. The pair states, however, should be strongly localized especially near defect positions and thus long relaxation times are expected for low energy two-

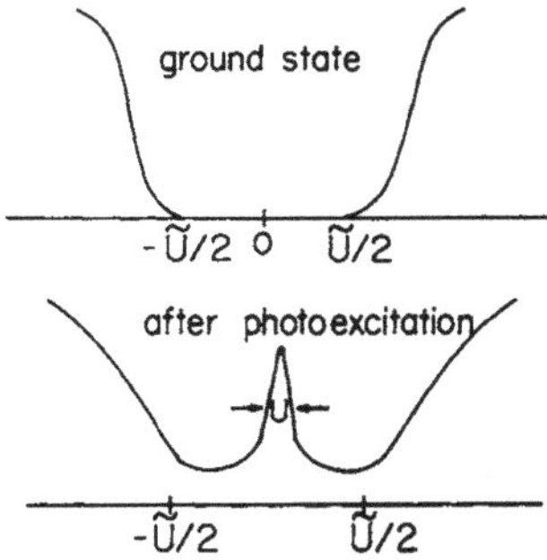

Fig. 3. The single-particle excitation spectra for a material in equilibrium (top sketch) and after considerable photo excitation (lower sketch). The gap fatiguing is due to the occupation of metastable two-particle states. Metastable *paramagnetic* states occur in a *possibly* narrow band (order U, the site Coulomb repulsion) near the Fermi level.

particle processes. Anderson has suggested [2] that out of equilibrium occupation of the pair states near E_F and the associated long times relaxation can provide explanations for photostructural effects including fatiguing and even switching phenomena.

It is certainly clear that any Fermi level smearing in the two-particle spectrum will induce lower energy absorption processes. After prolonged optical pumping, free electrons and holes will fall into available gap states, doubly occupying some two-electron centers above E_F. In addition, some centers will trap single particles which may relax to the effective energy $E_i = \varepsilon_i - \frac{1}{2}C_i$. If this particular center has an effective energy E_i which is smaller than the ground occupancy ($n = 0, 2$) in the same deformation configuration, the singly occupied state will be metastable. An explicit calculation using the same strong coupling model gives a band of paramagnetic metastable defects near the Fermi level:

$$E_F - \tfrac{1}{2}U_i < E_i < E_F + \tfrac{1}{2}U_i \,. \tag{4}$$

Note that the bandwidth is the Coulomb repulsion U_i and not the effective energy $\tilde{U}$. Thus if we presume that the Fermi level is pinned by 10^{19} states/cm^3 and take [5] a few tenths of an eV for the Coulomb term in amorphous chalcogenide glasses, we predict some 10^{18} spins/cm^3 should be observable in photo-induced ESR. A typical fatigued spectrum is shown in fig. 3. Thus we witness the reappearance of the missing gap states.

Acknowledgement

I wish to acknowledge helpful discussions with P. W. Anderson, R. Fisch, and F. D. M. Haldane.

References

[1] D. Weaire, Phys. Rev. Letters 26 (1971) 1541.
[2] P. W. Anderson, Phys. Rev. Letters 34 (1975) 953; J. Physique Colloq. 37, C4 (1976) 339.
[3] D. L. Stein, D. C. Licciardello and F. D. M. Haldane, to be published.
[4] D. C. Licciardello and D. J. Thouless, J. Phys. C11 (1978) 925.
[5] M. Kastner, D. Adler and H. Fritzsche, Phys. Rev. Letters 37 (1976) 1504.
[6] R. A. Street and N. F. Mott, Rev. Letters 35 (1975) 1293.

DO THE PROPERTIES OF SPIN GLASSES DEPEND ON THE SPIN DIMENSIONALITY?

Jacques VILLAIN

Département de Recherche Fondamentale,
Laboratoire de Diffraction Neutronique, Centre d'Etudes Nucléaires de Grenoble,
85 X, 38041 Grenoble Cedex, France

It is suggested that the spin dimensionality n is a much less important parameter for spin glasses than for ferromagnets. Experimental support as well as theoretical (mainly for $n = 2$) is discussed for the following properties: remanence, specific heat and eventual phase transition.

R. Balian et al., eds.
Les Houches, Session XXXI, 1978 – La matière mal condensée/Ill-condensed matter

1. Introduction

Consider a magnetic system, described for instance by the simple bilinear hamiltonian:

$$\mathscr{H} = -\sum_{ij} J_{ij} \boldsymbol{S}_i \cdot \boldsymbol{S}_j \,. \tag{1}$$

In the well understood case of a ferromagnet (or simple antiferromagnet) on a periodic lattice, the properties are well known to depend crucially on space dimensionality D and spin dimensionality n. For example:

(a) Long range order is possible at $T \neq 0$ is $D > D_0$, where:

$$D_0 = 1 \,, \quad \text{for} \quad n = 1 \,;$$
$$D_0 = 2 \,, \quad \text{for} \quad n \geqslant 2 \,.$$

(b) The specific heat of a simple antiferromagnet at low temperature T is

$$C \sim T^D \,, \qquad \text{if} \quad n \geqslant 2 \,;$$
$$C \sim \exp(-T/T_0) \,, \quad \text{if} \quad n = 1 \,; \quad T_0 \text{ is a constant.}$$

In the present communication, it is suggested that the effect of n may be much less important for spin glasses. There is some experimental support for this statement, as discussed in the next section.

2. Experimental facts

The most extensively studied spin glasses are dilute alloys: AuMn, CuMn, AuFe, ...

They are commonly believed to be accurately represented by a Heisenberg model. Their magnetic susceptibility is reasonably isotropic. However, they have certain properties of Ising spin glasses, namely:

– Remanence [1], which can hardly be explained if potential barriers do not exist in the configuration space. Such potential barriers do not exist in a normal Heisenberg ferro- or antiferromagnet (except for high lying topological excitations).

– Specific heat proportional to T at low temperature. According to Marshall [2] the magnetic specific heat of a spin glass per spin is:

$$C = \frac{\mathrm{d}W}{\mathrm{d}T} = \frac{\mathrm{d}}{\mathrm{d}T}\int_0^\infty \mathrm{d}^n HP(\boldsymbol{H})H\mathscr{L}(\beta H)$$

$$= -\frac{d}{\mathrm{d}T}\int \mathrm{d}^n HP(\boldsymbol{H})H[1 - \mathscr{L}(\beta H)]$$

$$\cong -\frac{\mathrm{d}}{\mathrm{d}T}\int T^{n+1}\,\mathrm{d}^n(\beta H)P(0)(\beta H)[1 - \mathscr{L}(\beta H)] \sim T^n ,$$

where $P(\boldsymbol{H})$ is the probability that the molecular field $\boldsymbol{H}$ has a given value and $\mathscr{L}$ is the Langevin function. Experimental data are consistent with the value $n = 1$.

There are two possibilities to explain this Ising behaviour:

(a) It can be due to anisotropic forces: (i) dipole forces are about 100 times weaker than RKKY exchange; (ii) random anisotropy due to pairs vanishes for strong dilution for a random alloy; (iii) cubic anisotropy seems to be the largest contribution, although a theory based on this effect meets serious difficulties. This will not be discussed here.

(b) The Ising behaviour can also be explained if a Heisenberg spin glass has potential barriers, or two-level systems (TLS) which make it look like an Ising spin-glass. This was first suggested by Anderson et al. [3,4]. The theory is much easier for XY spin glasses and is outlined in the next section. Heisenberg spin glasses are briefly discussed in sect. 4.

3. An *XY* spin glass is an Ising spin glass in disguise

3.1. Two-level systems (TLS) as a result of frustration

(a) We start with an extremely simple example: a set of 4 spins ("plaquette") with 4 interactions J_{12}, J_{23}, J_{34}, J_{41} (fig. 1). Classical spins are considered.

If $J_{12}J_{23}J_{34}J_{41} \geqslant 0$, the plaquette is not frustrated. The ground state is a "parallel" structure which minimizes the energy of each pair separately. The ground state is non-degenerate except under a uniform rotation of all spins, and this is a continuous degeneracy.

If $J_{12}J_{23}J_{34}J_{41} < 0$, the plaquette is frustrated [5]. For appropriate values of the interactions, the ground state is canted (fig. 1) and has (in

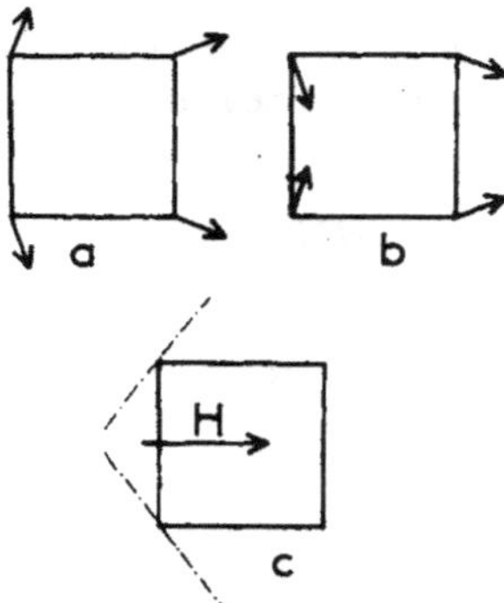

Fig. 1. Classical ground states of a single plaquette of XY spins. There is a two-fold degeneracy (a, b) in addition to the continuous degeneracy under rotation. (c) Shows the effect of a magnetic field plus a random anisotropy. If the spins have different easy axes (dashed lines), one of the states (here, state *a*) is favoured.

addition to the continuous rotation degeneracy) a two-fold degeneracy which can be characterized, for instance, by the sign of the vector product $(\boldsymbol{S}_1 \times \boldsymbol{S}_2)$.

Exercise

Prove that the necessary and sufficient condition for the ground state to be parallel (i.e., all spins parallel or antiparallel to a given direction) is that either:

$$J_{12}J_{23}J_{34}J_{41} > 0\,,$$

or

$$\frac{1}{|J_{12}|} > \frac{1}{|J_{23}|} + \frac{1}{|J_{34}|} + \frac{1}{|J_{41}|}\,,$$

if J_{12} is the smallest of all 4 interactions (in modulus).

(b) We consider now an infinite, two-dimensional system ($n = D = 2$) with short range interactions and one single frustrated plaquette. The number of antiferromagnetic interactions can be reduced by a Mattis transformation [5–7], but, even after this transformation, the system still contains an infinity of antiferromagnetic bonds which form a semi-infinite line (fig. 2). Sufficiently far from the frustrated plaquette, two neighbouring spins on different sides of the antiferromagnetic line have opposite directions. Along a path from one spin to the other, far enough from the frustrated cell and not crossing the antiferromagnetic line (dashed path on

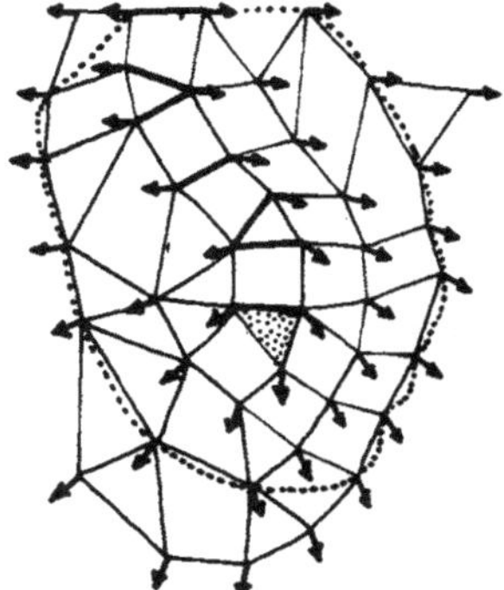

Fig. 2. Classical ground state of a two-dimensional XY model with one frustrated plaquette.

fig. 2) the spin direction rotates quasi-continuously by an angle π, or $-\pi$, which characterizes the two possible states of the system (which, of course, has also the continuous rotation degeneracy). This angle can be called vorticity, or chirality.

Thus, one isolated frustrated plaquette in a two-dimensional XY model is always a two-level system [8].

There is a remarkable analogy with the thermal, topological excitations of the XY ferromagnet, namely the "vortices" introduced by Kosterlitz and Thouless [9]. Here, however, one has half-vortices.

(c) Consider now an infinite, two-dimensional, XY magnet with two frustrated plaquettes. Their vorticities couple antiferromagnetically, i.e., they are opposite, because if they were identical, they would form a full vortex and the energy would be infinite (for an infinite volume). For opposite vorticities, the total energy (if the energy of each pair is measured from its ground state) is proportional to

$$J\int \mathrm{d}^2 r\,(\nabla\varphi)^2 \approx J\int_0^R 2\pi r\,\mathrm{d}r\,\frac{1}{r^2} \approx 2\pi J \log R\,,$$

where R is the distance between both frustrated cells, φ is the polar angle of the spin, J is the average modulus of the exchange interaction and the unit of length is the average interatomic distance [8,9].

If we want to extend these results to an arbitrary number of frustrated cells, and to three-dimensional media, it is convenient to use a modified XY model as explained below. The above considerations, however, suggest that the final results do not depend qualitatively on the modifications and oversimplifications which will be made.

3.2. *Quantitative theory* [10]

We consider two-dimensional, classical spins $S_i = (\cos\varphi_i, \sin\varphi_i)$.

In the usual XY model, the partition function and correlation functions depend on integrals containing

$$\exp(-\beta J\cos\psi)\,, \tag{2}$$

where J is the exchange constant for a given pair and ψ is the angle between the spins of this pair. In our modified model, this exponential is replaced by:

$$\text{const} \times \sum_{n=-\infty}^{\infty} \exp(-\beta g)\left(\psi - 2\pi n + \frac{1-\varepsilon}{2}\pi\right)^2, \tag{3}$$

where n is an integer, $\varepsilon = J/|J| = \pm 1$ and g is an appropriate, temperature dependent constant, determined in such a way that expression (3) resembles as much as possible the exponential (2). It should be recognized that this resemblance cannot be very good at very low temperature, in contrast with the ferromagnetic case [11], when it is sufficient to fit (2) near the maximum $\psi = 0$. However, (2) and (3) have the same periodicity and the same extrema.

The new partition function and correlation function are the same as those of the fictive hamiltonian [10]:

$$\mathscr{H} = \frac{1}{2}\sum_{i,j} g_{ij}\left[\varphi_i - \varphi_j - \frac{1-\varepsilon_{ij}}{2}\pi - 2\pi n_{ij}\right]^2, \tag{4}$$

which depends on unphysical, integer variables n_{ij}. This drawback is compensated by the fact that $\mathscr{H}$ depends quadratically on variables φ_i. Therefore, the continuous variables φ_i can be "eliminated" from the problem (i.e., their effect can be calculated in a straightforward way). This elimination procedure results in an effective interaction between the n_{ij}'s, just as elimination of phonons in superconductivity theory results in an effective attraction between electrons. Thus, the new problem (which contains all non-trivial features such as an eventual phase transition) is described by an effective hamiltonian which contains the n_{ij}'s only:

$$\mathscr{H} = \sum A_{ij}^{lm}\left(n_{ij} - \frac{1-\varepsilon_{ij}}{4}\right)\left(n_{lm} - \frac{1-\varepsilon_{lm}}{4}\right), \tag{5}$$

In order to be able to calculate coefficients A_{ij}^{ml} explicitly, additional simplifications are necessary:
- spins on a square lattice or a simple cubic lattice;
- nearest neighbour interactions;
- all g_{ij}'s equal.

Thus, the whole disorder is contained in the ε_{ij}'s.

We just mention that elimination of the φ_i's necessitates a little trick: the replacement of the integration:

$$\int_{-\pi}^{\pi} \mathrm{d}\varphi_i$$

by:

$$\frac{1}{2\sqrt{\pi\eta}}\int_{-\infty}^{\infty} \exp(-\eta\varphi_i^2)\,\mathrm{d}\varphi_i\,,$$

and one must take the limit $\eta \to 0$.

When one calculates the matrix $[A]$ in (5), it turns out that it has a lot of vanishing eigenvalues. In fact, variables n_{ij} are redundant and the hamiltonian can be expressed in terms of variables q_ρ associated to the plaquettes ρ of the lattice and defined by:

$$q_\rho = n_{ab} + n_{bc} + n_{cd} + n_{da} + \frac{1 - \varepsilon_{ab}}{4} + \circlearrowleft\,, \tag{6}$$

where a, b, c, d are the corners of the plaquette ρ, and

$$n_{ij} + n_{ji} = \varepsilon_{ij} + \varepsilon_{ji} = 0$$

for each bond i, j. It follows from (6) that: q_ρ is integer if the plaquette ρ is not frustrated; q_ρ is half-integer if the plaquette ρ is frustrated.

The final result [10] is an effective hamiltonian:

$$\mathscr{H} = \sum_{\rho \neq \rho'} V_{\rho\rho'} q_\rho q_{\rho'}\,, \tag{7}$$

where $V_{\rho\rho'}$ is essentially a D-dimensional Coulomb interaction:

$$V_{\rho\rho'} = -(\pi^2 g + 2\pi g \log|\boldsymbol{\rho} - \boldsymbol{\rho}'|) \qquad (D = 2)\,, \tag{8}$$

$$V_{\rho\rho'} = \frac{\pi g}{|\boldsymbol{\rho} - \boldsymbol{\rho}'|} \quad \text{(plaquettes parallel)} \qquad (D = 3)\,. \tag{9}$$

$$V_{\rho\rho'} = 0 \qquad \text{(plaquettes not parallel)}\,.$$

In addition the "charge" distribution should satisfy total neutrality (because of (6) and periodic boundary conditions)

$$\sum_{\rho} q_{\rho} = 0 , \tag{10}$$

and additional constraints for $D = 3$ [10]. Total neutrality has been taken into account in the derivation of (8). If total neutrality were not imposed, eq. (8) would describe a ferromagnetic interaction, in contrast with our statement of sect. 3.1(c).

The hamiltonian in terms of the Fourier transforms q_k reads:

$$\mathscr{H} = \sum_{k} \frac{4\pi^2}{k^2} g|q_k|^2 .$$

It is tempting to assume that at low temperature T, the $|q_k|$'s must be as small as possible, and the $|q_\rho|$'s must also be as small as possible, namely:

$$q_\rho = \tfrac{1}{2}\tau_\rho \tag{11}$$

where

$\tau_\rho = 0$ if ρ is not frustrated ,

$\tau_\rho = \pm 1$ if ρ is frustrated .

This speculation is correct for the XY ferromagnet [9,11–13]. For a spin glass, it is supported by the intuitive statement (sect. 3.1) that a frustrated plaquette is a two-level (but not more than 2!) system.

4. Heisenberg model

(a) Frustrated Heisenberg models can also have potential barriers. For instance, 2 neighbouring antiferromagnetic bonds in a ferromagnetic, triangular lattice constitute a TLS.

(b) For weak concentration of antiferromagnetic bonds the system is still ferromagnetic, magnons can be treated as harmonic and eliminated, and the resulting interactions between TLS is a dipole interaction, proportional to r^{-D}.

(c) In the spin glass case it is probably impossible to eliminate continuous degrees of freedom. There is no equivalent of formula (7).

(d) Frustration is not sufficient for appearance of TLS. A high degree of frustration is necessary.

5. Discussion

In sect. 1, we speculated that n is probably not a very important parameter in spin glasses. In fact we derived the equivalence of an XY spin glass with short range interactions with an Ising spin glass... with long range interactions! In the next sections, we shall argue that this difference is probably not essential. From now on, only XY spin glasses will be considered.

In addition, we started with a spin glass with both ferro- and antiferromagnetic bonds (Edwards–Anderson model) and the resulting Ising spin glass has well determined interactions, see formulae (8) and (9). In the new model, the spin glass character results from positional disorder of frustrated plaquettes. This difference is probably not too serious. It is, indeed, possible to find systems with antiferromagnetic interactions only, which are equivalent to Edwards–Anderson spin glasses [14].

Finally, the three-dimensional model is subject to constraints which make it closer to a six-vertex model than an Ising model.

In the next sections, we study the properties of the model (7).

6. Remanence

The fact that an XY or Heisenberg spin glass has certain remanence properties follows directly from the existence of potential barriers, demonstrated in sects. 3 and 4. The fact that interactions (8) or (9) are long range does not seem to be a problem in this case.

However, the remanence property that has been observed in practice concerns the magnetization. This type of remanence requires some anisotropy, because even if the q_ρ's are fixed, the demagnetizing field can produce a soft modulation of the φ_i's which is sufficient for the magnetization to vanish.

The present theory, however, explains why remanence is experimentally found to satisfy (when concentration is changed) scaling laws characteristic of $1/r^3$ interactions [1]. Indeed, according to our theory, remanence is essentially an effect of RKKY exchange interactions, and anisotropic interactions need not be proportional to $1/r^3$. We believe, in fact, that the main anisotropic force is the one-ion cubic anisotropy. Tholence and Tournier suggested that remanence is due to dipole forces (proportional to $1/r^3$) [1], but it seems unlikely that dipole interactions produce both the demagnetizing field and the anisotropy which opposes the demagnetizing field!

7. Low temperature specific heat

Marshall's calculation, outlined in sect. 2, agrees with experiment provided $n = 1$. The present theory explains why n should be taken equal to 1: in Marshall's calculation, the spins submitted to a molecular field H are not real Heisenberg spins, but fictive, Ising spins. In fact, (in the XY case), the relevant fictive spins are not the q_ρ's introduced in sect. 3.2, but *pairs* of q_ρ's (for $D = 2$). It is forbidden to reverse 1 single "charge" q_ρ since electric neutrality is imposed, but it is allowed to reverse two "charges" of opposite sign, which form a dipole moment μ. The two charges should not be too far from each other, because this would produce charge fluctuations of high energy. For $D = 3$, the charges form closed loops in the dual lattice [10]. It is also allowed to reverse all the charges on a given loop, and they also form something which looks like an electric dipole.

These ideas are apparently in good agreement with computer simulations of Walker and Walsted [15]: Marshall's argument cannot be applicable to real spins, but to two-level systems.

Exercise

Consider an Ising chain with nearest neighbour interactions J_n and probability distribution $p(J)$. Calculate the probability distribution $P(H)$ of the molecular field H seen by one real spin. Check that, if $p(0)$ is finite and different from 0, then $P(0) = 0$, in contradiction with Marshall's assumption. Check that Marshall's conclusion $C \sim T$ is nonetheless correct. What are the effective spins to be considered in Marshall's theory?

However, this is not the entire story. We have proved that it is reasonable to put the value $n = 1$ into Marshall's theory [2]. We shall now look for possible inconsistencies of Marshall's theory. This theory lies in the assumption $P(0) \neq 0$, and we have no means of checking this assumption (except computer simulation [15,16]). But, in addition, Marshall's theory requires that any number of dipoles μ submitted to a field smaller than T can be reversed *together*. This is obviously true for short range interactions, not for long range as argued below [16].

Consider two dipoles μ_1, μ_2 submitted in the ground state to molecular fields H_1^0, H_2^0 of order T or less. The ground state energy, if μ_1^0 and μ_2^0 are the ground state values, is:

$$W_{GS} = -\mu_1^0(H_1^0 - 2J_{12}\mu_2^0) - \mu_2^0(H_2^0 - 2J_{12}\mu_1^0) - 2J_{12}\mu_1^0\mu_2^0\,.$$

If both dipoles are reversed at the same time, the first two terms change their sign and the last term is not modified; the energy is:

$$W' = \mu_1^0(H_1^0 - 2J_{12}\mu_2^0) + \mu_2^0(H_2^0 - 2J_{12}\mu_1^0) - 2J_{12}\mu_1^0\mu_2^0 .$$

We now introduce our assumption that H_1^0 and H_2^0 are of order T or less:

$$W_{GS} = -\lambda T + 2J_{12}\mu_1^0\mu_2^0 ,$$

$$W' = \lambda T - 6J_{12}\mu_1^0\mu_2^0 .$$

Therefore, simultaneous reversal of both dipoles is possible if their interaction J_{12} is of order T or less. Assume the interaction to depend on distance as a power law:

$$|J(r)| \sim r^{-a} . \tag{12}$$

According to Marshall's assumption $P(0) \neq 0$, the average distance l between two neighbouring dipoles seeing a molecular field less than T is given by $1/l^D \approx T$ or $l \approx T^{-1/D}$. Insertion into (12) yields:

$$|J_{12}| \sim T^{a/D} .$$

As stated above, this must be of order T or less. We conclude that Marshall's argument is not inconsistent provided:

$$a \geqslant D .$$

In practice, the interaction J_{12} is of the dipole type, and proportional to r^{-D}. This is a marginal case [16], but apparently still favourable.*

We conclude that XY spin glasses may have a specific heat proportional to T at low temperature. We expect this to be true for Heisenberg spin glasses too. As discussed in sect. 4 the interaction between TLS in a disordered Heisenberg ferromagnet is also a $1/r^D$ interaction. In a spin glass, the exponent a might eventually be larger, but we do not see any mechanism that might lead to a lower value $a < D$.

Remark 1. The continuous, or transverse degrees of freedom of eq. (4) contribute a constant term to the specific heat in the classical limit, but quantization should lead to a T^D contribution.

* Numerical computations of ref. [16] seem to disagree with Marshall's assumption.

Remark 2. Kondo [17] has proposed another mechanism to explain the linear specific heat of a *metallic* spin glass. The specific heat per unit volume, however, is found to be independent of concentration and proportional to temperature as in Marshall's theory, and in agreement with experimental results. Experiments on insulating spin glasses would be of interest. It is rather surprising that both the Marshall and Kondo contributions have the same order of magnitude, since they depend on different quantities: Marshall's contribution depends on the intra-site s–d exchange integral J_0, and Kondo's contribution does not:

$C/K_B^2T \approx n(E_F)$ multiplied by a factor of order 1 to 10 say, for Kondo's contribution;

$C/K_B^2T \approx c/J \approx 1/J_0 n(E_F)$ for Marshall's contribution.

Remark 3. The square lattice considered in sect. 3.2 is appropriate for theory, but the network of frustrated plaquettes has certainly a lower frustration (with respect to the effective hamiltonian (7)) than the real network. Thus the specific heat, if calculated, would certainly be less than the experimental one.

Remark 4. The linear law for the specific heat is experimentally not too well satisfied at very low temperature [24].

8. Phase transition and spin glass order

(1) Our ambition is very limited. We shall not make any statement of the kind: "there is a spin glass transition" or "there is no spin glass transition". We assume the existence of such a transition in spin glasses with short range or RKKY or dipole interactions, and we make a few conjectures about it.

(2) The SG transition is quite different from the ferromagnetic transition of the XY model. The latter can be shown to be equivalent to a metal insulator transition [9,12,13]. This transition can only occur if the density of charges has no lower limit. The electrolyte obtained in sect. 3 is conducting at all temperatures [10,18]. This situation can be compared with the solidification of a molten salt (a random, charge-independent potential would be necessary for a complete analogy). The solid phase is still weakly conducting because of vacancies, etc.

(3) It has been speculated that the condition for spin glass ordering is [18–20]: $D > D_0$.

Anderson and Pond speculated that D_0 is different for $n = 1$ and $n = 2$ [18]. We support here the opposite speculation: D_0 is the same for $n = 1$ and $n = 2$, in agreement with Fish and Harris [19] and Bray and Moore [20].

(a) For ferromagnets, D_0 depends on n because the average square fluctuation of one spin in the spin wave approximation is proportional to:

$$\langle \delta S_i^2 \rangle \sim (n-1)T \int \mathrm{d}^D k/k^2 \,, \tag{13}$$

since $(n - 1)$ is the number of degrees of freedom. This diverges for $D \leqslant 2$ except if $n = 1$ (in this case, order is destroyed by solitons [21] for $D = 1$, not by spin waves).

In an XY spin glass, the average square fluctuation $\langle \delta\varphi_i^2 \rangle$ due to spin waves can be calculated from (4) and diverges for $D \leqslant 2$ exactly like (13). This is a very weak result since most of authors [18–20] expect $D_0 = 3$ or 4 for $n \geqslant 2$.

(b) If the XY model is replaced by the hamiltonian (7), Anderson's statement is equivalent to the statement that D_0 must be larger for an Ising model with Coulomb interactions (and additional constraints if $D \geqslant 3$ [10]) than with short range interactions. This is rather surprising: usually, long range interactions favour order. For instance, the Sherrington–Kirkpatrick spin glass has a phase transition. It will now be speculated that for $D = 2$, the Ising spin glass with Coulomb interactions has the same properties as the Ising spin glass with short range interactions. This statement is based on a renormalization group picture. If one does not want to use the $n \to 0$ replica trick, criticized by Blandin [22], the most accurate renormalization group technique for spin glasses is probably decimation [23], because no knowledge of the nature of the order parameter is required. Thus, one should like to express the free energy as a function of the τ_λ's, where the λ's designate some selected fraction (say $1/2^l$) of the frustrated plaquettes.

However, decimation is not appropriate for the description of Coulomb interactions. The best description of the long range part of the interactions is obtained if one introduces the charge Q_λ in the vicinity of the site λ

$$Q_\lambda = \sum_\rho A_{\lambda\rho} q_\rho \,,$$

where the positive coefficients $A_{\lambda\rho}$ must conserve the total charge, i.e.

$$\sum_\lambda A_{\lambda\rho} = 1 \,.$$

In addition A should vanish when increasing distance, for a distance of the order of the distance between 2 neighbouring λ's.

In principle, one should introduce renormalized dipole moments in addition to the renormalized charges but this complication will not be considered here.

We postulate a renormalized hamiltonian of the form

$$\mathscr{H} = -\sum g_{\lambda\mu}\tau_\lambda\tau_\mu + \sum V_{\lambda\mu}Q_\lambda Q_\mu + \sum \alpha_\lambda Q_\lambda^2 - \sum \gamma_\lambda\tau_\lambda Q_\lambda,$$

where $V_{\lambda\mu}$ is the Coulomb interaction and the other coefficients must be calculated by recursion formulae which have to be obtained by an appropriate generalization of Young's method [23]. Our statement is that screening necessarily forces α_λ to go to infinity and (probably) γ_λ to go to zero when l increases. One is left with the effective interaction $g_{\lambda\mu}$, which has no reason not to be short range.

References

[1] F. Holtzberg, J. L. Tholence, and R. Tournier, in: Amorphous Magnetism II, eds. R. A. Levy and R. Hasegawa (Plenum, New York, 1976) p. 155.
[2] W. Marshall, Phys. Rev. 118 (1960) 1519.
[3] P. W. Anderson, B. I. Halperin, and C. M. Varma, Phil. Mag. 25 (1972) 1.
[4] S. F. Edwards and P. W. Anderson, J. Phys. F5 (1975) 965.
[5] G. Toulouse, Commun. Phys. 2 (1977) 115.
[6] D. C. Mattis, Phys. Letters 56A (1976) 421.
[7] J. Villain, J. Phys. C10 (1977) 1717.
[8] J. Villain, 13th IUPAP Conf. on Statistical Physics; Ann. Israel Phys. Soc. 2 (1977) 625.
[9] J. M. Kosterlitz and D. J. Thouless, J. Phys. C6 (1973) 1181.
[10] J. Villain, J. Phys. C10 (1977) 4793; 11 (1978) 745.
[11] J. Villain, J. Physique 38 (1977) 385.
[12] J. M. Kosterlitz, J. Phys. C7 (1974) 1046.
[13] J. José, L. Kadanoff, S. Kirkpatrick, and D. Nelson, Phys. Rev. B16 (1977) 1217.
[14] J. Villain, Z. Phys. B. 33 (1979) 31.
[15] L. R. Walker and R. E. Walstedt, Phys. Rev. Letters 38 (1977) 515.
[16] S. Kirkpatrick and C. M. Varma, Solid State Commun. 25 (1978) 821.
[17] J. Kondo, Progr. Theoret. Phys. (Kyoto) 33 (1965) 575.
[18] P. W. Anderson and C. M. Pond, Phys. Rev. Letters 40 (1978) 903.
[19] R. Fisch and A. B. Harris, Phys. Rev. Letters 38 (1977) 785.
[20] A. J. Bray and M. A. Moore, J. Phys. C12 (1979) 1349.
[21] L. Landau and E. Lifshitz, Physique Statistique (Mir, Moscow, 1967) p. 577.
[22] A. Blandin, J. Physique 39 (1978) C6-1499.
[23] B. W. Southern and A. P. Young, J. Phys. C10 (1977) 2179.
[24] D. L. Martin, J. Physique 39, C6 (1978) 903.

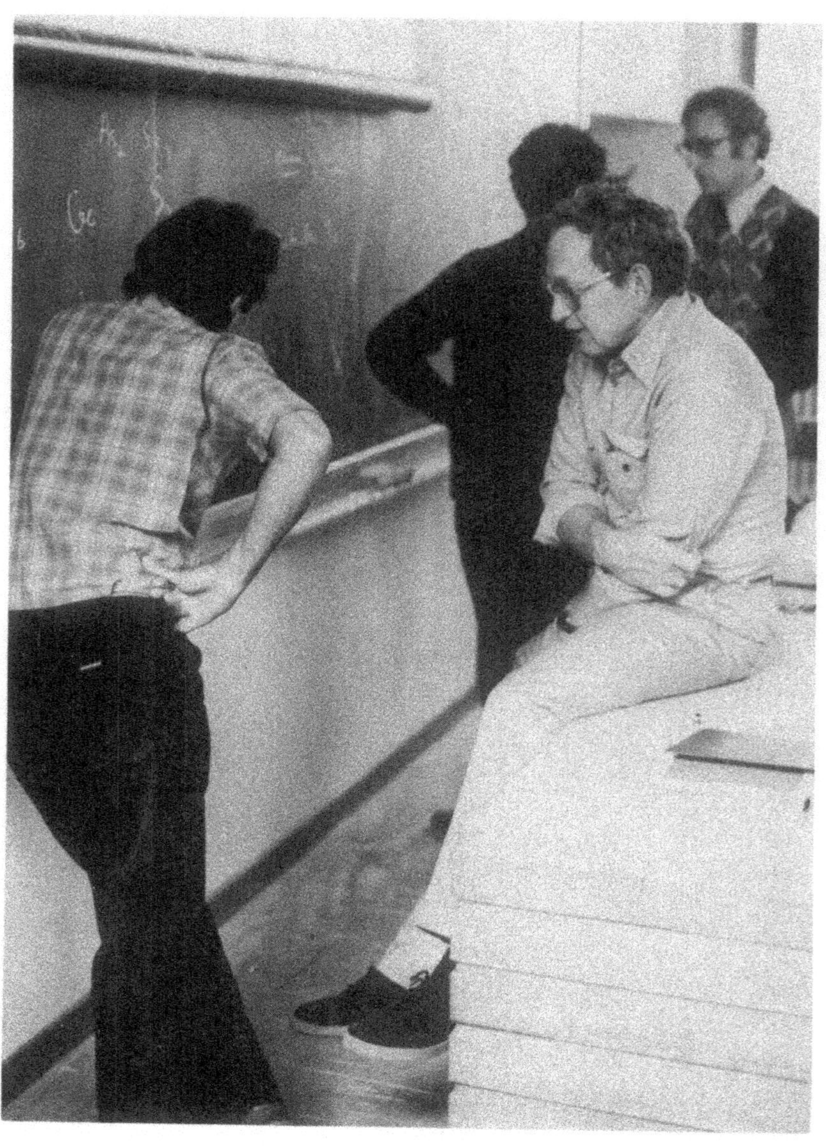

CONFORMATION DES POLYMÈRES EN SOLUTION

Gobalakichena SARMA

DPhSRM, CEA, Centre de Saclay, BP 2, 91190 Gif-sur-Yvette, France

The aim of these two lectures is to give a simple version of the recent theory of polymer configurations which was first initiated by P.-G. de Gennes for the dilute regime and extended by J. des Cloizeaux for the semi dilute regime. Some comparison is made with the early experimental work performed by the Polymer Group of Saclay under the direction of G. Jannink. More about these experiments, as well as a list of references, can be found in an article published in Macromolecules 8 (1975) 804. Complete information about the recent progresses can be obtained by writing to the Polymer Group of Saclay.

R. Balian et al., eds.
Les Houches, Session XXXI, 1978 – La matière mal condensée/Ill-condensed matter

1. Introduction qualitative au problème de la conformation des polymères

Le problème de la conformation des polymères en solution est un des problèmes de la physique de la matière condensée qui est resté longtemps incompris et qui a pris un nouvel essor grâce aux développements récents de la théorie des changements de phase.

Un polymère est une très longue macromolécule linéaire constituée de molécules liées par des liaisons covalentes. Typiquement sa masse moléculaire peut être de l'ordre de 10^6. Du point de vue de la configuration, on peut se représenter la macromolécule comme une très longue chaîne flexible formée

de quelque 10^4 maillons, chaque maillon ayant une dimension de quelques ångströms. La longueur du polymère est donc de l'ordre de quelques microns.

En solution diluée dans un solvant, le polymère prend une conformation de pelote embrouillée. Le volume occupé par le polymère est donc plutôt déterminé par le rayon de cette pelote que par sa longueur. Ce rayon R est appelé rayon de giration du polymère et c'est une quantité observable.

Un phénomène important dont il faut tenir compte dans les polymères est une interaction répulsive de cœur dur qui empêche deux molécules de venir au contact: c'est ce qu'on appelle *l'interaction de volume exclu*, qui revient à dire que la chaîne ne doit pas se recouper.

Configuration interdite

1.1. Le problème du polymère gaussien

On appelle polymère gaussien un polymère "idéal" où il n'y aurait pas d'effet de volume exclu: la chaîne a le droit de se couper. L'intérêt de cet exemple simple est qu'il est exactement soluble, et permet d'introduire et de comprendre des notions essentielles pour la suite.

Soit un polymère constitué de N maillons $\boldsymbol{a}_i$ de longueur a. Si toutes les configurations sont permises, il est clair que

$$\langle R^2 \rangle = \left\langle \left(\sum \boldsymbol{a}_i \right)^2 \right\rangle = Na^2 .$$

D'où le rayon de giration

$$\langle R^2 \rangle^{1/2} = N^{1/2} a ,$$

qui, on le voit, est beaucoup plus petit que la longueur du polymère. Pour une longueur de 10^4 Å, le rayon de giration n'est que de 100 Å.

Il est commode pour les calculs *d'incrire* le polymère sur un réseau. Une configuration de polymère est donnée par un problème de marche au hasard sur le réseau connaissant l'origine, l'extrémité et le nombre N de maillons élémentaires (voir la figure).

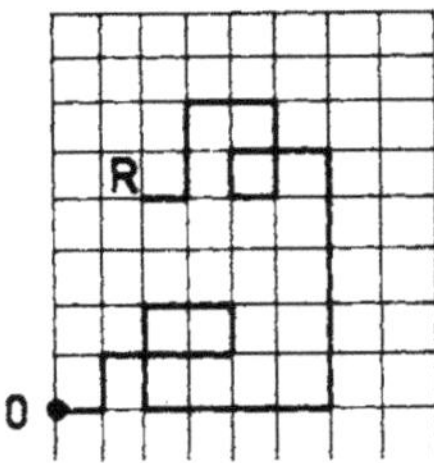

Soit $\Gamma(N, \boldsymbol{R})$ le nombre de configurations du polymère de N maillons partant de l'origine et aboutissant au point $\boldsymbol{R}$. Soit $\Gamma(N)$ le nombre total de configurations partant de l'origine

$$\Gamma(N) = \sum_{\boldsymbol{R}} \Gamma(N, \boldsymbol{R}) .$$

Il est clair que $\Gamma(N) = z^N$, où z est le nombre de voisins d'un site sur le réseau. On appelle $p(N, \boldsymbol{R})$ *la probabilité* de trouver l'extrémité en $\boldsymbol{R}$

$$p(N, \boldsymbol{R}) = \Gamma(N, \boldsymbol{R})/\Gamma(N) = \Gamma(N, \boldsymbol{R})/z^N .$$

Il est facile de calculer $\langle R^2 \rangle$ pour N fixé

$$\langle R^2 \rangle = \langle (\boldsymbol{a}_1 + \boldsymbol{a}_2 + \cdots + \boldsymbol{a}_N)^2 \rangle = Na^2 ,$$

car il n'y a pas de corrélation entre les $\boldsymbol{a}_i$, c'est-à-dire $\langle \boldsymbol{a}_i \cdot \boldsymbol{a}_j \rangle = 0$ pour $i \neq j$.

1.1.1. Calcul de $\Gamma(N, \boldsymbol{R})$

Il est clair que

$$\Gamma(N + 1, \boldsymbol{R}) = \sum_{\boldsymbol{R}' V \boldsymbol{R}} \Gamma(N, \boldsymbol{R}') ,$$

où $\boldsymbol{R}'$ est un site voisin de $\boldsymbol{R}$. D'où

$$z^{N+1} p(N + 1, \boldsymbol{R}) = z^N \sum_{\boldsymbol{R}' V \boldsymbol{R}} p(N, \boldsymbol{R}')$$

$$\Rightarrow p(N + 1, \boldsymbol{R}) - p(N, \boldsymbol{R}) = \frac{1}{z} \sum_{\boldsymbol{R}' V \boldsymbol{R}} [p(N, \boldsymbol{R}') - p(N, \boldsymbol{R})] .$$

En passant à la limite du continuum ($\boldsymbol{R}$ et N grands) ceci s'écrit:

$$\frac{\partial p}{\partial N} = \frac{1}{z} a^2 \nabla^2 p .$$

Définissons la transformée de Fourier en $\boldsymbol{R}$, $p(N, \boldsymbol{q})$,

$$\Rightarrow \frac{\partial p}{\partial N} = -\frac{q^2 a^2}{z} p .$$

D'où

$$p(N, \boldsymbol{q}) = \exp\left(-\frac{1}{z} N q^2 a^2 \right) \cdot$$

La constante d'intégration a été déterminée par la condition:

$$p(N, \boldsymbol{q} = 0) = \int \mathrm{d}\boldsymbol{R}\, p(N, \boldsymbol{R}) = 1 ,$$

qui exprime la certitude de trouver le N^{e} maillon quelque part dans l'espace. On en déduit

$$p(N, \boldsymbol{R}) \sim N^{-d/2} \exp(-R^2/2Na^2) ,$$

$$\Gamma(N, \boldsymbol{R}) \sim z^N N^{-d/2} \exp(-R^2/2Na^2) .$$

1.2. Remarques sur le problème du polymère gaussien

(a) On voit que $\Gamma(N, \boldsymbol{R})$ est une forme homogène généralisée ne faisant apparaître que la variable $R/N^{1/2}$, ce qui permet de retrouver facilement que $\langle R^2 \rangle \sim Na^2$.

(b) Posons $z^N = \exp(NT_c)$:

$$\Gamma(N, \boldsymbol{R}) = \exp(NT_c)\,, \quad N^{-d/2} \exp(-R^2/2Na^2)\,,$$

$$\Rightarrow \Gamma(N, \boldsymbol{q}) = \exp(NT_c) \exp(-Nq^2a^2)\,.$$

Définissons la transformée de Laplace:

$$\Gamma(T, \boldsymbol{q}) = \int_0^{+\infty} \mathrm{d}N \exp(-NT)\, \Gamma(N, \boldsymbol{q})\,,$$

on obtient:

$$\Gamma(T, \boldsymbol{q}) = \frac{1}{T - T_c + q^2a^2}\,.$$

Or ceci est la forme de la susceptibilité $\chi(\boldsymbol{q})$ d'un problème *magnétique* en théorie de Landau.

Ceci montre que si on définit une *température fictive* T comme *le potentiel chimique conjugué de* N, on obtient un problème critique où $N \to \infty$ correspond à $T - T_c \to 0$.

1.3. Phénoménologie des effets du volume exclu

On peut schématiser l'effet de l'interaction de volume exclu en interdisant dans la marche au hasard sur le réseau de repasser en un point déjà occupé: ceci est une approximation de haute température, car en ne tenant compte que du cœur dur répulsif, on ne calcule que l'entropie (nombre de configurations permises).

(a) *Le rayon de giration* sera plus grand car le volume exclu aura tendance à *dilater* la pelote. En particulier, à une dimension, le volume exclu imposerait $R = Na$, c'est-à-dire une chaîne parfaitement rigide. Il est donc logique de penser que

$$\langle R^2 \rangle \sim N^{2\nu} \qquad \text{où} \quad \nu > \tfrac{1}{2}\,,$$

$\nu = \frac{1}{2}$ correspondant au cas "gaussien". Pourquoi appeler cet exposant ν? Dans les problèmes de changement de phase on a une longueur de corrélation ξ qui diverge comme $\tau^{-\nu}$ ($\tau = (T - T_c)/T_c$). τ étant la variable conjuguée de N, il est donc naturel de penser qu'ici

$$\xi \sim \langle R^2 \rangle^{1/2} \sim N^\nu .$$

L'exposant $\nu = \frac{1}{2}$ pour $d > 4$ et on trouve la formule approchée $\nu = 3/(d+2)$ pour $d < 4$.

(b) Que devient $\Gamma(N, \boldsymbol{R})$? Il est naturel de supposer que c'est encore une forme homogène généralisée de R/N^ν. Donc

$$p(N, \boldsymbol{R}) \sim \frac{1}{N^{\nu d}} f(R/N^\nu) ,$$

le coefficient $1/N^{\nu d}$ étant imposé par la condition $\int \mathrm{d}\boldsymbol{R}\, p(N, \boldsymbol{R}) = 1$. D'où

$$\Gamma(N, \boldsymbol{R}) = \Gamma(N)\, N^{-\nu d} f(R/N^\nu) .$$

La quantité $\Gamma(N)$, qui est le nombre total de configurations de N maillons, est aussi modifiée par rapport au cas gaussien. Elle comporte encore un terme $(z^*)^N$ où $z^* < z$ car le nombre de voisins effectif est en quelque sorte diminué par l'effet de volume exclu, de plus on ne peut exclure une loi de puissance en N de la forme $N^{\gamma - 1}$ qui n'existe pas dans le cas gaussien où $\gamma = 1$. Il est ainsi raisonnable d'attendre pour $\Gamma(N, \boldsymbol{R})$ la forme suivante:

$$\Gamma(N, \boldsymbol{R}) \sim \exp(NT_c)\, N^{\gamma - 1 - \nu d} f(R/N^\nu) .$$

Cette forme, qui pour l'instant apparaît comme une hypothèse raisonnable, peut en fait être démontrée, et du même coup se trouveront justifiés les noms donnés aux divers exposants qui y figurent. Ceci sera le but du chapitre suivant.

2. Une version simple de la théorie actuelle de la conformation des polymères en solution

2.1. *Lemme géométrique*

Soit $\boldsymbol{S}$ un "spin" classique à n composantes, de norme $S^2 = \sum_\alpha S_\alpha^2 = n$. Ceci implique que

$$\langle S_\alpha^2 \rangle = 1 \quad \forall \alpha .$$

La contrainte de longueur fixée est une loi de probabilité dont la fonction caractéristique est donnée par la valeur moyenne *géométrique*

$$f(\boldsymbol{k}) = \left\langle \exp\left(\mathrm{i} \sum k_\alpha S_\alpha \right) \right\rangle .$$

Il est clair que $f(\boldsymbol{k}) = f(|\boldsymbol{k}|)$ est une fonction sphérique. Elle satisfait à l'équation différentielle

$$-\nabla^2 f = \left(\sum S_\alpha^2\right) f = nf\,,$$

qui s'écrit en coordonnées sphériques dans l'espace à n dimensions

$$-\left(\frac{\mathrm{d}^2}{\mathrm{d}k^2} + \frac{n-1}{k}\frac{\mathrm{d}}{\mathrm{d}k}\right) f = nf\,. \tag{1}$$

La solution de cette dernière équation est définie par deux conditions aux limites qu'on obtient par le développement de f pour k petit

$$f = 1 - \tfrac{1}{2}\left\langle\left(\sum k_\alpha S_\alpha\right)^2\right\rangle + \cdots\,,$$

$$\Rightarrow f = 1 - \tfrac{1}{2}k^2 + \cdots\,.$$

Or $f = 1 - \frac{1}{2}k^2$ est solution de (1) pour $n = 0$, donc pour $n \to 0$ on a:

$$f(k) = \langle \exp(\mathrm{i}\boldsymbol{k}\cdot\boldsymbol{S})\rangle = 1 - \tfrac{1}{2}k^2\,. \tag{2}$$

Notons que dans le cas général la fonction $f(k_\alpha) = \exp(i \sum k_\alpha S_\alpha)$ est la fonction génératrice des valeurs moyennes du type

$$\langle (S_1)^{p_1}(S_2)^{p_2}\cdots(S_n)^{p_n}\rangle\,,$$

où $p_1, p_2, \ldots, p_n$ sont des entiers positifs ou nuls. De telles valeurs moyennes sont visiblement proportionnelles au coefficient de $(k_1)^{p_1}(k_2)^{p_2}\cdots(k_n)^{p_n}$ dans le développement de f en puissances des k_α. Grâce au résultat (2), il est clair que pour un "spin" à *zéro composante*, toutes *ces valeurs moyennes sont nulles* sauf celles du type $\langle S_\alpha^2\rangle$ qui sont données par

$$\langle S_\alpha^2\rangle = 1\,.$$

2.2. *Le modèle "magnétique"*

Considérons des spins S_R à n composantes situés aux nœuds d'un réseau à d dimensions et couplés par une interaction K entre premiers voisins.

La loi de probabilité de ce système s'écrit:

$$\frac{\exp(-\beta\mathscr{H})}{Z} = \frac{1}{Z}\exp(\beta K)\sum_{R'VR} S_R S_{R'}\,. \tag{3}$$

Z est la fonction de partition, qui ne contient que des intégrales angulaires puisque la longueur des "spins" est fixée.

On peut développer Z sous la forme:

$$Z = \left\langle \prod_{RVR',\alpha} [1 + \beta K S_R^\alpha S_{R'}^\alpha + \tfrac{1}{2}(\beta K)^2 S_R^\alpha S_{R'}^\alpha + \cdots] \right\rangle ,$$

où RVR' désigne que R et R' sont des sites premiers voisins et $\langle\ \rangle_A$ désigne les valeurs moyennes angulaires. Dans la limite $n \to 0$, grâce au Lemme 1, il est inutile de pousser le développement plus loin.

Les différents termes contribuant à Z peuvent être représentés graphiquement. Un terme typique est une *boucle* sur le réseau correspondant à un produit de $\beta K S_R^\alpha S_{R'}^\alpha$, où l'indice α *est fixé.*

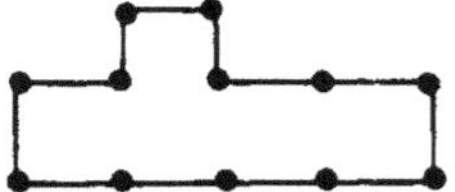

De plus, toute autre valeur moyenne géométrique que $\langle S_\alpha^2 \rangle$ étant nulle, on ne peut passer *qu'une fois* en un point du réseau; ainsi la *limite $n \to 0$ traduit le volume exclu*: un diagramme ne peut ni se recouper ni être coupé par un autre. La contribution d'une boucle est visiblement $(\beta K)^N$ où N est le nombre de maillons, mais comme cette boucle peut apparaître avec un indice de composante α quelconque, on obtient par sommation $n(\beta K)^N$ qui tend vers zéro quand $n \to 0$. On voit ainsi que

$$Z = 1 .$$

2.3. *Polymères en régime dilué*

En régime très dilué, on peut considérer le problème d'un seul polymère à volume exclu.

Le contenu essentiel du Lemme 1 est que les seuls diagrammes permis sont des diagrammes à volume exclu, c'est-à-dire qu'on ne peut passer plus d'une fois en un point du réseau.

Considérons maintenant la fonction de corrélation entre deux spins

$\boldsymbol{S}_{R_1}$ et $\boldsymbol{S}_{R_2}$ du "modèle magnétique" .

Puisque $Z = 1$, nous avons pour la fonction de corrélation pour la composante 1 l'expression

$$C_{11}(\boldsymbol{R}_1, \boldsymbol{R}_2) = \langle S_{R_1}^1 S_{R_2}^1 \rangle = \left\langle S_{R_1}^1 S_{R_2}^1 \prod_{RVR',\alpha} [1 + \beta K S_R^\alpha S_{R'}^\alpha + \cdots] \right\rangle .$$

Les différentes contributions à $C(\boldsymbol{R}_1, \boldsymbol{R}_2)$ peuvent être représentées graphiquement. La contribution de toute boucle étant réduite à zéro par sommation sur α, on voit que seuls restent les chemins de volume exclu joignant sur le réseau $\boldsymbol{R}_1$ à $\boldsymbol{R}_2$. Pour chaque maillon il faut utiliser le terme $\boldsymbol{S}^1_{\boldsymbol{R}}\boldsymbol{S}^1_{\boldsymbol{R}'}$, de sorte qu'il n'y a pas de sommation sur α.

Un tel graphe représente effectivement une configuration du polymère à extrémités fixées en R_1 et R_2. Chaque configuration est comptée une et une seule fois.

On obtient ainsi:

$$C_{11}(\boldsymbol{R}_1, \boldsymbol{R}_2) = \langle S^1_{\boldsymbol{R}_1} \cdot S^1_{\boldsymbol{R}_2} \rangle = \sum_N (\beta K)^N \Gamma(N, \boldsymbol{R}_1 - \boldsymbol{R}_2)\,, \tag{4}$$

où $\Gamma(N, \boldsymbol{R}_1 - \boldsymbol{R}_2)$ est comme au chapitre précédent le nombre de configurations d'un polymère de N maillons aux extrémités fixées en $\boldsymbol{R}_1$ et $\boldsymbol{R}_2$.

La relation (4), découverte par de Gennes, établit une *correspondance* entre le problème du polymère défini par $\Gamma(N, \boldsymbol{R})$ et le problème "magnétique" à zéro composante défini par la loi de probabilité (3) et les fonctions de corrélation de spins qui en résultent. Ce dernier problème, qui est un problème de changement de phases, est supposé connu et on en déduit $\Gamma(N, \boldsymbol{R})$.

La fonction de corrélation du problème magnétique est connue et se comporte au voisinage de T_c comme:

$$C_{11}(\boldsymbol{R}_1 - \boldsymbol{R}_2) \sim \frac{1}{|R_1 - R_2|^{d-2+\eta}} f\left(\frac{|\boldsymbol{R}_1 - \boldsymbol{R}_2|}{\xi}\right). \tag{5}$$

D'après (4) on peut écrire aussi:

$$C_{11}(\boldsymbol{R}_1 - \boldsymbol{R}_2) = \sum_N \left(\frac{K}{T_c}\right)^N \exp(-N\tau)\Gamma(N, \boldsymbol{R}_1 - \boldsymbol{R}_2)\,, \tag{6}$$

où $\tau = (T - T_c)/T$ apparaît une fois de plus comme la variable conjuguée de N. Rappelons que $\xi \sim |\tau|^{-\nu}$.

L'inversion de la transformée de Laplace (6) donne immédiatement

$$\left(\frac{K}{T_c}\right)^N \Gamma(N, \boldsymbol{R}) \sim N^{\gamma - 1 - \nu d} h\left(\frac{|\boldsymbol{R}|}{N^\nu}\right),$$

et enfin

$$\Gamma(N, \boldsymbol{R}) \sim \left(\frac{T_c}{K}\right)^N N^{\gamma-1-\nu d} h\left(\frac{|\boldsymbol{R}|}{N^\nu}\right). \tag{7}$$

Ceci résout le problème du polymère isolé. Les exposants dans la formule (7) sont *ceux du modèle magnétique isotrope à* $n = 0$ *composantes.* Nous confirmons ainsi l'approche intuitive du chapitre précédent, mais nous identifions en plus les valeurs des exposants. En particulier on aura (et on observe expérimentalement) en solution très diluée $R \sim N^\nu$ (R rayon de giration) où $\nu(n = 0) \sim 3/(d + 2) > \frac{1}{2}$ comme prévu.

Nous avons ainsi pu ramener le problème du polymère dans la limite $N \to \infty$ à un problème magnétique critique au voisinage de T_c.

2.4. Polymères en régime semi-dilué

Pendant longtemps, l'effet de l'augmentation de concentration du polymère dans le solvant restait incompréhensible. C'est des Cloizeaux qui devait récemment trouver la clé du problème, ce qui a provoqué en retour un rapide essor des expériences dans ce domaine.

Nous avons vu que le problème magnétique $n = 0$ résolvait le cas très dilué.

De même un problème magnétique $n = 0$ en présence *d'un champ appliqué* permet de comprendre le régime semi-dilué.

Problème N° 1: polymères semi-dilués.

Problème N° 2: problème magnétique $n = 0$ sous champ.

Nous montrerons que des relations permettent de passer du Problème N° 2 (supposé connu) au Problème N° 1. Nous établirons en particulier que la limite N grand (longueur des polymères) correspond au domaine critique du problème magnétique. Si le champ magnétique est appliqué suivant la direction (1) dans l'espace des spins à n composantes, la loi de probabilité des spins s'écrit:

$$\frac{\exp(-\beta\mathscr{H}_1)}{Z} = \frac{1}{Z}\exp\left\{\beta\left[K\sum_{R'VR} \boldsymbol{S}_R\cdot\boldsymbol{S}_{R'} + H\sum S_R^1\right]\right\}. \tag{8}$$

Nous avons vu plus haut que $Z(H = 0) = 1$. On peut donc écrire:

$$\frac{Z(H)}{Z(0)} = \left\langle \prod_{RVR',\alpha}(1 + \beta K S_R^\alpha S_{R'}^\alpha + \cdots)\prod_R(1 + \beta H S_R^1 + \cdots)\right\rangle_A .$$

Le développement en puissances de H ne contient à l'évidence que des puissances paires de H. Les boucles étant interdites, il ne reste que les graphes "polymères". Ainsi le terme en $(\beta H)^2$ somme les contributions de toutes les configurations *à un polymère* avec des extrémités $\boldsymbol{R}$ et $\boldsymbol{R}'$ arbitraires et un nombre N de maillons arbitraire, la contribution d'un tel graphe étant $(\beta H)^2 (\beta K)^N$.

En général un terme en $(\beta H)^{2p}$ est obtenu en traçant p graphes de polymères à *volume exclu* de longueurs $N_1, N_2, \ldots, N_p$.

Toute configuration est comptée une et une seule fois. Soit alors N_p le nombre de polymères, N_M le nombre *total* de maillons (ou de monomères), alors on a

$$\frac{Z(H)}{Z(0)} = \sum (\beta K)^{N_M} (\beta H)^{2N_P}\, U(N_M, N_P)\,, \tag{9}$$

où $U(N_M, N_P)$ est le nombre de configurations de N_P polymères contenant un nombre *total* N_M de monomères.

Ainsi, pourvu que $n = 0$ dans le problème magnétique, $Z(H)/Z(0)$ est *la grande fonction de partition* d'une assemblée de polymères de longueur variable, définie avec la convention que $2\log(\beta H)$ et $\log(\beta K)$ sont respectivement les potentiels chimiques conjugués du nombre de polymères N_P et du *nombre total* de monomères N_M.

Posant $K = 1$, ce qui revient à mesurer H et T en unités de K, on peut encore écrire (9) sous la forme

$$Z(H)/Z(0) = \sum \exp[2N_P \log H - (N_M + 2N_P)\log T]\,U(N_M, N_P)\,. \tag{10}$$

2.4.1. *Relations d'équivalence*

A gauche figureront les variables "polymère". A droite de même les variables "magnétiques". Soient ρ_P et ρ la concentration de polymères et de monomères (nombre par unité de volume), et soit $\Delta\mathscr{F} = \mathscr{F}(H, T) - \mathscr{F}(0, T)$ par unité de volume la variation d'énergie libre du système magnétique, on obtient alors grâce à (10)

$$2\rho_P = -H\frac{\partial}{\partial H}(\Delta\mathscr{F})\,,$$

$$\rho + 2\rho_P = T\frac{\partial}{\partial T}(\Delta\mathscr{F})\,,$$

$$\pi = -\Delta\mathscr{F} \qquad (\pi \text{ pression osmotique})\,.$$

Enfin la longueur moyenne des polymères est donnée par $N = \rho/\rho_P$. Remarquons que dans la limite N grand, ρ_P est négligeable devant ρ.

Le problème magnétique dépend de H et T dont l'élimination entre les 3 premières relations donnerait $\pi(\rho, \rho_P)$. En utilisant la 4e équation on obtiendrait $\pi(\rho, N)$ car expérimentalement la longueur N des polymères est fixée. Le problème est donc résolu si on élimine les variables "magnétiques".

Il est commode de décrire le problème magnétique plutôt en variables M, T (M étant l'aimantation). On y arrive par la relation

$$\Delta\mathscr{F}(H, T) = G(M, T) - HM\,,$$

avec $\partial G/\partial M = H$.

Négligeant ρ_P devant ρ, et remplaçant $T\,\partial/\partial T$ par $\partial/\partial\tau$ au voisinage du point critique ($\tau = (T - T_c)/T_c$) pour des raisons qui seront éclaircies plus loin, les relations d'équivalence peuvent s'écrire

$$\rho/N = \tfrac{1}{2}M\,\partial G/\partial M\,, \tag{11}$$

$$\rho = \partial G/\partial\tau\,, \tag{12}$$

$$\pi = -G + M\,\partial G/\partial M\,, \tag{13}$$

$$\rho/\rho_P = N\,. \tag{14}$$

Les relations (11) à (14) sont les relations fondamentales d'équivalence entre le problème magnétique et le problème des polymères. Supposons pour l'instant, sous réserve d'une démonstration ultérieure, que dans la limite de N grand on reste au voisinage de T_c. Nous pouvons alors utiliser la forme homogène généralisée pour $G(M)$

$$G(M) = |\tau|^{\nu d} g(x)\,, \quad x = M/|\tau|^{\beta}\,.$$

On en déduit sans calcul

$$\left.\begin{aligned} \rho/N &= |\tau|^{\nu d} g_P(x)\,, && (11\text{ bis})\\ \rho &= |\tau|^{\nu d-1} g_\rho(x)\,, && (12\text{ bis}) \end{aligned}\right\} \Rightarrow |\tau| N = g_\rho(x)/g_P(x)\,,$$

$$\pi = |\tau|^{\nu d} g_\pi(x)\,, \quad (13\text{ bis})$$

où $g_P(x)$, $g_\rho(x)$, $g_\pi(x)$ sont des fonctions universelles reliées à $g(x)$. De ces relations on déduit immédiatement l'équation d'état universelle

$$\pi N^{\nu d} = \phi(\rho N^{\nu d-1})\,, \tag{15}$$

où ϕ est une certaine fonction universelle. Cette équation d'état est valable pour N grand et pour un système de polymères à haute température (car nous n'avons pris en compte que le volume exclu, c'est-à-dire l'entropie).

Nous pouvons déjà voir l'existence d'une concentration de "crossover" (ou de changement de régime) définie par

$$\rho^* N^{\nu d - 1} \sim 1 . \tag{16}$$

2.4.2. Evolution en fonction de la concentration

Notre but est de décrire une situation de concentration variable, mais où la longueur N des chaînes est fixée.

L'équation $|\tau| N = g_\rho(x)/g_P(x)$, $x = M/|\tau|^\beta$, montre que cette condition impose une certaine relation entre τ et M dans le problème magnétique. Ceci définit dans le plan M, T une ligne "isométrique" dessinée en pointillé sur la figure. Le trait plein représente la courbe d'aimantation spontanée $M(T)$ du problème magnétique en champ nul.

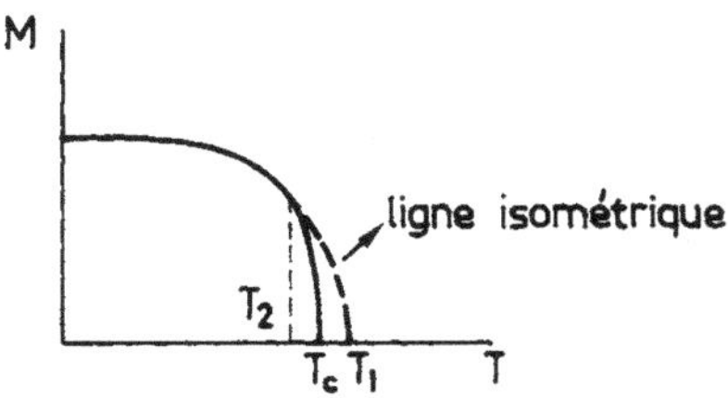

(a) Considérons le cas $x = M/|\tau|^\beta \ll 1$. Dans ce cas, G a un développement en puissances de x

$$G \sim \tfrac{1}{2}\tau^\gamma M^2 .$$

L'équation (11) donne $\rho/N = \frac{1}{2}\tau^\gamma M^2$ et (12) donne $\rho = \frac{1}{2}\gamma\tau^{\gamma-1}M^2$. Ainsi ρ *s'annule* au point $\gamma/\tau = N$ ce qui définit le point T_1 de la figure donné par $(T_1 - T_c)/T_c = \gamma/N$ ou

$$T_1 = T_c(1 + \gamma/N) .$$

Le point T_1 $(>T_c)$ est le point de départ de la ligne isométrique avec $\rho = 0$. Au voisinage de ce point nous avons donc une région qui correspond à $\rho \ll \rho^*$ et à un très proche voisinage de T_c. *C'est le régime dilué*. Il est immédiat que dans cette région, d'après (13), $\pi = \frac{1}{2}M\, \partial G/\partial M = \rho/N$ en première approximation

$$\pi = \rho/N \quad \text{pour} \quad \rho \ll \rho^* , \tag{17}$$

ce qui implique d'après (15), que

$$\phi(y) \sim y \quad \text{pour} \quad y \ll 1 .$$

La loi $\pi = \rho/N$ pour $\rho \ll \rho^*$ n'est autre que la loi de Mariotte ou des gaz parfaits.

(b) La ligne isométrique va ensuite dans une région $T \sim T_c$ et M fini, ce qui correspond à des concentrations $\rho \sim \rho^*$ (crossover).

(c) Enfin elle *rejoint* la courbe d'aimantation en champ nul pour $\tau < 0$ et $|\tau| \ll 1$ (nous allons le démontrer).

En effet, sur la courbe d'aimantation ($M \sim |\tau|^{\beta}$) on a $x = M/|\tau|^{\beta} = x_0$ qui est un nombre fixé. On a ici d'après (11 bis)

$$\rho \sim |\tau|^{\nu d - 1} g_\rho(x_0) \,.$$

On peut avoir simultanément (τ est négatif dans cette région)

$$|\tau| \ll 1 \quad \text{et} \quad |\tau| \gg 1/N \,,$$

ce qui implique $|\tau|^{\nu d - 1} \gg 1/N^{\nu d - 1}$, et donc $\rho \gg \rho^*$: nous sommes dans le *régime semi-dilué*, et comme prévu nous *restons dans le domaine critique* du problème magnétique.

De (12 bis) et (13 bis) nous tirons

$$\rho = |\tau|^{\nu d - 1} g_\rho(x_0) \,, \qquad \pi = |\tau|^{\nu d} g_\pi(x_0) \,,$$

ce qui donne la dépendance

$$\pi \sim A \rho^{\nu d/(\nu d - 1)} \quad \text{pour} \quad \rho \gg \rho^* \,,$$

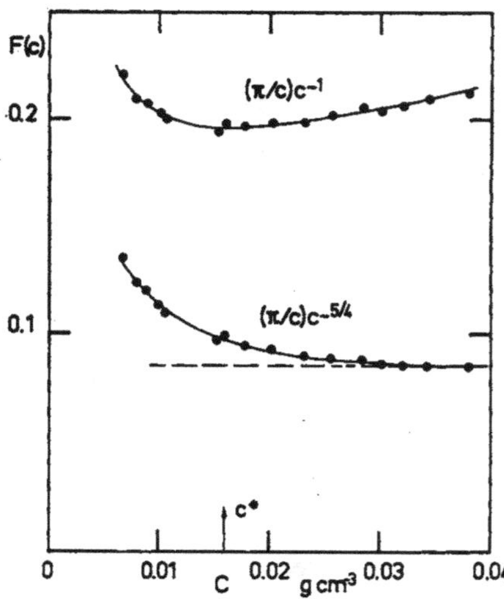

Fig. 1. $\pi \sim c^{9/4}$ pour $c \gg c^*$.

où A est une constante *indépendante de N*. Ceci implique d'après (15) que

$$\phi(y) \sim Ay^{\nu d/(\nu d-1)} \quad \text{pour} \quad y \gg 1 .$$

Ainsi le passage de $\rho \ll \rho^*$ (régime dilué) à $\rho \gg \rho^*$ (régime semi-dilué) est expliqué comme un crossover *par équivalence* avec un problème magnétique où on traverse T_c *tout en restant dans le domaine critique.*

Pour $d = 3$, on a $\nu \sim \frac{3}{5}$, ce qui prédit $\pi \sim \rho^{9/4}$ pour $\rho \gg \rho^*$. C'est ce que montre la figure 1. Notons en passant que toutes les courbes expérimentales proviennent du groupe des Polymères de Saclay.

2.4.3. Variation du rayon de giration

Le rayon de giration a aussi été mesuré expérimentalement comme fonction de la concentration. En régime dilué on a $R \sim N^\nu$. En régime semi-dilué ($\rho \gg \rho^*$) on retrouve $R \sim N^{1/2}$. Le crossover à $\rho = \rho^* = 1/N^{\nu d-1}$ implique que $R = N^\nu A(\rho/\rho^*)$ avec $A(0)$ fini ($\rho \ll \rho^*$). Si $\rho \gg \rho^*$, $A(\rho/\rho^*)$ se comporte comme $(\rho/\rho^*)^x$, ce qui donne

$$R = N^{\nu + x(\nu d-1)}\rho^x .$$

Comme $R \sim N^{1/2}$ on en déduit $x = -(\nu - 1/2)/(\nu d - 1)$ et enfin

$$R^2 = N\rho^{-(2\nu-1)/(\nu d-1)} , \quad \rho \gg \rho^* .$$

Pour $d = 3$, ceci donne $R^2 = N\rho^{-1/4}$, ce qu'on peut vérifier sur la courbe expérimentale de la figure 3.

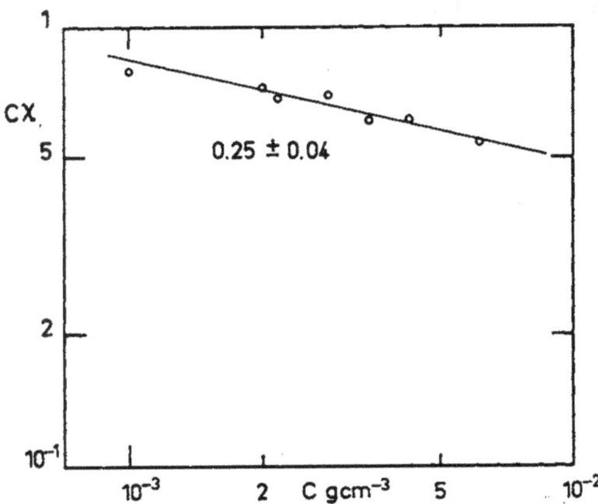

Fig. 2. Cette figure montre la variation expérimentale de $c(\partial c/\partial \pi) \sim c^{-1/4}$ pour $c \gg c^*$.

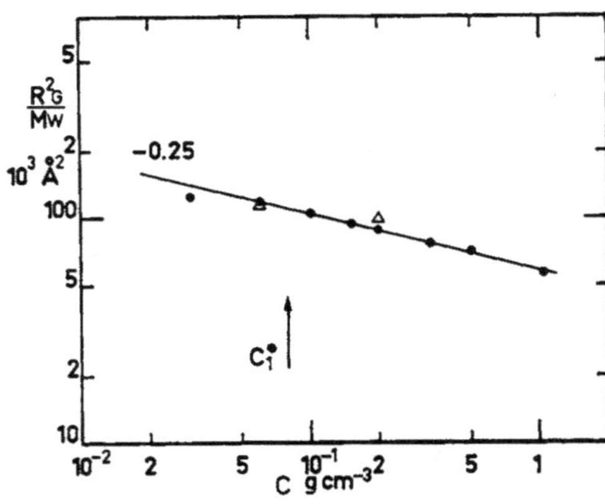

Fig. 3. $R^2/N \sim c^{-1/4}$.

2.5. *Conclusion*

Nous voyons que du point de vue théorique il a été possible d'établir une correspondance générale entre le problème de la configuration des polymères et un problème de changement de phase magnétique.

Les récents progrès de la théorie des changements de phase nous ont permis d'expliquer beaucoup de propriétés de ces polymères et d'en prédire qui ont été confirmées par l'expérience.

Mais le champ des problèmes de polymères reste encore largement ouvert à la fois à l'expérience et à la théorie.

NEUTRONS ET COHÉRENCES

Jacques JOFFRIN

Institut Laue–Langevin, 156 X, 38042 Grenoble Cedex, France

Contents

R. Balian et al., eds.
Les Houches, Session XXXI, 1978 – La matière mal condensée/Ill-condensed matter

1. Introduction

Les neutrons sont des particules de masse $m_n = 1.67 \times 10^{-27}$ kg, de spin $\frac{1}{2}$ et de moment magnétique dipolaire $\psi_n = -1.91$ magnetons nucléaires. Le neutron a une charge nulle, mais bien des expériences de physique fondamentale cherchent à déterminer son moment électrique dipolaire; on sait seulement que celui-ci est inférieur à 3×10^{-24} cm; si l'on parvenait à gagner ne serait-ce qu'un facteur 10 en précision on serait en mesure de tester plusieurs théories sur les interactions faibles et les courants neutres.

Les grandeurs cinématiques qui caractérisent le neutron sont: la quantité de mouvement $\boldsymbol{p}$, le vecteur d'onde $\boldsymbol{k} = \boldsymbol{p}/\hbar$, la vitesse $\boldsymbol{v} = \boldsymbol{p}/m_n$, la longueur d'onde $\lambda = 2\pi/k$, et l'énergie cinétique $E = p^2/2m_n$.

Une population de neutrons en équilibre thermique avec un modérateur à température ordinaire a une répartition d'énergie maxwellienne dont le maximum est situé à 25 meV (4×10^{-18} J), correspondant à une longueur d'onde d'environ 1.8 Å. Les neutrons sont à la fois des objets d'étude et des instruments d'étude. Si bizarre que cela paraisse on ne connait pas exactement la durée de vie du neutron; son moment dipolaire électrique est seulement borné supérieurement. Bref, comme particule le neutron est encore l'objet d'études.

La grande gamme d'énergie (ou de longueur d'onde) dans laquelle on sait le produire sous forme de faisceaux collimatés intenses en fait un instrument très versatile pour toutes les études de physique nucléaire ou de physique de la matière condensée. Les raisons qui en ont fait un instrument d'analyse sont les suivantes.

(I) L'absorption des neutrons par la matière est très petite pour presque tous les corps. Les exceptions sont fameuses et bien connues: l'He^3 par exemple et cela interdit les études à très basse température des phases suprafluides et limite celles à température plus élevée (1 K!) à des examens en surface du liquide ou du solide He^3; le Cd ce qui en fait inversement un bon écran; le Gd et quelques isotopes rares.

Bref les neutrons sont peu absorbés; on peut donc travailler sans difficulté sur des échantillons massifs. Par comparaison, les rayons X où les particules chargées, (électrons ou ions par implantation) pénètrent beaucoup moins; les phonons eux-mêmes, à moins d'atteindre de très basses fréquences sont très rapidement thermalisés. Seuls les photons, dans le cas des cristaux ou des matériaux transparents peuvent avoir des sections d'absorption notablement plus petites.

(II) Corrélativement, l'interaction des neutrons avec la matière est faible; de sorte que dans toutes les expériences de diffusion on se contente souvent de négliger la contribution des diffusions multiples; les résultats sont interprétés dans le cadre d'une approximation de Born au premier ordre.

Il y a des cas où les corrections sont nécessaires: détermination des facteurs de structure d'un liquide, d'un amorphe; cristallographie de précision; mais enfin, on sait faire, et les corrections ne sont pas plus/moins indispensables que celles auxquelles on est conduit en spectroscopie optique par battement de photons (expérience de démixion critique, élasticité des gels) en effet Brillouin ou en RX.

(III) L'énergie des neutrons thermiques est du même ordre que celle des excitations de toute nature que l'on rencontre en physique de la matière condensée; leur vecteur d'onde est une bonne mesure des dimensions de la zone de Brillouin. Cela laisse espérer une capacité d'analyse de la zone de Brillouin qui est dans ce domaine sans comparaison. Si l'on ajoute que l'énergie des neutrons avec lesquels on peut travailler s'étend de presque 1 eV jusqu'à environ 10^{-7} eV on a là un instrument assez unique.

Dans le domaine inélastique on est à l'heure actuelle en mesure d'apprécier des excitations de quelques centaines de meV (excitations moléculaires, excitations d'impuretés dans un champ cristallin) ou d'une fraction de micro-eV (relaxations de spins, mouvements dispersifs de polymères en solution, vibrations internes de l'hémoglobine, rotations moléculaires génées).

Je ne crois pas que dans un proche avenir il soit possible d'augmenter encore cette résolution du côté des basses fréquences; en particulier, il est exclu de pouvoir concurrencer la spectroscopie optique de haute résolution qui permet d'atteindre 100 c/s. Mais par contre le joint est fait avec ce que l'on sait déduire de la mesure des temps de relaxation de RMN ou RPE.

Il faudrait ajouter que par rapport à d'autres techniques, les règles de sélection ne sont jamais une grosse difficulté; on s'en affranchit aisément en choisissant une configuration de travail convenable.

(IV) Le neutron a un spin $\frac{1}{2}$, c'est un énorme avantage s'il est correctement exploité; on va le voir, la diffusion du neutron n'est pas uniquement de type nucléon–nucléon, mais aussi de type magnétique; de sorte que la diffusion d'un faisceau de neutrons polarisés est un instrument irremplaçable en cristallographie pour l'étude des structures magnétiques, en chimie du solide pour l'étude de la liaison chimique lorsqu'elle est polarisée par un radical ou un atome magnétique, en physique du solide pour l'étude de la densité de spin ou de la densité électronique dans un alliage.

Le but de ces quelques notes n'est pas de faire une présentation de la diffusion des neutrons dans sa généralité mais de mettre en évidence quelques propriétés qu'on peut regrouper sous le titre de "cohérence" dans le but très pédagogique de faire une comparaison avec la cohérence en optique ou en acoustique et même de faire les analogies nécessaires avec les rayons X.

2. Diffusion élémentaire

Les neutrons, dans leur interaction avec d'autres noyaux, peuvent être soit absorbés, soit diffusés.

Absorbés, ils produisent des réactions nucléaires via l'excitation de toute une série de niveaux: il y a décomposition de l'état excité par émission d'autres particules ou par fission. C'est un domaine encore fort actif mais qui n'est pas dans l'axe de ce séminaire; disons seulement que pour des neutrons thermiques, c'est-à-dire loin des pics de résonance, cette absorption est faible mais croît comme $1/v$.

Diffusés, ils donnent lieu à des phénomènes cohérents ou incohérents suivant que les noyaux diffusants ont ou n'ont pas un degré de liberté interne: le spin ou l'isotope. La section efficace de diffusion pour un neutron thermique ne dépend pas de son énergie.

La section efficace de diffusion cohérente par le noyau est définie par la relation:

$$\sigma_{\mathrm{coh}} = 4\pi b^2 , \tag{1}$$

où b est ce qu'on appelle la longueur de diffusion. La définition de b est que l'onde diffusée par le noyau s'écrit

$$\psi = \exp(\mathrm{i}k_i z) - \frac{b}{r}\exp(\mathrm{i}k_i r) ; \tag{2}$$

b peut être positif ou négatif et cela a bien des conséquences comme on le verra plus loin. Numériquement b est de l'ordre de 10^{-12} cm pour tous les noyaux. Le potentiel d'interaction correspondant à (2) est défini par

$$V(\boldsymbol{r}) = \frac{2\pi\hbar^2}{m_{\mathrm{n}}} b\delta(\boldsymbol{r}) . \tag{3}$$

La section efficace incohérente au contraire varie notablement d'un isotope à un autre produisant un effet de contraste important dont le plus fameux et le plus employé est celui relatif à l'hydrogène et au deuteron; il a été exploité soit pour distinguer les objets organiques dans un solvant (eau légère ou eau lourde) soit pour deuterer ces mêmes objets (deutération partielle ou totale des chaînes de polymères). Cet effet qui est exploité systématiquement est lié au mécanisme de diffusion qui est du type noyau–noyau. Il se distingue radicalement de celui qui gouverne la diffusion des RX, des photons, des électrons ou des phonons qui toujours relèvent de la densité électronique. C'est bien pour cette raison que la substitution isotopique n'a aucune influence sur les RX, qui ont d'ailleurs bien du mal à localiser les atomes pauvres en électrons (cristallographie des protéines).

L'existence du spin du neutron, ou éventuellement la possibilité de travailler avec des neutrons polarisés, montre qu'une deuxième interaction entre en jeu: l'interaction magnétique qui est de nature très classique; c'est une interaction dipolaire magnétique entre le mouvement du neutron et le champ magnétique local $\boldsymbol{H}$ produit par les électrons magnétiques de l'atome diffusant. A la différence de l'interaction noyau–noyau, l'interaction magnétique n'est plus ponctuelle; il s'introduit donc de manière naturelle un facteur de forme qui dans le cas le plus simple est la transformation de Fourier de la densité de spin de l'atome, et qui dans les cas où le magnétisme orbital est mélangé avec le magnétisme de spin, est relativement compliqué. Bref, il est possible de définir, comme précédemment une longueur de diffusion magnétique b_{H}, positive ou négative telle que la longueur de diffusion totale par un atome soit, suivant que le spin du neutron par rapport au spin du noyau est parallèle ou antiparallèle,

$$b_{\text{totale}} = b \pm b_{\mathrm{H}} . \tag{4}$$

La diffusion magnétique des neutrons a des caractères qui la rendent très semblable à celle des rayons X: un facteur de forme dépendant de la distribution de la densité de spin; en optique ou en acoustique, au contraire le facteur de forme entre peu ou pas en ligne de compte dans la mesure où les longueurs d'ondes sont grandes devant les objets diffractants.

Il existe plusieurs autres origines à la diffusion des neutrons qui tiennent compte d'interactions plus faibles comme par exemple le couplage électromagnétique qui résulte du mouvement d'un dipole magnétique dans un champ électrique ou dans un gradient de champ électrique ayant son origine dans la distribution de charges électriques des électrons ou du noyau. Mais il est une dernière interaction que je voudrais mentionner car elle a un caractère assez spectaculaire sur la réalisation des interféromètres à neutrons que je décrirai dans la suite: c'est l'interaction gravitationnelle.

Deux termes entrent en jeu: si V_g est l'interaction du neutron avec le champ de gravité on aura:

$$V_g = mgh + \hbar\Omega\cdot(\boldsymbol{r}\boldsymbol{k})\,.$$

Le premier terme est le potentiel du neutron dans le champ gravitationnel d'accélération g et h la hauteur dans ce champ; le deuxième terme tient à la rotation de la terre: Ω est la vitesse angulaire et $\boldsymbol{r}$ la distance au centre de rotation. Mettons des ordres de grandeur sur ces différentes contributions.

La première, l'interaction nucléaire, donne pour un neutron dans un matériau où N est le nombre de noyaux par unité de volume:

$$V_{\text{Noyau}} \simeq (2\pi\hbar^2/m)Nb\,;$$

on obtient $V_N \simeq 10^{-6}$ eV. L'interaction magnétique V_M est de l'ordre de 10^{-7} eV dans un champ de 10^4 Gauss ce qui rend bien

$$V_M \simeq V_N\,.$$

Le premier terme de V_g pour $h = 10$ cm, ce qui est une distance typique pour faire une expérience, est de l'ordre de 10^{-8} eV alors que le dernier, beaucoup plus petit, vaut 10^{-10} eV.

Une conséquence immédiate de la longueur de diffusion, tout au moins après un petit calcul simple, est l'existence d'un indice de réfraction pour les neutrons. L'analogie avec l'optique ou l'acoustique se poursuit longtemps même si l'on introduit la diffusion magnétique, électromagnétique, gravitationnelle. On montre en effet que l'indice "neutronique" est donné par

$$n = 1 - \lambda^2 Nb_{\text{totale}}/2\,, \qquad (5)$$

où λ est la longueur d'onde des neutrons dans le vide, et N est le nombre de centres diffuseurs indépendants par unité de volume.

Il résulte de (5) que si b_{tot} est positive (ou b seulement, comme c'est le cas pour la majorité des noyaux hormis quelques éléments comme l'hydrogène, le titane ou le manganèse), l'indice est plus petit que 1 et il y a réflexion totale pour les neutrons provenant d'un indice 1 (air ou vide). De toutes façons, l'indice étant très peu différent de 1, l'angle de réflection totale θ_t est donné par

$$\theta_t = (Nb_{\text{tot}}\lambda/\pi)^{1/2}\,, \qquad (6)$$

mais il est très petit par comparaison avec l'optique ou l'acoustique. A titre d'exemple, pour le nickel on a:

$$\theta_c = 1.7 \times 10^{-3}\,\lambda\,, \tag{7}$$

où λ est exprimé en Å. θ_c est donc de l'ordre de 10 minutes pour des neutrons thermiques. Cet effet très simple a été utilisé pour fabriquer des guides de neutrons dont le principe est analogue à celui des guides de lumière ou aux guides à ondes acoustiques de surface. En jouant sur la longueur de diffusion magnétique on peut de même préparer des guides qui transmettent une polarisation de spin et pas l'autre.

3. Interféromètre à neutrons; cohérence

Un interféromètre à champ d'ondes, qu'il s'agisse d'optique, d'acoustique, ou de neutronique combine toujours deux ondes dont l'une sert de référence et l'autre est affectée par une interaction de la particule avec un diffuseur.

Si ψ est l'onde de référence et $V\psi$ l'onde diffusée un interféromètre est composé de sorte que l'on construise une onde $\psi + V\psi$ dont on mesure l'intensité

$$I = \langle \psi \pm V\psi | \psi \pm V \rangle\,.$$

Dans la mesure où l'on peut négliger l'absorption, V est unitaire et peut s'écrire sous la forme

$$V = A \exp(\mathrm{i}\phi)\,.$$

Lorsque ψ est normé convenablement

$$I = 1 \pm \langle A \cos\phi \rangle\,.$$

Mesurer I c'est donc mesurer ϕ ou V.

Deux exemples simples: si l'interaction V correspond à la traversée d'une plaque parallèle d'épaisseur d d'indice n, on a évidemment

$$V = \exp[\mathrm{i}2\pi d(n-1)/\lambda]\,;$$

si l'interaction vient du passage du neutron dans une zone de champ magnétique H, sur une distance d, à la vitesse v, on a:

$$V = \exp\left(\mathrm{i}\boldsymbol{\sigma}\cdot\boldsymbol{H}\cdot\frac{\gamma d}{V}\right),$$

où γ est le rapport gyromagnétique du neutron.

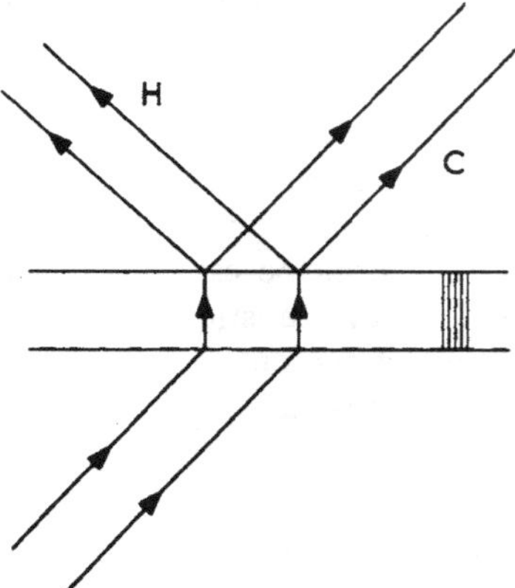

Fig. 1.

En optique ou en acoustique, les indices sont notablement différents de 1 ; par ailleurs, on dispose de sources cohérentes de photons ou de phonons et cela rend très aisé la réalisation d'hologrammes qui incorporent en plus une information sur les propriétés spatiales du diffuseur ; c'est d'ailleurs ce qui les rend attractifs.

En neutronique, comme en RX rien d'analogue n'existe et il faut utiliser une conséquence de la théorie dynamique de la diffraction, l'effet Bormann, pour parvenir à préparer deux faisceaux cohérents séparés spatialement à partir d'une source incohérente.

Lorsqu'un cristal parfait est attaqué par un faisceau de neutrons ou de RX en position de Bragg on montre en effet que dans la géométrie de Laue, l'énergie a tendance à cheminer parallèlement aux plans cristallins et à ressortir de la plaque cristalline sous forme de deux faisceaux cohérents avec incidence de Bragg (voir fig. 1).

En théorie cinématique de la diffraction, il y a deux nappes de dispersion D_1 et D_2 qui se coupent en O : ce sont deux sphères centrées en deux nœuds du réseau réciproque de rayon k_i. Dans le vide ou dans le cristal (en oubliant la différence d'indice qui compliquerait la figure sans amener de physique nouvelle) i_0 est le faisceau incident, h_0 le faisceau réfléchi. Si S est la trace de la surface du cristal, les directions de propagation du rayon sont perpendiculaires à D_1 et D_2. En théorie dynamique, D_1 et D_2 donnent naissance à deux nouvelles nappes (1) et (2). (1) et (2) résultent de D_1 et D_2 par couplage avec la densité nucléaire du cristal dans le cas des neutrons ou avec la densité électronique dans le cas des RX (voir fig. 2).

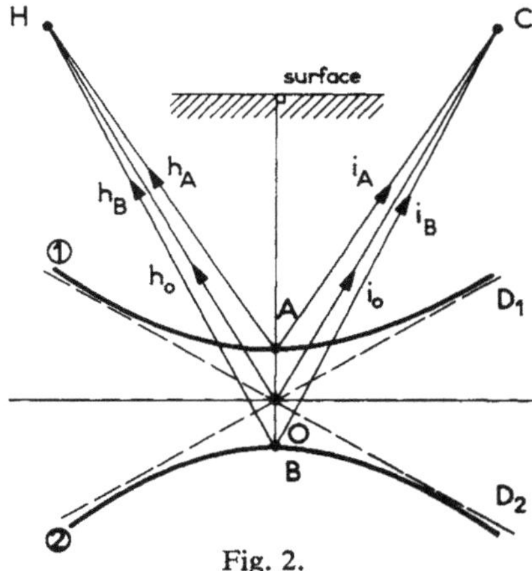

Fig. 2.

Comme les composantes des vecteurs d'ondes parallèles à S se conservent, (invariance par translation le long de S) les vecteurs d'ondes des faisceaux dans le cristal sont dirigés suivant $\boldsymbol{i}_B$, $\boldsymbol{h}_B$, $\boldsymbol{i}_A$, $\boldsymbol{h}_A$; mais les directions de propagation des faisceaux sont perpendiculaires aux nappes (1) et (2) en A et B, c'est-à-dire perpendiculaires à S, c'est-à-dire encore, parallèles au plan des strates.

Sur la face de sortie, les faisceaux $\boldsymbol{h}_A$, $\boldsymbol{h}_B$ et $\boldsymbol{n}_A$, $\boldsymbol{l}_B$ se recomposent pour donner les faisceaux $\boldsymbol{H}$ et $\boldsymbol{C}$ de la figure 1 de même orientation que $\boldsymbol{h}_0$ et $\boldsymbol{i}_0$.

Pour parvenir à observer un tel effet l'angle d'incidence doit être celui de Bragg à quelques secondes d'arc près; c'est aussi la distance angulaire typique entre $\boldsymbol{h}_A$ et $\boldsymbol{h}_B$. Quand on écarte le faisceau incident de la direction $\boldsymbol{i}_0$ on a une oscillation de l'intensité des faisceaux C et H; de même en modifiant l'épaisseur de la lame en incidence $\boldsymbol{i}_0$ on observe des oscillations (pendellösung) dont la périodicité est proportionnelle à la distance AB.

Fabriquer un interféromètre consiste donc à recombiner les faisceaux C et H pour en faire les deux bras; c'est ce qu'on fait au moyen des deux lames (2) et (3) (voir fig. 3). On mesure les intensités des faisceaux i_3 ou h_3 qui résultent des interférences des deux faisceaux cohérents $\overleftrightarrow{h_1 h_2}$ ou $\overleftrightarrow{i_1 i_2}$. Il suffit alors d'interposer dans l'un des bras une différence de "trajet optique" pour modifier l'intensité détectée en h_3 ou i_3 et observer un déplacement des franges de l'interféromètre.

L'art est difficile: il faut éviter la moindre vibration qui brouille les franges, éviter les écarts de température, collimater les faisceaux, travailler à la seconde d'arc sur trois lames successives... Mais on y arrive:

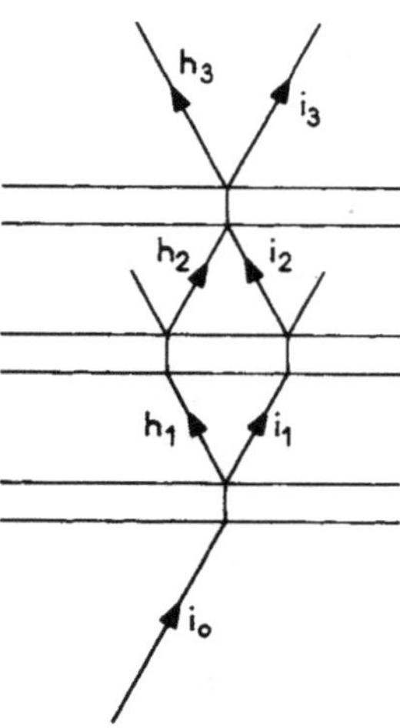

Fig. 3.

"le jeu vaut la chandelle" car l'ordre d'interférence d'un tel instrument est énorme (trajet optique 5 à 10 cm; longueur d'onde 2 Å).

Quelles informations obtient-on? Une première série d'expériences correspond aux mesures précises de longueur de diffusion: en introduisant une lame d'égale épaisseur dans l'un des bras on obtient b, éventuellement b_H si on travaille en neutrons polarisés; le corps étudié peut être à volonté solide, liquide ou gazeux: en particulier pour une détermination ou plutôt pour un test des forces nucléaires élémentaires, il y a un grand intérêt à mesurer correctement les longueurs de cohérence relatives aux noyaux à trois ou quatre nucléons He^3, He^4, T.

Dans le domaine de la physique de la matière condensée, on peut penser à observer l'évolution de la cohérence en fonction de l'inhomogénéité de l'échantillon: au cours d'une précipitation par exemple où il y a en plus une diffusion aux petits angles qui diminue le contraste de l'interférence.

Références

U. Bonse et M. Hart, Z. Physik 188 (1965) 154; 194 (1966) 1.
W. Bauspiess, U. Bonse et H. Rauch, Nucl. Instr. Methods 157 (1978) 495.

AN INFORMAL CRASH COURSE

ON ERGODIC THEORY
(From a topologist's standpoint)

Valentin POÉNARU

Département de Mathématiques, Faculté des Sciences d'Orsay
Université Paris XI, 91405, Orsay, France

Notes written by Françoise Axel and Nicolas Rivier.

Contents

R. Balian et al., eds.
Les Houches, Session XXXI, 1978 – La matière mal condensée/Ill-condensed matter

1. Historical introduction to ergodic theory, and some examples

1.1. Introduction

The subject of these very informal lectures, will be the "statistical behaviour of dynamical systems". We will try to discuss those patterns of behaviour of dynamical systems which correspond roughly to the idea that "any point goes everywhere", or "points are indistinguishable" or "homogeneous chaos". These patterns will be characterized by the adjectives below (which will be properly defined later)

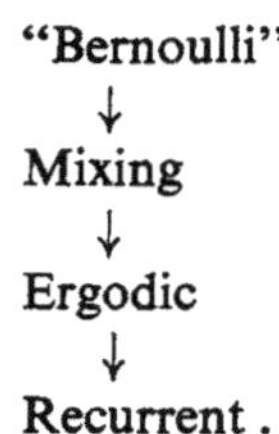

To our dynamical systems, we will attach a numerical invariant "the entropy", which will be defined in two contexts: topology and probability theory.

We use the words "dynamical system" in four ways, schematically represented in the diagram below.

	Continuous time	Discrete time
Classical systems having smooth or continuous properties		
Abstract systems (stochastic models)		

We will give only few proofs but *many* handwaving arguments!

Preliminary remark. The expression "almost all" refers to the Lebesgue measure: that is: "except for a set of measure zero" in the situation of interest. Recall that a set of measure zero is contained in an arbitrarily small open set. "Almost all" refers to something sure in probability theory which cannot see a set of measure 0. At this level there is a sharp contrast between topology and probability theory: Any two (closed, connected) n-manifolds become the same after one removes appropriate sets of "measure 0".

The original aim of ergodic theory was to reconcile the irreversible character of classical thermodynamics with the reversible character of classical mechanics.

Let $\mathscr{H}(q_1, \ldots, q_n, p_1, \ldots, p_n)$ be a time independent hamiltonian. Here, the phase space X is an open set of $2n$ space.

One can write the equations of motion which describe the evolution of the system and represent this evolution by a smooth operator T_t acting on phase space, and such that

$$\left.\begin{aligned} T_0 &= \mathbf{1} \\ T_{t+t'} &= T_t T_{t'} \\ T_{-t} &= T_t^{-1} \end{aligned}\right\} \begin{array}{l} \text{the } T\text{'s form a continuous group} \\ \text{(can be defined for discrete time too)} \\ \text{also called } \textit{dynamical group}. \end{array}$$

From now on, the dynamical system is defined by X, the function $\mathscr{H}(q, p)$ and the group T.

The Liouville measure of a subset A is a volume in phase space

$$\mu(A) = \int_A dq_1 \ldots dq_n \cdot dp_1 \ldots dp_n .$$

This turns out to be invariant by the dynamical group T_t:

LIOUVILLE'S THEOREM. $\mu(T_t A) = \mu(A)$.

Since $\mathscr{H}(p, q)$ does not depend on time the dynamical system will stay on a surface of equal energy $\mathscr{H}(p, q) = c^{st}$ which is a smooth manifold of dim $2n - 1$ of phase space (that is for almost all values of the energy). Suppose also this is compact, which is indeed the case for the ideal gas in a finite container. Then one can redefine as follows the notion of volume:

$$d\omega = d\sigma / |\text{grad}\, \mathscr{H}| , \quad \text{where } d\sigma = \text{“surface element”} ,$$

is invariant by the dynamical group T_t (acting on the equal energy surface).

1.2. Poincaré's theorem and the ergodic hypothesis

Let the surface be $\Omega = \mathscr{H}^{-1}(c^{st})$; we normalize so that $\mu(\Omega) = 1$. The reversible character of mechanics is dramatically epitomized by Poincaré's recurrence theorem or "l'éternel retour":

Let $A \subset X$ have a volume $\mu(A) > 0$. For almost all points in A, $x \in A$, $\exists t$ arbitrarily large, such that $T_t x \in A$ (i.e., if a sufficiently long time is given, one point of phase space will return within a small neighbourhood of its initial position).

Imagine a two compartment container such that at $t = 0$ only one compartment is full of molecules moving at random and the barrier is open. The whole situation at $t = 0$ is just *one point* in the phase space, and what Poincaré's theorem tells us, is that if one waits long enough, there will be a time when most molecules will be back in B. This forms an apparent "paradox": nobody has ever seen this kind of thing happen!

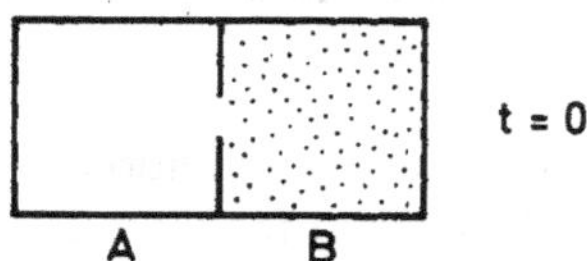

Weak form of Poincaré's theorem. Let X be a space with a probability μ (that is a measure of total measure 1), let T be a transformation: $X \to X$ which is invertible and preserves the measure:

$$\mu(TA) = \mu(A),$$

let $A \subset X : \mu(A) > 0$ then

$$\exists x \in A, \exists n,$$

integer arbitrarily high, such that: $T^n x \in A$.

Proof. Let us form the sequence:

$$A, TA, \ldots, T^{n-1}A, T^n A, \ldots .$$

Because the volume of X is finite and equal to 1 and the volumes of A, TA, etc. are finite, >0, and all equal, there must be at least one non-empty intersection of two elements of the sequence: then,

$$T^p A \cap T^q A \neq 0 \to T^{p-q} A \cap A \neq 0 .$$

In this way we have produced one n such that $T^n A \cap A \neq \emptyset$. But the same argument produces infinitely many such n's. Q.E.D.

Boltzmann's answer to the paradoxes consisted, basically, of the following four points which we will discuss one by one.

(1) The container situation is a highly unusual one, where our system consists of *very many* molecules, and that portion of the phase space which corresponds to "all molecules being in B", has *very small* volume. This is, of course, a very reasonable assumption.

(2) Fix the phase space X, a set $A \subset X$, and some point $x \in X$; let $[0, \tau]$ be a time interval and define $t(x, A, \tau)$ as the total time that x spends in A between instants 0 and τ, that is the measure of the following set: $\{u, T_u x \in A, 0 \leqslant u \leqslant \tau\}$. The following limit (provided it exists) can be thought of as representing the "relative time x spends in A":

$$\lim_{\tau \to \infty} t(x, A, \tau)/\tau .$$

Boltzmann conjectured that this limit does indeed exist. Much later, about 1930, Birkhoff proved that the limit exists for almost all x's. (This is called the *ergodic theorem.*) A slightly weaker version (von Neumann) says that the limit exists in the sense of mean squares. We will give this proof a bit later.

(3) DEFINITION. A system is said to be *indecomposable* or *ergodic* if it has no proper invariant subset; or: given (X, μ), T, our X is indecomposable or ergodic $\Leftrightarrow \nexists A \subset X$: $\mu(A) \neq 0$ and $\mu(A) \neq 1$, such that $T_t A = A$.

Boltzmann made the following

ERGODIC HYPOTHESIS. A perfect gas (many molecules and elastic shocks) is ergodic. (This is a very difficult question which has been proved only recently by Sinaï [6].) Contrast this with the 3-body problem (few particles and no shocks), which is *not* ergodic.

(4) Finally, Boltzmann claimed that (2) + (3) imply the following equality:

$$\lim_{\tau \to \infty} t(x, A, \tau)/\tau = \mu(A) \quad \text{(for almost all } x\text{'s)} .$$

So the average time which x spends in A is equal to the volume of A. In particular if $\mu(A)$ is very small the chance of seeing $T^n x$ in A is very small too. This is Boltzmann's answer to the paradox.

Proof of (4) (granted (2) and (3)). Given X and a probability μ on X (which will mean some Lebesgue type measure normalized so that $\mu(X) = 1$), and $A \subset X$ let us suppose time is discrete and T is an invertible transform preserving μ. Let us define the characteristic function for a measurable set A:

$$\chi_A(x) = \begin{cases} 0 & \text{if} \quad x \notin A \\ 1 & \text{if} \quad x \in A \,. \end{cases}$$

For $x \in X$ we construct the sequence of points in X: $x, Tx, \ldots, T^{n-1}x, T^n x, \ldots$.

Question. Which elements of this sequence belong to A? These are the n's such that $\chi_A(T^n x) = 1$. Let us look at

$$\lim_{n\to\infty} \frac{1}{n+1} [\chi_A(x) + \chi_A(Tx) + \cdots + \chi_A(T^n x)] = h(x) \,.$$

Fix A, and let us assume this limit exists, for almost all x; we call it $h(x)$ but it is not necessarily independent of x. Note that this is exactly Boltzmann's

$$\lim_{\tau\to\infty} \frac{t(x, A, \tau)}{\tau} \quad \text{(in the discrete context)} \,.$$

CLAIM. *This limit is $\mu(A)$ if T is ergodic.* (Lebesgue integral is used from now on.)

We will check convergence later. Once it is granted, one notes the following:

(a) $h(x)$ is measurable which means that $\forall \alpha \in \mathbf{R}$, $h^{-1}[-\infty, \alpha]$ is measurable [this is the set of all the points x such that $h(x) \leq \alpha$].

(b) Apply T to the sequence: and since there are no edge effects $\Rightarrow$

$$h(Tx) = h(x) \,, \quad (h \text{ is "}T\text{-invariant"}) \,.$$

(c) All the sets $h^{-1}[-\infty, \alpha]$ are invariant under T, hence h is (almost everywhere) constant.

Proof of (c). Assume this were not the case, hence there would be two disjoined sets of values for h,$\mathscr{E}$ and $\mathscr{F}$ such that $\mu(h^{-1}(\mathscr{E})) > 0$, $\mu(h^{-1}(\mathscr{F})) > 0$. But

$$h^{-1}(\mathscr{E}) \cap h^{-1}(\mathscr{F}) = h^{-1}(\mathscr{E} \cap \mathscr{F}) = \emptyset \,,$$

and so one would have disjoint proper invariant subsets: that is there would exist a proper invariant subset $\Rightarrow$ no ergodicity $\Rightarrow$ $h(x) = c^{st}$ except for subsets of probability 0. This constant can be computed:

$$\int_{\text{all space } X} h(x) = h_0 \int_X 1 = h_0$$

$$= \int \lim_{n \to \infty} \frac{1}{n+1} [\chi_A(x) + \chi_A(Tx) + \cdots + \chi_A(T^n x)] .$$

But the integral sign can go inside the bracket (Lebesgue), and

$$\int \chi_A(x) = \mu(A) , \qquad \int \chi_A(Tx) = \int \chi_{T^{-1}A}(x)$$

$\Rightarrow$ the limit is $\mu(A)$.

Von Neumann's convergence in quadratic mean. (Birkhoff's theorem is slightly sharper and takes longer to prove.) Let $f, g \in L^2(X, \mu)$, be complex valued, one defines the scalar product

$$\langle f, g \rangle = \int f\bar{g};$$

$L^2(X, \mu)$ is a Hilbert space of L^2 integrable functions. Let $T: X \circlearrowleft$ be a transform of X.

The operation T will be replaced by a unitary operator on $L^2(X, \mu)$, where by definition: $(Uf)(x) = f(Tx)$ (or possibly $f(T^{-1}x)$). Since T preserves the measure, U is a unitary operator (or, in the continuous case $(U_t f)(x) = f(T_t x)$, U_t is a one-parameter-group of unitary operators). Note this passage

$$\begin{pmatrix} T \\ \text{Classical dynamical} \\ \text{system (non-linear)} \end{pmatrix} \xrightarrow[\text{baby ``quantization''}]{} \begin{pmatrix} U \\ \text{linear (unitary) operator} \\ \text{on Hilbert space} \end{pmatrix} .$$

Apply the spectral theorem for unitary operators:

$$U = e^{iA} , \qquad A = \int_{-m}^{+m} \lambda \, dE_\lambda$$

$\Bigg[$or Stone's theorem (for 1-parameter groups)'

$$U_t = e^{itA} , \qquad A = \int_{-\infty}^{+\infty} \lambda \, dE_\lambda \Bigg] ,$$

and recall that the expression

$$\frac{1}{N}\sum_{0}^{N-1} e^{inx} = \begin{cases} 1 & \text{if } x = 0, 2\pi, \text{ etc.} \\ 0 & \text{otherwise}. \end{cases}$$

If one computes the ergodic sum

$$\lim_{n\to\infty} \frac{1}{n+1}\left[\chi_A(x) + \chi_A(Tx) + \cdots + \chi_A(T^n x)\right],$$

outside the eigenvalue $\lambda = 0$ the integrals are oscillating very fast and everything cancels out. Thus:

$$\lim_{N\to\infty} \frac{1}{N+1}\left(\sum_{0}^{N} \chi_A(T^p x)\right) = \lim_{N\to\infty} \frac{1}{N+1}\left[\sum_{0}^{N} U^p \chi_A\right]$$
$$= E_0\chi_A \in L^2(X, \mu).$$

The second limits converge in the Hilbert space sense, i.e. in quadratic mean.

Exercise

Show that T ergodic

$\Leftrightarrow \forall f \in (X, \mu)$

$$\lim_{N\to\infty} \frac{1}{N+1}\left(\sum_{0}^{N} f(T^n x)\right) = \int f(y)\, d\mu(y),$$

and similarly for dynamical system with continuous time. (In some sense this kind of equality is really a version of the law of large numbers. . . .)

Note. Notion of *isomorphism* for abstract dynamical system.

To begin with, there is a notion of *isomorphism* for probability spaces: X^1, X^2. This means there is a 1-to-1 map φ

$$X^1 - \tilde{X}^1 \xrightarrow{\varphi} X^2 - \tilde{X}^2,$$

where $\mu(\tilde{X}^1) = \mu(\tilde{X}^2) = 0$, such that φ (and φ^{-1}) transforms measurable sets into measurable sets and preserves the measure.

According to a theorem of Rohlin all "reasonable" probability spaces (reasonable means without atoms plus some more technical conditions) are isomorphic to [0, 1] (with the Lebesgue measure).

Two measure-preserving transformations are, by definition, isomorphic, if there is an isomorphism in between the underlying probability spaces, carrying one transformation into the other.

Very often, in the sequel, we will not distinguish between isomorphic objects (spaces, transformations, . . .).

1.3. Examples of ergodic and non-ergodic systems

(a) Boltzmann asserted that an ideal gas (of hard spheres in a hard container) is ergodic.

(b) A symmetrical top rotating about its axis of symmetry is non-ergodic.

(c) The two-body and three-body problems are non-ergodic. (The corresponding systems have stability properties which prevent them from being ergodic.)

Roughly speaking, any very stable system is non-ergodic. *A system which has an invariant function apart from the energy is non-ergodic.* But it is possible to restate the problem on a phase space of lesser dimensionality in this case, and the new problem might be ergodic.

(1) *Recurrence.* (Warning: recurrence can be a purely topological notion, but we interpret it here probabilistically.)

(Mild) recurrence: $\forall A$ such that $\mu(A) > 0$, and *almost all* $x \in X$,

$$\exists n \quad \text{such that} \quad T^n x \in A\,.$$

Remark. Ergodic $\Rightarrow$ recurrent (*the converse is not true*). (Suppose (ad abs.) $\exists \mu(A) > 0$ and $\mu(B) > 0$ such that $\forall x \in B$, $\forall n T^n x \notin A$. Then $\bigcup_{-\infty}^{+\infty} T^n B$ is a proper invariant set, etc.)

(2) *Mixing.* Expresses the notion of independence in a probability space (statistical independence)

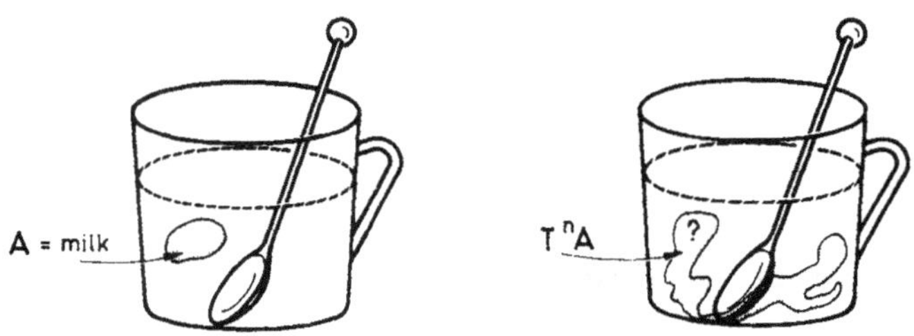

Start with 10% of milk. Stir. After n stirs (n large enough), expect physically that the proportion of milk in *any* part B of the cup be 10%.

Let us recall the notion of *statistical independence*:

with (X, μ) and $A, B \subset X$.

DEFINITION. $\mu(A \cap B) = \mu(A)\mu(B) \Leftrightarrow$ "A and B are independent". This means:

$$\frac{\mu(A \cap B)}{\mu(B)} = \mu(A).$$

(That is, the probability of finding A inside B equals the probability of finding A in general.)

DEFINITION. A dynamical system is *mixing* if, $\forall A, B$

$$\lim_{n = \infty} \mu(T^n A \cap B) = \mu(A)\mu(B).$$

PROPOSITION. Mixing $\Rightarrow$ ergodic (the converse is not true).

Proof. Suppose (ad abs.) T *not* ergodic $[X = A \cup B,\ A \cap B = \emptyset,$ $\mu(A) > 0$, $\mu(B) > 0$, and $T^n A = A]$. Then, $\mu(T^n A \cap B) = 0$, whereas

$$\mu(A)\mu(B) \neq 0. \qquad \text{Q.E.D.}$$

Exercise

Rotate a circle by $2\pi\theta$. Show that if θ is irrational, the transformation is ergodic. [(*Hint*. Use Fourier analysis.) T is *not* mixing:

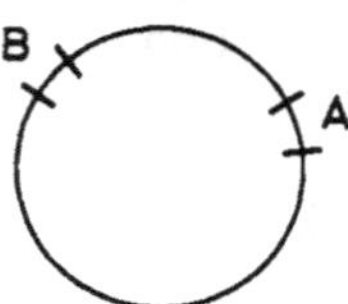

sometimes $T^n A \cap B = 0$, sometimes $\mu(T^n A \cap B) > \varepsilon > 0$ for fixed $\varepsilon > 0$.]

1.4. More on discrete dynamical systems on the circle

Although the circle is, topologically speaking, a trivial object, dynamics on the circle is a highly non-trivial topic.

Let $T: S^1 \to S^1$ be an orientation-preserving C^∞-diffeomorphism of the circle. We can lift T to the universal covering of S^1

$$\begin{array}{ccc} R & \xrightarrow{\tilde{T}} & R \\ {\scriptstyle p}\downarrow & & \downarrow{\scriptstyle p} \\ S^1 & \xrightarrow{T} & S^1 \end{array}, \quad \text{where } p(x) = e^{2\pi i x},$$

Observe that $\tilde{T}(x + 1) = \tilde{T}(x) + 1$, hence

$$\tilde{T}(x) = x + \varphi(x),$$

where $\varphi(x)$ is *periodic.* We can consider φ as being a smooth function on S^1, which will be denoted by Φ.

Poincaré proved that the following limit exists and is independent of x:

$$\lim_{n\to\infty} \frac{1}{n} \tilde{T}^n(x) = \rho(T), \quad (x \in R).$$

This is called the *rotation number of* T, and it is well-defined in R/Z. It can be easily shown that $\rho(T)$ is, in fact, the following ergodic sum:

$$\rho(T) = \lim \frac{1}{n} \sum_{k=0}^{n-1} \Phi(T^k(y)), \quad (y \in S^1)$$

$$= \lim \frac{1}{n} \sum_{k=0}^{n-1} (\tilde{T}^{k+1}(x) - \tilde{T}^k(x)).$$

So $\rho(T)$ is somehow the "average step of $\tilde{T}$".

Exercises

(1) If $T = R_\alpha$ (which denotes the rotation of angle $2\pi\alpha$), then $\rho(T) = \alpha$.

(2) If $d\mu$ is a T-invariant measure on S^1, then

$$\rho(T) = \int_{S^1} \Phi(y)\, d\mu(y).$$

Poincaré also proved that $\rho(T)$ is rational iff T has periodic points. Later Denjoy showed that if $\rho(T) = \alpha \in$ irrationals, then T is *topologically* conjugate to R_α. This means that there exists a homeomorphism $h = h_T$ such that $T = h \circ R_\alpha \circ h^{-1}$. As Denjoy showed, h is not even necessarily of class C^1 (although T and R_α are C^∞).

Very recently Herman proved the following conjecture of Arnold: for *almost* any irrational $\alpha \in R/Z$, a diffeomorphism T such that $\rho(T) = \alpha$ is *smoothly* conjugate to R_α.

One of the intermediate steps in Herman's proof deserves special attention. Let T be such that $\rho(T)$ = irrational. Herman showed that there is no proper T-invariant *Lebesgue-measurable* subset of S^1.

This is an ergodic theorem in an unusual setting, since the Lebesgue-measure itself is *not* T-invariant (unless T is a rotation). One should not think that ergodic theory has to be necessarily confined to measure preserving transformations (although this is, of course, the most usual setting for it).

2. A lecture on abstract dynamical systems

2.1. Some definitions

(1) *Probability space.* It is a set X where we have chosen a collection Σ of subsets which are the *events*

$$\Sigma = \{\text{collection of subsets of } X\} \quad \text{(it does not necessarily include all the subsets)}.$$

such that

- if $A \in \Sigma$ then $C_A \in \Sigma$ (where C_A is $X - A$);
- if $A_1, \ldots, A_n, \ldots$ is a (finite or countable) sequence, then

$$\bigcup_i A_i \in \Sigma, \qquad \bigcap_i A_i \in \Sigma.$$

Σ thus defined is a *σ-algebra* of sets.

Now add a probability law defined as follows:

- $$\Sigma \supset A \to \mu(A) \in [0, 1] \quad \text{with} \quad \mu(\phi) = 0, \quad \mu(X) = 1,$$
- if $A_1, A_2, \ldots, A_n, \ldots$ is an, at most, infinitely countable sequence such that $A_i \cap A_j = \phi$ when $i \neq j$, then

$$\mu\Big(\bigcup_i A_i\Big) = \sum_i \mu(A_i).$$

Then:

Axiom. If A is any set in X and $A \subset B$ with $B \in \Sigma$ and $\mu(B) = 0$

$$\Rightarrow A \in \Sigma \quad \text{and} \quad \mu(A) = 0$$

(in particular A is an acceptable event).

What is the classical model behind this?

Let $X = [0, 1]$ and Σ_0 a slightly smaller collection of subsets of X, called the Borel sets. Σ_0 is anything which contains open and closed sets, stable with respect to countable $\cap$, countable $\cup$, and C.

This collection is not big enough: one has to add what Lebesgue has called sets of measure 0. That is: add any set $E \subset A$ such that A is Borel and $\mu(A) = 0$ (and define $\mu(E) = 0$).

(2) *Independent events.* Let A and B be events, that is measurable sets. They are called "independent" if

$$\mu(A \cap B) = \mu(A)\mu(B)\,.$$

(3) *The notion of isomorphism of probability spaces.* Let us have some probability space X^1, with Σ^1 and μ^1, and another probability space X^2, with Σ^2 and μ^2.

They are *isomorphic* if

$$\exists A_1 \subset X^1 \quad \text{such that} \quad \mu(A^1) = 0$$

$$\exists A_2 \subset X^2 \quad \text{such that} \quad \mu(A^2) = 0\,,$$

and there is a one-to-one correspondence

$$X^1 - A^1 \xrightarrow{\varphi} X^2 - A^2\,,$$

which preserves the measures μ_1 and μ_2, and transforms (acceptable) sets from Σ^1 into acceptable sets of Σ^2 so that $\mu^1(A) = \mu^2(\varphi A)$.

In fact there is only one "reasonable" probability space, it is isomorphic to [0, 1] with Lebesgue measure. ("Reasonable" includes a number of conditions like nonexistence of atoms; and our statement corresponds to a theorem of Rohlin.) So there is, basically, *only one probability space*. But this space supports an infinitely rich variety of dynamical systems.

Remark. No Topology is involved here.

(4) (*Finite*) *random variable.* It is a function f defined on [0, 1] which takes finitely many distinct values $(a_1, \ldots, a_n)$

$$[0, 1] \xrightarrow{f} (a_1, \ldots, a_n)\,,$$

such that $\{f^{-1}(a_i)\}$ is a measurable set. So this is just a finitely-valued Lebesgue measurable function.

(5) *Independent random variables.* We consider here only finite-valued functions. Then:

$$f\colon [0,1] \to (a_1, \ldots, a_n)$$

$$g\colon [0,1] \to (b_1, \ldots, b_m)\,,$$

are independent if $f^{-1}(a_i)$ and $g^{-1}(b_j)$ are independent for all i and j. Then

$$\int f \cdot g = \int f \cdot \int g\,.$$

2.2. Tossing the coin: the oldest model for "Bernoulli"

Let us take a coin with faces H and T ("heads" and "tails") and assume that time is discrete and goes only one way. We have:

$$t = 1 \qquad 2 \qquad 3 \qquad 4 \quad \ldots$$

tossings are

$$\begin{array}{cccc} a_1 & a_2 & a_3 & a_4 \\ = H \text{ or } T\,, & H \text{ or } T\,, & H \text{ or } T\,, & \ldots\,, \end{array}$$

hence

$$X = \{a_1, a_2, \ldots, a_n, \ldots\}$$

and

$$a_i \in \{H, T\}\,.$$

Call $H_i(T_i)$ the "cylindrical" set of X defined by $a_i = H(T)$.

The following conditions define the model:

(a) The coin is *fair*, consequently for the measure:

$$\mu(H_i) = \mu(T_i) = \tfrac{1}{2}\,;$$

(b) The tossings are *independent* (see definition above), for example if at moment i, I have H; if at moment j, I have T; if at moment k, I have H; then

$$\mu(H_i, T_j, H_k) = \tfrac{1}{8}\,.$$

More generally, a subset of X where finitely many tossings (l) have been fixed is a cylindrical set with l coordinates fixed: its measure (or probability) is $(\frac{1}{2})^l$.

(c) Define the smallest σ algebra that contains the cylindrical sets (as defined above), stable for $\cap$, $\cup$, and $\complement$, let us call it Σ_0. As seen above, we must add to Σ_0 certain sets of measure 0, which are not a priori in it: they are sets A such that: $A \subset B$, $B \in \Sigma_0$, and $\mu(B) = 0$; then $\mu(A) = 0$, $A \in \Sigma$ and A is acceptable.

Then:

THEOREM. This model (X, μ) is isomorphic to $[0, 1]$ with the Lebesgue measure.

The *proof* runs as follows: let $t \in [0, 1]$. We shall write it in binary notation:

$$t = \frac{\varepsilon_1(t)}{2} + \frac{\varepsilon_2(t)}{2^2} + \cdots + \frac{\varepsilon_n(t)}{2^n} + \cdots$$

[$\varepsilon_n(t)$ is the nth digit of t in binary notation].

LEMMA. The $\varepsilon_n(t)$ are independent random variables ("$\forall i \forall j \varepsilon_i(t)$ and $\varepsilon_j(t)$ are independent random variables" (see definition above)).

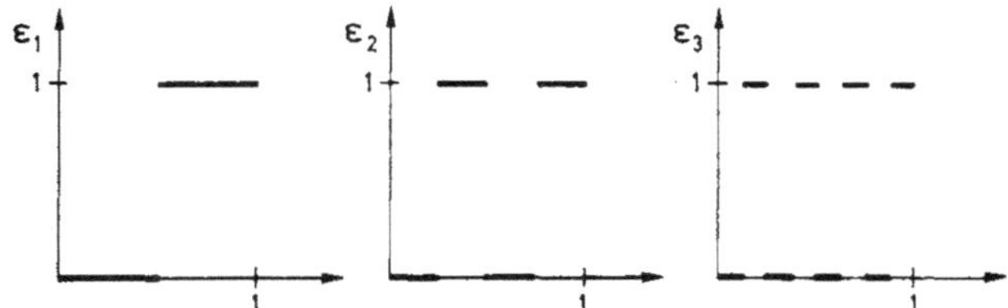

We see that $\varepsilon_1: [0, 1] \to (0 \text{ or } 1)$ is a random variable; the measure of $\varepsilon_1^{-1}(0)$ and $\varepsilon_1^{-1}(1)$ exists and is $\frac{1}{2}$ in both cases.

ε_1 and ε_2 are independent. For example $\varepsilon_1^{-1}(0)$ and $\varepsilon_2^{-1}(0)$ are independent because $\mu[\varepsilon_1^{-1}(0) \cap \varepsilon_2^{-1}(0)] = \frac{1}{4} = \mu[\varepsilon_1^{-1}(0)]. \mu[\varepsilon_2^{-1}(0)]$, etc., ... More generally, let

$$H_i = \text{subset of } [0, 1], \quad \text{where} \quad \varepsilon_i(t) = 0,$$

$$T_i = \text{subset of } [0, 1], \quad \text{where} \quad \varepsilon_i(t) = 1.$$

We can make the following dictionary, where a number in binary notation corresponds to the tossing of the coin (to a point in X)

$$0 \leftrightarrow H$$

$$1 \leftrightarrow T$$

$$\varepsilon_i \leftrightarrow a_i$$

$$[0, 1] \supset H_i \leftrightarrow H_i \subset X$$

$$[0, 1] \supset T_i \leftrightarrow T_i \subset X.$$

The sets of H_i, T_i of $[0, 1]$ *or* of X determine *all* the other measurable sets and their measures exactly in the same manner. This allows us to continue our dictionary, as follows:

measurable set $\leftrightarrow$ event

(Lebesgue) measure $\leftrightarrow$ probability .

2.3. *"Bernoulli"*

The space of "Bernoulli" is defined very much like the space of coin tossings, except that "time" runs now through all the values..., -2, -1, 0, 1, 2,... (of course the difference is only notational).

2.3.1. *Preliminary remark*

Let X be a probability space (X, μ) and T an isomorphism of X into itself. (That is an invertible measure-preserving transformation T; this situation corresponds to an abstract dynamical system with discrete time.) In analogous fashion, give (Y, ν) and U. As usual, all is well defined except for some sets of measure 0.

Recalling the notion of isomorphism of probability spaces given above, we shall say that the transformations T and U are *isomorphic* if we can find isomorphisms which are identical on the vertical lines and one has:

$$\begin{array}{ccc} (X,\mu) & \xrightarrow{T} & (X,\mu) \\ {\scriptstyle I}\downarrow & & \downarrow{\scriptstyle I} \\ (Y,\nu) & \xrightarrow{U} & (Y,\nu) \end{array} \quad \text{hence} \quad T = I^{-1}UI .$$

From now on, we shall disregard differences between two isomorphic dynamical systems.

2.3.2. *Definition of "Bernoulli"*

Let us give ourselves a "probability vector" $(p_1, p_2, \ldots, p_n)$:

$$0 \leqslant p_i \leqslant 1 , \qquad \sum_i p_i = 1 ,$$

and n abstract symbols $a_1, a_2, \ldots, a_n$ (these correspond to a number n of states). The space will be

$$X = \{\text{all the sequences}, \ldots, x_{-2}, x_{-1}, x_0, x_1, x_2, x_3, \ldots, x_l, \ldots\} ,$$

where $x_i \in \{a_1, a_2, \ldots, a_n\}$.

Let us fix x_i to be a_j ($x_i = a_j$). Then, by definition, the probability for the event $x_i = a_j$ is p_j that is $\mu(x_i = a_j) = p_j$ (in other words, the coin is now biased by the p_j's and has n faces).

By definition of the probability space (X, μ) [of the Bernoulli transformation $T = B(p_1, p_2, \ldots, p_n)$], such events are independent that is

$$\mu(x_{i_1} = a_{i_1}, x_{i_2} = a_{i_2}, \ldots, x_{i_k} = a_{i_k}) = p_{i_1}, p_{i_2}, \ldots, p_{i_k} .$$

(This corresponds to the idea that the various tossings of our n-faced unfair coin are *independent*.)

Now, once the measure for cylindrical sets has been established we can construct all the σ-algebras of measurable sets (and their measures). The Bernoulli transformation $B(p_1, \ldots, p_n) = T$ is a transformation $X \to X$ that shifts indices by one: it is *the flow of time.*

2.3.3. Relationship between the baker's transform and "Bernoulli"

(1) Let us recall schematically the steps of the baker's transform: it is a transformation of the square onto itself (not well defined on some sets of measure 0)

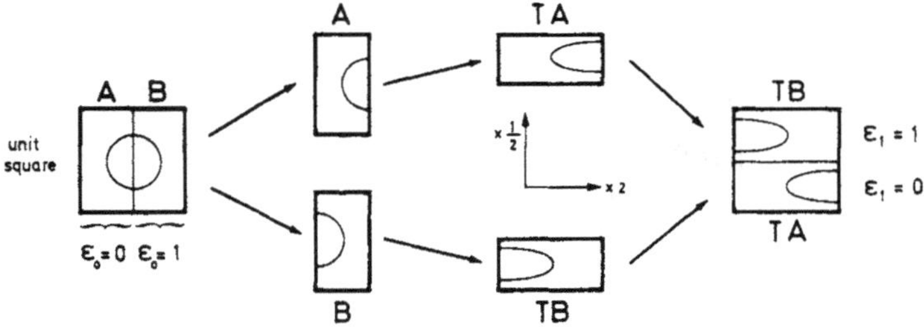

This baker's transform T is isomorphic to $B(\frac{1}{2}, \frac{1}{2})$.

Sketch of proof.

t ↑ → s

Let us write the point $x \in$ unit square in binary digits ε_i in this coordinate system. Then

$$s(x) = 0\,, \quad \varepsilon_0\varepsilon_{-1}\varepsilon_{-2}\varepsilon_{-3}\cdots$$

$$t(x) = 0\,, \quad \varepsilon_1\varepsilon_2\varepsilon_3\cdots .$$

Let X be the space of $B(\frac{1}{2}, \frac{1}{2})$ and

$$f\colon X \ni x \mapsto (s(x), t(x)) \in \text{unit square}$$

f is an isomorphism of $(X, \mu) \mapsto$ (square, Lebesgue measure)

(cf. the relationship between coin tossing and Lebesgue measure seen above).

$B(\frac{1}{2}, \frac{1}{2})$ is a transformation of (X, μ) into itself, the baker's transform is a transform of the unit square into itself. One has:

$$s(Tx) = 0, \varepsilon_1\varepsilon_2 \cdots$$

$$t(Tx) = 0, \varepsilon_0\varepsilon_{-1}\varepsilon_{-2} \cdots,$$

that is:

$$\text{baker of } \begin{pmatrix} 0, \varepsilon_0\varepsilon_{-1}\varepsilon_{-2}\cdots \\ 0, \varepsilon_1\varepsilon_2\cdots \end{pmatrix} \xrightarrow[\text{is the point}]{} \begin{pmatrix} 0, \varepsilon_{-1}\varepsilon_{-2}\varepsilon_{-3}\cdots \\ 0, \varepsilon_0\varepsilon_1\varepsilon_2\cdots \end{pmatrix}.$$

We see that with our example

A | B		
	ε_0 means first digit in $s \rightarrow$	$\varepsilon_0 = 0$ for A $\varepsilon_0 = 1$ for B
$\uparrow$ $\varepsilon_0 = 0$ $\uparrow$ $\varepsilon_0 = 1$	ε_0 also means first digit in t after transformation $\uparrow$	$\varepsilon_0 = 1$ for TB $\varepsilon_0 = 0$ for TA.

You can check that the same is true for the ε of any index so what we have found is that our isomorphism f carries T into the baker's transformation:

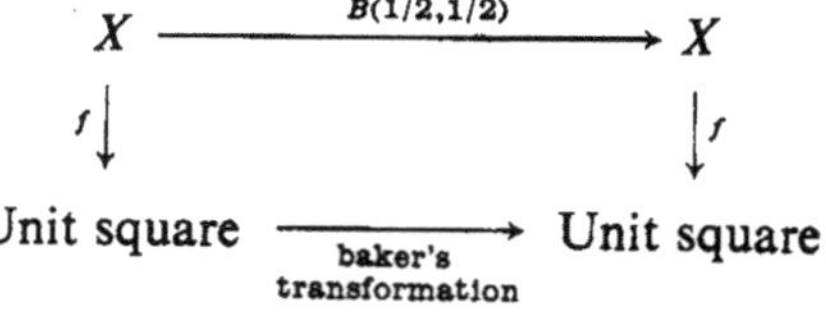

(2) *Generalized baker's transform.* Suppose I have $B(p_1, \ldots, p_n)$. It can be *realized* geometrically as follows, by a generalized baker's transformation

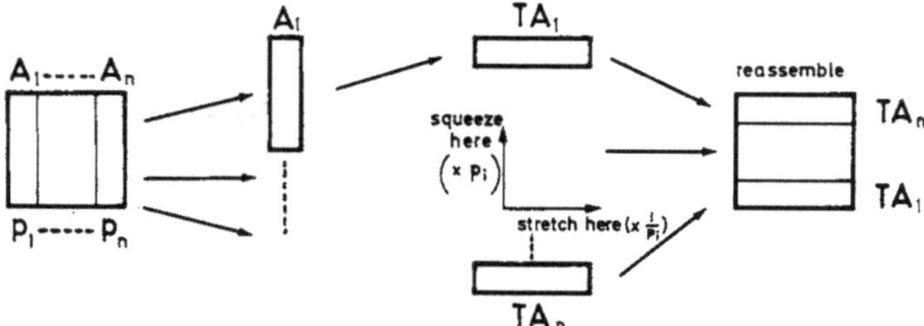

Recall that underlying space of Bernoulli is always isomorphic to [0, 1] Lebesgue interval.

THEOREM. Bernoulli is mixing, hence it is ergodic.

Proof. Let us take two sets A and B; their *symmetrical difference* is defined as

$$A \triangle B = (A - B) \cup (B - A) .$$

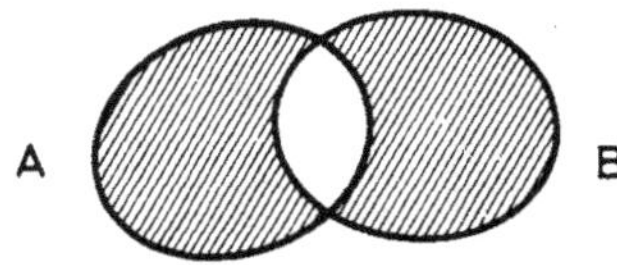

If $\mu(A \triangle B)$ is small, then A and B are "almost" the same.

Let X be the underlying space of the Bernoulli shift. Let $A, B \subset X$, and both be measurable. We want to show that:

$$\lim_{n\to\infty} \mu(T^n A \cap B) = \mu(A)\mu(B) .$$

Any measurable set can be approximated as closely as one wishes by a finite union of cylindrical sets (see above). Approximated means that chosen ε and given A, I can find a set C_1 which is a finite union of cylinders such that $\mu(A \triangle C_1) < \varepsilon$:

$$\forall \varepsilon > 0 , \quad \exists C_1 : \mu(A \triangle C_1) < \varepsilon .$$

We can do the same thing for B. That is we have

C_1 approximates A better than ε, i.e. $\mu(A \triangle C_1) < \varepsilon$

C_2 approximates B better than ε, i.e. $\mu(B \triangle C_2) < \varepsilon$

hence for any integer n:

$$\mu(T^n A \triangle T^n C_1) < \varepsilon$$

(T maps everything, preserving the triangle sign $\triangle$ and the measure). Since the C_i are a finite union of cylinders, that is, only a finite number of fixed coordinates come in, for a high enough power of T these coordinates can be made disjoint. This means:

$$\exists N , \quad \forall n > N : T^n \text{ (coordinates of } C_1) \cap \text{(coordinates of } C_2) = \emptyset .$$

Then: if $n > N$, $T^n C_1$ *and* C_2 *are independent*, that is

$$\mu(T^n C_1 \cap C_2) = \mu(C_1)\mu(C_2) .$$

Let $n > N$, then

$$|\mu(T^n A \cap B) - \mu(A)\mu(B)| \leqslant |\mu(T^n A \cap B) - \mu(T^n C_1 \cap C_2)| + |\mu(A)\mu(B) - \mu(C_1)\mu(C_2)| .$$

The second term is less than 2ε, and one can prove that the first is less than 2ε too:

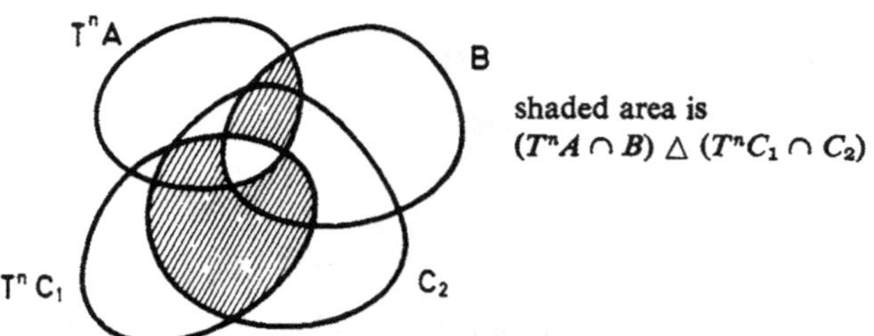

The drawing shows that

$$(T^n A \cap B) \triangle (T^n C_1 \cap C_2) \subset (T^n A \triangle T^n C_1) \cup (B \triangle C_2) .$$

Hence

$$\begin{aligned} |\mu(T^n A \cap B) - \mu(T^n C_1 \cap C_2)| &\leqslant \mu[(T^n A \cap B) \triangle (T^n C_1 \cap C_2)] \\ &\leqslant \mu(T^n A \triangle T^n C_1) + \mu(B \triangle C_2) \\ &\leqslant 2\varepsilon . \end{aligned}$$

Finally:

$$\forall \varepsilon, \exists N, \forall n > N, \quad |\mu(T^n A \cap B) - \mu(A)\mu(B)| \leqslant 4\varepsilon .$$

Hence T is mixing, hence it is ergodic. Q.E.D.

To summarize. The Bernoulli shift:

- n-faced unfair coin with independent tossings, and probabilities $(p_1, p_2, \ldots, p_n)$ $\sum_i p_i = 1$;
- a geometrical model is the generalized baker's transform;
- it is a discrete transformation = the flow of time.

2.4. Bernoulli processes as a special case of Markov processes

(1) *Definition of a Markov process* or "how do letters follow each other in a Russian poem?". Let us give ourselves "states" or boxes $a_1, \ldots, a_n$ (for example a molecule will jump from box a_i to box a_j) and transition

probabilities p_{ij}; $(0 \leqslant p_{ij} \leqslant 1)$ p_{ij} is the probability to pass from state a_i to state a_j. What are reasonable conditions on this idea?

CONDITION 1. $\forall_i \sum_j p_{ij} = 1$.

This summarizes the following intuitive statements:

- "from a_i one has to go somewhere";
- "if $j \neq k$, the passages $a_i \to a_j$, $a_i \to a_k$ cannot occur at the same time".

CONDITION 2. One gives oneself (what is in fact a stationary state of the situation) a set of numbers

$$p_1 \cdots p_n\,, \quad 0 \leqslant p_i \leqslant 1\,, \qquad \sum_i p_i = 1\,,$$

such that $\forall_j \sum_i p_i p_{ij} = p_j$ (this is the stationarity condition). (This means that if we assign to the state a_i probability p_i to occur, this is an equilibrium situation with respect to our law of transition.)

What is the probability that two given letters will follow one another in a poem by Pushkin? Let us consider the following model:

$$X = \{\ldots, x_{-1}, x_0, x_1, x_2, \ldots\}\,,$$

where $x_i \in (a_1, \ldots, a_n)$; let us think of a point in X as being a "molecule" (or "letter") which *at time i is in the state x_i*.

Let the initial state be a_{i_1}, there are transitions to $a_{i_2}, a_{i_3}, \ldots, a_{i_p}$. Let $E_m\,(a_{i_1}, a_{i_2}, \ldots, a_{i_p})$ be the cylindrical set defined by the conditions

$$x_m = a_{i_1}\,, \qquad x_{m+1} = a_{i_2}, \ldots, x_{m+p-1} = a_{i_p}\,.$$

We define a probability (measure) μ on X as follows:

$$\mu(E_m(a_{i_1}, \ldots, a_{i_2}, \ldots, a_{i_p})) = p_{i_1} p_{i_1 i_2} p_{i_2 i_3}, \ldots, p_{i_{p-1} i_p}\,,$$

meaning: our system is at time m in a_{i_1}, then $a_{i_1} \to a_{i_2}$, etc., ..., then $a_{i_{p-1}} \to a_{i_p}$ and these events are *independent*.

We define $T : X \to X$ as for the Bernoulli process (this is just: "going into the future by one unit of time"), T is a measure preserving isomorphism.

(2) Bernoulli is a special case of Markov, with $p_{ij} = p_j$ (the probability of going from $a_i \to a_j$ depends *only* on a_j).

Markov is not necessarily ergodic. The possible ergodicity of Markov is characterized by the fact that "any state communicates with any other state"

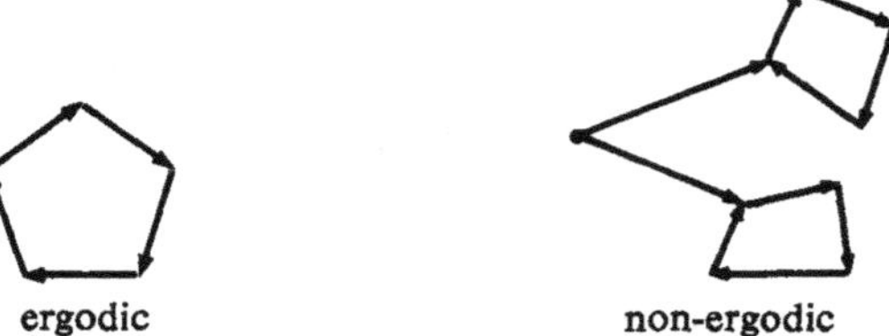

This can be seen directly by looking at the matrix $\{p_{ij}\}$. It is a "discrete property" of the matrix.

(3) THEOREM (without proof) (Ornstein).

Markov + mixing ↔ Bernoulli.

Bernoulli flows. By definition, an abstract flow $(X, \mu) \supset T_t$ is a "Bernoulli flow" if every individual transformation is Bernoulli. Anticipating on the next section: by use of entropy methods, Ornstein proved that (up to isomorphism and change of time-unit).

"*There is only one Bernoulli flow*". (We will see that there are infinitely many Bernoulli shifts.) In the next chapter this (unique) Bernoulli flow will be presented in the shape of the geodesic flow on a negatively curved manifold.

2.5. Entropy

It can be defined in a probabilistic context (then it is connected with the usual thermodynamics) and in a topological context.

Reference (for thermodynamical interpretation): Wehrl Rev. Mod. Phys. (April 1978).

2.5.1. *Quantity of information in a message* (Shannon, 1948)

(1) Let X be a probability space.

If somebody says "event A has happened" what amount of information does one get?

Assume that this amount of information depends only on the probability of the event A, that is:

I("event A has happened") $= I(p)$, then:

(a) $I(0) = \infty$.

(b) $I(1) = 0$.

(c) Since $p(A \cap B) = p(A) + p(B)$,
if A and B are independent, then $I(p \cdot q) = I(p) + I(q)$.

$\Rightarrow$ I is a logarithmic function $I(p) = -\log_2 p$ (up to a choice of unit).

(2) Entropy for a finite partition of (X, μ).
A finite partition $\mathscr{P}$ of (X, μ) is breaking X into a union of disjointed events:
$X = A_1 \cup A_2 \cup \cdots \cup A_n$
$A_i \cap A_j = \emptyset$
$\mu(A_i) = p_i$, $\sum p_i = 1$.
By definition the entropy of $\mathscr{P}$ is $H(\mathscr{P}) = -\sum_i p_i \log_2 p_i$ (=the average amount of information) if $p_i = 0$ we consider the product $p_i \log_2 p_i$ to be zero)

- if n is fixed, H is maximum when
  $p_1 = p_2 = \cdots = p_n = 1/n$;
- if X is broken into two pieces, one being $\emptyset$, one being X then
  $H = 0$ (or: if any of the $p_i = 1$).

Remark that *all this is static*: the partition is fixed, nothing refers to a dynamical system.

Now we want to define entropy for a continuous transformation in the topological contexts; afterwards we come back to the stochastic context.

2.5.2. Topological entropy

Let X be a topological space, take it compact for simplicity, and define a metric: hence there exists a distance. Consider a continuous transformation

$$T : X \to X,$$

we want to attach its "topological entropy" $h(T)$ to T, that will measure how complicated the *dynamics* of T is

(1) Let $\mathscr{U}$ be a finite open covering of X

$$X = \mathscr{U}_1 \cup \mathscr{U}_2 \cup \mathscr{U}_3 \cdots \cup \mathscr{U}_p$$

$\mathscr{U} = (\mathscr{U}_1, \ldots, \mathscr{U}_p)$, where $\mathscr{U}_i$ is open.

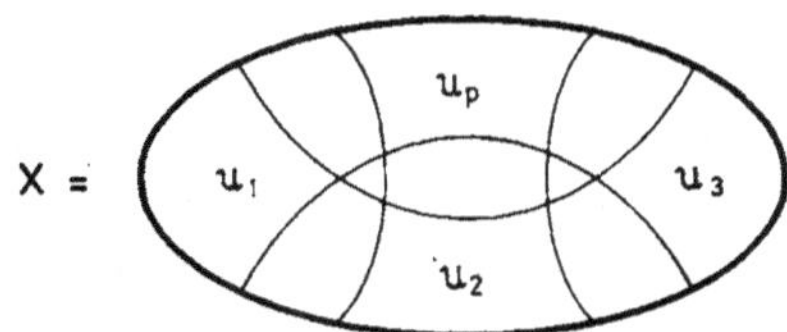

Consider $T^0, T^1, T^2, \ldots, T^{n-1}$ and let $x \in X$ be any given point of X; in time it has an "n-history"; that is, $x = T^0 x \in \mathscr{U}_{i_0}$, $Tx \in \mathscr{U}_{i_1}$, $T^2 x \in \mathscr{U}_{i_2}$, $\ldots$, $T^{n-1}x \in \mathscr{U}_{i_{n-1}}$ where the $\mathscr{U}_{i_k}$'s are not uniquely defined (this is the trajectory of x). Note that x is such that

$$x \in \mathscr{U}_{i_0} \cap T^{-1}\mathscr{U}_{i_1} \cap T^{-2}\mathscr{U}_{i_2} \cap \cdots \cap T^{-(n-1)}\mathscr{U}_{i_{n-1}},$$

this is the generic element of a new open covering which we denote by $\mathscr{U} \cap T^{-1}\mathscr{U} \cap \cdots \cap T^{-(n-1)}\mathscr{U}$.

$\mathscr{U}_{i_0}\mathscr{U}_{i_1}\cdots\mathscr{U}_{i_{n-1}}$ is a word describing the history of point x. Let $N(\mathscr{U}, T, n)$ be the minimum number of words (sequences) of the form $\mathscr{U}_{i_0}\mathscr{U}_{i_1}\cdots\mathscr{U}_{i_{n-1}}$ sufficient to describe *all* n-histories from 0 to $n - 1$, for *all* $x \in X$. [Another way of putting things is to say, for a covering $\mathscr{U}$ we call $N(\mathscr{U})$ the minimal number of elements in $\mathscr{U}$ sufficient to cover X then

$$N(n, \mathscr{U}, T) = N(\mathscr{U} \cap T^{-1}\mathscr{U} \cap \cdots \cap T^{-(n-1)}\mathscr{U}).]$$

Let p be the number of sets in the original covering of X, it is easy to imagine the worst case

$$N(\mathscr{U}, T, n) \leqslant p^n = 2^{n \log_2 p}$$

one usually will expect something like

$$N(\mathscr{U}, T, n) \sim 2^{\lambda n} \quad \text{(asymptotically)}.$$

Such a λ is by definition

$$\lambda = \overline{\lim_{n\to\infty}}\, (1/n) \log_2 N(\mathscr{U}, T, n) = h(T, \mathscr{U}),$$

which is the "*entropy of T with respect to the partition* $\mathscr{U}$" (the bar denotes the upper limit).

What is $h(T, \mathscr{U})$ measuring? If I take "$2^{h(T\mathscr{U})}$ symbols", then words of length n written with these symbols are asymptotically enough to represent all histories on time interval $(0, n - 1)$. Or, another way to put it: binary words of "length hn" represent "roughly" all histories on $(0, n - 1)$.

(2) Topological entropy of T. This was for a fixed covering. If I take finer and finer coverings, then by definition

$$h(T) = \sup_{\mathscr{U}} h(T, \mathscr{U}).$$

THEOREM. Let $\mathscr{U}_1, \mathscr{U}_2, \ldots, \mathscr{U}_n, \ldots$ be a sequence of covers, and denote by $|\mathscr{U}_i|$ of a cover to be the largest diameter of an open set belonging to $\mathscr{U}_i$. It can be shown that if $|\mathscr{U}_i| \to 0$, the sequence $h(T, \mathscr{U}_i)$ converges to $h(T)$ (note that this limit is not necessarily finite).

2.5.3. Back to probability theory

Let now $(X, \mu) \rightleftarrows T$ be an invertible measure-preserving transformation. If $\mathscr{P}$ is a partition of (X, μ) one defines the (probabilistic) entropy of T with respect to $\mathscr{P}$:

$$h(\mathscr{P}, T) = \overline{\lim}(1/n)H(\mathscr{P} \cap T^{-1}\mathscr{P} \cap \cdots \cap T^{-(n-1)}\mathscr{P}),$$

and the (probabilistic) *entropy of* T:

$$h(T) = \sup_{\mathscr{P}} h(\mathscr{P}, T).$$

A very vague way of stating one of Shannon's theorems is to say that, in this context, binary words of length hn represent "roughly" all n-histories. So there is a (vague) dictionary:

topology		probability theory
$\mathscr{U}$	$\leftrightarrow$	$\mathscr{P}$
$\log N$	$\leftrightarrow$	H.

Examples (for h_{top})

$$(1)\begin{cases} X \xrightarrow{id} X \quad h(id) = 0 \\ X \xrightarrow{T} X \quad T \text{ is distance preserving (i.e. isometry e.g. rotation)} \\ \qquad\qquad h(T) = 0. \\ T \text{ invertible (homeomorphism)}, \quad h(T^n) = |n| h(T) \\ T \text{ periodic}, \quad h(T) = 0. \end{cases}$$

(2) $SL(2, \mathbf{Z})$ is the group of linear maps of the plane onto itself such that

$$\det\begin{pmatrix} a & b \\ c & d \end{pmatrix} = 1,$$

with $a, b, c, d \in \mathbf{Z}$. These matrices preserve the natural Bravais lattice in $\mathbf{R}^2$ with integral coefficients $\Rightarrow \begin{pmatrix} a & b \\ c & d \end{pmatrix}$ defined on a torus, where it preserves the natural notion of area.

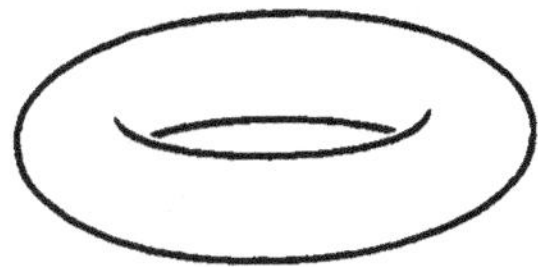

Let

$$T = \begin{pmatrix} a & b \\ c & d \end{pmatrix} \in SL(2, \mathbf{Z});$$

let us look at the eigenvalues λ_1, λ_2.

(α) λ_1, λ_2 complex conjugate $\Rightarrow$ T is periodic, hence T is non-ergodic, $h(T) = 0$ ("elliptic "case).

(β) $\lambda_1 = \lambda_2 = \pm 1$ ("parabolic" case) in Jordan form, the matrix looks like $\begin{pmatrix} 1 & a \\ 0 & 1 \end{pmatrix}$; this can be interpreted as follows: one cuts the surface along a simple closed curve C that does not separate the torus in two and glues back the two ends after a rotation of 2π a (Dehn's twist). The entropy is zero ($h(T) = 0$).

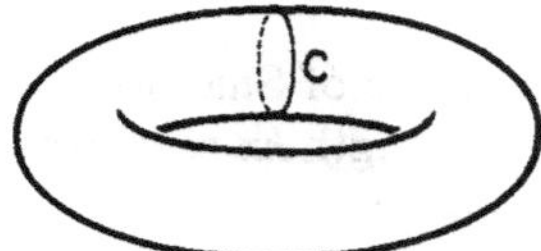

(γ) The interesting case, it is the "hyperbolic" case (*Anosov*) where λ_1, λ_2 are real, $\lambda_1 \neq \lambda_2$, $\lambda_2 = 1/\lambda_1$, $0 \leqslant |\lambda_1| \leqslant 1$ (both are irrational).

Let us consider the corresponding eigenvectors in the plane ($\mathbf{R}^2$), and the families of lines they generate. They project down to the torus with *dense* image; each family is globally preserved but in one direction one stretches in the other one, one expands ($\lambda_2 = 1/\lambda_1$).

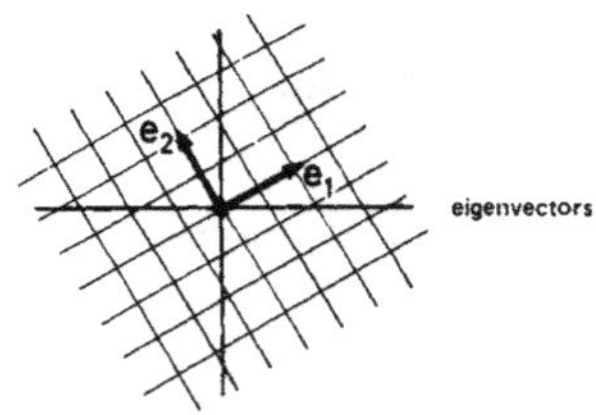

This looks like the baker's transform and in fact it can be shown it is Bernoulli, hence ergodic (notice that each of the families of lines in the torus looks like a smectic liquid crystal without defects). One also has

$$h_{\text{top}}(T) = h_{\text{proba}}(T) = -\log_2|\lambda_1| .$$

2.5.4. Relationship between topological entropy and homology

I recall that the first homology group H_1X is:

$$H_1 = \pi_1/[\pi_1, \pi_1] .$$

It is, by definition, abelian. Let T be a continuous map: $X \to X$ then

$$T_*X : H_1X \to H_1X$$

is linear. Let us consider the eigenvalues of T_*, and let the largest be λ: Then it can be shown that

$$h(T) \geqslant \log_2 \lambda .$$

This is part of a general "entropy conjecture" due to Shub.

Let T be a continuous map of a compact space $T : X \to X$. If one puts on that space a probabilistic measure which is invariant by T (which one can always do) we can compute the probabilistic entropy of T with respect to μ and it can be shown that

$$h_{\mathrm{top}}(T) = \sup_{\mu} h_{\mathrm{proba}}(T, \mu) .$$

Back to Bernoulli. It can be shown that for a Bernoulli transformation the (probabilistic) entropy has a very simple expression

$$h_{\mathrm{proba}}(T) = -\sum_{i=1}^{n} p_i \log p_i ,$$

where $T = B(p_1, \ldots, p_n)$.

The following is a big theorem due to Kolmogoroff, Sinaï and Ornstein:

THEOREM. *Two Bernoulli transformations are isomorphic* iff *they have the same (probabilistic) entropy.*

3. Classical systems

In the last section we have talked about "abstract" dynamical systems, which means measure preserving transformations in probability theory. Now we will talk about "classical" systems which means smooth invertible transformations (i.e. diffeomorphisms of smooth manifolds) or smooth flows (on smooth manifolds), from the standpoint of ergodic theory.

3.1. Curvature

We will very briefly review the notion of curvature (which has already been mentioned during this Summer School, in the context of gauge theories). Because of lack of time we will concentrate mainly on *gaussian curvature* (surface), but most of this and the next paragraph remains true mutatis mutandis in the context of *Riemannian curvature* (dimension $\geqslant$ 3). We will consider a surface M^2 (=2-dimensional smooth manifold) endowed with a Riemannian metric

$$ds^2 = \sum_{i,j=1}^{2} g_{ij}(x)\, dx^i\, dx^j .$$

This leads to the concepts of *geodesic* (shortest or "straight" line) and of *parallel transport* of a (tangent) vector along a smooth path L.

This is defined, first for L = geodesic

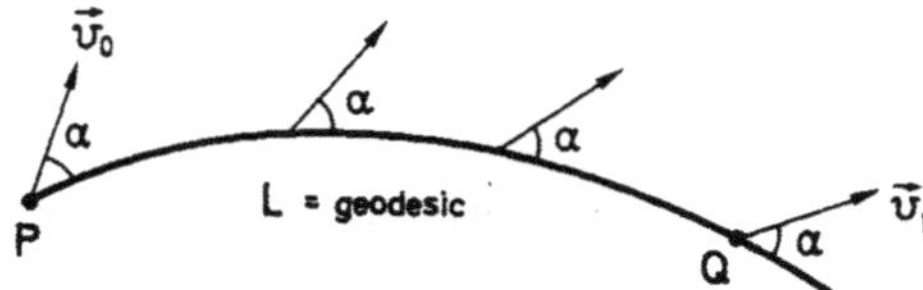

The result of parallel transport of v_0, from P to Q, along L.

Then for L = broken geodesic path = $\{L_1, L_2\}$

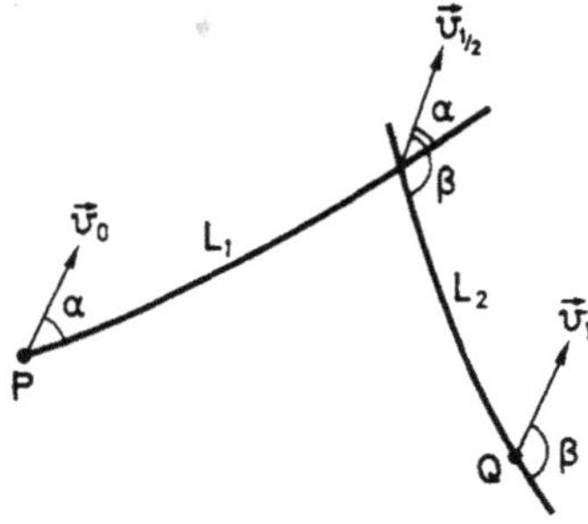

Finally, by going to the limit, for an arbitrary smooth L, as for example along a circle in $\mathbf{R}^2$

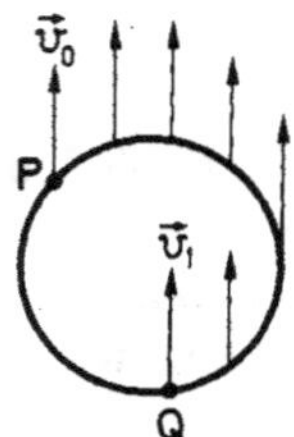

Notice that the angle of the transported vector with the circle L is *not* constant.

On our surface M^2 take a point p and a small simple closed *oriented* curve Γ, going around p and enclosing an area A. We will consider the parallel transport of a vector ($V_0 \mapsto V_1$) along Γ. There are three cases:

(1) The flat (or "euclidean", or "parabolic") case, where one comes back exactly with the same vector ($V_0 = V_1$)

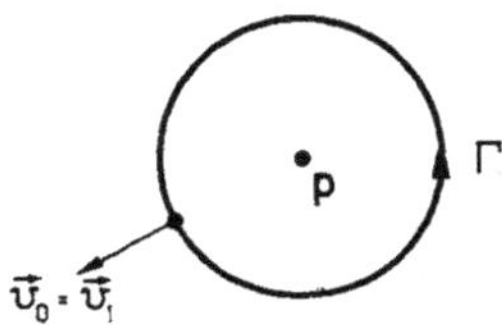

Here, by definition, the *curvature* $K = 0$.

(2) The "elliptic" case: the vector moves forward; by definition the curvature is positive.

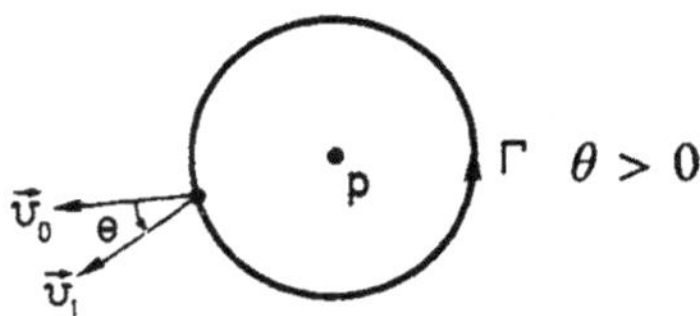

(3) *The "hyperbolic" case* (this is really the interesting one, from our standpoint). The vector lags behind, and by definition we have *negative curvature* ($K < 0$).

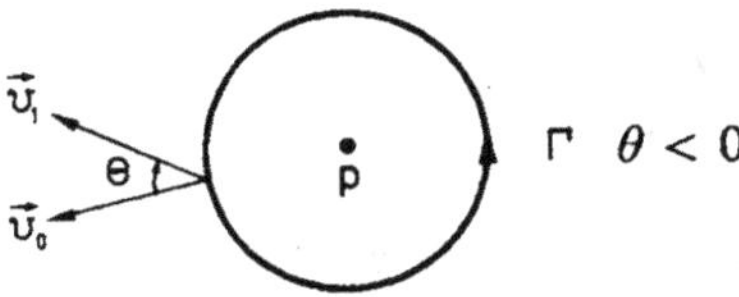

The formal definition of curvature (which is a real number, with *sign*, attached to p): divide θ by A, or more exactly:

$$K = \lim_{A \to 0} \frac{\text{angle } (V_0, V_1) \text{ with its sign}}{\text{area of } A}.$$

Exercise

Take for Γ a (small) geodesic triangle

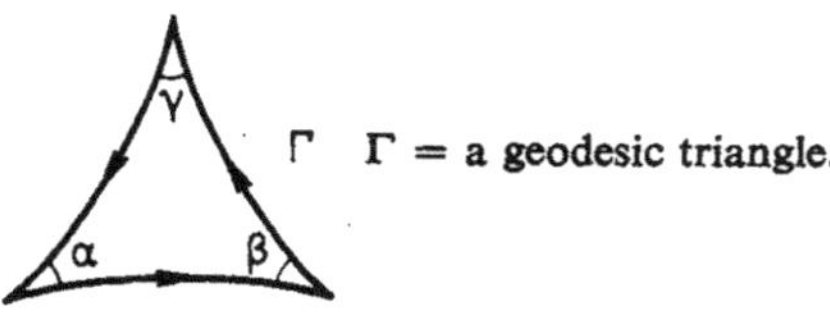

Then

$$\theta = (\alpha + \beta + \gamma) - \pi \,.$$

So, in the flat case, we have the euclidean formula $\alpha + \beta + \gamma = \pi$, and, in general, curvature measures the amount by which our geometry fails to be euclidean. In the hyperbolic case $\alpha + \beta + \gamma$ is always *less* than π.

Let us take a closer look at our three cases.

(1) The "flat" case is (locally) like our euclidean plane. The length of the circle is a linear function of its radius ($L = 2\pi R$)

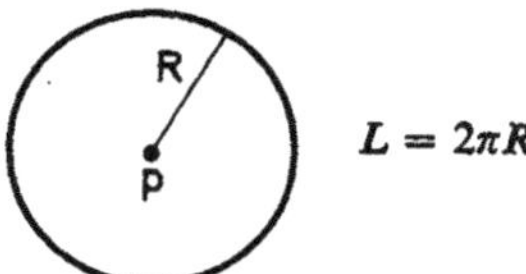

A discrete model: put 6 equilateral triangles together, like this:

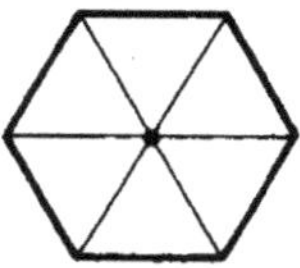

(2) The "elliptic" case is locally like a cup

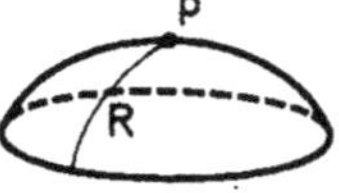

L is small with respect to R.

Here the periphery grows slower with respect to R (or even decreases). Spheres are exactly the cases $K =$ constant > 0. A discrete model: put together 5 equilateral triangles.

(3) *The "hyperbolic" case looks locally like a saddle*

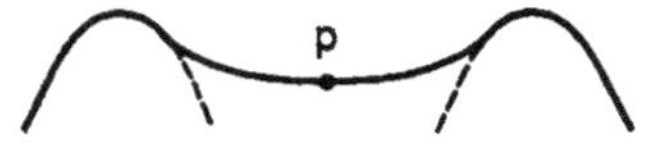

The "periphery" (length of L) grows *exponentially* with R. A discrete model: put together 7 equilateral triangles.

A brief comment on dimensions $\geqslant 3$ (Riemann). These notions (and what comes next) can be generalized for manifolds of higher dimensions but curvature is no longer a number but a tensor. Let, ξ, η be two distinct tangent vectors at p and $V^2(\eta, \xi)$ a small piece of geodesic surface around p, spanned by η, ξ. We consider

$$K(\eta, \xi) = K \quad (\text{of } V^2(\eta, \xi) \text{ at } p)$$

and, by definition if $K(\eta, \xi) < 0$ for all η, ξ, then M^n has *negative curvature.*

Brief comments on hyperbolic geometry (and its beauty). Because of time limitations we will restrict ourselves to dim $= 2$, $K = -1$ ("the hyperbolic or Lobatchevskyan plane"). But most of our comments extend, mutatis mutandis to higher dimensions and variable negative curvature. Here are some facts:

(1) Because the length of the circumference is abnormally large (when compared to the radius), if one shoots straight from p_0 to two far away points P, Q, the shortest path from P to Q looks rather like this:

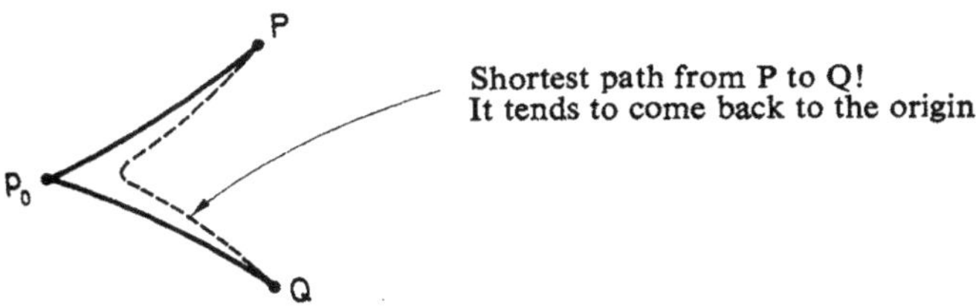

(2) Random walk is highly *non*-recurrent (it goes to a precise point at ∞ with probability (1)). Hence heat diffuses very fast, or a drunkard left loose gets immediately lost. This is in contrast to euclidean $\mathbf{R}^1$ or $\mathbf{R}^2$ where brownian motion is recurrent. (It is non-recurrent only in euclidean $\mathbf{R}^n$ for $n \geqslant 3$.)

(3) An infinitely long stick occupies only a *finite* portion of the horizon.

(4) Shape determines size (i.e. two similar objects are actually congruent; this is a general fact in constant non-zero curvature (Gauss–Bonnet)).

(5) There are infinitely many regular lattices. The hyperbolic plane H^2 has a much richer symmetry structure than the euclidean plane. This has to do with the fact that its group of isometries $SL(2, \mathbf{R})$ is much subtler than the group of euclidean translation + rotations in $\mathbf{R}^2$.

Some of these hyperbolic lattices have been used by the Dutch artist M. Escher in his etchings, via the "Poincaré model for the hyperbolic plane". (In this model, H^2 = the interior of the unit disk, geodesics = circles orthogonal to the unit circle, non-euclidean angles = euclidean angles, ...)

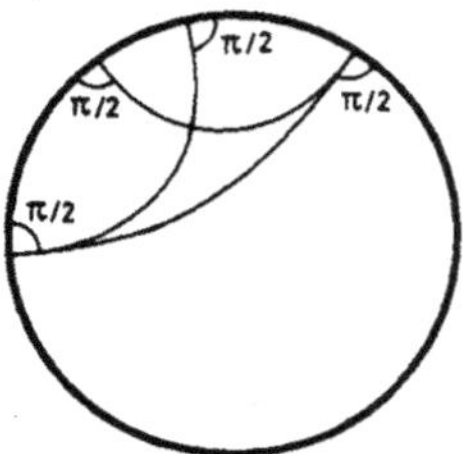

3.2. *The geodesic flow*

It is defined by a particle moving *freely* (by inertia in the absence of external forces of friction), on a manifold M^n, equipped with a metric $ds^2 = \sum g_{ij}\, dx^i\, dx^j$. The particle moves along *geodesics*, with a constant speed ($\|\text{speed}\|$ = constant = 1, say). The *phase space* consists therefore of all points in M^n with all tangent vectors of unit length on M^n

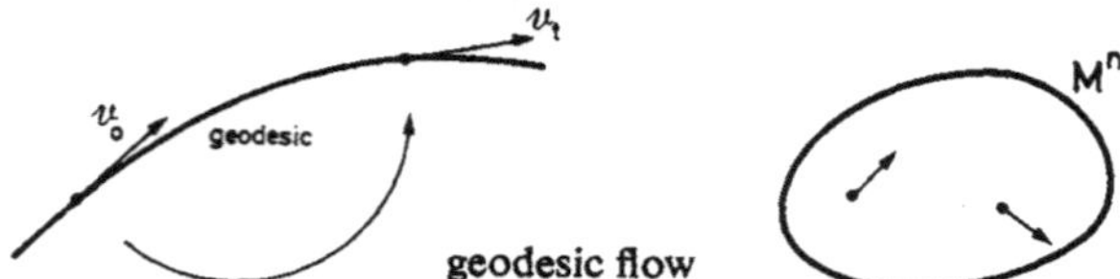

Note that there is only *one* geodesic through each point along a given tangent.

Thus: geodesic flow is $X \circlearrowleft T_t$. This is a classical system, the point moves on a smooth manifold and the transformation T_t is smooth, and also measure preserving (it preserves the volume in phase space – Liouville's theorem). Is it ergodic? (For $K = 0$ the geodesic flow is non-ergodic (obvious: plane). For $K > 0$, the geodesic flow is non-ergodic (sphere: it is very easy to separate great circles into two parts. Flow is hence not ergodic).

THEOREM. If $K < 0$ (M^n compact – for technical reasons) the geodesic flow is *Bernoulli*, hence ergodic. (Anosov (proof of ergodic), Sinaï, Ornstein (proof of Bernoulli for $n = 2$, stated for general n).)

Thus geodesics on a manifold with $K < 0$ behave like *tossing the coin.* To see that, let us go back into history and consider Hadamard's remarkable (*hyperbolic*) *pair of pants.*

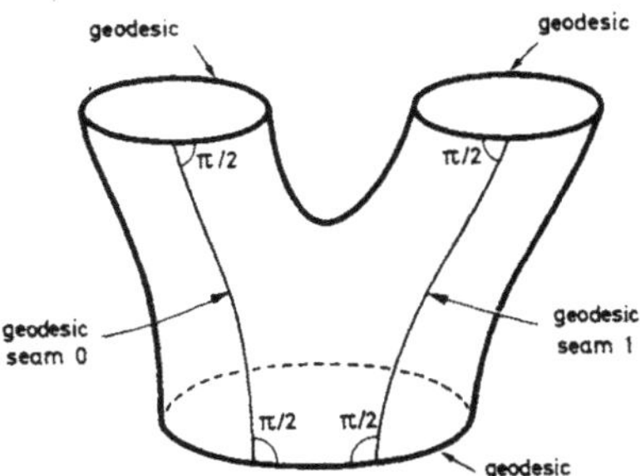

Hadamard proved that any random (infinite) sequence

$$\ldots 0\,0\,1\,0\,1\,1\,1\,0\,1\,1\,1\,1\,0\,0\,1\,0\,1\,0 \ldots,$$

can be realized by some geodesics starting from p on the pair of pants (the digits are written in the order in which the geodesic crosses the seams).

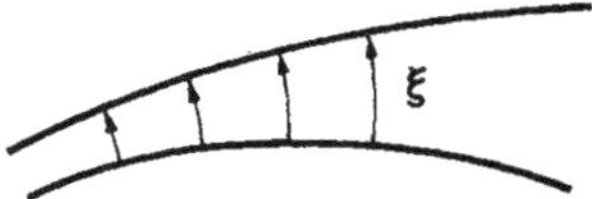

Heuristic idea of the proof that the geodesic flow is Bernoulli. Let us first look at the geodesics close to a *given* geodesic Γ. In a first approximation, they differ by a vector field ξ for which a differential equation can be written:

$$D^2\xi/Dt^2 = -\text{grad}\, U.$$

D^2/Dt^2 is the covariant derivative (associated with parallel transport of ξ), and U is a *potential* (called potential of the geometry) $U \sim \frac{1}{2}K\langle\xi, \xi\rangle$ (where $\langle\ ,\ \rangle$, denotes the scalar product).

This is the generalization of the equation of motion of a linear oscillator. For $K > 0$, the situation (the oscillations) is stable, for $K < 0$, it is highly unstable.

In configuration space (not phase space)

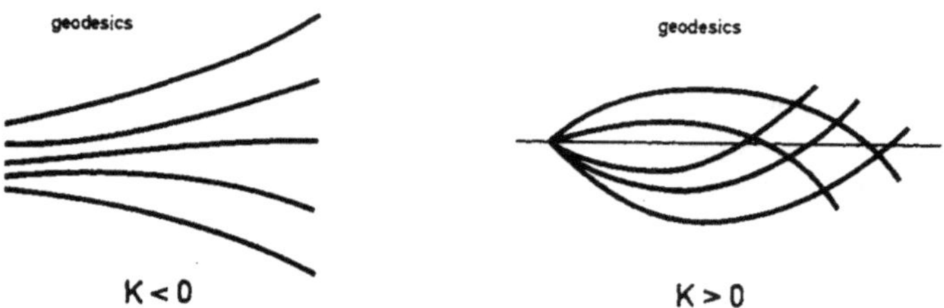

For $K < 0$, the geodesics go far away from each other in an exponential manner (a point stressed above without proof).

After these general comments, we return to $K < 0$. The geodesics (solution of the above "harmonic" equation), neighbouring a given geodesic Γ, split into two classes, those (G_u, u standing for "unstable") which go away exponentially from Γ, and those (labelled G_s, s = "stable") approaching Γ as $t \to +\infty$.

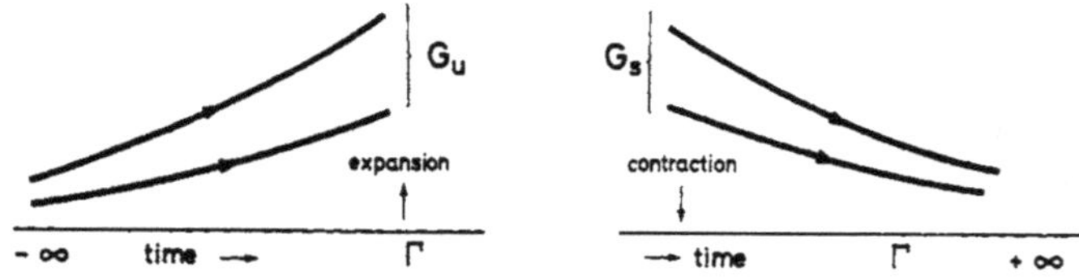

(Recall that the figures above are drawn in the configuration space, not in phase space, which, in the neighbourhood of Γ, splits up as indicated.)

The phase space is therefore $G_u \oplus G_s$, in the neighbourhood of our given geodesic $\Gamma = G_u \cap G_s$. When t increases the G_u are expanded and the G_s are contracted in a fashion reminiscent of the *baker's transformation*.

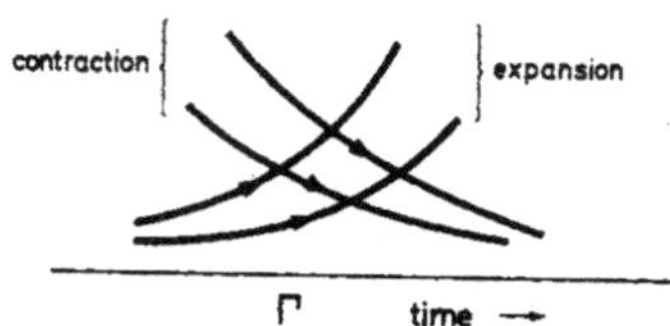

This is the heuristic way of understanding the correspondence between $K < 0$ geometry and Bernoulli transformation (tossing the coin).

3.3. Is the "ideal gas" ergodic? (Boltzmann's original problem)

We consider a collection of hard spheres in a box (regularly shaped). All shocks (between the spheres and between the spheres and the boundary) are elastic.

THEOREM (*Sinaï* (Ornstein)). *If the box is rectangular (or if it is a* 3d *torus), the motion is Bernoulli and hence ergodic.*

This is – to say the least – a difficult theorem, for which all details of the proof have, precisely, not yet been worked out. But the heuristic idea of the proof goes as follows. (For the argument to work, it is essential that there should be many particles and elastic shocks.)

Consider two hard spheres A and B, only. Let A be fixed in the physical space, in which B moves (we recall that the space (the box) is convex). We represent this space by a straight line

and think of it as made of two sheets, like this

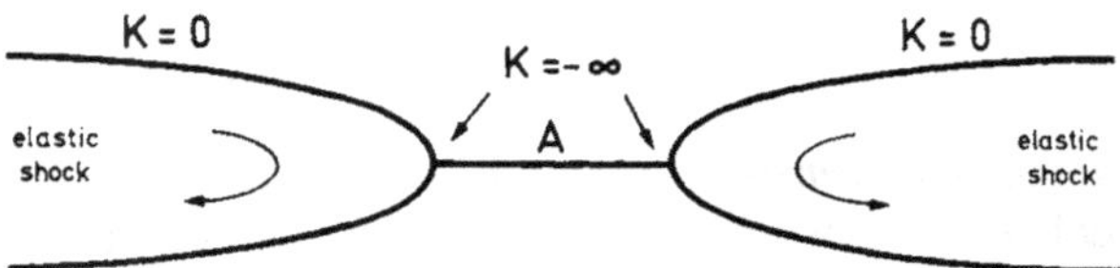

B is moving on one or the other sheet before and after its collision with A.

The physical space is saddle-like around the boundary of A (which can be visualized as the hole in the middle of a torus, or looks like a torus.) The curvature of the physical space vanishes except in the neighbourhood of A where it is (strongly) negative. It can be represented by a (negative) Dirac delta function. We end up with enough negative curvature (albeit *concentrated*) and the arguments used to show that the geodesic flow on a negatively curved manifold is Bernoulli can be, somehow, extended to this more general case. Recall that there is only *one* Bernoulli flow (up to isomorphism), which *is* therefore the geodesic flow constructed here. Recall also Hadamard's pair of pants, for which any arbitrary sequence of binary digits can be realized exactly by *one* particular geodesic flow.

This idea of *concentrating* the negative curvature appears also in another context as we shall see (Teichmüller's theory). (All this has been limited to a quadratic metric $ds^2 = \sum g_{ij}\,dx_i\,dx_j$, which restricts the hamiltonian systems under consideration.)

3.4. *Discrete dynamical systems in* dimension 2 (genus > 1)

The key words, here, will be *isotopy* and *foliation* (on a manifold of dimension 2). A two-dimensional manifold M^2 looks locally like the euclidean plane $\mathbf{R}^2$. But it is nice to think of a manifold as an *arena*, on which things happen. "Things happening", in this paragraph, will be (discrete) dynamical systems and *foliations*. When endowed with a (non-singular) foliation $\mathscr{F}$, the manifold looks locally like $\mathbf{R}^2$ with horizontal lines and this notion of horizontality is preserved when passing from one local coordinate chart to another (a horizontal line from one chart is still horizontal in the other chart).

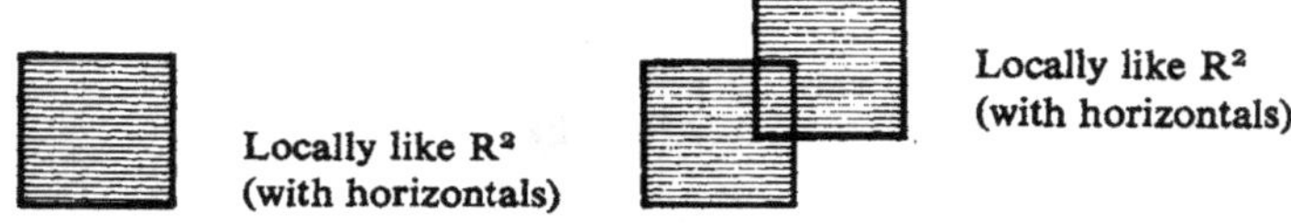

M^2 is covered with *leaves* (horizontal lines), each leaf is a smooth, one-dimensional object. Two leaves are either the same object or they never touch each other (the leaves are disjoint).

Examples

(i) Consider two foliated annuli

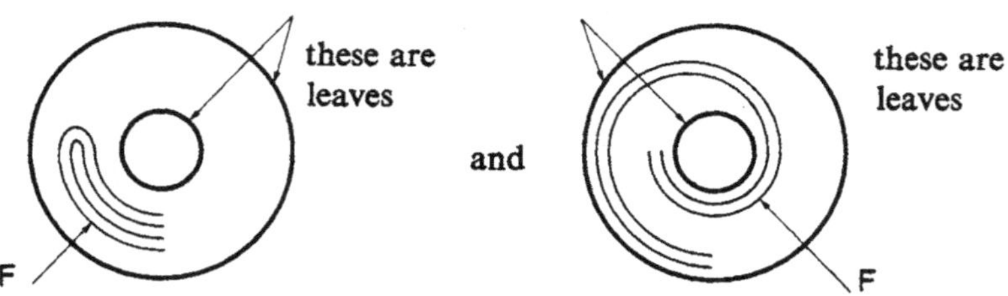

By glueing these annuli together one gets a torus T^2, *with a foliation.*

However, this foliation *cannot* be oriented coherently. Indeed, let us unwrap the annuli like this:

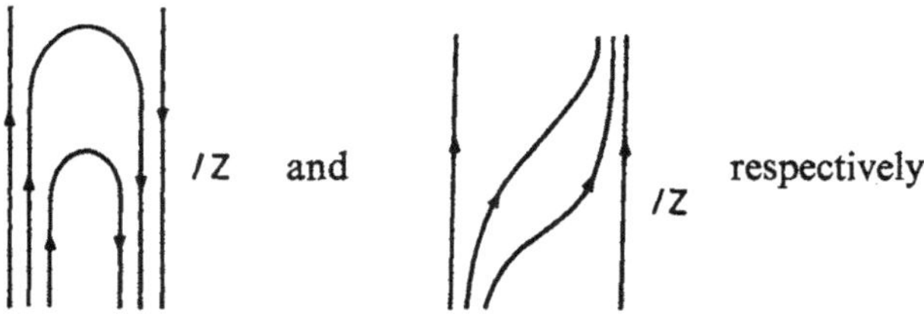

(we divide by the unit vertical translation using the periodicity of the whole situation and we get the annuli). The arrows on the right-side edges are *inconsistent*.

(ii) Consider a system of differential equations

$$\mathrm{d}x/\mathrm{d}t = f(x, y)\,, \qquad \mathrm{d}y/\mathrm{d}t = g(x, y)$$

where f and g are doubly-periodic and $f^2 + g^2$ never vanishes. The orbits can be naturally considered as a foliation on the torus and this kind of foliation can be naturally oriented by the arrow of time.

We introduce now the basic notion of *measured foliation*. Here there is a notion of distance in between two (neighbouring) leaves and this notion is conserved when passing from one local chart to another.

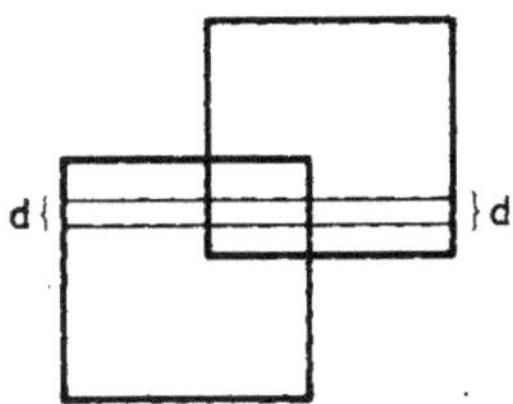

The fact that such a distance can be defined consistently is an intrinsic topological property of the foliation ("invariant by isotopy"). The foliation in example (i) *cannot be measured. Smectic* liquid crystals offer a natural example of measured foliations.

On T^2 the *linear foliations* are measured (and up to isotopy-notion defined below) these are the only ones. (Indeed, the torus can be represented as a Bravais lattice with periodic boundary conditions, and a measured foliation will be a collection of equidistant, parallel lines on this lattice in an arbitrary direction, and topologically speaking *all* measured foliations of it look like this.)

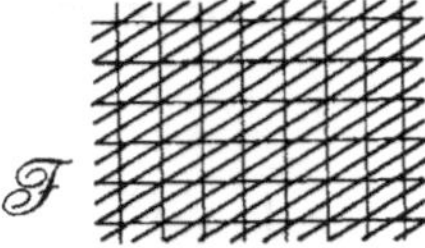

From now on we consider a manifold of genus $g > 1$, M_g, for example:

Since $\chi(M_g) \neq 0$, in order to construct a measured foliation, one has then to accept *singularities*, e.g.

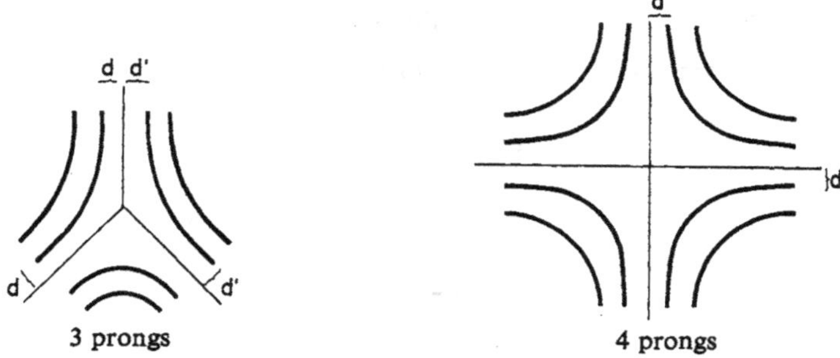

Remark that one cannot put arrows consistently, not even locally if the number of prongs is odd. (The odd-prongs singularities are not even *locally orientable*.)

Remark. The following class of singularities,

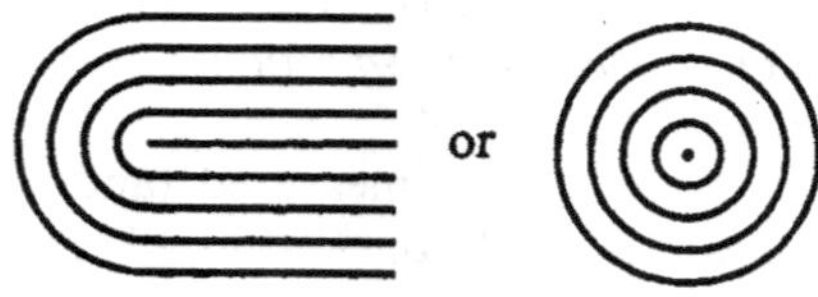

which occurs in condensed matter physics, are *not* considered in our topological discussion, for technical reasons. Compare the singularities drawn above with the figures in Kleman's book (on defects and textures).

Short digression on dimension 3. Let V be a vector field defined in a region of $\mathbf{R}^3$ and let $\mathcal{N}$ be the family of normal planes to V

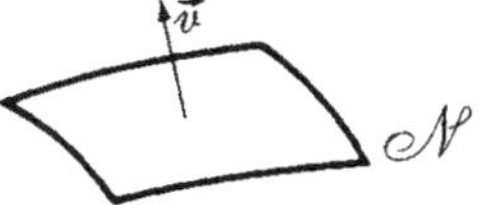

Then $\mathcal{N}$ comes to form a foliation *iff* $V \cdot \mathbf{rot}\, V = 0$ (scalar product). Moreover $\mathcal{N}$ comes to form a *measured* foliation *iff* $\mathbf{rot}\, V = 0$.

Consider now two discrete dynamical systems of the same manifold

$$M^n \overset{T_0}{\underset{T_1}{\rightrightarrows}} M^n$$

(recall that T_0, T_1 are invertible smooth transformations with smooth inverses). By definition T_0 and T_1 are *isotopic* if one can join them by a continuous (or smooth) one-parameter family of discrete dynamical systems

$$M^n \xrightarrow{T_t} M^n, \quad t \in [0, 1]$$

Notice that "T_0 and T_1 isotopic" $\Rightarrow$ "T_0 and T_1 are homotopic". In dimension $\geqslant 3$ the converse is generally false. It is, however, true in dimension 2.

We come now to *Thurston's theorem* which describes discrete dynamical systems on closed 2-manifolds of genus > 1, *up to isotopy*. This is a vast generalization of the situation on the torus T^2 (and infinitely harder to prove). Recall that up to isotopy any discrete dynamical system on T^2 is given by a matrix

$$\begin{pmatrix} a & b \\ c & d \end{pmatrix} \in \mathrm{SL}(2, \mathbf{Z}).$$

There are three cases: the elliptic one (imaginary eigenvalues), parabolic ($\lambda_1 = \lambda_2 = \pm 1$) and hyperbolic (or Anosov) where the eigenvalues are real and distinct.

THURSTON'S THEOREM. Let M_g^2 be a closed surface of genus $g > 1$. Any discrete dynamical system on M_g^2 is isotopic to one of the three models described below (and these cases are mutually disjoined):

(a) elliptic case;

(b) hyperbolic case (or "pseudo-Anosov");

(c) parabolic case (or "reducible case").

The description after isotopy is:

(a) *Elliptic case.* There is a hyperbolic metric ($K = -1$) on M_g^2 such that T is an isometry. Such a T preserves the corresponding (hyperbolic) area, is periodic, hence *non*-ergodic, and the entropy is

$$h_{\text{top}}(T) = h_{\text{proba}}(T) = 0 .$$

(b) *Hyperbolic case* (this is really the interesting one). What does such a T do? M_g^2 is covered by two measured foliations

(u stands for "unstable" and s stands for "stable").

They cut each other transversally (and have the same singular points)

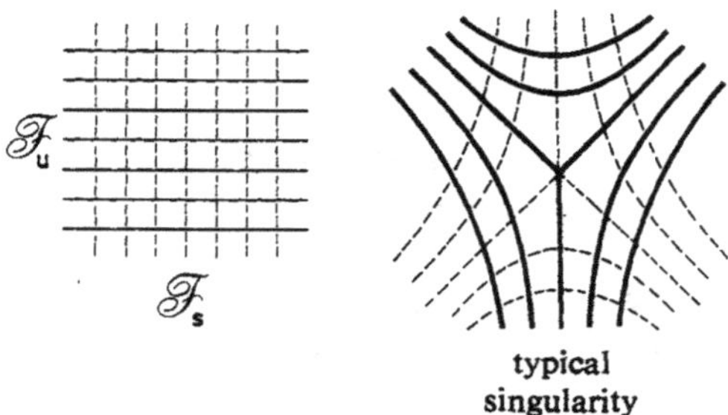

and the surface is covered by the two foliations. What does T do? T preserves both drawings

(1) black line $\to$ black line
dashed line $\to$ dashed line

$$\begin{matrix} T(\text{——}) = \text{——} \\ T(---) = --- \end{matrix} \quad \text{or} \quad \begin{cases} T \text{ preserves the leaves of } \mathscr{F}_u \\ T \text{ preserves the leaves of } \mathscr{F}_s , \end{cases}$$

but

(2) $\exists$ a number λ, $0 < \lambda < 1$, such that

$$\begin{cases} T \text{ shrinks distances in between leaves of } \mathscr{F}_s \text{ with factor } \lambda \\ T \text{ expands distances in between leaves of } \mathscr{F}_u \text{ with factor } 1/\lambda \end{cases}$$

$\Rightarrow$ there exists a notion of invariant *volume* for T (or invariant "area").

Hence one can talk about probabilistic and topological entropy of T. They are equal:

$$h_{\text{proba}}(T) = h_{\text{top}}(T) = -\log_2 \lambda .$$

T realizes the minimum topological entropy in its isotopy class. Also

T is Bernoulli → ergodic .

Minimum. Let $T' : M_g^2 \to M_g^2$ such that T' is isotopic to T (homotopic). T' can be continuously joined to T provided T is hyperbolic

$$h_{\text{top}}(T') \geqslant h_{\text{top}}(T)$$

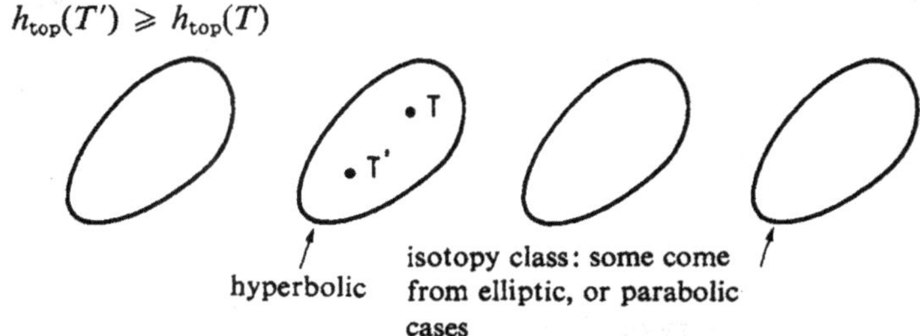

All this has been described on closed manifolds, but it makes sense on manifolds with boundaries too, and the typical *parabolic* system (which is never ergodic) is the mixture of hyperbolic and/or elliptic, like this one:

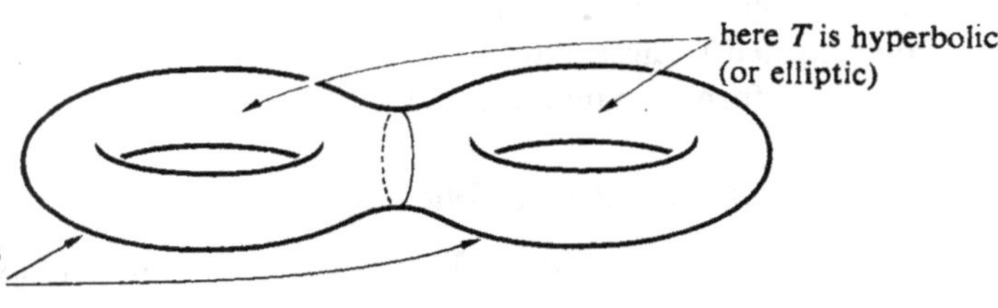

Non-linear spectral analysis. (THURSTON): any discrete dynamical system T on a surface M_g^2, has a finite sequence of "eigenvalues": $1 \leqslant \lambda_1 < \lambda_2 < \cdots < \lambda_k$.

They have the following properties: Take any simple closed curve (non-homotopic to 0) on M_g^2, call it α, and some fixed hyperbolic metric ($K = -1$). There is a unique geodesic in the class of α, and its length will be denoted by $\|\alpha\|$. Then

$$\lim_{n=\infty} \sqrt[n]{\|T^n{}_\alpha\|} = \text{one of the } \lambda_i\text{'s} .$$

T is isotopic to a pseudo-Anosov $\Leftrightarrow k = 1$ and $\lambda^{-1} = \lambda_1 > 1$. These λ's (which are very hard to hit upon; they have subtle arithmetical properties) are somehow "a spectrum for M_g^2".

Last comments. Cover a surface M_g^2 with two transverse measured foliations, $\mathscr{F}$ and $\mathscr{F}'$. Locally, the following drawings are cartesian systems of coordinates

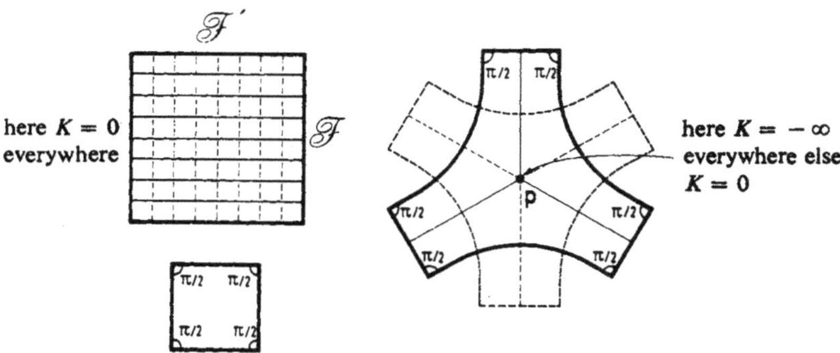

They put a flat geometry on the surface M_g^2, everywhere except at p, where $K < 0$ is concentrated (a Dirac delta function in p). Theoretically, this is somehow similar to the ideal gas method (see above). (The right-hand side figure is an hexagon with $\sum$ (angles) $= 3\pi$, while the euclidean hexagon has $\sum$ angles $= 4\pi$, $\Rightarrow$ inside the hexagon some *negative* curvature is concentrated.) These kinds of ideas come from the Teichmüller theory (quasi-conformal mappings, . . .)

One should consider the following kind of relationship:

Theory of manifolds $\leftrightarrow$ Theory of dynamical systems
(classical, but then also abstract . . .) .

On one hand: "manifolds are *arenas* on which things happen, and these things are dynamical systems". But also: "$(n - 1)$ manifolds + dynamical system" is roughly speaking an "n-manifold" (think of glueing together two n-manifolds with diffeomorphic boundaries by some $T: \partial M \xrightarrow{\approx} \partial N \ldots$).

References

[1] M. Kac, Statistical Independence in Probability, Analysis and Number Theory, Carus: Math. Monography of the MAA (John Wiley).

[2] P. Billingsley, Ergodic Theory and Information (John Wiley, 1965).

[3] V. I. Arnold and A. Avez, Problèmes ergodiques de la mécanique classique (Gauthier-Villars, 1967); also in English translation: Benjamin, 1968.

[4] P. S. Shields, The Theory of Bernoulli Shifts (Univ. of Chicago Press, 1973).
[5] D. Ornstein, Ergodic Theory, Randomness and Dynamical Systems (Yale Univ. Press, 1974).
[6] Y. Sinaï, Introduction to Ergodic Theory, translated from Russian (Princeton Univ. Press, 1976).
[7] W. Thurston, On the Geometry and Dynamics of the Diffeomorphisms of Surfaces (to appear).
[8] Séminaire d'Orsay sur les difféomorphismes des surfaces et les espaces de Teichmüller, d'après W. Thurston (to appear in book form; eds. A. Fathi, F. Laudenbach and V. Poénaru).
[9] M. R. Herman, Sur la conjugaison différentiable des difféomorphismes du cercle à des rotations (to appear).